SOLID-EARTH
SCIENCES AND SOCIETY

Committee on Status and Research Objectives in the Solid-Earth Sciences:
A Critical Assessment
Board on Earth Sciences and Resources
Commission on Geosciences, Environment, and Resources
National Research Council

NATIONAL ACADEMY PRESS
Washington, D.C.   1993

NATIONAL ACADEMY PRESS ● 2101 Constitution Avenue, N.W. ● Washington, D.C. 20418

NOTICE: The project that is the subject of this report was approved by the Governing Board of the National Research Council, whose members are drawn from the councils of the National Academy of Sciences, the National Academy of Engineering, and the Institute of Medicine. The members of the committee responsible for the report were chosen for their special competences and with regard for appropriate balance.

This report has been reviewed by a group other than the authors according to procedures approved by a Report Review Committee consisting of members of the National Academy of Sciences, the National Academy of Engineering, and the Institute of Medicine.

Support for this study was provided by the W. M. Keck Foundation, the G. Unger Vetlesen Foundation, and the National Academy of Sciences' Arthur L. Day Fund and Maurice Ewing Earth and Planetary Science Fund.

Library of Congress Cataloging-in-Publication Data

National Research Council (U.S.). Committee on the Status and Research Opportunities in the Solid-Earth Sciences.
    Solid-Earth Sciences and Society/Committee on the Status and Research Opportunities in the Solid-Earth Sciences, Board on Earth Sciences and Resources, Commission on Geosciences, Environment, and Resources, National Research Council.   p.   cm.
    Includes bibliographical references and index.
    ISBN 0-309-04739-0
    1. Earth sciences—United States.   I. Title.
QE47.A1N38   1993
550'.973—dc20   92-41781
        CIP

Cover art by Y. David Chung. Cover design by Rumen Buzatov. Chung and Buzatov are graduates of the Corcoran School of Art, Washington, D.C. In 1988, Chung won the Mayor's Art Award for Outstanding Emerging Artist and has exhibited widely throughout the country, including the Studio Museum in Harlem and the Whitney Museum of American Art, New York City.

The cover includes many artistic depictions, both ancient and modern, of the solid-earth sciences. At the center is a subduction zone—high mountains, deep-sea trenches, and volcanic activity. The frog dropping a ball into the mouth of a dragon is part of an ancient Chinese seismometer that indicated earthquake direction. The Mariner spacecraft, used to study Mars, represents our new abilites to view the Earth and other planets on different scales. On the spine is an Armillary sphere used in Renaissance Europe as a way of depicting the Earth at the center of the universe. On the back cover are representations of a mid-ocean ridge, an offshore oil derrick, the center section of an early Mayan calendar, and plate movements off the east coast of Africa.

Printed in the United States of America

NATIONAL RESEARCH COUNCIL

2101 CONSTITUTION AVENUE   WASHINGTON, D. C. 20418

OFFICE OF THE CHAIRMAN

Breakthroughs in scientific understanding during the past quarter century as well as innovative technologies for gathering and organizing large amounts of information are expanding the frontiers of knowledge in the earth sciences at an accelerating pace. Basic research has increased our understanding of the origin and internal workings of our planet, of the processes that modify our landscape, and of the evolution of life during times of quite different global environments. A new approach to studying earth processes, in which the earth is viewed as an integrated, dynamic system rather than a collection of isolated components, has emerged.

Solid-Earth Sciences and Society explores these important new directions in earth sciences research and examines how they can enhance society's ability to make wise decisions on resource development, waste disposal, environmental protection, natural hazards reduction, and land use. The report, which reflects a long-term effort by a diverse expert committee, presents a vision of this rapidly changing field: its scope and goals, its emerging research issues, and its scientific contributions and applications.

We have reached a critical time in the solid-earth sciences. Many in the professional community are shifting their focus from exploring for and developing resources to addressing environmental and social problems on global as well as regional scales. Others are working to maintain the research base and acquire the new knowledge upon which the applications are built. Solid-Earth Sciences and Society recommends priorities for future research and discusses the scientific challenges facing our society. It should prove helpful to the research community, to practitioners, to educators, to students, and to all of us with an interest in the earth sciences.

We are particularly indebted to the W. M. Keck Foundation, without whose support this would not have been possible.

Frank Press
Chairman

THE NATIONAL RESEARCH COUNCIL IS THE PRINCIPAL OPERATING AGENCY OF THE NATIONAL ACADEMY OF SCIENCES AND THE NATIONAL ACADEMY OF ENGINEERING

TO SERVE GOVERNMENT AND OTHER ORGANIZATIONS.

Prologue

The GOAL is:

to understand the past, present, and future behavior of the whole earth system. From the environments where life evolves on the surface to the interaction between the crust and its fluid envelopes (atmosphere and hydrosphere), this interest extends through the mantle and the outer core to the inner core. A major challenge is to use this understanding to maintain an environment in which the biosphere and humankind will continue to flourish.

SOCIETAL CHALLENGES FOR EARTH SCIENTISTS

The solid-earth sciences are essential to:

■ provide sufficient resources—e.g., water, minerals, and fuels;
■ cope with hazards—e.g., earthquakes, volcanoes, landslides, tsunamis, and floods;
■ avoid perturbing geological environments—e.g., soil erosion, water contamination, improper mining practices, and waste disposal; and
■ learn how to anticipate and adjust to environmental and global changes.

RESEARCH FRAMEWORK

The information needed to achieve the goal of the solid-earth sciences and to meet the societal challenges derives from research that can be described conveniently in a matrix of four *objectives* and five *research areas*. *Research opportunities* can be located within the elements of this matrix, and from these have been selected research topics of top priority and high priority.

Objectives

The following four *objectives* are derived from the challenges facing society in which fundamental understanding of the solid-earth sciences plays a primary role:

■ *Understand the processes involved in the global earth system, with particular attention to the linkages and interactions between its parts (the geospheres)*
■ *Sustain a sufficient supply of natural resources*
■ *Mitigate geological hazards*
■ *Minimize and adjust to global and environmental change*

Research Areas

The following research areas provide promise of achieving the scientific goal:

■ *Global paleoenvironments and biological evolution*
■ *Global geochemical and biogeochemical cycles*
■ *Fluids in and on the Earth*
■ *Dynamics of the crust (oceanic and continental)*
■ *Dynamics of the core and mantle*

EARTH SYSTEM SCIENCE

The goal represents an integrated approach to the study of the earth system, requiring interdisciplinary investigations of the geology, physics, chemistry, and biology of the whole Earth, because all parts of the Earth are interconnected through geological, geophysical, and geochemical processes, some of which are monitored by biological activity near the surface.

Attainment of the specific objectives may be greatly enhanced by more complete understanding of processes occurring on a global scale. Boundaries between basic and applied solid-earth sciences are artificial.

This process-oriented, integrated global approach should be incorporated into revised earth science curricula in universities and schools. There are also educational opportunities in redefined engineering geology.

COMMITTEE ON STATUS AND RESEARCH OBJECTIVES IN THE SOLID-EARTH SCIENCES: A CRITICAL ASSESSMENT

PETER J. WYLLIE (*Chairman*), California Institute of Technology
PHILIP H. ABELSON, American Association for the Advancement of Science
SAMUEL S. ADAMS, Minerals Consultant, Lincoln, New Hampshire
CLARENCE R. ALLEN, California Institute of Technology
G. ARTHUR BARBER, Minerals Consultant, Denver, Colorado
ROBIN BRETT, U.S. Geological Survey
ROBERT A. BERNER, Yale University
JOHN D. BREDEHOEFT, U.S. Geological Survey
ROBERT G. COLEMAN, Stanford University
BRUCE R. DOE, U.S. Geological Survey
CHARLES L. DRAKE, Dartmouth College
LARRY W. FINGER, Carnegie Institution of Washington
WILLIAM L. FISHER, Texas Bureau of Economic Geology
ALEXANDER F. H. GOETZ, University of Colorado
ALLEN W. HATHEWAY, University of Missouri at Rolla
JOHN D. HAUN, Barlow & Haun, Inc.
JAMES F. HAYS, National Science Foundation
WILLIAM J. HINZE, Purdue University
RAYMOND JEANLOZ, University of California, Berkeley
MARVIN E. KAUFFMAN, National Science Foundation
JUDITH T. PARRISH, University of Arizona
CHARLES T. PREWITT, Carnegie Institution of Washington
LEE R. RUSSELL, ARCO Oil and Gas Company
STANLEY A. SCHUMM, Colorado State University
BRIAN J. SKINNER, Yale University
STEVEN M. STANLEY, The Johns Hopkins University
DONALD L. TURCOTTE, Cornell University
KARL K. TUREKIAN, Yale University
ROBERT E. WALLACE, U.S. Geological Survey
DANIEL F. WEILL, National Science Foundation
ROBERT E. ZARTMAN, U.S. Geological Survey

Staff

LALLY A. ANDERSON, Staff Assistant
KEVIN C. BURKE, Scholar-in-Residence
CATHERINE MacMULLEN, Consultant
THOMAS M. USSELMAN, Senior Staff Scientist

The National Academy of Sciences is a private, nonprofit, self-perpetuating society of distinguished scholars engaged in scientific and engineering research, dedicated to the furtherance of science and technology and to their use for the general welfare. Upon the authority of the charter granted to it by the Congress in 1863, the Academy has a mandate that requires it to advise the federal government on scientific and technical matters. Dr. Frank Press is president of the National Academy of Sciences.

The National Academy of Engineering was established in 1964, under the charter of the National Academy of Sciences, as a parallel organization of outstanding engineers. It is autonomous in its administration and in the selection of its members, sharing with the National Academy of Sciences the responsibility for advising the federal government. The National Academy of Engineering also sponsors engineering programs aimed at meeting national needs, encourages education and research, and recognizes the superior achievements of engineers. Dr. Robert M. White is president of the National Academy of Engineering.

The Institute of Medicine was established in 1970 by the National Academy of Sciences to secure the services of eminent members of appropriate professions in the examination of policy matters pertaining to the health of the public. The Institute acts under the responsibility given to the National Academy of Sciences by its congressional charter to be an adviser to the federal government and, upon its own initiative, to identify issues of medical care, research, and education. Dr. Kenneth I. Shine is president of the Institute of Medicine.

The National Research Council was organized by the National Academy of Sciences in 1916 to associate the broad community of science and technology with the Academy's purposes of furthering knowledge and of advising the federal government. Functioning in accordance with general policies determined by the Academy, the Council has become the principal operating agency of both the National Academy of Sciences and the National Academy of Engineering in providing services to the government, the public, and the scientific and engineering communities. The Council is administered jointly by both Academies and the Institute of Medicine. Dr. Frank Press and Dr. Robert M. White are chairman and vice chairman, respectively, of the National Research Council.

Contents

Preface

The Committee on Status and Research Objectives in the Solid-Earth Sciences was charged by the Board on Earth Sciences and Resources with preparing a comprehensive and critical review of the current state of the science, to identify opportunities for research during coming decades, and to consider the issue of establishing priorities. The study was supported by a major grant from the W. M. Keck Foundation of Los Angeles, together with grants from the G. Unger Vetlesen Foundation of New York, and the National Academy of Sciences' Arthur L. Day Fund and Maurice Ewing Earth and Planetary Science Fund. In addition, the following scientific societies provided support: American Association of Petroleum Geologists; American Institute of Professional Geologists; American Geological Institute; Association of American Geographers; Association of American State Geologists; Association of Earth Sciences Editors; Geological Society of America; Society of Economic Geologists, Inc.; Society for Sedimentary Geology/SEPM; and Society of Vertebrate Paleontology. No funds for the study were specifically requested from federal agencies.

COMMITTEE PROCESS

The committee began its work with a series of meetings during the summer of 1988 to plan its approach and to initiate the study. There was a conscious decision to organize the report around two principal themes: (1) basic understanding of solid-earth processes and their interaction with other parts of the earth system and (2) societal issues in which the solid-earth sciences provided significant information in the decision-making process. There was no attempt to organize the study on a disciplinary basis or to review the appropriateness of specific federal agency programs.

The committee formed 21 panels to help synthesize the vast body of earth science knowledge on specific societal issues or related to a few subdisciplines. The panels and their membership are included in Appendix B; over 150 earth scientists were involved in this process. The panels, through several individual meetings, produced draft reports that provided a major input to the report. Because of differences in approach and content among the panel reports, there are no plans to issue those draft materials.

In addition to input provided by individual committee members and drafts from the panels, the committee was aware of the findings and conclusions of the many recent reports produced by the National Research Council, by various federal agencies, and by other consortia and planning groups. These are listed as a bibliography at the end of the appropriate chapters. These materials provided a second major input to the committee's deliberations. As many of the committee and panel members had participated in preparing one or more of these previous reports and long-range plans, this experience helped to put the discussions and the possibilities in a broad and informed perspective.

Input from other members of the earth science community was sought in several ways. A questionnaire about priorities in the earth sciences was distributed to the councils of 40 national societies representing the spectrum of the pure and applied earth sciences. A written solicitation for suggestions was published in *EOS*, and a letter was widely circulated to individual scientists from society membership lists and through chairmen of earth science and selected engineering departments in North American universities. Eighty-three responses were received. Presentations of the work of the committee were made at three national meetings: a lecture at the Geological Society of America in 1989, and focused symposia at the American Association for the Advancement of Science's annual meeting in 1990 and the American Geophysical Union's meeting in spring 1990. Open discussion sessions were held at the latter two symposia. When writing was under way, many individual earth scientists were approached to provide a paragraph or page on specific topics that were not covered adequately through the other input processes.

The committee held a series of meetings and workshops to consider the various inputs and to design the report. Several editorial subgroups were formed; these subgroups held several meetings and prepared drafts of the specific chapters, largely from the source materials. The draft chapters were circulated to the full committee for comment. Based on the comments, revisions were made and redistributed to the committee. At a final workshop, the committee discussed the report as a whole and reached consensus on the various priorities.

The committee's overall approach was to determine the most important earth processes and then to consider what methods and facilities would be most effective in providing answers to the process-oriented problems. The committee recognized that in a field as diverse as the solid-earth sciences research must advance on a broad front. In preparation for, and during the final workshop in October 1991, the committee members reviewed a draft of the report and selected a top-priority research topic in each of eight areas, together with supporting research programs and the infrastructure required for implementation of the programs.

STRUCTURE OF THE REPORT

This report covers the workings of the whole Earth, concentrating on solid-earth processes and their influence on and interactions with human society. The table below illustrates the report's structure, which consists of seven chapters following the Executive Summary. A global overview of the present status of the solid-earth sciences is presented in the first chapter. The second chapter deals with the processes on Earth that are driven by internally generated heat; it includes some discussion of plate tectonics, volcanoes, earthquakes, the origin and history of the continents, and where research is going in all those areas. The third chapter complements the second by addressing two processes driven by the solar energy that falls on the surface—erosion and deposition—and portrays how they have operated since the beginning of earth history. It also tells the story of life on Earth and of its evolution as revealed by fossil

Chapter	Executive Summary		
1	Essay	**Global Overview**	
2	Essay	**Understanding Our Active Planet**	Research Opportunities
3	Essay	**The Global Environment and Its Evolution** (near-surface processes)	Research Opportunities
4	Essay	**Resources of the Solid Earth** (water, minerals, fuels)	Research Opportunities
5	Essay	**Hazards, Land Use, and Environmental Change**	Research Opportunities
6	Essay	**Ensuring Excellence and the National Well-Being**	Recommendations
7	**Research Priorities and Recommendations**		Recommendations
Appendix A	Data Base of Federal Programs and Their Budgets		

organisms, including such issues as catastrophic extinction. The fourth chapter treats resources—land, water, and mineral deposits, including metals, oil, gas, and coal—in light of the two previous chapters. The fifth chapter deals with hazardous phenomena such as earthquakes, volcanoes, and unstable land surfaces. It also describes how human beings interact with the Earth, changing its environment both locally, as in urban pollution, and globally, as in the composition and temperature of the ocean or atmosphere. The sixth chapter looks at how the solid-earth sciences are practiced, where research is going, what demographic changes are happening, the nature of changes in instrumentation and data handling, and the international role of the solid-earth sciences. The seventh chapter summarizes the goals of the solid-earth sciences, the research opportunities, the facilities required, and the priorities and ends with a list of recommendations.

Each chapter is introduced by an essay. At the ends of Chapters 2 through 5, specific research opportunities related to the topics addressed are summarized in a research framework. At the ends of Chapter 6 and 7, recommendations are listed. These front and back portions of each chapter are shaded in the table, and highlighted in the report. Appendix

A gives information on the budgets of federal government research programs.

There are several different ways to sample the report:

■ The Executive Summary conveys the essentials.

■ The Executive Summary and Chapter 1 give a global overview of the whole volume.

■ The shaded essays provide the essence of each chapter without the detail.

■ The shaded research opportunities outline important research without the detailed background.

■ Chapter 7 gives a detailed discussion of opportunities and priorities.

■ The bodies of Chapters 2 through 6 provide a more technical treatment of the fields.

SOLID-EARTH
SCIENCES AND SOCIETY

Executive Summary

The solid-earth sciences* address the planet we live on, the continents and ocean basins from which we derive our mineral and energy reserves, the rocks from which soils to raise crops are derived, and the rock formations where we dispose of most of our waste products. Earth scientists analyze the physical and chemical processes that link all of these domains and those of the Earth's interior. They characterize the internal and external energy systems that drive and have driven these fundamental processes. They unravel the record of life on the planet and interpret the changing environments in which biological evolution has proceeded to its present state. Understanding our Earth has now become essential to humankind's existence.

Currently, the expanding world population requires more resources; faces increasing losses from natural hazards; and contributes to growing pollution of the air, water, and land. The activities of humans and their consequences are now comparable in magnitudes and rates as perturbations of the Earth's environment to many natural processes. Many of these human perturbations are not beneficial to life on the planet.

Human societies face momentous decisions concerning their control of many future activities that require understanding the Earth. The issues include atmospheric changes, environmental degradation, vulnerability of populated sites to natural disasters, soil erosion, contamination of water supplies, provision of adequate supplies of energy and mineral resources, weighing the potential of nuclear power, and the destruction of species. The rates of changes have become so rapid that these issues cannot be ignored any longer if the Earth is to be managed as a sustainable habitat. To accomplish sustainability will require all of our scientific understanding of the natural materials and processes, particularly the material and energy transfers linking the geosphere, hydrosphere, atmosphere, and biosphere. Life prospers or fails at the surface of the Earth where these environments intersect. The slower geological processes, which have created the life-productive, quality environments, are complex and sensitive. Human decisions that must be made, including analysis and prediction of change, will require reliable knowledge based on profound understanding of the Earth's interconnected systems.

Attaining that fundamental understanding is the primary objective of the solid-earth sciences. There

> *The study of the whole earth system provides a research framework essential to the solution of global problems.*

---

*Note on terminology: We use the term *solid-earth sciences* to specifically apply to terra firma—the solid surface and the planet's interior; the term includes geology (and all of its subdisciplines) along with significant portions of geophysics and geochemistry. *Earth science* (also geoscience) refers to all of the disciplines that study the planet and includes oceanography; atmospheric science; hydrology; and parts of ecology, biology, and solar-terrestrial physics. *Earth system* is used in reference to all of these disciplines and emphasizes their interactive processes.

1

are many major challenges facing solid-earth scientists as they serve societal needs. Prominent among these are:

■ to provide sufficient resources—for example, water, minerals, and fuels
■ to cope with hazards—for example, earthquakes, volcanoes, landslides, tsunamis, and floods
■ to avoid perturbing geological environments—for example, soil erosion, water contamination, improper mining practices, and waste disposal; and
■ to learn how to anticipate and adjust to environmental and global changes.

An important force driving earth science research is human curiosity regarding our origins, evolution, and the processes that shape our environments.

> *Pure and applied earth sciences are intimately interwoven.*

Programs designed to improve the human condition, whether related to resources, hazards, or environmental change, depend on the results of basic research aimed at expanding our understanding of the Earth's processes. Therefore, the **GOAL OF THE SOLID-EARTH SCIENCES** is:

to understand the past, present, and future behavior of the whole earth system. From the environments where life evolves on the surface to the interaction between the crust and its fluid envelopes (atmosphere and hydrosphere), this interest extends through the mantle and the outer core to the inner core. A major challenge is to use this understanding to maintain an environment in which the biosphere and humankind will continue to flourish.

New concepts and methodologies are emerging that permit the synthesis of solid-earth science data on a global scale. The new capabilities allow construction of testable models of interaction among the many subsystems that form the whole earth system. This global view was heralded by the plate tectonics revolution, which recognized that material making up the rigid outer plates comes from the interior at suboceanic spreading centers, is modified at the surface, and either returns to the interior at subduction zones or is added to the continents. Current research into the interconnected systems aims at developing an understanding of convection in the solid interior, the specific plate-driving mechanism, and the connection between convection and the hydrosphere and biosphere, including long-term atmospheric and oceanic changes.

PRESENT STATE OF THE SOLID-EARTH SCIENCES

Twenty-five years ago our understanding of the global system was revolutionized by plate tectonics and the recognition of a highly mobile outer shell of the Earth. This breakthrough, as well as the continued demand for water, mineral, and energy resources, led to a surge in the number of qualified researchers. These researchers have access to advanced instrumentation in laboratories, in the field at the Earth's surface on land and sea, and in aircraft and in space. Computational capabilities have revolutionized the handling of the vast amounts of data generated in earth science research and facilitated the rapid construction and testing of sophisticated models.

Although recent years have, in some sense, been the best of times with the introduction of new concepts and sophisticated observing systems, the increasing calls for guidance or predictions or solutions for major societal problems require an even more dedicated cadre of earth scientists and facilities to meet the challenges of the next century. The expectation continues that the earth science community in the United States will play a leadership role in global scientific cooperative research. This report addresses the research areas, their applications, and the personnel and facility requirements for the U.S. earth sciences to fulfill national and international expectations.

During most of the twentieth century, the mineral and energy extractive industries employed most of the geologists and geophysicists and commanded a large fraction of the basic and applied research conducted. Because they are fundamentally cyclical industries, their support of research has waxed and waned with economic fluctuations. The resource industries are now restructuring, and there is a growth of employment and research opportunities related to environmental matters and engineering geology, including hydrology and waste isolation. These areas may become the dominant opportunities for future geoscientists. Indeed, the environment between the solid and fluid geospheres is recognized as a major challenge in the solid-earth sciences. The recommendations and priorities presented in this report focus strongly on the problems of the solid-earth sciences and the opportunities for understanding that they provide.

PRIORITIES

Priority Themes: Objectives and Research Areas

The range and scale of research opportunities far exceed the financial and personnel requirements that could reasonably be available. Therefore, priorities are given to guide the allocation of available resources. The only reliable prediction about where scientific breakthroughs can be expected is that such predictions will fail to anticipate discoveries that emerge unexpectedly from investigations. It is therefore important to maintain some level of research activity across the entire field and to ensure that the system is creative and responsive to new ideas and techniques. The committee structured its priorities on the basis of four broad objectives and the major research areas that support them. This framework of objectives and research areas was the basis for the committee's consideration of priorities in the solid-earth sciences.

The following four *objectives* are derived from the challenges facing society in which fundamental understanding of the solid-earth sciences plays a primary role:

A. **Understand the processes involved in the global earth system, with particular attention to the linkages and interactions between its parts (the geospheres)**
B. **Sustain sufficient supplies of natural resources**
C. **Mitigate geological hazards**
D. **Minimize and adjust to the effects of global and environmental change**

The committee selected the following five *research areas* that will provide the understanding needed to address the above objectives:

I. **Global paleoenvironments and biological evolution**
II. **Global geochemical and biogeochemical cycles**
III. **Fluids in and on the Earth**
IV. **Dynamics of the crust (oceanic and continental)**
V. **Dynamics of the core and mantle**

These research areas all relate to the dynamic behavior of the earth system, but they emphasize different time scales, processes, and environments, and they progress from the surface downward into the core. These research areas also provide much of the scientific basis for the objectives.

The objectives and research areas identified in this report are used in two different ways. First, they can be used as the axes of a 4 × 5 matrix to provide detail about solid-earth science research. The entries in the matrix could be current research projects, recommended research topics, or federal funding of research, among others. Indeed, we use the matrix in all these ways. Second, understanding the processes (Objective A) in each of the five research areas, plus the other three objectives, can form an eight-item list, which we designate as *priority themes*. These priority themes (Table 1) are all important areas for research in the solid-earth sciences that provide the promise of achieving the scientific goals, and we use these themes as a first step toward defining research priorities.

Selection of High-Priority Research Opportunities

In selecting the highest-priority research opportunities for each of the priority themes, the committee concentrated on processes rather than disciplines. A rather comprehensive list of *research opportunities* was identified; this list is given in Table 2. From the great variety of worthy research topics, a limited set of research opportunities was selected for each priority theme. This set represents a second stage in the selection of research priorities. Detailed discussion by the committee developed a remarkable degree of consensus concerning the high-priority research opportunities.

The selected top-priority research opportunity for each priority theme is given in the research recommendations below. There are supporting and supplementary research programs associated with each top-priority research selection as well as other high-priority research programs. For each top-priority selection, two additional high-priority research subjects are listed (with three for resources), most of which could compete strongly for the top position. The importance of maintaining research on a broad front cannot be overemphasized.

TABLE 1 The Objectives and Research Areas Used Throughout the Report[a]

Objectives

A. To Understand the Processes in All Research Areas

To understand the origin and evolution of the Earth's crust, mantle, and core and to comprehend the linkages between the solid earth and its fluid envelopes and the solid earth and the biosphere. We need to maintain an environment in which the biosphere and humankind can flourish without risk of mutual or shared destruction.

B. To Sustain Sufficient Supplies of Natural Resources

To develop dynamic, physical, and chemical methods of determining the locations and extent of nonrenewable resources and of exploiting those resources using environmentally responsible techniques. The question of sustainability, the carrying capacity of the Earth, becomes more significant as the resource requirements grow.

C. To Mitigate Geological Hazards

To determine the nature of geological hazards, including earthquakes, volcanic eruptions, tsunamis, landslides, soil erosion, floods, and materials (e.g., asbestos, radon) and to reduce, control, and mitigate the effects of these hazardous phenomena. It is important to consider risk assessment and levels of acceptable risk.

D. To Minimize and Adjust to the Effects of Global and Environmental Change

To mitigate and remediate the adverse effects produced by global changes of environment and changes resulting from modification of the environment by human beings. These latter changes may necessitate changes in human behavior. In order to predict continued environmental changes and their effects on the Earth's biosphere, we need the historical perspective given by reconstructed past changes.

Research Areas

I. Global Paleoenvironments and Biological Evolution

To develop a record of how the Earth, its atmosphere, its hydrosphere, and its biosphere have evolved on all time scales from the shortest to the longest. Such a record would provide perspective for understanding continuing environmental change and for facilitating resource exploration.

II. Global Geochemical and Biogeochemical Cycles

To determine how and when materials have moved among the geospheres crossing the interfaces between mantle and crust, continent and ocean floor, solid earth and hydrosphere, solid earth and atmosphere, and hydrosphere and atmosphere. Interactions between the whole solid-earth system and its fluid envelopes represents a further challenge. Cycling through the biosphere and understanding how that process has changed in time is of special interest.

III. Fluids in and on the Earth

To understand how fluids move within the Earth and on its surface. The fluids include water, hydrocarbons, magmas rising from great depths to volcanic eruptions, and solutions and gases distributed mainly through the crust but also in the mantle.

IV. Crustal Dynamics: Ocean and Continent

To understand the origin and evolution of the Earth's crust and uppermost mantle. The ocean basins, island arcs, continents, and mountain belts are built and modified by physical deformations and mass transfer processes. These tectonic locales commonly host resources introduced by chemical and physical transport.

V. Core and Mantle Dynamics

To provide the basic geophysical, geochemical, and geological understanding as to how the internal engine of our planet operates on the grandest scale and to use such data to improve conditions on Earth by predicting and developing theories for global earth systems.

[a]Sequence implies no ranking. All of the research areas and Objectives B, C, and D are treated as priority themes; Objective A involves understanding the processes in each of the research areas, and so was not itself designated as a separate priority theme.

PLANNING FOR THE FUTURE

Personnel Requirements

Geologists and other solid-earth scientists have played a pivotal role in sustaining societal growth for the past century. Through their efforts great deposits of underground water, minerals, and energy resources have been found and made available. The composition and dynamics of the solid earth have been explored, leading to insights of scientific, aesthetic, and economic value. Interactions with the defense community have been organized around a common interest in the physical nature of the Earth, a concern over ensuring adequate supplies of strategic materials, and activities ranging from develop-ment of navigation systems for submarines by mapping the magnetic field to detection of underground nuclear explosions.

If advances in the solid-earth sciences are to meet evolving societal needs, there must be a sufficient number of well-qualified professionals. Employment projections indicate that opportunities in the earth sciences are growing, with particular emphasis on issues of engineering geology, groundwater, the siting of waste repositories, and environmental cleanups.

Education Requirements

During the latter half of the 1980s, there was a nationwide decline in enrollment in university earth science courses and in the number of undergraduate earth science majors. This is a major concern at a time when the need for earth science expertise is increasingly recognized. The content of earth science curricula has lagged behind the rapidly changing societal concerns and employment opportunities, and major revisions in how the earth sciences are defined and taught are now being made.

General introductory courses at universities should be among the best that an earth science department offers. For many students, such courses may be their only planned exposure to the earth sciences. These courses must both educate future citizens and attract potential earth science majors. Earth science departments also need to collaborate with education departments in designing programs for future precollege teachers of the earth sciences.

Two significant changes in curricula are likely to develop in the 1990s. First, there will be a greater need for courses preparing students for growth in such areas as hydrology, land-use planning and engineering geology, environmental and urban geology, and waste disposal, which will result from society's need to devise environmentally sound ways of exploiting mineral and energy resources. Second, the conventional disciplinary courses will be supplemented by more comprehensive courses in earth system science that emphasize the interrelationships and feedback processes and the involvement of the biosphere in geochemical cycles. For both undergraduate and graduate earth science students, flexibility, versatility, and a firm foundation in basic sciences are crucial. Fundamental principles must be emphasized, because the narrow focus of initial job training will eventually become obsolete as national needs change.

Facilities and Equipment

In recent years technological progress has transformed the analytical tools available to earth scientists. Although the course of science is largely unpredictable, the needs of a discipline for instruments and facilities must be planned in advance. New instruments and facilities are needed to advance the highly promising research areas identified in this report.

The approach taken by the committee was to determine the most important earth processes and then to consider what methods and facilities would be most effective in providing answers to the process-oriented problems. No attempt was made to prioritize facilities and equipment; this will be highly dependent on the priority themes being addressed. The needed facilities and instruments are discussed under the Research Opportunities sections in Chapters 2 through 5 and are summarized in Chapter 6. Actual equipment ranges from large platforms (such as space satellites and drilling vessels) through supercomputers and laboratory experimental equipment (such as large-volume, high-pressure apparatus) to sensitive analytical instruments (such as ion microprobes), a host of smaller laboratory instruments, and field equipment (such as digital seismometers). Many of the priority themes share the need for certain instrumentation and facilities. This multiple use is one of the criteria that might be used in decisions about prioritization of equipment purchase, given a budget allocation. Another is the novelty of the research and exploration made possible by the equipment.

Data Gathering and Handling

The earth sciences, along with many other fields, are experiencing revolutions in data handling and computing. There are enormous amounts of information available to solid-earth scientists in the form of maps, text, physical samples, aerial and space-based imagery, well logs, potential field data, and seismic data, for example. The acquisition, retention, dissemination, and use of data have changed because of the rapid development of the computer. The growth and use of digital data have overwhelmed more traditional methods of data management. Within the profession, coordination of the retention and distribution of data is currently limited. Incompatible data formats, lack of knowledge about the existence of data, proprietary and national security concerns, and the lack of available archives all limit the potential use of data in solving important problems.

The traditional focus has been on the scientific effort, not data management. This emphasis is appropriate and should continue; however, ways of managing and resources for handling data acquired through scientific studies should be considered in the early stages of planning experimental programs and investigations.

Greatly improving the availability and utility of solid-earth science data requires a national solid-earth science data policy or set of guidelines to address a wide range of issues. Such a policy should deal with issues of data conversion (from analog to digital form), data rescue, incentives for data retention and dissemination, data-base standardization, exchange formats, data directories or catalogs, in-

TABLE 2 Research Opportunities

Research Areas	Objectives	
	A. Understand Processes	B. Sustain Sufficient Resources Water, Minerals, Fuels
I. Global Paleoenvironments and Biological Evolution	■ Soil development and contamination ■ Glacier ice and its inclusions ■ Quaternary record ■ Recent global changes ■ Paleogeography and paleoclimatology ■ Paleoceanography ■ Forcing factors in environmental change ■ History of life ■ Discovery and curation of fossils ■ Abrupt and catastrophic changes ■ Organic geochemistry	■ Mineral deposits through time
II. Global Geochemical and Biogeochemical Cycles	■ Geochemical cycles: atmospheres and oceans ■ Evolution of crust from mantle ■ Fluxes along ocean spreading centers and continental rift systems ■ Fluxes at convergent plate margins ■ Mathematical modeling in geochemistry	■ Organic geochemistry and the origin of petroleum ■ Microbiology and soils
III. Fluids in and on the Earth	■ Analysis of drainage basins ■ Mineral-water interface geochemistry ■ Pore fluids and active tectonics ■ Magma generation and migration	■ Kinetics of water-rock interaction ■ Analysis of drainage basins ■ Water quality and contamination ■ Modeling water flow ■ Source-transport-accumulation models ■ Numerical modeling of the depositional environment ■ In situ mineral resource extraction ■ Crustal fluids
IV. Crustal Dynamics: Ocean and Continent	■ Landform response to change ■ Quantification of feedback mechanisms for landforms ■ Mathematical modeling of landform changes ■ Sequence stratigraphy ■ Oceanic lithosphere generation and accretion ■ Continental rift valleys ■ Sedimentary basins and continental margins ■ Continental-scale modeling ■ Metasomatism and metamorphism of lithosphere ■ State of the crust: thermal, strain, stress ■ Convergent plate boundary lithosphere ■ History of mountain ranges: depth-temperature-time ■ Quantitative understanding of earthquake rupture ■ Rates of recent geological processes ■ Real-time plate movements and near-surface deformations ■ Geological prediction ■ Modern geological maps	■ Sedimentary basin analysis ■ Surface and soil isotopic ages ■ Prediction of mineral resource occurrences ■ Concealed ore bodies ■ Intermediate-scale search for ore bodies ■ Exploration for new petroleum reserves ■ Advanced production and recovery methods ■ Coal availability and accessibility ■ Coal petrology and quality ■ Concealed geothermal fields
V. Core and Mantle Dynamics	■ Origin of the magnetic field ■ Core-mantle boundary ■ Imaging the Earth's interior ■ Experiments at high pressures and temperatures ■ Chemical geodynamics ■ Geodynamic modeling	

TABLE 2 Research Opportunities (continued)

Objectives

Research Areas	C. Mitigate Geological Hazards Earthquakes, Volcanoes, Landslides	D. Minimize Global and Environmental Change Assess, Mitigate, Remediate
I. Global Paleoenvironments and Biological Evolution		■ Environmental impact of mining coal ■ Past global change ■ Catastrophic changes in the past ■ Solid-earth processes in global change ■ Global data base of present-day measurements ■ Volcanic emissions and climate modification
II. Global Geochemical and Biogeochemical Cycles	■ Seismic safety of reservoirs ■ Precursory phenomena and volcanic eruptions ■ Volume-changing soils	■ Earth-science/materials/medical research ■ Biological control of organic chemical reactions ■ Geochemistry of waste management
III. Fluids in and on the Earth		■ Isolation of radioactive waste ■ Groundwater protection ■ Waste disposal ■ In situ cleanup of hazardous waste ■ New mining technologies ■ Waste disposal from mining operations ■ Disposal of spent reactor material
IV. Crustal Dynamics: Ocean and Continent	■ Earthquake prediction ■ Paleoseismology ■ Geological mapping of volcanoes ■ Remote sensing of volcanoes ■ Quaternary tectonics ■ Densifying soil materials ■ Landslide susceptibility maps ■ Preventing landslides ■ Dating techniques ■ Real-time geology ■ Systems approach to geomorphology ■ Extreme events modifying the landscape ■ Geographic information systems ■ Land use and reuse ■ Hazard-interaction problems ■ Detection of neotectonic features ■ Bearing capacity of weathered rocks ■ Urban planning: underground space ■ Geophysical subsurface exploration ■ Detection of underground voids	
V. Core and Mantle Dynamics		

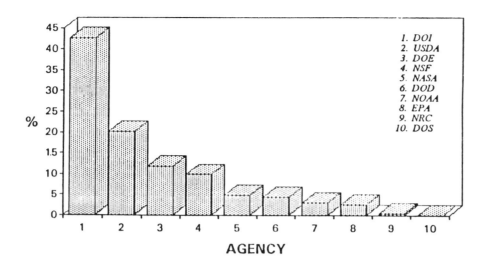

1. DOI
2. USDA
3. DOE
4. NSF
5. NASA
6. DOD
7. NOAA
8. EPA
9. NRC
10. DOS

FIGURE 1 Estimated percentages of total federal agency expenditures (fiscal year 1990) in the solid-earth sciences; details are given in Appendix A.

formation systems research, and training of students and professionals in data management.

Funding for Priority Themes

Total federal funding for solid-earth science activities in fiscal year 1990 was on the order of $1,368 million (see Appendix A). Figure 1 shows the relative percentages of support by the different departments and agencies. Details (in relation to the priority themes) of the types of activities that were supported are given in Appendix A.

Considering the overall distribution of federal support in Appendix A across the priority themes, there do not appear to be significant gaps, which indicates that the existing national research structure is working reasonably well. Many of the priority themes are already well established. New funding for modern equipment, technical support staff, and adequate ongoing operating support is required to sustain important progress on most of the themes.

Future science planners might benefit by keeping in view the overall distribution of funds among overlapping and interlocking priority themes. The selections and funding base are considered to be the starting points for priority considerations for the next decade of research. Each federal agency should do what it can to increase support in those high-priority areas within its domain, developing a

schedule to bring the appropriate equipment and facilities into operation as funding permits. For several programs, closer coordination among agencies would surely be fiscally sound as well as beneficial to scientific inquiry.

State geological surveys had a combined budget of $133 million in 1991, a base equal to about 10 percent of the federal support; these expenditures tend to focus on site-specific issues. The petroleum and mining industries have typically spent an amount equivalent to about a quarter of the federal total on solid-earth science research (usually resource focused but with an increasing amount of effort devoted to environmental issues).

Global Collaboration

In 1982 the National Science Board pointed out that "maintaining the vigor of the U.S. research effort requires a broad, worldwide program of cooperation with outstanding scientists in many nations." The solid-earth sciences, by their very nature, have always had a global orientation, but a special opportunity is at hand for international activities. It is now irrefutable that humans are altering the environment of the Earth at a rapid rate. A growing concern of all nations is the influence of increasing populations on the environment. Understanding and addressing global changes demand

> *International scientific cooperation is needed to understand global earth systems.*

international cooperation and planning. The earth sciences, through their emphasis on international research and communication, can help show the way toward a new era of global cooperation in science and societal relations. Solid-earth scientists will play increasingly important roles in assembling crucial data required to make provident decisions on an ever increasing range of issues. The United States, as a member of the global community, has a responsibility to aid in resolving such problems in the interest of all humanity.

In addition, many geological processes that are known but imperfectly displayed in the United States must be studied in other countries in order to be understood. For example, the principles of metallogenesis, tectonism, and crustal evolution that are applied to geological studies in the United States are derived from observations made throughout the world. This leads to refinements in our understanding of earth processes.

RECOMMENDATIONS

Recommendations for action in the areas affecting the solid-earth sciences—education, research support, and the national approach to both—are presented below. **The committee's overarching recommendation, which is basic to all its other suggestions, is that there should be a commitment within the United States to earth system science.** Knowledge of the interrelationships among the solid earth, its fluid envelopes, and the biosphere is crucial to humankind's continued well-being.

Education Recommendations

The continued vitality of the solid-earth sciences is critically dependent on a continuous supply of well-prepared geoscientists. Chapter 6 presents a number of specific actions for the graduate, undergraduate, and secondary-school levels. Three recommendations for college curricula merit special attention:

1. Conventional disciplinary courses should be supplemented with more comprehensive courses in earth system science. Such courses should emphasize the whole Earth, interrelationships and feedback processes, and the involvement of the biosphere in geochemical cycles.

2. New courses need to be developed to prepare students for increased employment and research opportunities in such areas as hydrology, land use, engineering geology, environ-mental and urban geology, and waste disposal. Such courses will be necessary to prepare students for changing careers in both the extractive industries and environmental areas of the earth sciences. No longer are these two areas separate, as mineral and energy resources need to be exploited in environmentally sound ways.

3. Colleges and universities should explore new educational opportunities (at both the undergraduate and graduate levels) that bridge the needs of earth science and engineering departments. This need arises from the growth of problems related to land use, urban geology, environmental geology and engineering, and waste disposal. The convergence of interests and research is striking, and the classical subject of "engineering geology" could become a significant redefined area of critical importance for society.

Research Recommendations

The committee discovered a remarkable degree of consensus when it selected the top-priority research opportunities for each of the priority themes; the eight themes are listed below and are summarized in Table 3. Each has two high-priority research opportunities associated with it under the same priority theme (with a third for resources). In many cases they were strong contenders for the top-priority position. These high-priority selections follow the top-priority selections.

Priority Theme A-I

Top priority: There should be a coordinated effort to understand how the Earth's environment and biology have changed in the past 2.5 million years. The current research activities of many federal agencies bear on this issue, and international involvement would be appropriate as well.

High-priority topics are:

■ characterization of the environmental and biological changes that have taken place over the past 150 million years, since the oldest preserved oceans began to evolve, and

■ exploration of environmental and biological changes prior to 150 million years ago.

Priority Theme A-II

Top priority: The earth sciences need to establish how global geochemical cycles have oper-

TABLE 3 Summary of the Top-Priority and High Priority Research Opportunities for Each Theme

Top Priorities	High Priorities
A-I. Global Paleoenvironments and Biological Evolution **The Past 2.5 Million Years**	■ The past 150 million years ■ Prior to 150 million years ago
A-II. Global Geochemical and Biogeochemical Cycles **Biogeochemistry and Rock Cycles Through Time**	■ Construct models of the interaction between cycles ■ Establish how geochemical cycles operate in the modern world
A-III. Fluids in and on the Earth **Fluid Pressure and Fluid Composition in the Crust**	■ Fluid flow in sedimentary basins ■ Microbial influences on fluid chemistry
A-IV. Crustal Dynamics: Ocean and Continent **Active Crustal Deformation**	■ Landform responses to climatic, tectonic, and hydrologic events ■ Understanding crustal evolution
A-V. Core and Mantle Dynamics **Mantle Convection**	■ Origin and variation of the magnetic field ■ Nature of the core-mantle boundary
B. To Sustain Sufficient Natural Resources **Improve the Monitoring and Assessment of the Nation's Water Quantity and Quality**	■ Sedimentary basin research ■ Thermodynamic and kinetic understanding of water-rock interaction and mineral-water interface geochemistry ■ Energy and mineral exploration, production, and assessment strategies
C. To Mitigate Geological Hazards **Define and Characterize Regions of Seismic Hazard**	■ Define and characterize areas of landslide hazard ■ Define and characterize potential volcanic hazards
D. To Minimize and Adjust to the Effects of Global and Environmental Change **Develop the Ability to Remediate Polluted Groundwaters, Emphasizing Microbial Methods**	■ Isolation of toxic and radioactive waste ■ Geochemistry and human health

ated through time. This information, which is essential to working out how the earth system operates, is now a realistic target that could be achieved by coordinating a number of federal programs and current national and international activities.

High-priority topics are:

■ construction of models of the interaction between biogeochemical cycles and the solid earth and climatic cycles and
■ establishment of how geochemical cycles operate in the modern world.

Priority Theme A-III

Top priority: The earth sciences need to take up the challenge of investigating the three-dimensional distribution of fluid pressure and fluid composition in the Earth's crust. The instrumental, observational, and modeling capabilities that exist within various federal programs can be effectively focused on this problem. International coordination is important.

High-priority topics are:

■ modeling fluid flow in sedimentary basins and
■ improved understanding of microbial influences on fluid chemistry, particularly groundwater.

Priority Theme A-IV

Top priority: There should be coordinated and intensified efforts to understand active crustal deformation. The opportunity exists to revolutionize current knowledge of this area, which is vital not only to the solid-earth sciences but also to the missions of several federal agencies and various state and international bodies.

High-priority topics are:

■ exploration of landform response to climatic, tectonic, and hydrologic events and
■ increased comprehension of crustal evolution.

Priority Theme A-V

Top priority: An integrated attack needs to be mounted to solve the problems of understand-

ing mantle convection. Seismic networks, satellite data, high-pressure experiments, magnetic observatories, geochemistry, drilling, and computational modeling can all be brought to bear. Again, federal, national, and international organizations will be involved.

High-priority topics are:

■ establishment of the origin and temporal variation of the Earth's internally generated magnetic field and

■ determination of the nature of the core-mantle boundary.

Priority Theme B

Top priority: A dense network of water quality and quantity measurements, including resampling at appropriate intervals, should be established as a basis for scientific advances. Coordination of federal and state agencies that have programs in the field will be needed.

High-priority topics are:

■ sedimentary basin research, particularly for improved resource recovery;

■ improvement of thermodynamic and kinetic understanding of water-rock interaction and mineral-water interface geochemistry; and

■ development of energy and mineral exploration, production, and assessment strategies.

Priority Theme C

Top priority: There should be an effort to define and characterize regions of seismic hazard. Because many people (and much property) in the United States are endangered by earthquakes, improved understanding of seismic occurrences is a pressing need. This issue is important to the missions of several federal agencies and to organizations ranging from local to international.

High-priority topics are:

■ definition and characterization of areas of landslide hazard and

■ definition and characterization of potential volcanic hazards.

Priority Theme D

Top priority: The earth sciences need to develop the ability to remediate polluted groundwaters on local and regional scales, em-

phasizing microbial methods. Coordination of local, industry, state, and federal activities will enhance the potential for success, and international involvement would be desirable.

High-priority topics are:

■ secure the isolation of toxic and radioactive waste from household, industrial, nuclear-plant, mining, milling, and in situ leaching sources and

■ investigation of the relationship between geochemistry and human health.

General Recommendations

Recommended priorities for research will need to be developed within the existing complex structure in which federal agencies, most with highly specific missions, interact with universities, industry, and each other. These groups should also be interacting with professional societies, state and local agencies, other nations, and international organizations. The recommendations that follow are intended to provide guidance for the diverse communities involved in research and practice in the solid-earth sciences in the coming decade.

RECOMMENDATION 1. There should be a major commitment to the study of the whole earth system, emphasizing interrelationships among all parts of the Earth. The recommended commitment should be akin to the space missions that have revolutionized our understanding of other planets in the past two decades. We are able for the first time to recognize features associated with the internal evolution of our planet, the actual heterogeneities that drive the geological processes of the Earth. Thus, we are at the threshold of a new and fundamental understanding of global geological phenomena. To be effective, a "Mission to Planet Earth" must be a visionary and broad-ranging study of our entire planet, from core to crust. At least four elements are widely recognized as being crucial to this program: (1) the need for global observations, including those based on space technologies and international collaborations; (2) the development and application of novel instrumentation; (3) the utilization of new computer technologies; and (4) a commitment to support advanced training.

RECOMMENDATION 2. High priority should continue to be given to the best proposals from individual investigators. The intellectual resources represented by members of the earth science community are our most valuable asset. The U.S. scientific and industrial population may receive less support in some areas than its international competitors, but it does not suffer from lack of imagination. Core support for individual investigators is the best way to ensure that the diversity of ideas and approaches that are at the root of American inventiveness remains a strong feature of the U.S. effort.

RECOMMENDATION 3. The newest tools for data acquisition need to be made available for use in earth science research. Advanced instrumentation is urgently needed for experiment and analysis in the laboratory and for deployment in space (on satellites), at sea (on research vessels and on the sea bottom), in aircraft, and on land (in networks and in boreholes and movable arrays).

RECOMMENDATION 4. Opportunities for the integration and use of observations and measurements from advanced space-borne instruments in solid-earth geophysics and geology should continue to be made available. The opportunity for increased understanding of the continents using an integrated approach with remote sensing, field, laboratory, and other data (e.g., seismic) is extraordinary. Remote sensing data should be incorporated and used as a standard field geology tool, throughout the undergraduate curriculum and especially in field geology courses. At the graduate level, research should address geological problems aided by remote sensing methods rather than consider remote sensing as a separate discipline.

RECOMMENDATION 5. There is an essential need for the production and availability of interactive data banks on a national level within the earth sciences. With new methods of digital acquisition, handling, and archiving, and with the growth in the use of geographic information systems along with the Global Positioning System, there are major opportunities to apply the computer revolution to the solid-earth sciences. It is time to integrate the vast amounts of solid-earth science data in nondigital form, like maps, with the exponentially growing digital data sets. National coordination of data-handling services, retrieval procedures, networking, and dissemination practices is required to improve access to the wealth of data held by government, industrial, and academic organizations. This will ensure its best use in understanding the Earth, in sustaining resources, in

mitigating impacts of hazards, and in adjusting to environmental change.

RECOMMENDATION 6. Efforts need to be made to expand earth science education to all. Citizens need to understand the earth system to make responsible decisions about use of its resources, avoidance of natural hazards, and maintenance of the Earth as a habitat. School systems must respond to this need. At the university level, curricula should be adjusted to meet the needs of contemporary society while maintaining excellence at the professional level.

RECOMMENDATION 7. Research partnerships involving industry-academia-government are encouraged to maximize our understanding of the Earth. Cooperative multidisciplinary investigations that pool intellectual resources residing in government, academic, and industrial sectors can produce more comprehensive research efforts. The primary objectives of governmental, industrial, and academic groups are diverse. The breadth of disciplines that collectively exist within groups spans our science, but each has its own primary research objectives. Each sector has much expertise to offer that would make it possible to capitalize on the complementary nature of collaboration. The solid-earth sciences stand to gain immeasurably if these three major research communities establish forward-looking cooperative programs.

RECOMMENDATION 8. Increased U.S. involvement in international cooperative projects in the solid-earth sciences and data exchange are essential. Increased understanding of the Earth as a system requires that regional problems be looked at from an international perspective. Cooperative programs involving both nongovernmental international science programs and individuals should be strengthened. Groups involved in U.S. foreign policy decisions should be aware of the importance of the earth sciences in global agreements about issues such as waste management, acid rain, hazard reduction, energy and mineral resources, and desertification. New linkages between the West, the former Soviet Union, and Eastern Europe present a timely opportunity for U.S. scientists to join with scientists from those countries in data collection and data sharing to increase knowledge of earth systems. Such cooperation with other countries also can be an important tool in U.S. foreign policy.

1

Global Overview

ESSAY ON THE EARTH SCIENCES

This is a particularly opportune time to assess the state of the earth sciences. New concepts arising from breakthroughs of the past quarter century as well as innovative technologies for gathering and organizing information are expanding the frontiers of knowledge about the earth system at an accelerating pace. Research ranges from the atomic scale of the scanning-tunneling microscope and ion microprobe to the global scale of worldwide seismic networks and images produced from data gathered by orbiting satellites. Earth scientists are constructing a comprehensive picture of the Earth, one that interrelates the physical, chemical, geological, and biological processes that characterize the planet and its history (Plate 1). Scientific advances and opportunities abound at the same time that environmental and resource problems affecting the world's population have become international political issues.

Human societies face momentous decisions in the next few years and decades. Issues such as atmospheric changes, nuclear power, hazardous wastes, environmental degradation, overpopulation, soil erosion, water quality and supplies, and the destruction of species cannot be ignored. The decisions eventually made about these environmental issues, including prediction of changes, must be based on reliable knowledge—knowledge that comes from understanding the earth system and its history.

Living on Earth

The more we have learned about the Earth, the more we have come to appreciate the many ways in which it is suited to life. The Earth is just the right distance from the Sun to have surface temperatures optimal to sustain living things. Its vast oceans have remained liquid, neither boiling into the atmosphere nor freezing into a solid block of ice, since shortly after its formation some 4.5 billion years ago. It is the only planet in the solar system that exhibits plate tectonics, which recycles nutrients and other materials essential to life through the interior of the planet and back to the surface. The Earth is unique in sustaining an atmosphere that is one-fifth oxygen—oxygen that was generated over eons by single-celled organisms and that, in turn, spurred the evolution of multicellular organisms.

More particularly, the Earth is a congenial home for humans. All of the materials we use in our daily lives come from the Earth—fuels, minerals, groundwater, even our food (through the intermediaries of soil, water, and fertilizer). We have a strong psychological affinity for certain places on the Earth, for the regions where we grow up and live and for the wild and beautiful tracts, whether preserved in parks or apparent in our everyday surroundings. But we have altered the surface extensively during our occupation—erecting structures, clearing forests, damming rivers. Despite our activities, the Earth continues its normal path of inexorable change—through violent paroxysms such as earthquakes and landslides or through the slow, steady erosion of soils and coastlines.

The Earth is a resilient planet. Long after human beings have vanished from the planet, its basic cycles will persist. The largest man-made structures will erode and disappear, radioactive materials that have been gathered will decay, man-made concentrations of chemicals will disperse, and new species will evolve and perish. The oceans, atmosphere, solid earth, and living things will continue to interact, just as they did before humans appeared on the scene.

But some earth systems are very fragile. We dispose of our wastes in the same sedimentary basins that supply us with the bulk of our groundwater, energy, and mineral resources. Through our social, industrial, and agricultural activities, we are changing the composition of the atmosphere, with potentially serious effects on climate and on terrestrial and marine ecosystems. The human population is expanding into less habitable parts of the world, which increases vulnerability to natural hazards and strains the biological and geological systems that sustain life.

If present trends continue, the integrity of the more fragile systems on which human societies are built cannot be assured. The time scale for the breakdown of these systems may be decades or it may be centuries, but we cannot continue to use the planet as we have been using it. Present trends need not continue, because we are unique among the influences that affect earth systems: we have the ability to decide among various courses, to weigh the pros and cons of alternative actions, and to behave accordingly.

The history of the geological sciences offers many reasons to be optimistic. One of the triumphs of twentieth-century science and technology has been the worldwide identification and extraction of energy and mineral resources, an activity that has brought an increased standard of living to an expanding human population. The geological sciences have demonstrated ways of maintaining water quantity and quality, disposing of wastes safely, and securing human structures and facilities against natural hazards. Essentially, the geological sciences teach us about the nature of the Earth and about our role on it.

Understanding the Earth

The process of understanding the Earth has just begun. If human beings are to survive and prosper for more than a moment in geological time, an understanding of the intricacies of interacting earth systems is a necessity. We must find and develop the resources needed to sustain and improve the human condition. We must also preserve and improve the environment on which essential and aesthetic human needs depend. We need to know enough to predict the beneficial and destructive consequences of our actions.

Among the questions now confronting solid-earth scientists are the following: How did the Earth form, and what accounts for the composition of its various components? How does the core interact with the mantle, and how does the mantle drive plate tectonics? How have the continents evolved? How has the solid earth influenced the course of biological evolution and extinctions, and how have these processes affected the Earth? How have the ocean and atmosphere changed over time, and how can they be expected to change in the future as human influence increases? How can wastes be safely isolated in geological repositories? How are the small-scale movements of the surface related to tectonic processes? None of these questions can be answered without contributions from several subdisciplines of the solid-earth sciences. Thus, interdisciplinary interactions will continue to grow in importance as the earth sciences advance.

Because its concerns are with global features and phenomena, the earth sciences are intrinsically an international undertaking. Basic field data must be gathered from diverse regions of the Earth, and earth scientists must conduct experiments on a global basis to accurately determine its composition and dynamics. International cooperation and exchange among scientists in different countries are essential.

Quantitative models have been developed for a number of earth processes. For instance, numerical computer simulations have been devised for mantle convection, for the evolution of sedimentary basins, for surface processes such as erosion, for fluid flow, and for rupture processes in earthquakes. At the same time, solid-earth scientists are relying on increasingly sophisticated instrumentation to expand and to keep up with the data needed for future discoveries and for models that assess and predict earth processes. Even such traditionally descriptive subdisciplines as mapping and paleontology are becoming increasingly quantitative with the advent of digital analyses and computerized data bases.

Predicting the Earth's Future

Better understanding of the way natural processes operate over time, whether quantitative or descriptive, leads to more accurate prediction of the future effects of those processes. For instance, a greater understanding of surface processes enhances the ability to predict such events as floods, landslides, and subsidence, which assault human structures and facilities. This enhanced ability is the basis for land-use planning, enabling society to minimize the costs and problems associated with geological hazards. The potential losses due to earthquakes and volcanic eruptions have stimulated research on the condition of and changes in the state of the crust before such events, in the hope that similar changes may be used for future predictions.

Prediction in the earth sciences requires an understanding of the role that past events have had in shaping our planet and an accurate projection of data-based interpretations into the future. Field work will remain the principal source of ground truth observations, while laboratory work will calibrate those observations. At the same time, theory will produce generalizations that can be extrapolated to unexplored regions, to different scales, and to events occurring at different times.

Greater predictive abilities should enable earth scientists to tell how human influences will shape the future. Earth systems undergo natural fluctuations on which the changes caused by human influences are superimposed. Distinguishing between natural fluctuations and human-induced changes in earth systems remains an essential task.

An important question is to what extent geological changes, whether driven by natural processes or human intervention, exhibit chaotic behavior. If chaos dominates, the ability to predict many earth phenomena may be limited. But even chaotic systems are subject to statistical predictions related to the size of the events that may be expected and to their average frequency.

THE EARTH AND ITS COMPONENTS

Development of the geological sciences has been marked by two seemingly contradictory but in fact complementary trends. Earth scientists have become ever more aware of the great complexity of the Earth and its constituents; certain individual minerals, for example, are among the most complex inorganic substances known. At the same time, earth scientists are uncovering the principles that lie behind the Earth's bewildering diversity—unifying concepts that bring coherence to a welter of seemingly unrelated observations.

One way to approach the planet's apparent complexity is to look upon the Earth as a system of interacting components. Viewed in its entirety, the earth system is driven by energy from two sources: (1) inside the Earth from primordial energy remaining from accretion and by radioactive decay of elements and (2) from the Sun (Figure 1.1). As it moves through a homogeneous system, energy tends to reorganize that system into components and then to drive physical and chemical interactions among the components. In this way the Earth has been organized into a system of interacting components on a variety of spatial scales. As with any

classification, the boundaries between these components are not everywhere sharp; for instance, the Earth contains liquids (in the form of magmas,

FIGURE 1.1 Earth system processes are driven by both internal and external (solar) energy. From NASA *Earth System Science* report (1986).

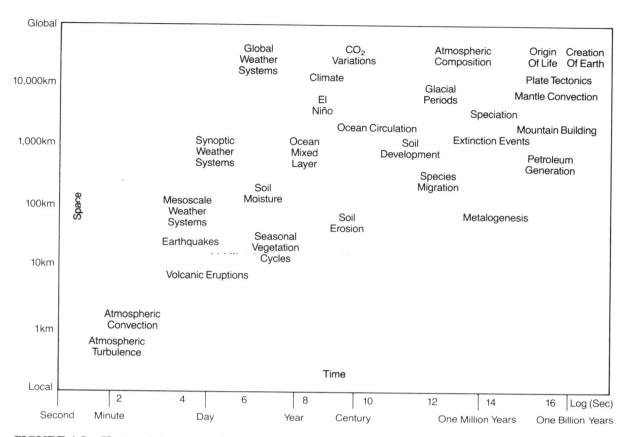

FIGURE 1.2 Characteristic space and time scales of earth system processes. From NASA *Earth System Science* report (1986).

groundwater, hydrocarbons, and so on) and gases (Figure 1.2).

The Earth can be organized into temporal as well as spatial components. Geologists demonstrated the great antiquity of the Earth in the eighteenth and nineteenth centuries by analyzing the layers, or strata, of rock formed from sedimentary deposits. Since then, geologists have continued to refine and elaborate the geological time scale through increasingly sophisticated dating and stratigraphic techniques.

The atmosphere and oceans mix too rapidly to carry information about times long past except in their bulk compositions. But some of the rocks in the continents have remained virtually unchanged since their formation billions of years ago. Thus, events that took place 3.5 billion years ago are frozen like a snapshot in the rocks of the oldest parts of continents. Geologists examine these rocks for clues about conditions when the rocks were formed and changes they underwent thereafter. In this way geologists have been able to chart past changes in temperature, sea level, atmospheric and oceanic composition, volcanism, plate tectonics, the evolu-

tion of life, the Earth's magnetic field, sedimentation rates, and glaciation (Plate 2). Earth scientists study this record to learn how the components of the Earth behaved and interacted in the past, to understand how they interact now, and to predict how they will interact in the future.

Geology is a science in which time plays an especially crucial role. This dependence on time adds a unique dimension to geological phenomena. Every rock, every mountain, every volcano is the product of a particular circumstance. Similarly, every period of earth history is in some way special because of the ever-changing combinations of variously interacting spatial and temporal components, including the remarkable consequences of evolving life.

A further complicating factor is the broad range of scales over which geological events occur. The spatial components range from individual mineral grains to objects the size of the Earth and larger, as in the case of the magnetic field. The temporal components range from seconds—the duration of an earthquake—to billions of years. Understanding geological processes therefore involves enormous extrapolations of experimental and theoretical knowledge.

Until recently, the geological sciences dealt with the great complexity of the Earth largely by focusing on specific spatial, temporal, or compositional regimes. Subdisciplines studied phenomena that were largely compartmentalized, and influences from outside the domain of study were greatly simplified or ignored. As a result, the various subdisciplines of the field—geology, geochemistry, geophysics, paleontology, geohydrology, geoengineering—were neatly pigeonholed and had little need or incentive to communicate or work with one another.

During the past 25 years, that fragmentation of the geological sciences has been disappearing under the influence of several momentous developments. The result has been a profound change in the way geoscientists study the Earth.

UNIFYING FORCES IN THE GEOLOGICAL SCIENCES

Many advances have simultaneously contributed to a general unification of the geological sciences, but three in particular stand out: the plate tectonic revolution, the enhanced ability to produce images of both the surface and the interior, and the increasing recognition of humanity as a geological agent.

Plate Tectonics

In the 1960s the geological sciences experienced a conceptual revolution that continues to affect the field today. Traditionally, most geologists analyzed the history of the Earth primarily in terms of vertical movements: mountains emerged from a buckling crust and were eroded, sea levels rose and fell, whole areas of continents were uplifted. But a quarter century ago a series of developments in marine geology and paleomagnetism—the record of the magnetic signals preserved in the rocks—resulted in a radically new picture of the Earth. This new picture acknowledged the essential role of large-scale horizontal movements throughout the Earth's evolution as well as that of vertical movements.

This concept of plate tectonics has endured through two decades of scientific scrutiny and is now regarded as an established fact, a situation nearly unthinkable in 1960. Scientists now know that the crust is composed of about a dozen major (Plate 3) and several minor plates that constantly move and jostle each other in response to movements in the underlying mantle. Where two plates converge, one may override the other, and the leading edge of the lower plate may melt as it reaches greater depths or may produce melting in the overlying mantle. This convergence creates the oceanic trenches and zones of coastal volcanoes seen on the western edge of South America, in the Southwest Pacific, and elsewhere. If neither plate sinks, collision creates a wrinkled mountain belt, such as the Himalaya or the Urals. Where plates merely sideswipe each other, the boundary shears laterally, as happens along California's San Andreas fault. In each type of plate boundary zone, earthquakes occur when the plates bind, build up stress, and suddenly slip free. A plate may break apart; two observable examples are a rift across a volcanic center such as the Red Sea and a rift through the interior of the plate related to stresses on a distant plate boundary, as at Lake Baikal.

The plate tectonic model enables geoscientists to synthesize formerly separate and enigmatic facets of crustal processes. This ongoing synthesis helps explain the distribution and timing of mountain building, igneous activity, earthquakes, sedimentary basins, ore deposits, and other features of both practical and theoretical import. Plate tectonics also establishes a new view of earth history, in which the plates have moved with time, with consequent effects on the hydrosphere, atmosphere, and biosphere.

Plate tectonics relates activity within the Earth that is directly associated with the movements of plates to characteristics and changes of the surface. For example, midplate hotspots, which are responsible for such features as the volcanic activity at Yellowstone Park and the Hawaiian Islands, are caused by plumes of hot material rising from within the mantle, perhaps from as deep as the core-mantle boundary. Plate tectonics also offers a natural framework for geochemical cycles, which involve the transfer of elements among the various envelopes that form the earth system.

The revolutionary theory of plate tectonics is comparable in power and elegance to the Copernican theory of the sixteenth century, to Newton's theory of gravitation in the seventeenth century, to the establishment of atomic theory at the beginning of the nineteenth century, to the theory of evolution in biology later in that century, and to the development of quantum mechanics and relativity in physics in the twentieth century. Plate tectonic theory is the newest of these major advances in human understanding, and as a result many of its implications have yet to be worked out. Fundamental questions about plate tectonics remain unanswered. The theory is better at explaining the large-scale features of the Earth than the smaller ones. It suggests the locations of some natural resources but not others. And although the source of the energy that drives the plates is assumed to be convection in the mantle, this is still unproven. These questions are fundamentally important, not only for understanding the functioning of the Earth but also for applying the theory to practical problems such as resource exploration or the prediction of natural hazards.

Images of the Earth

Research in the geological sciences is a data-intensive activity. Many advances in the discipline have occurred when new observations became available or when innovative technologies allowed the gathering of new kinds of data. At the same time, recently revised ideas about the Earth have spurred the development of original instruments and techniques specifically designed to test those ideas.

Measurements of the Earth can today be made with unprecedented levels of accuracy and sensitivity. Just a few decades ago, classical wet chemical methods typically allowed the detection of elements in natural materials at a level of about one part per thousand. Today, mass spectrometers, electron microprobes, and particle accelerators detect elements and isotopes at levels of one part per billion or better from sample areas of just a few microns. Such levels of sensitivity have tremendously increased our ability to date earth materials and analyze the changes they have undergone by measuring isotopic and elemental compositions.

The traditional forms of gathering geological data are field observations and the preparation of geological maps, which remain a fundamental enterprise. Maps of the surface and near surface are critical in locating and assessing fuels, minerals, and other resources; in analyzing environmental conditions and geological hazards; and in reconstructing the geological history of an area. Maps are the basic tools that geologists use in approaching specific problems (Plate 4). The ability to use maps effectively is greatly enhanced by geographic information systems (GIS), a method of analyzing and integrating digitial spatial data of diverse geological and geographical nature.

Geologists have also developed methods for producing images of the Earth's interior. Drilling can probe a few kilometers into the crust to produce measurements and samples of material; beyond a dozen or so kilometers, indirect geophysical methods must be used. Seismic waves, generated either by earthquakes or human activity, are the most effective means of producing images of the deep layers. Geophysicists can detect variations in the velocity of reflected and refracted seismic waves from the surface to its core. These variations (Plate 5) are caused by changes in the density of the materials within the Earth. Laboratory measurements of the physical and chemical properties of earth materials supplement seismic images to model the structure, composition, and dynamics of otherwise inaccessible portions of the Earth. In addition, geochemists sample materials that originated in the crust and mantle and develop experiments that test the range of conditions that led to their formation; both produce further hypotheses about the nature of the Earth's interior.

A new source of striking images began to be produced in the 1960s with the launch of the first satellites designed to observe the Earth and return data to ground receivers. Satellite monitoring offers an integrated and efficient method of gathering information on a global basis. Visual images are one form of remote sensing obtained from orbiting platforms, but satellites also monitor the Earth at nonvisible wavelengths, gauge the gravity and magnetic fields, measure the altitude of the land and sea precisely, and perform a number of other data-gathering operations.

Global positioning systems use satellites to measure the location of receivers on the Earth to an accuracy of a few centimeters. This technique has enabled earth scientists to measure the relative velocities of the plates, determine how the velocities change where plates come together, and assess the rigidity of plate interiors. In the future, global positioning systems may be used to measure changes in the ground surface before earthquakes, volcanic eruptions, or other natural hazards occur. These measurements may lead to reliable predictions of potential disasters and timely warnings to endangered populations.

Observations of other planets by interplanetary probes also have shed light on the history and nature of the Earth. The other planets in the solar system bear striking resemblances and notable dissimilarities to the Earth. Our neighbors in the solar system—the Moon, Venus, Mars, and Mercury—are analog of the Earth, frozen in different states of evolution. Investigations of these planets on an integrated global scale have encouraged earth scientists to adopt a similar approach to our own.

Variations in the Earth's Orbit

It was recognized in the nineteenth century that variations in the Earth's orbit would cause changes in incoming solar radiation that could be important in controlling ice ages. Theoreticians first calculated how these variations would interact, and the study of deep-sea sediments has yielded persuasive evidence that the recurring ice ages are indeed closely associated with orbital cycles. The importance of orbital variations in controlling the climatic changes of the past 2 million years has proved revolutionary. The critical studies of sediments from deep-sea cores were published less than 20 years ago, and since then application of the idea of orbital control has been applied to studies of ice cores and cave deposits. Researchers now have a yardstick to apply to the complicated record of the most recent past, which is helping to make their interpretations more quantitative. Problems being addressed currently include the extent to which evidence of orbital control can be recognized in older parts of the geological record and how orbital variations (which act directly on insolation) have affected not only such variables as sea-surface temperature but also atmospheric trace gas concentrations.

Humankind as a Geological Agent

The geological sciences have traditionally been responsible for finding and maintaining adequate supplies of fuels, minerals, and nonmineral resources at reasonable prices. The continuing ability to meet this demand depends on basic and applied research in the geosciences. But as human endeavors put pressure on the carrying capacity of settled regions, different components of the Earth's systems are affected. This report therefore does not limit its consideration of resources to the traditional meaning of the word.

Groundwater is an essential resource. In parts of the world the supply of groundwater is dwindling because it is being withdrawn from underground aquifers faster than it is being replenished. Accessible and stable places to live are also a resource that human populations are rapidly consuming. Soil erosion is accelerating in many areas around the world; for every pound of food consumed in the world, an average of 7 pounds of soil is lost to erosion. A habitable environment is a resource that pollutants released to the air, water, or solid earth can compromise. Biological species of the world are also a resource, but human activities have caused

extinctions at an accelerating rate. If current trends persist, a large proportion of the species existing today may disappear during the next few decades.

The consequences of human activity are a recent factor needed to understand earth systems. Before the control of fire, the environmental effects of human beings were comparable to those of other species. Then crop cultivation, animal domestication, and the subsequent appearance of urban civilization introduced a new set of forces onto the Earth. Today, human beings are changing basic earth processes in ways they have never been changed before because of the acceleration and concentration of the effects.

Each person in the United States uses an average of 16 metric tons—about 35,000 pounds—of minerals and fossil fuels each year. This amount includes only the use of materials; it does not include the material moved during the construction of homes, parking lots, factories, dams, and so on. On a worldwide basis, the human population uses nearly 50 billion metric tons of material each year. That is more than three times the amount of sediment transported to the sea by all the rivers of the world.

Clearly, humankind has become a geological agent that must be taken into account in considering the workings of the earth system. If future generations are to have resource supplies in the full sense of the word, decisions must be made within a context that considers the Earth as a total entity. The geological sciences offer information that will be invaluable in evaluating trade-offs and in balancing competing demands.

GOALS OF THE EARTH SCIENCES: RESEARCH FRAMEWORK

There are two fundamental reasons for pursuing the earth sciences. One is to learn more about the world where we live, to satisfy a basic human curiosity about our surroundings and our relationship to them. The other is that research in the geosciences can be used to improve the human condition. Geological knowledge is essential for making decisions regarding the use and preservation of resources or the protection of human life and habitats from the effects of natural disasters.

Stated this way, the distinctions between intellectual curiosity and utilitarian concern may appear fairly sharp, but in fact the two categories overlap extensively. Not all scientific studies have immediate applications, but experience shows that few findings in the geosciences remain unapplied for long. Today's theoretical science may very well be tomorrow's applied science. The process also works in reverse: applied studies, particularly through their effects on technology, make fundamental contributions to research on the nature and history of the planet.

The geological sciences draw on tools and knowledge developed in several other scientific disciplines, especially chemistry, physics, biology, and mathematics. At the same time, geological research has contributed concepts and techniques to these other disciplines. To take just one example, the structural determination of high-temperature superconductors drew heavily on mineralogical principles, and the perovskite structure of these superconductors is similar to the mineral structure of large parts of the mantle.

Because of its data-intensive nature, the earth sciences have been marked by a particularly close relationship between science and technology. As earth scientists have sought to increase their observational powers, they have developed or refined several devices that were then applied in many other fields, such as x-ray diffraction; electron and ion microprobes; and high-temperature, high-pressure equipment. Similarly, work in the geosciences has produced a number of materials that have found applications far beyond their original realm, including Pyrex glass, zeolite catalysts, and synthetic crystals.

In organizing this report, the committee made extensive use of the close links between basic and applied components of the geosciences. In addition to considering research on the Earth and its components, this report considers the field within the context of problems related to society's needs. The basic science appears within these sections to reflect considerations of how it might contribute to more applied problems.

This organization should not be interpreted as an indication of the relative importance of basic and applied research. Research must not be limited to short-term directed investigation. If it were, scientific innovation and serendipitous discovery would seldom occur, and the plethora of intellectual challenges that provide new dimensions for future scientific opportunity would disappear. A balance must be maintained between basic and applied science, so that each symbiotically nourishes the other.

Goal, Objectives, and Research Themes

Recent research and discoveries have made it possible to consider the Earth as a set of interrelated systems. The concept of plate tectonics provides a grand example of the planet as an integrated system, every part functioning to some degree separately but ultimately dependent on all others. New probes of the interior have reinforced the notion of an internal engine that drives geological processes. The Earth operates as a thermodynamic engine that generates flows and stresses and engenders geochemical cycles. The surface topography is roughly determined by internal movement, and its detailed architecture is sculpted by the action of fluids driven by energy from the Sun with the aid of gravity and tides. The near-surface chemistry involves interaction between the oceans, atmosphere, and fluids from the crust and mantle. The sinking of plates at subduction zones, the slow thermal convection of the mantle, and the volcanism associated with hotspots result in an exchange of materials between the surface and deep interior. The multidisciplinary research themes described in this report reflect a new awareness of the many chemical and thermal exchanges characterizing earth systems.

The **GOAL OF THE SOLID-EARTH SCIENCES** is:

to understand the past, present, and future behavior of the whole earth system. From the environments where life evolves on the surface to the interaction between the crust and its fluid envelopes (atmosphere and hydrosphere), this interest extends through the mantle and the outer core to the inner core. A major challenge is to use this understanding to maintain an environment in which the biosphere and humankind will continue to flourish.

The **OBJECTIVES** associated with this goal are:

A. **Understand the processes involved in the global earth system, with particular attention to the linkages and interactions between its parts (the geospheres)**
B. **Sustain sufficient supplies of natural resources**
C. **Mitigate geological hazards**
D. **Minimize and adjust to the effects of global and environmental change**

The following **RESEARCH AREAS** were selected for a dual purpose—they provide comprehensive coverage of the whole earth system, and they identify those processes and topics that offer the promise of achieving the scientific goal:

TABLE 1.1 Solid-Earth Science Research Framework

Objectives

Research Areas	A. Understand Processes	B. Sustain Sufficient Resources—Water, Minerals, Fuels	C. Mitigate Geological Hazards—Earthquakes, Volcanoes, Landslides	D. Minimize Global and Environmental Change—Assess, Mitigate, Remediate
I. Global Paleoenvironments and Biological Evolution	I-A	I-B	I-C	I-D
II. Global Geochemical and Biogeochemical Cycles	II-A	II-B	II-C	II-D
III. Fluids in and on the Earth	III-A	III-B	III-C	III-D
IV. Crustal Dynamics: Ocean and Continent	IV-A	IV-B	IV-C	IV-D
V. Core and Mantle Dynamics	V-A	V-B	V-C	V-D
	Facilities-Equipment-Data Bases Education: Schools, Universities, Public			

I. **Global paleoenvironments and biological evolution**
II. **Global geochemical and biogeochemical cycles**
III. **Fluids in and on the Earth**
IV. **Dynamics of the crust (oceanic and continental)**
V. **Dynamics of the core and mantle**

These research areas all represent major processes in the evolution of the Earth. They address exploration of the unknown, data collection, and theoretical modeling, and they offer the prospect of breakthroughs arising from new techniques, new instruments, and new concepts. They are all multidisciplinary: none is independent of the others. For example, the lithosphere is the rigid outer layer of the Earth, and its dynamic behavior relates to that of the mantle. The crust, an assemblage of rocks with a variety of compositions, is a blanket on top of the lithosphere, and movements of the crust cannot be considered independently of lithospheric dynamics. Geochemical cycles transfer material within and between the several geospheres; a better understanding of connections between major cycles in which the fluid envelopes dominate and those in

which the interior characteristics dominate is a conspicuous need. Much of the material transfer in geochemical cycles is accomplished by fluid transport involving either silicate melts or aqueous solutions. Important investigations of biogeochemical cycles consider the interaction between solid earth and fluid envelopes. A major factor in changing the paleoenvironments that influence biological evolution is plate tectonics, which is associated directly with mantle convection and lithosphere dynamics. Interconnections are legion, and most frontier research topics relate to more than one theme.

Research Framework

The objectives and research areas of the solid-earth sciences form a matrix (Table 1.1) that provides a **RESEARCH FRAMEWORK**, with the objectives (A-D) placed along the top and the research areas (I-V) defining the side of the matrix. (The numbering of the latter carries no implication for priorities; it merely reflects the progression deeper into the Earth.) The matrix has obvious parallels with Plates 1 and 2 showing versions of the whole earth system: the sketch of the globe with its interior exposed and the block diagram illustrating the relationships among the various geospheres. This matrix also outlines the organization of the report, as discussed below.

Every one of the research areas I-V not only requires the further understanding (basic science) represented by Objective A but also has applications to one or more of the societal challenges (applied science) represented by Objectives B, C, and D. Research projects with specific applications can be located in boxes I-B to V-D. Facilities, education, and other topics related to the infrastructure of science are shown below the matrix.

Most of the research discussed in this report fits somewhere into the matrix. The matrix displays the broad spectrum of earth sciences in a manner that avoids putting scientific merit into direct competition with societal needs. It also permits consideration of priority recommendations according to various criteria and from differing viewpoints. Priority is in the eyes of the beholder; different agencies have different missions, and they will enter the matrix at different points. However, the continuity between pure and applied science becomes obvious. Many scientific projects can have multiple objectives and themes, and this factor needs to be considered in setting a priority agenda.

Priority Theme Selection

In every scientific field, subdisciplinary committees and workshops have generated long lists (and short lists) of high-priority research programs. The contributors to this report are all too familiar with these efforts. From many possibilities, the committee selected five research areas and four objectives that include the top-priority scientific issues for understanding the solid Earth, for discovering and managing its resources, and for maintaining its habitability. These constitute the research framework (Table 1.1) that serves to simplify concepts of classification and relationship. Each of the major objectives is most appropriate for particular research areas, but all are relevant to more than one. Indeed, the potential that a research program will contribute to more than one objective or area is one of the criteria to be used in the selection of priorities.

These research areas and objectives consolidate into eight *priority themes*. (As mentioned earlier, Objective A, comprehensive understanding, is considered to be implicit in the other objectives.) The aims of these priority themes are summarized in Table 1.2.

The selection of these priority themes represents a preliminary stage in prioritization. If funding can be obtained only in increments, the facilities required to implement these themes need not be established in any particular sequence. If sufficient funding were available, simultaneous construction of these facilities would be a wise investment for the nation. However, it is apparent that not all research programs in this array can be fully funded simultaneously.

Despite current limitations on funds, it is most important to provide the type of support that encourages continuing engagement of the best young minds in postdoctoral independent research. One of the strengths of the American research enterprise is the independence granted to researchers who have fresh ideas and are in the prime of their intellectual careers. But it is widely felt that the system is in jeopardy. The problem is variously ascribed to inadequate funding for the sciences, to a surplus of scientists for the available funds, or to the expenditure of too high a proportion of available research funds on megaprojects. (These issues are discussed more fully in Chapter 7.) Whatever the reason, it is clear that priorities must be established.

Predicting the next scientific breakthrough is difficult. The only prediction that is likely to hold true is that all other predictions will be "scooped" by some unanticipated discovery. And great discoveries may spring from anywhere; this was illustrated when paleomagnetism and magnetic anomalies turned out to be crucial to the recognition and formulation of the theory of plate tectonics. Any serendipitous advance in instrument design or invention may reveal a deeper level of understanding.

Despite the apprehension of some scientists concerning the definition of scientific programs defined by committees, funding agencies must have guidelines. The priority themes outlined above (and detailed in the next section) are consistent with similar topics selected through the past decade by many committees and workshops for their high research promise. Because the topics are broad, additional evaluation of research frontiers and priorities within each theme is required. It is necessary to determine the instrumental resources required for success and to establish a procedure for determining priorities according to the overall funding available. This must involve considerations of timing and of a logical sequence in which facilities and equipment should be made available to optimize applications of developing technologies. Connections between objectives and research areas within the Research Framework may help to guide some redistribution of effort. The detailed review of the priority themes that follows, with consideration of the personnel,

TABLE 1.2   The Objectives and Research Areas Used Throughout the Report[a]

Objectives

A.   To Understand the Processes in All Research Areas

To understand the origin and evolution of the Earth's crust, mantle, and core and to comprehend the linkages between the solid earth and its fluid envelopes and the solid earth and the biosphere. We need to maintain an environment in which the biosphere and humankind can flourish.

B.   To Sustain Sufficient Supplies of Natural Resources

To develop dynamic, physical, and chemical methods of determining the locations and extent of nonrenewable resources and of exploiting those resources using environmentally responsible techniques. The question of sustainability, the carrying capacity of the Earth, becomes more significant as the resource requirements grow.

C.   To Mitigate Geological Hazards

To determine the nature of geological hazards, including earthquakes, volcanic eruptions, tsunamis, landslides, soil erosion, floods, and materials (e.g., asbestos, radon) and to reduce, control, and mitigate the effects of these hazardous phenomena. It is important to consider risk assessment and levels of acceptable risk.

D.   To Minimize and Adjust to the Effects of Global and Environmental Change

To mitigate and remediate the adverse effects produced by global changes of environment and changes resulting from modification of the environment by human beings. These latter changes may necessitate changes in human behavior. In order to predict continued environmental changes and their effects on the Earth's biosphere, we need the historical perspective given by reconstructed past changes.

Research Areas

I.   Global Paleoenvironments and Biological Evolution

To develop a record of how the Earth, its atmosphere, its hydrosphere, and its biosphere have evolved on all time scales from the shortest to the longest. Such a record would provide perspective for understanding continuing environmental change and for facilitating resource exploration.

II.   Global Geochemical and Biogeochemical Cycles

To determine how and when materials have moved among the geospheres crossing the interfaces between mantle and crust, continent and ocean floor, solid earth and hydrosphere, solid earth and atmosphere, and hydrosphere and atmosphere. Interactions between the whole solid-earth system and its fluid envelopes represents a further challenge. Cycling through the biosphere and understanding how that process has changed in time is of special interest.

III.   Fluids in and on the Earth

To understand how fluids move within the Earth and on its surface. The fluids include water, hydrocarbons, magmas rising from great depths to volcanic eruptions, and solutions and gases distributed mainly through the crust but also in the mantle.

IV.   Crustal Dynamics: Ocean and Continent

To understand the origin and evolution of the Earth's crust and uppermost mantle. The ocean basins, island arcs, continents, and mountain belts are built and modified by physical deformations and mass transfer processes. These tectonic locales commonly host resources introduced by chemical and physical transport.

V.   Core and Mantle Dynamics

To provide the basic geophysical, geochemical and geological understanding as to how the internal engine of our planet operates on the grandest scale and to use such data to improve conditions on Earth by predicting and developing theories for global earth systems.

[a]Sequence implies no ranking. All of the research areas and Objectives B, C, and D are treated as priority themes; Objective A involves understanding the processes in each of the research areas, and so was not itself designated as a separate priority theme.

---

facilities, and equipment required to implement the best programs, should lead to appreciation of which research should be expanded first. These matters are taken up again in Chapter 7.

PRIORITY THEMES: RESEARCH AREAS

The priority themes can be related to form together in a continuous story. The Earth operates as a thermal engine, driven by primordial heat from accretion, by the continuing decay of radioactive material dispersed through the interior, and by energy generated by the gravitational accumulation of the inner core from the outer core. The Earth is hot and continuously releases that heat into the cold abyss of space. Cooling is efficiently accomplished by thermal convection within the molten core and the solid mantle, with its volcanic by-products, and out through the crust and atmosphere. The brittle crust is moved and dislocated in association with the internal convective disturbances. The solid surface is a major boundary in this dynamic system at which energy from the Sun is imported to the planet. For this reason, understanding the surface processes that affect humankind—landslides, earthquakes, volcanoes, ocean cycles, and basin development, with their characteristic accumulation of resources—requires knowledge of the whole dynamic earth system and all of its interrelations. Even the deep core affects the surface and is therefore important to so-

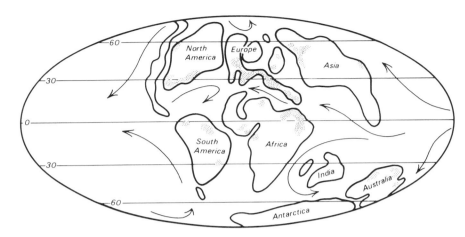

FIGURE 1.3 Distribution of laterite (shaded) on land masses and estimated major ocean surface currents about 80 million years ago. Laterite are soils that form under relatively uniform maritime conditions on windward sides of continents within the humid intertropical, or tropical forest, zone. From *Climate in Earth History*, National Academy Press (1982).

ciety. For example, convection in the molten iron-alloy core produces a magnetic field that protects most organisms from lethal solar radiation, provides a basis for navigation, and occasionally disrupts electromagnetic communication systems.

Priority Theme A-I: Global Paleoenvironments and Biological Evolution

Ongoing programs have reconstructed climatic and oceanographic conditions during ancient intervals that were much different from those of today. These varying conditions range from intervals when the deep sea was much warmer than at present and its oxygen content was depleted to times when vast ice sheets spread to regions far from the poles. Efforts to piece together the kaleidoscopic movements of the continental lithosphere, from climate change evidence found in the rock record and from mountain-building and ocean-forming processes, are progressing rapidly. Researchers have established the relative positions of most of the larger parts of the continents with respect to each other and with respect to the North and South poles for the past 500 million years.

This record of moving plates provides a meaningful framework for the study of paleogeography, paleobiology, paleoclimatology, and paleoceanography. This framework has so revolutionized these fields that they are, for all intents and purposes, new and budding disciplines. Paleogeography produces maps that show not only the relative positions of continental masses but also the locations of high mountain ranges and shallow seas. Finer details, gleaned from paleoclimatic evidence, indicate specific environments such as deserts, coal swamps, and glacial ice (Figure 1.3). Paleoenvironmental maps represent beautiful reconstructions from scientific detective work, and they have become im-

portant guides in exploring for minerals and energy resources. These reconstructed maps lead explorers to potential sites of valuable concentrations, formed during continental development, of material such as petroleum and phosphates. The discovery of the submarine hydrothermal vents found at oceanic spreading centers and the associated mineral deposits has revolutionized interpretations of the origins of many ancient ore concentrations.

One important endeavor is unraveling the record of past continental movements. Another is constructing circulation models for ancient oceans and atmospheres. Reconstruction of such circulation models may illuminate distinctive transitions from one stable environment to another, as well as the forcing factors responsible for such transitions. New approaches to the study of global geochemical cycles contribute to these reconstructions of large-scale environmental change. Researchers are only beginning to appreciate how much life-supporting oxygen and heat-trapping carbon dioxide have fluctuated.

The Earth's dynamic environments form the backdrop against which the history of life unfolds. The rock record is a vast repository of information documenting natural processes of trial and error. These natural proving grounds can be considered experiments that offer lessons about the relationship between life and environments—lessons that will serve humans well in their attempt to confront global change in the decades ahead. The most important of these lessons will come from understanding the events of the past 2.5 million years. The record of the most recent climatic lurch—when deglaciation accelerated around 15,000 years ago—lies right at the surface. The evidence is in the landscape, on the ocean floor, and within the top layers of the remnant ice sheets. Pertinent information is garnered from ice and ocean cores, tree rings,

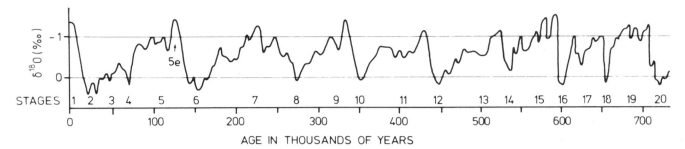

FIGURE 1.4 Oxygen isotope variations of oceanic plankton, which is a surrogate for climatic fluctuations, over the past 700,000 years. The stages are those reflecting glacial conditions. Note the rapid increase over the past 15,000 years reflecting the transition from the most recent glaciation to the present interglacial. From *Climate in Earth History*, National Academy Press (1982).

fossil pollen, fossil soils, loess deposits, lake varves, relict shorelines, ancient river channels, and flood deposits. These data are used to build and test computer models of global circulation patterns. Detailed analyses of these data and isotope dating combine with field documentation of landscape evolution to promote recognition of geological thresholds that mark profound changes from one climatic state to another.

Sedimentary rocks from more ancient intervals also attest to periods of global warming and consequent environmental change. Many of the most exciting conceptual developments in modern paleontology are based on explanation of rates, trends, and patterns in the evolution, migration, and extinction of species—all of which can be interpreted only in a context of environmental change. The evolutionary process is becoming more intelligible because of advances in the documentation, resolution, and management of fossil data (Plate 7). Mass extinctions, including the catastrophe that destroyed the dinosaurs, are among the many important subjects in the study of environmental change.

Global environmental changes are of three kinds. First are the secular changes, such as biological evolution. Since the first single-celled organism exchanged chemical compounds with its surroundings, and persisted and even thrived through that exchange, biota have affected the environment in myriad ways. The changes caused by recent human endeavor are not unnatural, but the intensity of those changes, the concentrations and rates, can cause problems. Second are the cyclic changes, such as day following night, ocean basins opening and closing, and mountains being built and eroded. Familiarity with cyclic patterns allows earth scientists to interpret repeating sequences observed in the geological record. Tides, seasons, floods, droughts, sea level changes, and glacial ages all leave distinc-

tive patterns indicating cycles (Figure 1.4). Secular and cyclic changes continue together. Thus, despite the recurrence of patterns, nothing returns exactly to its original state. The third kind of change is catastrophic. These abrupt changes have occasionally punctuated the record with highly unusual events such as the extraterrestrial impact that led to the demise of the dinosaurs.

Understanding earth history must be grounded in reliable chronologies of events. In the decades ahead, solid-earth scientists will benefit from new methods of dating ancient materials and events—the exploitation of previously ignored fossil taxa and the application of new isotopic techniques or of chemical signals that record unique global events in the rocks. A few years ago no one would have envisioned recent radiometric triumphs: dating zircon grains billions of years old with a precision of a few million years, or dating coral skeletons nearly two centuries old with a precision of just a few years.

Priority Theme A-II: Global Geochemical and Biogeochemical Cycles

Geochemical cycles—the cyclic movement of chemical elements, nuclides, and compounds through the great reservoirs of the Earth constituted by the biosphere, atmosphere, oceans, sediments, crust, mantle, and core—have received recent attention because of advances in analytical techniques, data-handling, and modeling capabilities. The reservoirs may be very large, on the scale of the whole Earth, sequestering elements for a billion years, or they may be as small as a backyard, exchanging material with the surroundings over a few days. Plate tectonic cycles, which involve exchange of material between the mantle and the crust, operate on a time scale of hundreds of millions to billions of

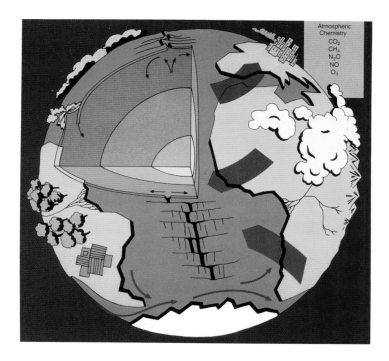

PLATE 1 Schematic view of the earth system. The successive geospheres are shown from the center to the outside of the solid earth: the inner and outer cores, the mantle, the lithosphere (the outer shell of the mantle with its blanket of crust, oceanic, and continental), the hydrosphere (dominated in terms of quantity by the ocean), and the atmosphere; the biosphere is represented by the trees, the crop-growing fields, and the occupants of the city. The magnetosphere extends beyond the limits of the atmosphere. The processes represented are convection in the mantle, seafloor spreading forming a mid-ocean ridge and reshaping the ocean floor, subduction (lithosphere sinking back into the mantle) recycling elements through the interior, volcanism associated with mantle plumes and with sinking lithosphere, the use of earth resources for construction of cities, urban growth, photosynthesis by terrestrial vegetation, ocean circulation (arrows for ocean currents) around the polar icecap, evaporation and precipitation of water into and from the atmosphere, and circulation of the atmosphere in winds. From NASA *Earth System Science* report (1986).

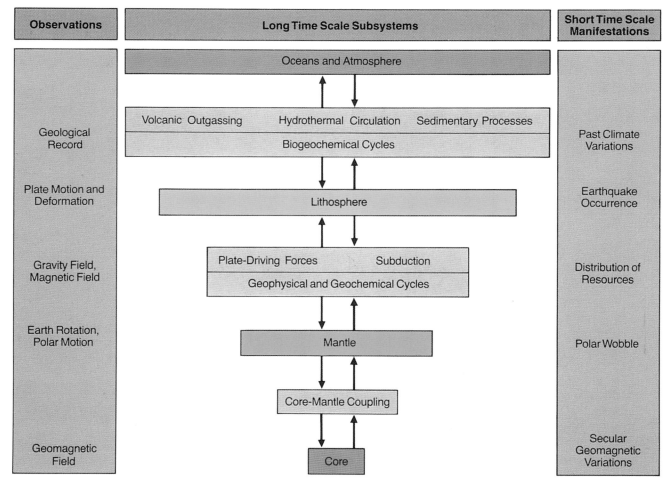

PLATE 2 Schematic of solid-earth processes. From NASA *Earth System Science* report (1986).

PLATE 3 Crustal plate boundaries shown on a global relief map. Earthquakes of magnitudes greater than 5 for the period 1980 to 1990 are shown in red. Photo courtesy of the NOAA National Geophysical Data Center.

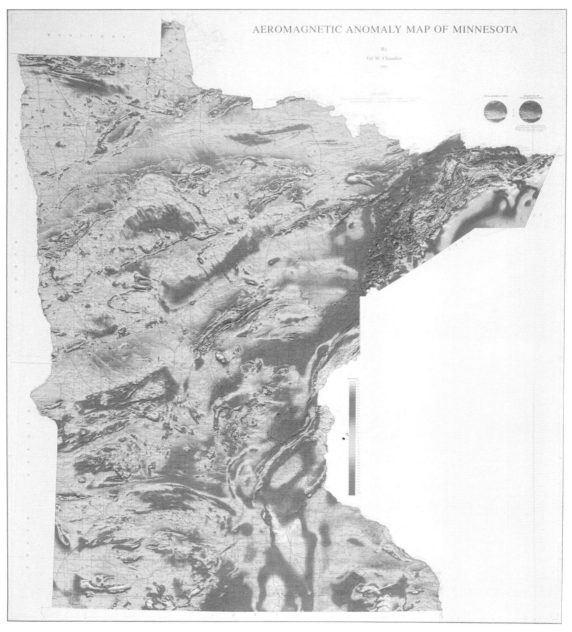

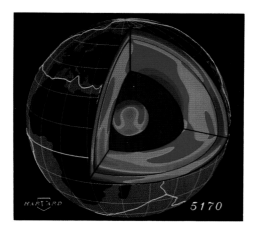

PLATE 5 The view of Earth in the picture is a perspective roughly from the altitude of a geostationary satellite, about 36,000 km. The cut into the interior shows a composite image synthesized from seismic tomographic mapping. Results from seismic tomography are used to answer fundamental questions about the evolution and present-day dynamics of Earth. Velocities slower than average are shown in orange, faster than average in blue. Figure courtesy of Adam Sziewonski, Harvard University; computer graphics by John H. Woodhouse.

PLATE 6 Black smoker (hydrothermal vent) and vent shrimp at the TAG hydrothermal field, mid–Atlantic Ridge (26° N, 45° W). Photo courtesy of Peter A. Rona, National Oceanic and Atmospheric Administration.

PLATE 7 Fossil evidence of evolutionary juvenilization in ammonites, which are relatives of the pearly nautilus that died out with the dinosaurs. The two large speciments include a male (smaller of the two) and a female of the Upper Cretaceous species, *Scaphites whitfieldi*. The small specimen, which retains some of the mother-of-pearl inner layer of its shell, represents the species, *Scaphites coloradensis*. Evolution produced the small 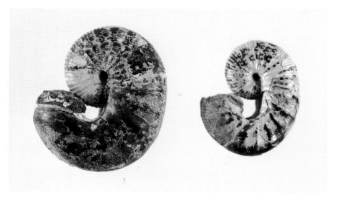 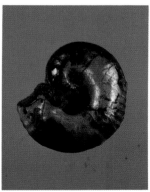 species from a population of the larger one by terminating growth at a small size while preserving many juvenile features. The parent and daughter species ended up living side-by-side in shallow seas that flooded the western interior of North America. Photos courtesy of Neil H. Landman, American Museum of Natural History.

PLATE 4 *(Opposite)* The 1991 aeromagnetic anomaly map of Minnesota shown as colors and a shaded relief image (illuminated from the north). Total intensity aeromagnetic data measured in nanoteslas (nT), and the range from low (−1,300 nT) in blue to high (+700 nT) in light pink. The strongly magnetic mid-continent rift extends from the Iowa border northward under Minneapolis-St. Paul. This survey was funded from state cigarette taxes. Plate courtesy of the Minnesota Geological Survey.

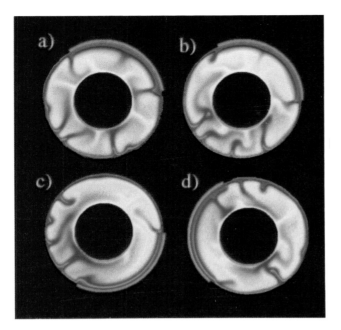 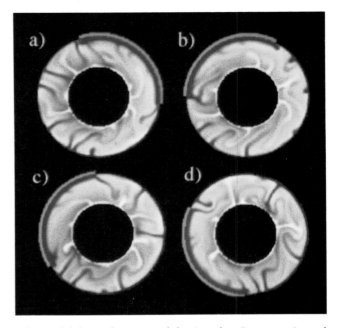

PLATE 8   Models of mantle convection at a series of different times. The models have a large nonsubducting plate (i.e., a continental plate on the surface); plate velocity and position is completely dictated by the dynamics of the convecting system; and there is a full two-way dynamic feedback between the plate and convection. The green strip shows the location of the 300-km thick continental lithosphere. The motion of the plate and the corresponding effect and consequent temperature variations are shown. The yellowish-white areas are hottest and the blue areas colder. In both simulations, plate velocities can reach 5 cm/year. Left: The plate is stationary and positioned toward the upper right. From a to b the temperature under the plate builds up. From b to c the plate moves rapidly off the hot areas in a counter-clockwise direction. Right: Again time increases progressively from a to d and the plate moves in a counter-clockwise direction; from c to d the plate is mostly stationary. Photos courtesy of Michael Gurnis, The University of Michigan.

PLATE 9   Mount Rainier as seen from Tacoma, Washington. Mount Rainier has been inactive in historical times but is still considered to be hazardous in a highly populated area because of the threat of future volcanic eruptions and the potential for massive mudflows. Photo courtesy of Lyn Topinka, USGS.

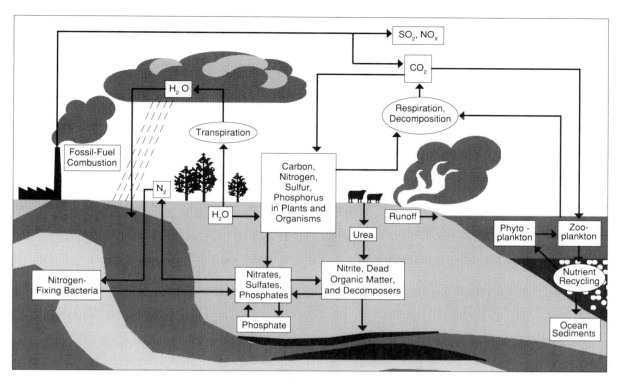

FIGURE 1.5 Movements of key elements (carbon, nitrogen, sulfur, phosphorus, and others) through the earth system's biogeochemical cycles. From NASA Earth System Science report (1986).

years. Sedimentary cycling through the land, oceans, and atmosphere operates over hundreds of thousands to hundreds of millions of years. Ocean cycles involving the biological transfer of nutrients between deep and shallow waters extend over decades to millennia. And atmospheric cycles involving input of gases to, and removal from, the atmosphere have a time scale of days to centuries.

Material transport involving the solid earth, the atmosphere, the hydrosphere, and the biosphere is biogeochemical cycling (Figure 1.5). The principal elements concerned are hydrogen, oxygen, carbon, nitrogen, phosphorus, and sulfur. These nutrient elements are required by organisms and are obtained by them directly from the atmosphere and hydrosphere. The atmosphere and hydrosphere, in turn, obtain carbon, nitrogen, phosphorus, and sulfur from interaction with the solid earth. Rocks yield these elements by chemical weathering and by exchanges between water and bottom sediments or volcanic rocks in the oceans. To complete the cycle, the organisms themselves give up these elements to the hydrosphere, atmosphere, and solid earth either as part of their life processes or after death by bacterial decomposition. Rocks affect life, but also life affects rocks. For example, both bacteria and organic constituents can affect chemical weathering

and the soft-tissue remains of ancient organisms are sources for coal and oil and gas deposits. The main elements of the global carbon cycle are given in Figure 1.6.

Humanly altered biogeochemical cycles modify the composition of the atmosphere. For example, the addition of carbon dioxide and sulfur-containing gases to the atmosphere by the burning of coal and oil results in greenhouse warming and acid rain. The burning of coal and oil short-circuits a natural geological process, chemical weathering, that normally proceeds slowly. Humans produce in decades what would take nature thousands to millions of years to produce. Another example is the phosphorous cycle. Phosphorus is ultimately derived from rocks by chemical weathering, but it is rapidly cycled within reservoirs such as soils, lakes, and estuaries. These sensitive environments, which support vital biological activity, overdose on concentrations of fertilizers and detergents containing phosphorus. They choke to death when oversupplied with what is normally a limiting nutrient.

These rapid cycles involving the biosphere, hydrosphere, and atmosphere are linked through the crust to slower cycles involving the interior. The slower cycles occur along ocean spreading centers where new material from the interior reacts with

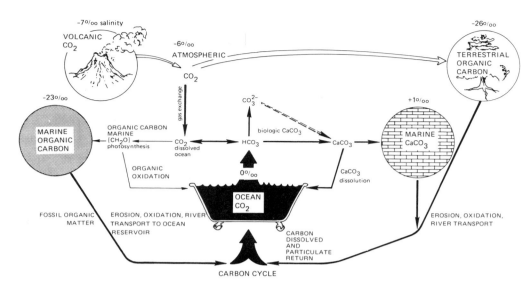

FIGURE 1.6 The main elements of the global carbon cycle. From *Climate in Earth History*, National Academy Press (1982).

ocean water and in subduction zones where the lithosphere—after interaction with the atmosphere, hydrosphere, and biosphere—sinks back into the interior and releases volatiles. The chemical reactions characterizing these environments circulate and remobilize water, carbon, nitrogen, phosphorus, sulfur, and all of the other elements that are so important in the biosphere.

Isotopic studies continue as essential tools for investigation of global geochemical cycles. Researchers use various stable isotopes, such as those of carbon, oxygen, and sulfur, to assess biogeochemical cycles. Ratios of stable isotopes of carbon and oxygen, for example, provide important chemical tracers. They may provide answers to fascinating problems such as whether diamonds contain organic carbon that has been recycled into the interior.

New techniques using radioactive isotopes date critical events during biogeochemical cycles. Radiogenic isotopes can also be used to quantify the size and mean age of major global reservoirs. The information emerging from these studies suggests a new way to understand earth systems.

The ultimate goal of studies of geochemical cycling is reconstruction of the historical development of the Earth and its environment. Geochemical cycling is a unifying concept that brings together results from studies of various earth features, including mantle evolution, global tectonics, rock-water interaction, paleoclimatology, and organic evolution. Compilation of geological history and study of modern processes and their rates permit

mathematical modeling of these cycles. These models provide a means of evaluating the quantitative significance of specific earth processes during selected intervals of the past. Such modeling can provide a basis for making future predictions.

Priority Theme A-III: Fluids in and on the Earth

Fluids dominate the redistribution of mass and energy through the earth system. Their role is prominent in the generation, migration, and eruption of magmas; in the chemical alteration of masses of rock during metamorphism and mountain building; in the mechanical transport of material on the surface; in the formation of ore deposits; and in the migration and entrapment of oil and gas within sedimentary basins. Water is the active agent in many of the mechanical, chemical, and thermal processes operating within the crust and on the surface. An understanding of fluid flow is needed to predict the movement of contaminants in groundwater aquifers. Fluids also play a role in triggering earthquakes and landslides.

Chemical differentiation of the crust, hydrospheres, and atmosphere involves volcanic activity, beginning with partial melting of the mantle and ending with eruption of material at the surface. The chemistry of magmas is controlled not only by the conditions of melting but also by the physics of rock-fluid systems—a realization that has renewed interest in the physics of magmatic processes. Analyses of physical data, and of the requirements for

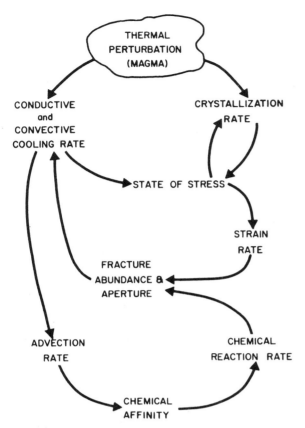

FIGURE 1.7 Relationships among principal fluid transport rates and their products. Arrows depict directions of energy, mass flow, and feedback effects. From *The Role of Fluids in Crustal Processes*, National Academy Press (1990).

accumulation, flow, and escape of melts from within a rock matrix, are beginning to produce coherent explanations. In addition to experimental studies of physical properties at high pressures and temperatures, several theoretical developments have clarified the properties of flow through a deformable rock matrix and have defined conditions for fluid migration by percolation, diapirism, or through cracks (Figure 1.7). These investigations have only begun; coupling of the laws that govern fluid dynamics in rocks is truly a frontier.

One of the most remarkable discoveries of recent years was that of the submarine hydrothermal vents at spreading centers, where ocean water is transformed chemically by heat and interaction within the basaltic crust (Plate 6). The emerging hot springs sustain colonies of huge tubeworms and other creatures without photosynthesis. Abundant sulfide minerals precipitate from the mineral-saturated flow. These discoveries have affected many hypotheses, including those about the chemical composition of ocean water and the formation of

mineral deposits. Estimates show that an amount of water equal to the volume of the whole ocean cycles through the oceanic crust in about 10 million years.

Tectonic, hydrological, chemical, and biological processes interact at the surface with a complexity that has hampered accurate understanding of the system as a whole. The reactions that occur at these interfaces produce the surface environment. They play critical roles in determining the quality of fresh water supply, the development of soils and the distribution of nutrients within the soils, the integrity of underground waste repositories, the genesis of certain types of ore and hydrocarbon deposits, and the geochemical cycling of elements. Bacterial activity appears to be an essential component in many processes that contribute to the evolution of the environment. This recent realization has led to the preliminary development of biological remedies for water pollution and methods for waste treatment.

Priority Theme A-IV: Crustal Dynamics: Ocean and Continents

The notion of terra firma is deeply ingrained in the human subconscious, and yet testimonials to the Earth's vigorous dynamics litter the landscape. An abundance of diverse landforms and surface environments, from spectacular mountains to desert valleys, from rocky coastlines to verdant tropical islands, demonstrate ongoing adjustments to the planet's surface. These alterations proceed at a tremendous range of rates; some are detectable only with sensitive instruments, some are catastrophic. Processes such as basin subsidence, mountain uplift, and continental drift occur at rates of only centimeters per year or less—slowly enough to sustain the sense of a solid earth but persistent enough to completely reorganize the surface in a few million years. Other processes are readily perceptible, such as the violent volcanic eruptions of Mount Pinatubo or Mount St. Helens, earthquake ruptures that displace high rock volumes by many meters within minutes, or mud flows and avalanches that can outrace a sprinting human. The disorientation and stark terror experienced by many survivors of strong earthquakes result from the violation of basic instincts when the solid earth reveals its inner turmoil.

Throughout earth history continents have assembled, broken up, and reassembled from different fragments—repeatedly flooded by the waters of the ocean and bombarded with objects from outer space. These surface reconfigurations are manifesta-

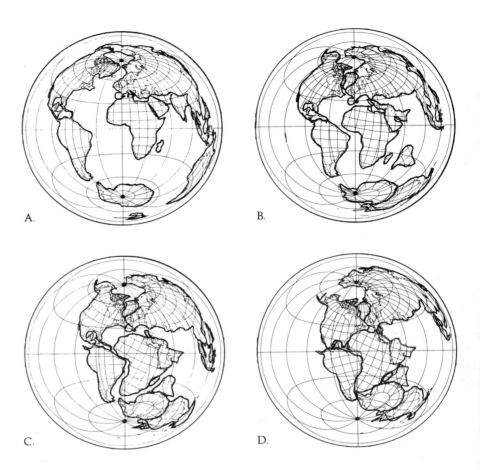

FIGURE 1.8 Approximate distribution of continental areas between the breakup of the supercontinent Pangea (about 180 million years ago) and the present.

tions of interior processes. All of the deep oceanic areas are underlain by a thin crust formed within the past 200 million years. The comparatively simple oceanic crust is made of basalt erupted in a submarine environment at magmatic spreading centers. The oceanic crust then migrates toward subduction zones where it eventually plunges back into the mantle. In contrast, the continental crustal areas are more complex; their growth patterns extend over 4 billion years. Earth scientists agree that for a short period about 250 million years ago, nearly all the continental material was consolidated into a great landmass, the supercontinent Pangea (Figure 1.8). They are not sure whether similar supercontinents existed before Pangea because the record of continental history deteriorates with increasing age. The superior quality of the most recent record is a major reason why researchers of continental dynamics concentrate on active processes. Perhaps even more important is that active processes—earthquakes, volcanic eruptions, elevation and depression, erosion and sedimentation—can be studied using methods that cannot be used to study processes that no longer operate.

Understanding continental crustal evolution re-

quires explanation of long-term deformation mechanisms. This would assist the development of realistic plans to mitigate future disasters caused by short-term crustal deformation. Geologists, geodesists, and geophysicists are working on problems of the deformation occurring between and within continental plates. Their questions address the geometries and histories of structures at the margins of plates, the relationships between forces in the crust and upper mantle, and the tectonic nature and evolution of the now-stable interiors of the continents.

The processes of plate deformation leave evidence in the rocks. Rocks of mountain chains are commonly metamorphosed—recrystallized into new minerals—by the physical changes experienced during the burial, strain, and heating that occur during mountain building. Experimental petrologists have devised methods to determine the stability fields for minerals and mineral assemblages in terms of pressure and temperature, providing the framework for evaluation of the depths and temperatures reached by rocks during deformation (Figure 1.9). Our understanding of these processes has involved the combined approaches of field and structural geol-

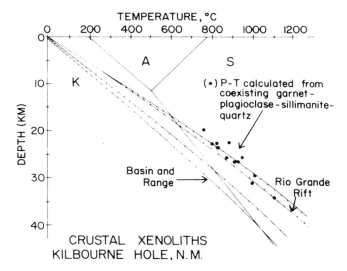

FIGURE 1.9 Temperature-depth plot showing conditions as recorded by coexisting garnet and plagioclase in Kilburne Hole (Arizona) granulites. The aluminum silicate stability fields are shown as reference (K: kyanite; A: andalusite; S: sillimanite). From *Continental Tectonics*, National Academy Press (1980).

ogy, calibrated by experimental petrology and thermodynamics.

The discoveries in the European Alps, China, and Norway of mineral assemblages stable only at very high pressures confirm that some continental rocks have been recrystallized locally at depths of at least 100 km and temperatures of 800°C or more. Continental material is too buoyant to remain at this depth, and rapid uplift with concomitant erosion must have occurred. These observations present two challenges. The first is to explain how the continental rocks arrived at such depths. The second is more difficult: that is, how did the rocks return to the surface fast enough to escape the further metamorphism that can destroy assemblages formed at depth?

Researchers are using two approaches for introducing the dimension of time into depth-temperature analyses. The first enlists the power of computer modeling to project the behavior of a rock mass subjected to pressure and temperature changes and influenced by myriad variables, including fluid flow. The second examines isotopic geochronology and thermobarometric pressure values available from radial growth patterns of minerals and from their inclusions. These new techniques, which combine mineralogy with geochemistry, yield direct estimates for the rates of geological processes associated with mountain building. The rates of these processes complement the record of mountain erosion found in sedimentary basins. Geochronologists are now capable of dating rocks and land surfaces with a resolution that was not previously possible. They can confidently estimate how long a mountain range has been elevated and how fast it is being eroded away.

It is becoming apparent that the continents cannot be treated as simple rigid blocks. The rheological properties of the crust need to be established by a combination of field observations, laboratory experiments, and measurements of active crustal deformation rates, which can now be made with the aid of satellites. Large rotations occur during continental deformation. Studies of the paleomagnetic records in rocks have indicated rotations of 90° in 20 million years within belts 200 km wide. Measurements using global positioning systems have confirmed the evidence of rotation during short (several years) observation terms.

Seismic reflection methods, developed for oil exploration (still its predominant use), when used to study the structure of the continents, have enriched our evidence of the processes operating close to the surface. Several remote sensing techniques, particularly seismology, have produced remarkable advances in theory and data analysis, and exploitation of images showing structure through the entire thickness of the crust and into the mantle continually supplies new data and raises new questions.

Over the past decade, researchers have recognized structural styles associated with large-magnitude extension of the crust. Extensional strains in the continental lithosphere appear to be accommodated in the upper crust by areas of normal faulting called detachment systems (Figure 1.10). Detachment systems adjust to the separation of relatively undeformed crustal blocks, with individual faults slicing through the upper 15 km of the crust and having displacements that measure in tens of kilometers. The results of continental extension are broad areas of mid-crustal rocks veneered by regionally subhorizontal detachments. These detachments have accommodated the removal of upper crustal rocks from the region of extension. Such extended domains exist in the Basin and Range Province of the western United States. They are 100 to 200 km wide and are interspersed with relatively unextended crustal blocks. Over the next decade, interdisciplinary study of the lithosphere in regions of extension should produce significant advances in understanding deformational processes in the continental lithosphere.

While geophysicists, geochemists, experimental petrologists, and structural geologists study the processes that build continents, complementary re-

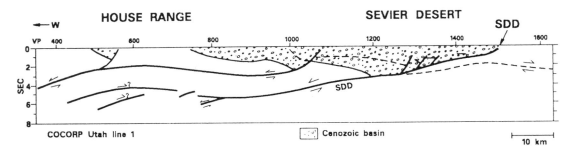

FIGURE 1.10 Interpretation of a seismic reflection line from the eastern Basin and Range over the Sevier Desert basin and House Range showing major low-angle normal faults (e.g., Sevier Desert detachment, SSD). From J. H. McBride, 1991, *Tectonics 10*, 1065–1083; © American Geophysical Union).

search investigates the processes that destroy mountain ranges and fill basins. Physical and chemical weathering promotes erosion of uplifted masses. Streams and rivers entrain eroded sediments and carry massive amounts of material into downstream basins. Accumulating basin deposits sink into the crust, while weathered highlands rebound, relieved of the weight. Researchers studying the Mississippi-Missouri River drainage basin and the Mississippi delta find that while the delta continues to sink, the amount of sediment delivered by the river has decreased because of upstream dams (Figure 1.11). In this case, human intervention is disrupting the balance of crustal dynamics.

Priority Theme A-V: Core and Mantle Dynamics

The internal structure of the Earth has been studied in recent years from several very different perspectives: seismic waves generated by both earthquakes and explosions have shown how the structure varies, geochemical analysis of mantle-derived rock reveals variations in the mantle's composition, and meteorites that are fragments of planetary objects supply information about other mantles, similar to Earth's. Analyses of rare gases from the mantle further describe its history. Each investigative approach contributes to the overall body of knowledge about the deep interior (Figure 1.12).

Heterogeneities within the inner core, the mantle, and the crust are revealed through seismic tomography, in which seismic waves projected through the Earth's layers are built into a three-dimensional image. The speed and paths of seismic waves are indicative of physical properties that vary according to differences in composition, temperature, amount of melting, and other factors. The resulting seismograms permit a depiction of the structure and prop-

erties along the path of the wave. The three-dimensional images (see Plate 5) constructed from individual seismic records show trends and discontinuities—such as the discontinuity found at 670 km—with remarkable resolution. These seismic tomographic images showing the three-dimensional structure of the mantle are now possible because of advances in seismic instrumentation and the availability of powerful computers capable of handling data from hundreds of thousands of seismograms. Heterogeneities are commonly interpreted as evidence of density changes, which can be correlated approximately to parts of the mantle that sink because of high density and rise because of low density. This is indicative of convection within the mantle (Plate 8). An important question concerns whether the mantle convects in one or two shells between the crust and the core. Solving the convection question is one of the challenges for the 1990s.

Whereas seismic tomography provides a picture of the locations of heterogeneities in the mantle, geochemical analyses of basaltic lavas and other igneous rocks from the mantle provide information about the compositions and histories of the heterogeneous rock masses involved in convection. The heterogeneities are distinct on both small and large scales, and their development has taken billions of years—perhaps as long as half the Earth's age.

There is a clear chemical distinction between the mantle source rocks that supply lava to mid-oceanic ridges and those that supply oceanic island volcanoes associated with hot spots, such as Hawaii. Up to five different sources are required to account for the compositions of ocean island basalts. Isotopes and trace elements in the lavas serve as tracers for detecting the recirculation of continental and oceanic sedimentary rocks into the mantle. One provocative question is whether—and if so, how—subducted lithosphere is able to penetrate through the 670-km discontinuity within the mantle.

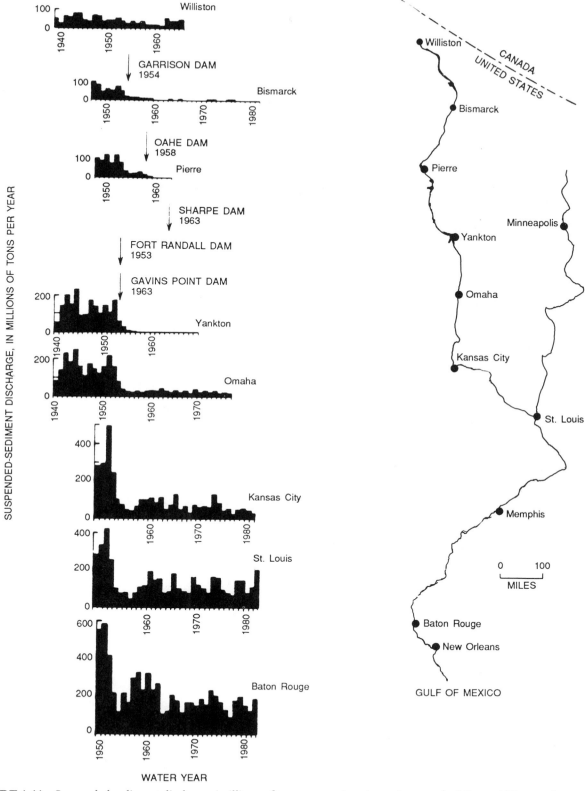

FIGURE 1.11 Suspended sediment discharge (millions of tons per year) at six stations on the Missouri River and two stations on the Mississippi River showing how the construction of reservoirs reduced downstream sediment loads by about half during the period 1939 and 1982. After R. H. Meade and R. S. Parker, 1985, *U.S. Geological Survey Water Supply Paper 2275.*

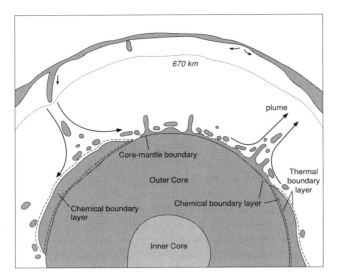

FIGURE 1.12 Core-mantle transition zone features based on geophysical data. A heterogeneous chemical boundary layer is embedded in a thermal boundary layer produced by a large contrast in temperature between the core and mantle. Large-scale mantle circulation transports chemical heterogeneities to the base of the mantle and returns them to shallower depths by entrainment. The dotted line represents the base of the upper mantle. Modified from T. Lay, 1989, *EOS: Trans. Am. Geophys. Union 70*, 49; © American Geophysical Union.

High-pressure experimental devices, such as large presses, small diamond anvils, and dynamic shock-wave apparatus provide key data on the properties of material from the core and mantle. Until recently, experiments determining the temperatures necessary for mantle melting had reached pressures corresponding to depths of only 100 to 200 km. Improvements now provide detailed melting and other physical property determinations at pressures that correspond to the 670-km seismic discontinuity. Less detailed measurements are now possible at pressures corresponding to depths approaching that of the mantle-core boundary.

As investigations continue to decipher mantle processes, information about the core builds toward a comprehensible picture. The melting curve of iron has been measured in shock-wave apparatus at pressures corresponding to those at the core. These results provide a basis for estimating the temperature at the transition between the molten outer core and the solid inner core. From that estimate, temperature calculations can be extrapolated both outward to the core-mantle boundary and inward to the Earth's center.

Other high-pressure experiments have assessed the compositional and physical state of the molten core. These results, combined with seismic data that reach to the core-mantle boundary region, feed speculation about a relationship between core-mantle interactions and magnetism. Changes in the main magnetic field, known to be generated in the core, remain a major unsolved problem.

PRIORITY THEMES: OBJECTIVES

The broad objective of earth system science is to understand the intricate processes that make this planet what it is: a world balanced among so many interacting factors that the life-supporting outer layer suggests an incredible series of coincidences. Understanding the interacting factors and incredible coincidences is the challenge facing earth system science. The excitement and importance of this challenge are enhanced by the immediate benefits to society to be derived from that understanding. Comprehension of earth system processes applies directly to the specific objectives of maintaining resource supplies; preventing damage caused by geological disasters; and assessing, mitigating, and remediating effects of global environmental change.

Military analysts sometimes claim that superior technology won World War II and has so far prevented World War III. The wars of the twenty-first century may be of a nonmilitary nature. The enemies may be the modern Four Horsemen—overpopulation, disease, environmental degradation, and resource exhaustion. The preemptive strikes of the next decades could originate from the bastions of technology and in the trenches of research. The winning strategy may depend on an integrated understanding of the earth system—its organic and inorganic components and their interactions.

Priority Theme B: Sustaining Resource Supplies

The resources needed in ever-growing quantities to sustain civilization are land, water, energy, and minerals. Resource geologists are seldom concerned with materials and processes active beyond 10 km into the Earth. This is the shallow crust. Further division separates into the top 10 m, where soil develops and surface water saturates; intermediate depths from 10 to 100 m, where groundwater and surface mining dominate; and 100 m to 10,000 m—the domain of petroleum production, deep mining, and deep groundwater. The distinction is arbitrary and the boundaries gradational, but the different problems of determining structure, understanding processes, and methods of investigation have enough in common to provide a certain unity to each of the three areas. Basin analysis is emerging as a central field within the shallow crust. Characteristics of the sedimentary rocks within a basin can be used to detect groundwater, petroleum, coal, and ore deposits. Distinct sedimentary packages also provide clues about the way fluid movement has cemented or altered parent rocks.

Researchers now use isotopic methods to determine the ages of surfaces and of shallow materials. These dates influence interpretations of landscape development. For example, soils are complex systems supporting interactions among various earth components. They contain water, which typically cycles on seasonal and annual time scales; organic materials, which flourish over decades to centuries; and silicate minerals, which persist for thousands to millions of years. Soils have always been difficult to study because of the temporal ranges involved. During the past decade, the use of cosmogenic nuclides to determine dates for soil material and processes and of analytical methods that can handle the mass of data, has advanced the fruitful dialogue between geologist and agronomist.

The Earth's landscapes consist of nested drainage basins on horizontal scales ranging from that of a freshet network on a hillside to the vast Amazon Basin. Relating these networks and their evolution to the sedimentary basins that receive eroded material remains a major focus of earth science. Geomorphologists hope that improved temporal resolution will distinguish between the evidence of steady basin evolution and that of catastrophic events.

Surface water and groundwater are absolutely vital resources. But pressure from human activities threatens their availability. Areas of necessary research range from modeling water flow and the kinetics of water-rock interaction to determining surface-water quality and monitoring groundwater contamination on a national scale. The disciplinary division between solid-earth science and hydrology fades on close inspection of these research areas; both are components of the larger earth system science.

Fossil fuels are abundantly available but sporadically distributed within the crust. Intense exploration over the past century in the conterminous United States has found most of the major oil fields. Production from these fields has reduced the reserves of much of the primary recoverable petroleum. While domestic exploration continues, the main priority involves increasing production from existing fields. Success depends on multidis-

ciplinary efforts to understand reservoir characteristics, especially heterogeneity. New extraction techniques will endeavor to recover petroleum resources more thoroughly and more efficiently.

Outside the conterminous United States, the petroleum industry still focuses on new reserves in undiscovered oil fields. Modern exploration methods use an integrated approach. This approach first establishes the tectonic environments of basin development (Figure 1.13) and then interprets local depositional, structural, and thermal evolution. This research forms a background for studying potential oil and gas generation within the basin, to assessing possible fluid migration and related diagenetic change, and finally to defining the types of traps in which oil and gas should be located.

Coal is abundant in the United States, but exactly how much can be developed is not clear. It presents significant environmental problems in development and use. Both open-pit and underground mines require the removal and disposal of massive amounts of rock. Burning coal that is rich in sulfur contributes to acid rain. And the combustion of coal and other fossil fuels produces carbon dioxide, which can intensify the atmospheric greenhouse effect. Solving the environmental problems of coal use while maintaining its economic viability is a key challenge for the future.

Coal originates with a peat-rich material and passes through lignitic, subbituminous, bituminous, and anthracitic phases. During these phases the fixed carbon content increases and the percentage of volatile material and moisture decreases. Coal is composed of woody and waxy organic material and is likely to generate gas on burial. Recent research has shown that in some environments and at low latitudes coals are more prone to yield oil. The oil- and gas-producing potential of coals is one of the most significant exploration frontiers at present.

Nuclear power plants represent an efficient energy source, relying as they do on concentrated and enriched uranium for fuel, but problems with dangerous by-products inhibit development of the industry. Solving these problems through containment or chemical neutralization remains the primary task in this area. Earth scientists can contribute to this work through investigations of fluid-rock interactions.

Uranium concentrations occur widely in sedimentary basins, and the fluid dynamics and chemical interactions of aqueous fluids in the subsurface play a preeminent role in their formation. Most of the commercial uranium in the United States comes

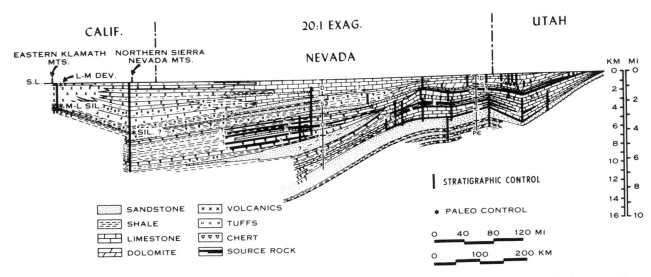

FIGURE 1.13 Stratigraphic cross section across the Great Basin. From *Continental Margins*, National Academy Press (1979).

from sedimentary deposits in stream channel sandstone deposits of the Colorado plateau, Rocky Mountain intermontane basins, and the Texas coastal plain. These deposits originated as disseminated uranium was taken into solution and transported in oxygenated groundwater. Where the water came in contact with reducing agents such as carbonaceous logs, leaf material, or various sulfides, the uranium precipitated. Anaerobic sulfate-reducing bacteria living on the organic material may intensify the deposition. The chemistry and dynamics of groundwater flow in basins are central to understanding the mechanisms of uranium deposition. Uranium deposits are most commonly strip mined, although subsurface hard-rock mines and in situ subsurface leaching also exist.

During the 1960s and 1970s, theories about the origin of ore deposits were overhauled by the advent of plate tectonics and by the discovery of metalliferous brines in geothermal systems such as the Salton Sea and black smokers at ocean spreading centers (see Plate 6). Nearly every type of mineral deposit can now be explained by known processes, rather than by ore-forming mechanisms of debatable nature. Replenishment of dwindling available metal resources is promised by the ability to describe and model the processes and to predict probable locations. For example, recognition that the chemistry of igneous intrusions is closely related to plate tectonic processes has led to understanding and additional discoveries of porphyry-copper and porphyry-molybdenum ore bodies. Besides plate tectonic reconstructions, modern mineral exploration also integrates the understanding of volcanology, geothermal studies, geochemistry, geophysics, and economic geology in the successful search for ore deposits.

A picture is emerging of the characteristics of fluid flow in sedimentary basins. New hydrologic research in basin analysis focuses on understanding the geological mechanisms controlling basin-scale fluid flow. This research is also applied toward more efficient and successful exploration for energy and mineral resources. The resource requirements of the planet grow as developing nations increase their consumption of mineral resources while at the same time so do the industrialized nations. Despite these increases, imminent resource depletion is unlikely because major conceptual and technological leaps during the past decade have greatly enhanced our capabilities to seek and develop these resources. The challenge of resource sustainability is discussed further in Chapter 4.

Priority Theme C: Preventing Damage from Geological Hazards

Human society is a part of the biosphere, which exists within the zone where the solid earth interacts with its fluid envelopes. This zone is perched between the two engines of mantle convection and solar energy that drive geological processes. The more dynamic surface manifestations of those processes can destroy parts of the biosphere, including its human inhabitants and their structures. As society has expanded its population and increased its use of resources, the vulnerability of its encounters with dangerously vigorous processes has also increased. The term "geological hazards" is used for these perfectly normal adjustments, which have been typical of the Earth's surface since long before even ancestral humans arrived on the scene.

Natural events that are considered hazards include instability of the ground supporting human activities, eruption of lava initiated by processes deep within the Earth, and behavior of water in the hydrosphere above and below ground; even potential impacts by asteroids or comets should be considered hazards. Some human activities can also induce hazards. The consequences of resource acquisition, urban growth, and waste disposal accelerate the rates of change in natural processes. This acceleration may gain a momentum in landscape degradation and climate change that cannot be checked.

Classification of a geological event as a hazard indicates a potential danger or risk. The hazard may pose a relatively minor risk that will have a minimal effect, or it may be a potential catastrophe that involves great damage and loss of life. Most geological hazards can be avoided, or at least tempered, by taking a few responsible measures. These include proper land-use planning, appropriate construction practices, building of containment facilities such as dams, stabilization of landslide-prone slopes, and development of effective prediction and public warning systems. Already measures of this kind have reduced human suffering from geological hazards in most parts of the world.

In the United States, research into the disaster potential of areas threatened by earthquake and volcanic eruption has been spurred by such natural events. The seismic activity common to the San Andreas fault zone jolted the nation to awareness with the San Fernando earthquake of 1971 and the Loma Prieta earthquake of 1989; most recently, the 1992 Landers and Big Bear earthquakes amplified this awareness. The eruptions of Mount St. Helens in 1980 and Mount Pinatubo (Philippines) in 1991 brought the dangers of volcanic hazards (Plate 9) to the attention of scientists, politicians, and citizens. But in addition spectacular geological disasters, there are geomorphic hazards. These are the slow progressive landform changes that evolve into hazards involving costly preventive or corrective measures. For example, the slow shift of a meandering river may undermine a bridge's foundation, eventually causing the structure to collapse; or steady deposition within a harbor area can require extensive dredging to maintain safe accessibility for large container vessels.

Recording, processing, and interpreting geological changes in time frames of seconds to weeks (termed realtime geology) is an exciting new approach that promises to help mitigate the effects of natural disasters. Traditionally, solid-earth scientists have concentrated their studies on time frames of millions or billions of years, and yet many processes take place rapidly. For example, to predict—with sufficient time to avert disaster—when an earthquake will happen, when a volcano will erupt, or when a landslide will occur, data on strain accumulation and release must be gathered quickly and continuously, and an analysis of the data must be performed as rapidly as possible. Researchers in real-time seismology hope to construct networks that monitor the direction and strength of the most destructive earthquake waves and warn threatened communities. Those communities could shut down utilities before the shaking begins; fire threats from broken gas pipes and electricity lines could be reduced, and communication links could be preserved to aid recovery efforts. Orderly procedures could be taken to secure nuclear power plants, and centralized computer facilities could take actions to avoid massive disruptions. Successful networks that track and predict tsunami dangers have been in operation for decades.

Priority Theme D: Assessing, Mitigating, and Remediating Effects of Environmental and Global Change

The environment changes progressively as continents drift, ocean basins open and close, climates fluctuate, and ocean currents adjust accordingly. Geological evidence demonstrates that rates of environmental change vary and that some catastrophic events cause abrupt changes. Humans have continually influenced their local environment, as most living things do, but in the past these disturbances have been minor. However, these influences are increasing. Ecological repercussions traditionally control population pressures, in human as well as nonhuman communities. Now humans have become major agents of environmental—and geological—change. Society has reached such density that its effects are concentrated: humans now move and use more solid material each year than is transported from continents to oceans in all of the world's rivers.

Problems are seldom as simple as they appear. In the beginning of the twentieth century, urban planners were delighted at the prospect of a hydrocarbon-powered transportation system dominated by horseless carriages as an escape from the waste

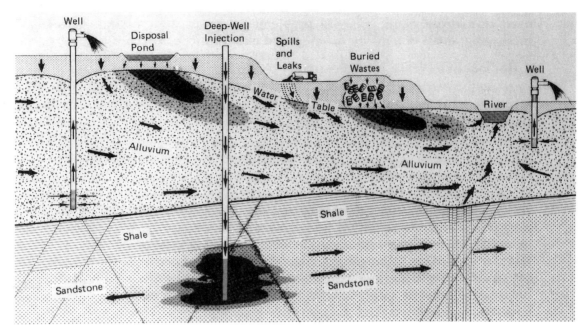

FIGURE 1.14 Schematic for possible groundwater contamination from a variety of waste disposal practices. From *Groundwater Contamination*, National Academy Press (1984).

discharged by thousands and thousands of horses. What would they think of today's smog, waste motor oils, and mountains of used tires and junked autos or of the negative balance of payments due to importation of the oil that literally drives this burdensome waste?

The growing effects of an industrial society on the environment and on geological cycles necessitate close monitoring of human activities (Figure 1.14). This attention will lead to an understanding of cause and effect, so that precisely what perturbs the system can be discovered and factored into future practices. Not only can the material controlled by human society disrupt physical cycles by increasing mass transfer; it can also disrupt chemical cycles. The conventional cycles of biogeochemical processes concentrate elements into living creatures and into ore deposits and other geological features. Now these processes follow new paths of chemical migrations and concentrations, forming compounds deliberately generated by manufacturing and inadvertently generated by the disposal of materials. Many of these concentrated compounds are toxic. Effective remediation of toxic compounds is possible by controlling the biogeochemical environments at disposal sites.

People all over the Earth are moving from less inhabited rural areas to the environs of often overcrowded cities in hopes of a better livelihood. Wastes produced in and around cities further com-

promise the quality of surrounding regions. At the same time there is a strong economic force demanding urban renewal, or recycling of crowded urban space. Structures for human habitation, transport, or manufacturing are made taller and heavier, and they may be built on old foundations that cannot support the newly designed structure.

Geology is the main interconnecting element between the craft of civil engineering and the intricacies of nature, and, in that capacity, geology is used to accommodate the demands and to reduce the damages of societal growth. Every effort toward improvement of the human condition necessarily results in some effect on the environment. Geological conditions control the maintenance and influence the quality of human life in both developing and developed regions. Geological understanding not only identifies the potential of major natural disasters but it can also warn of activities that could result in minor disasters or in the waste of resources.

The surface is molded by two competing, but usually sluggish, forces. One force levels and smooths the landscape through erosion and deposition, and the other creates topographic relief by uplift. Humans evolved as bystanders in this competition, but increasingly have acquired an instrumental role—destroying forests and plowing fields, enhancing erosion, creating reservoirs that interrupt sediment migration, and generating massive amounts of carbon dioxide that influence global

TABLE 1.3 Approximate Percentages of Expenditures Keyed to the Research Framework of the Federal Agencies for Fiscal Year 1990[a]

Objectives

Research Areas	A. Understand Processes	B. Sustain Sufficient Resources— Water, Minerals, Fuels	C. Mitigate Geological Hazards— Earthquakes, Volcanoes, Landslides	D. Minimize Global and Environmental Change—Assess, Mitigate, Remediate
I. Global Paleoenvironments and Biological Evolution	2	<1	<1	1
II. Global Geochemical and Biogeochemical Cycles	4	20	—	1
III. Fluids in and on the Earth	2	12	<1	3
IV. Crustal Dynamics: Ocean and Continent	19	22	4	6
V. Core and Mantle Dynamics	4	—	<1	—

[a]One percent of the total of $1,368 million is about $13 million (see Appendix A).

climate patterns. Predictions of global change based on historical observations consider only a brief period of geological time; the context for understanding how the environment changes as a result of human activities will be provided only thorough analysis and interpretation of the enduring geological record. While geologists traditionally have based their working hypotheses on the idea that the present is the key to the past, scientists investigating possible environmental changes recognize that the past may well be the key to the future.

RESEARCH SUPPORT

While the federal government is the source of the largest part of U.S. research expenditures in the solid-earth sciences, the petroleum and mining industries are significant contributors in certain areas. Future researchers will benefit from examining past allocation trends.

Federal Funding

Total federal funding for solid-earth science activities in fiscal year 1990 was on the order of $1,368 million (see Appendix A). The details (in relation to the Research Framework) of the types of activities that were supported are given in Appendix A, and their estimated relative percentage is shown in Table 1.3. Given the diversity of agencies and accounting methods, there is some uncertainty about what matrix box is most appropriate for some of the research funds, but the broad picture is valid.

Considering the overall distribution of federal support across the Priority Themes in the Research Framework, there do not appear to be significant gaps, which indicates that the existing national research structure is working reasonably well. Many of the priority themes are already well established. New funding for modern equipment, technical support staff, and adequate ongoing operating support is required to sustain important progress on most of the themes.

Industry Support of University Research

The petroleum industry has traditionally supported hydrocarbon research that involves theoretical and, more particularly, applied geology, geophysics, and geochemistry. Likewise, so has the mining industry. It is hard to assign a meaningful dollar cost to all this research. One rough guide might be this: the seismic exploration industry worldwide is expected to rise to about $5 billion by the mid-1990s. If about 1 percent of this sum goes to related earth science research, industry support would be about $50 million. Other estimates indicate that a range of $100 million to $275 million is expended annually on oil and gas research in the United States in both the public and private sectors. Although most of the research is in-house, both mining and petroleum industries historically have supported research projects in university departments and collaborated in research with federal agencies (e.g., Bureau of Mines and Department of Energy).

Mining industry support of university research typically involves funding graduate student work in the field or laboratory, summer or interim employment of students, consulting arrangements with faculty, and direct grants. During the fiscal decline of the mining industry in the early and mid-1980s, this support diminished considerably as companies cut back on research and exploration activities as well as on geoscientific personnel. In recent years a growing proportion of the supported research has been in the area of low-temperature, heavy metal geochemistry—a reflection of concern with waste management. At the same time, support for basic research in ore-forming processes and igneous petrology has declined.

The petroleum industry currently supports university research through granting foundations in the form of doctoral and master's fellowships, direct faculty support, and grants for equipment and laboratories. At the same time, many companies are providing support directly through their research and operating subsidiaries, either through membership in industrial consortia or through direct funding of research by faculty and students. Additional research funding is handled by trade associations, such as the American Petroleum Institute and the American Gas Association. The industry-supported Petroleum Research Fund of the American Chemical Society has played an important role for decades. A wide variety of university programs have been encouraged through these means, ranging from basic research in petrology, paleontology, and sedimentology to technologies for reservoir characterization, enhanced oil recovery, and seismic signal processing. Petroleum industry support of environmental research is growing. Particular emphasis is being placed on the disposal of solid and liquid wastes and on groundwater management.

The main thrust of oil and gas company research is naturally toward the development of technology and science that can be directly applied to exploration for and development of oil and gas. If an application cannot be defined, support for a research project is unlikely to be granted. It should be noted, however, that a surprising number of research programs pursued by industry have led to significant bodies of fundamental knowledge, which in turn have supported societal endeavors quite apart from the search for energy resources.

An early example is the study of sedimentary processes and sedimentation of the past 10,000 years, which led to funding of American Petroleum Institute Project 51 for study of the northwest Gulf of Mexico. This project was a pioneering effort in basin-wide sedimentology and ecology, which set the stage for subsequent work on deltas and shorelines around the world. Since then, comparable studies of sedimentology have moved offshore to investigate the processes of continental slopes and submarine fans. The results of these endeavors, many of which were supported by industry consortia, are applicable to environmental engineering and to national defense considerations.

The use of regional seismic data to unravel the stratigraphic history of basins, which was pioneered by industry, has led to the idea that global sea-level change controls patterns of sedimentation within basins. These studies are leading to a critical reassessment of the various controls on the development of continental shelves and slopes, submarine fans, and carbonate platforms. Each of these depositional environments offers a rich source of information about the sedimentary record of the past and about the continental margins of the present.

While the basis for digital analysis and processing of seismic signals originated in the university community, intensive development of the technology has been carried out by the petroleum industry. Many of the techniques are now used in seismology and also in the defense and medical fields. Furthermore, the need for high-speed signal processing spawned the development of array processors, now widely used for computations of many kinds.

The occurrence, nature, and diagenesis of organic materials in sedimentary rocks is another area of science that incubated in the applications laboratories of the petroleum and coal industries. Investigation of the burial and preservation of organic-rich sediments in modern environments such as the Black Sea illuminates episodes in the past when oxygen-depleted waters occurred in isolated basins. Evidence suggests that sometimes oxygen depletion spread widely over shallow shelf seas. No modern analogs exist for the latter conditions, and the study of these events—which gave rise to a large part of the world's oil resources—aids in evaluation of the processes that control oceanic circulation and lead to relatively rapid changes in the makeup of life in the seas.

Bacterial and thermal alteration of buried organic material forms oil, natural gas, and coal. The by-products include organic acids, which have an important influence on the reactions between rocks and the subsurface brines contained within their pore space. These reactions affect the processes of lithification, by which friable and plastic sediments are converted to rock, and of metalliferous ore formation.

INSTRUMENTATION, COMPUTATIONAL CAPABILITY, AND DATA MANAGEMENT

Facilities and Instruments

Many advances in the earth sciences have followed advances in instrumentation. In every decade since the 1890s some major change in instrumentation, beginning with the earliest seismographs and the discovery of x-rays and radioactivity, has resulted in major repercussions for the earth sciences. Progress in scientific instrumentation has accelerated at a remarkable rate. The instruments in modern laboratories include those that can produce images of materials on the atomic scale, determine crystal structure under immense pressure (Figure 1.15), and chemically analyze the composition of less than one-billionth of a gram of material.

A paradoxical situation has arisen. On the one hand, the earth sciences now depend on many more laboratory scientists than 25 years ago. New instruments yield more kinds of data of greater precision than ever before. And many more questions are posed and answered at greater rates. On the other hand, the necessary instruments are more expensive and they become obsolete more quickly. Practicing scientists tend to see the glass as half empty, concentrating on the competition, expense, and obsolescence, while administrators and funding agencies consider the glass half full, regarding the numerical increases as proof of the discipline's health. The instruments used by laboratory earth scientists seldom cost as much as $1 million, although some essential data can only be acquired using expensive facilities such as accelerators, mass spectrometers, or synchotrons.

Success in obtaining some of the more expensive instruments often comes from collaboration, con-

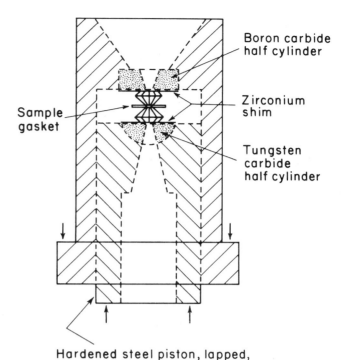

FIGURE 1.15 Schematic of a high-pressure diamond anvil; research using similar apparatus have brought many aspects of mineralogy and geophysics together to understand mantle and core properties and dynamics.

sortia, matching institutional and state funds, and other varieties of cost sharing. Within and among institutions, a variety of ingenious schemes for sharing operating costs have developed, while efficiency of use is maintained by the continuing support of competent staff—a policy that allows sharing of equipment requiring highly trained operators. A small number of private foundations with an interest in the development of the earth sciences help by supplementing and complementing public funds.

There is a widespread perception that the overall condition and availability of research instrumentation are rapidly deteriorating. One specific reason for this perception is that it is now 20 years since the solid-earth sciences community began to optimize laboratory instrumentation, at the time of the Apollo program.

The most costly instrumentation in the solid-earth sciences, as in all earth system science, is mounted on spacecraft and on ocean-going vessels. The use of manned and unmanned aircraft as instrument platforms plays a relatively small role at present, except for a continuing commitment from the National Aeronautics and Space Administration to develop space instruments by testing on aircraft. This wide global perspective is growing at the same time that internationally cooperative involvement has increased in programs addressing earth science systems.

Most ocean-going research vessels used for solid-earth sciences research are also used for oceanographic and other earth system science research. The Ocean Drilling Program, which costs about $45 million a year (mainly from the National Science Foundation with a little less than half of the funds come from collaborating countries), represents the largest single current operation in the solid-earth sciences. (Its role is discussed in Chapters 2 and 3.) Its unique contributions to an understanding of how the Earth works are clear. The Continental Scientific Drilling Program is smaller than the Ocean Drilling Program but promises to provide complementary information about continental processes.

Solid-earth scientists have diverse interests in measurements involving space. These extend from the three high-precision distance-measuring techniques—very long baseline interferometry (VLBI), satellite laser ranging (SLR), and global positioning systems (GPS)—through altimetry and magnetometry to remote sensing of the land, especially that using high-spatial-resolution methods such as that available from Landsat and the Systeme Pour l'Observation de la Terre (SPOT). The ambitious plans for an Earth Observing System (EOS) lie at the core of the U.S. global change program.

Theoretical models of earth processes will become increasingly important for solving many problems that simply could not be tackled previously. These include problems that are among those of greatest concern to society. Finding the solutions will require the best research data bases, the fastest computers, and highly imaginative people.

Earth scientists have already developed quantitative models for many earth processes. Numerical computer simulations have been devised for mantle convection, the evolution of sedimentary basins, the concentration of oil and gas resources, climate change, and surface processes such as erosion, fluid flow, and rupturing of the crust in earthquakes. These models are tools that have been used to allow scientists to rapidly test hypotheses about earth processes. The necessary answers involve understanding the complex interactions of many processes. This requires the computational power to handle large numbers of calculations and large amounts of data.

The development of high-performance computing and communications is currently one of the Presidential Initiatives in the United States. It may be time to consider whether a dedicated high-speed computational facility for solid-earth scientists is necessary or whether improved access to current and developing facilities would be more helpful. Whatever route is taken, it is clear that many solid-earth problems will make high demands on available computational facilities.

Molecular geochemistry is one of the exciting earth science frontiers opened up by advances in theoretical chemistry, numerical algorithms, and data-handling capabilities. Calculations are based on quantum mechanical equations governing the bonding and energetics of atoms in minerals and fluids. These calculations provide substantial insight into our understanding of mineral composition and structures and aqueous solution behavior under a wide variety of pressure and temperature conditions. The prognosis for future increases in computational power at both the supercomputer level and the workstation level promises to establish these methods as a major requirement for theoretical studies of solid-earth processes.

Data Handling

Science is experiencing a variety of distinct revolutions in data handling. Very large amounts of data can now be generated very quickly, and much of it needs to be archived. Spatial data amenable to geographic information systems (GIS) characterizes the solid-earth sciences. For much of the United States digital data sets are commonly available for such basic information as topography. That level of coverage, however, is uncommon over much of the Earth. Ironically, higher-resolution digitized topographic data are available for Venus than for Earth, partly because Venus has no ocean but also because of the lack of military and political sensitivities. The same may soon be true of Mars.

In general, the digital revolution is making great changes in the way geological data are used. Geological maps for many areas are stored in digital format, and hard copies are generated as needed. The practical problems of cities and counties that require geological and geographical information are being addressed as GIS allows easy integration of diverse spatial relationships.

The World Data Centres, established as an archive for information acquired since the International Geophysical Year, have worked well as custodians and monitors of geophysical data for nearly 30 years. Global change studies, involving a wider range of data gleaned from ecological and sociological sources, will modify the traditional way that these centers operate.

Study of the solid earth requires careful archiving of and wide access to regional subsurface information such as seismic reflection data, well logs, core samples, and cuttings. At present, these are all handled in different ways, and much of the data are proprietary in nature. Current systems are working well, but increased access will become necessary as societal needs evolve.

Storage of rocks, fossils, and minerals is the responsibility of museums because proper curation depends on the specialized knowledge of properly trained personnel. The rising base costs of museums are forcing managing boards to take a hard look at how useful museums are to the supporting community. For solid-earth scientists, well-run museums are essential. Fossil collection in particular provokes conflicts among scholars, museums, commercial collectors, and others.

As we look at future programs in global change, geological hazards and their reduction, as well as programs using space-based techniques, a recurring central theme is the need to establish effective data management schemes. Essentially all of our programs are data intensive and depend on digital data. Thus, if they are to be successful, management procedures must be established for organizing the digital data chain from capture to analysis and for making those data available.

Rapidly expanding quantities of earth science data, increased awareness within the community of digital methods, improved technology in both hardware and software, and the necessity for multicomponent data-set synthesis will continue to apply pressure for improved efficacy of data management. Digital earth science data should be available at an international level to broaden the research base for all scientists.

EDUCATION AND EMPLOYMENT

There is a national crisis in science education. The problems are magnified for the earth sciences, to which few students are exposed prior to college. College enrollments in science are decreasing while job opportunities expand—in new areas for which few adequate college programs exist. An aging college faculty and a low level of both public and decision-maker awareness about earth science issues compound the problem. The geosciences profession continues to have to cope with the peculiar problem that a significant part of the population of the United States considers the Earth to be no more than a few thousand years old. Many citizens flatly reject biological evolution, particularly as it applies to the human species. This situation is not likely to change, but its consequences recur prominently from time to time. These conditions contrast with the ever-increasing need for knowledge about Earth because of the growing worldwide consequences of human activities. Sensitivity to this educational challenge is spreading throughout the profession, and energetic initiatives are coming from professional societies, teaching faculty, and other groups. These developing initiatives should enhance general appreciation for the earth sciences and improve the educational status of the discipline.

The Earth sciences are taught in fewer than 5 percent of the nation's high schools. There are few qualified earth science teachers for kindergarten through grade 12 because of the paucity of teaching opportunities. Not only is the opportunity missed to excite youngsters about the Earth, and ultimately to attract a number of them to the profession, but the opportunity to generate interest in science in general is wasted. Study of the earth sciences offers the prospect of focusing students' attention on a subject they are familiar with and already have questions about, such as the local landscape. An innovative approach would begin with local field trips, including the study of biology, which ultimately would lead to explanations involving the study of chemistry and physics. This type of program could serve as a general introduction to science in the context of the real world.

A related concern is a general lack of awareness of the earth sciences in societal issues, a situation that aggravates the educational problem. The average citizen has a need for better appreciation of natural phenomena, resources, and environmental-economic concerns to make informed decisions about protection from earthquakes and floods; about maintaining sources of water, minerals, and energy; and about the threats from acid rain, hazardous wastes, and global change.

The education problem is severe in the solid-earth sciences because of fluctuations in the demand for geologists as well as the severely decreased student enrollments. Historically, the domestic petroleum industry has been the greatest single employer of geologists, geophysicists, and geochemists. As a result of the dramatic decline in petroleum prices in the early 1980s, following aggressive hiring by the industry during the 1970s, many geoscientists found themselves without jobs. At the same time, another major earth science employment segment, the mining industry, was in the midst of a long period of depressed commodity prices. So thousands of mineral resource geoscientists also were without jobs. Many individuals in both fields resorted to early retirement, nongeological employment, or marginal industry employment at levels far below their capacities, and their talents were forever lost to science. This represents a waste of scientific resources orders of magnitude greater than the resources lost by an oil spill, yet no newspaper headlines heralded the event. There is no science equivalent of an Environmental Protection Agency or the National Guard to remedy, or even recognize, a disastrous brain drain.

Enrollment in undergraduate geoscience programs plummeted from 28,000 in 1982 to 9,000 in 1988 and is now rising only very slowly. In the early 1980s, while petroleum-related jobs became scarce, environmental legislation addressing waste disposal sites was enacted and strictly enforced, which opened employment opportunities for individuals with appropriate training. Employment projections indicate a sixfold increase in many geoscience areas during the remainder of the century for Superfund hazardous waste sites alone, and there is no question that employment opportunities in the earth sciences are growing. They are shifting from concentration on resource problems toward hydrology, rock-fluid relationships, and associated problems.

Against these brightening prospects, however, where will the replacement faculty and other geoscientists come from in light of the present extraordinarily low enrollments? If the petroleum industry rebounds from its decline, how will its demands further stress the labor pool? Supply and demand have been out of phase for decades in the geosciences.

Women and minorities are not common among earth science professionals. Some estimates place their involvement as less than 5 percent. Surely opportunities exist to make the earth sciences more intriguing and rewarding for this majority, women and minorities, of the North American population.

Universities and colleges might wish to examine their undergraduate curricula in the earth sciences to see whether they adequately train future professionals and whether they provide scientific awareness for the lay citizen. Introductory geology courses, which represent the only exposure that most students have to the earth sciences, should be stimulating and challenging. Comprehensive courses in earth system science, emphasizing interrelationships and feedback processes, as well as the involvement

of the biosphere in geochemical processes, would appeal not only to life and earth scientists but also to a wider range of curious students. Professional programs must increase offerings in emerging fields and might consider phasing out less useful requirements. Growth in both employment and research opportunities can be expected in such areas as hydrology, land use, engineering geology, environmental and urban geology, and waste disposal. The interests of earth science and engineering departments converge in these fields, and opportunities to link the two should be explored. Unfortunately, there are few qualified scientists to teach these programs at the graduate level, because those who are qualified are attracted by the higher incomes of the industrial and consulting fields.

The greatest potential for enhancing earth science education lies with the federal agencies that have programmatic and legislated authority to do just that. Over $360 million per year is spent on science and engineering education by the National Science Foundation and the Department of Energy; the U.S. Geological Survey, which works mainly at the precollege level through development of publications and increasing services to teachers. Other federal organizations have expressed an interest in the education process, including the Department of Defense, the National Oceanic and Atmospheric Administration, the National Aeronautics and Space Administration, and the Environmental Protection Agency. These organizations appreciate the pending problems in the availability of earth scientists and the need for more public awareness of science issues. The need is critical now, as the Department of Energy initiates its multidecade cleanup of waste sites throughout the country, with many projects heavily dependent on earth science knowledge.

When there are opportunities to offer expert advice on issues such as hazard mitigation, the sustainability of life, resource exploration and depletion, and waste disposal, the solid-earth sciences community should be ready.

INTERNATIONAL SCOPE

Earth system science is an intrinsically international undertaking. The global character of geological processes translates into a realization that, for example, many tectonic and sedimentary processes are best illustrated outside our national borders; that volcanic and other hazards are shared by all peoples; that environmental degradation does not stop at political boundaries; and that climatic and sea-level

TABLE 1.4 Organization of the Report

Overall perspective: Chapter 1
Conclusions, priorities, recommendations: Chapter 7

Facilities, Equipment, Data Bases: Chapters 6 and 7
Education: Schools, Universities, Public: Chapter 6
Human Resources, Professionals: Chapter 6

Objectives

Research Areas	A. Understand Processes	B. Sustain Sufficient Resources— Water, Minerals, Fuels	C. Mitigate Geological Hazards— Earthquakes, Volcanoes, Landslides	D. Minimize Global and Environmental Change—Assess, Mitigate, Remediate
I. Global Paleoenvironments and Biological Evolution	Chapter 3	Chapter 3 Chapter 4	Chapter 3 Chapter 5	Chapter 3 Chapter 4 Chapter 5
II. Global Geochemical and Biogeochemical Cycles	Chapter 2 Chapter 3	Chapter 2 Chapter 3 Cahpter 4	Chapter 3 Chapter 5	Chapter 2 Chapter 3 Chapter 4 Chapter 5
III. Fluids in and on the Earth	Chapter 2 Chapter 3 Chapter 4	Chapter 3 Chapter 4	Chapter 3 Chapter 5	Chapter 3 Chapter 4 Chapter 5
IV. Crustal Dynamics: Ocean and Continent	Chapter 2 Chapter 3	Chapter 3 Chapter 4	Chapter 2 Chapter 5	Chapter 5
V. Core and Mantle Dynamics	Chapter 2	Chapter 3 Chapter 4	Chapter 5	Chapter 5

changes occur all over the world. Basic field data must be gathered from diverse regions of the Earth, and earth scientists must conduct their experiments on a global basis to accurately determine the composition and dynamics of the whole Earth. International cooperation and exchange among geoscientists in different countries are thus essential. The successes of a variety of programs of international scope, such as the Ocean Drilling Program and global seismic networks, attest to their value to the earth sciences.

In 1982 the National Science Board concluded that the United States was at a critical point in its international scientific relationships. The technical lead had been lost in many fields at the very time when the global nature of scientific problems was becoming more apparent. The loss of scientific leadership translates directly into lost technological edge and lost trade advantages, as well as lost science.

Economic and scientific common sense dictates that the interests of the United States in the global economy and in the global scientific community are best served by strong international science programs. Compelling arguments can be presented that in the future U.S. energy and minerals resources will be drawn from overseas to an even greater degree. Because of this dependency, information about the amount and distribution of global resources will be required for a range of policy decisions.

ORGANIZATION OF THE REPORT

The chapters in this report are keyed to the research framework, as can be seen in Table 1.4. Following the definition of goals and objectives and the Global Overview of the priority themes in this chapter, Chapter 2, Understanding Our Active Planet, concentrates on the Earth's internal workings. Research into the relationship between the core and mantle, and the dynamic convective motions within the mantle, has advanced rapidly. New results are bringing researchers closer to understanding how internal processes affect events at the Earth's surface: earthquakes and volcanoes in the short term, and plate tectonics and mountain building in the long term.

Chapter 3, The Global Environment and Its Evolution, is concerned with the crust and its interaction with the fluid envelopes, which is driven by solar energy. Past environmental changes and their effects on life can be determined from the study of sedimentary rocks, which contain the record of ancient environments, including information about the composition of the atmosphere and oceans through time. Studies of continental, oceanic, and atmospheric processes have been revitalized by the integration of this compositional information with interpretations of plate tectonic movement and the resulting changes in distribution of continents and ocean basins. The record of rocks and fossils also documents evolution and extinction and provides an intellectual framework for understanding humanity's place in nature.

Chapter 4, Resources of the Solid Earth, includes environmental concerns related to resource extraction and use. Groundwater, and the flow and reaction of fluids within the crust, is a pervasive theme associated with most geological resources. The report therefore considers water as well as the traditional earth resources of minerals and fossil fuels.

Chapter 5, Hazards, Land Use, and Environmental Change, addresses the geological problems facing humankind and involves research throughout most of the matrix. Topics include hazards caused by earthquakes, landslides, and volcanic eruptions; environmental changes associated with urban growth, engineering, and agricultural practices; and some modern consequences of the long-term global change reviewed in scientific terms in Chapter 3.

Chapter 6, Ensuring Excellence and the National Well-Being, reviews topics such as education, demographics, instruments and facilities, the revolution in data gathering and handling, and international collaboration.

Chapter 7, Research Priorities and Recommendations, tackles the problem of determining science priorities for the whole array of research activities covered in Chapters 2 through 6 and the sources of federal funding for solid-earth science research. The top-priority and high-priority research selections and the facilities required for their implementation are outlined.

There are two informative short-cut paths between Chapters 1 and 7, highlighted for easy identification. Essays at the beginnings of Chapters 2 through 6 provide the sense of the science in each chapter. At the end of each chapter, the most promising research opportunities are summarized in research frameworks without background detail (these are also collected into a single table in Chapter 7).

2

Understanding Our Active Planet

ESSAY: THE DYNAMIC EARTH

Understanding earth processes requires broad and eclectic thinking. The earth system is complex, with open channels between interacting boundaries the norm rather than the exception. Many researchers think of the solid Earth as an engine driven by radioactive decay, while others expand this view to include the whole earth system and consider the added processes driven by solar energy. Others see the Earth as a system of geochemical cycles with interchanges spanning ranges of time and space that extend back to the birth of the solar system. Finally, some scientists regard the planet as a series of concentric domains with ill-defined layers distinguished by the transfer of mass and energy.

The Earth is all of these and more. The accelerated understanding of the earth system that characterized the past few decades is attributable to problem-solving strategies based on integration of these various interpretations. Contributions from geochemistry support theories developed from seismological data, structural geology depends on investigations in physics, and organic chemistry offers potential explanations for problems encountered in both resource extraction and waste management. Since their adoption of this expanded tool kit for investigating the implications of plate tectonics, earth scientists have made unprecedented progress.

The Earth began over 4.5 billion years ago with the accretion of material orbiting around the Sun, supplemented by the capture of other bodies from intersecting orbits. Early in the process of consolidation, proto-Earth collided with a Mars-sized body and the material from both reorganized into the Earth-Moon system. Soon—in a geological sense—after that event, convection cells became established within the Earth's mantle, a crust developed, free water entered the atmosphere, plates diverged, ocean basins evolved, and mountains rose through tectonic forces at work along plate boundaries. These distinctive earth phenomena are both cause and effect in a multiscaled arrangement of interacting processes.

From the perspective of geochemical cycles, there are two end-member processes, differentiation and mixing; two end-member domains, the exterior Earth environment and the interior; and two end-member time frames, hundreds of million of years and the instant. Within each end-member pair, there is a continuum of possibilities. Generally, surface domain processes occur very quickly and interior processes endure over long intervals, although there are exceptions. Some continental material has endured for billions of years near the surface, and mantle plumes may erupt at the surface with no detectable warning, after migrating from the core-mantle boundary over mere millions of years. And while differentiation and mixing of large volumes continue for eons within the mantle, incremental changes within small volumes can take place quickly both at the surface and within the interior.

The domains that extend above and beneath the surface contain the Earth's fluid envelopes. Water vapor in the atmosphere condenses and falls as rain. At the surface, water weathers the rocks physically and chemically: physically by impact and by freeze-thaw action and chemically by solution and the introduction of ions that foster reactions with rock minerals. Particles and solutions from crustal rocks wash downstream and enter the great water reservoirs of river, lake, ocean, and groundwater—settling out as detrital sediment and as precipitates.

At ocean spreading centers and in volcanic environments, water may aid in the precipitation of mineral concentrations that become valuable resources when discovered in accessible terrain. Magmas and other fluids that move through the crust have the potential of becoming significant sources of minerals and energy. Along subduction zones, hydrated crust and water-saturated ocean-bottom sediments descend into the interior beneath a mantle wedge that extends over the sinking plate. At high temperature and pressure these rocks dehydrate, which leads to melting. Volatile-rich magmas rise and interact with crustal rocks to generate the type of gas-charged magmas that erupt at the surface with devastating violence. The volatility is most pronounced along Cordilleran arcs of continents; the volcanoes along the South and North American Cordillera erupt in explosions that may literally blow them apart, as Mount St. Helens did in 1980.

Volcanoes that build over hot spots, such as those in the Hawaiian Islands, erupt magmas that flow rather placidly because of their chemical makeup. They contain smaller proportions of silica, and gases escape readily. By the time the magma reaches the surface it is less explosive and sticky, so it flows easily. Eruptions from hot spots produce large volumes of basaltic lava spreading over extensive areas in layered sheets that may accumulate to great thicknesses; the Hawaiian volcanoes reach heights over 9,000 m above the deep ocean floor and nearly 4,000 m above sea level.

Hot spots originate deep within the Earth. They are the result of plumes that reach the surface after rising through mantle and crust. The source of these plumes may be within the mantle or from the mantle-core boundary. They may originate at both depths, and researchers are not yet able to recognize evidence that could characterize distinct sources.

Researchers are also investigating the possible causes of mantle plumes. Physical anomalies along the core-mantle boundary and chemical anomalies attributable to recycled surface material are two intriguing possibilities. Whatever the cause or the source of mantle plumes, they bring to the surface basaltic lavas with chemical clues about the deep mantle material from which they were extracted.

The movement of plumes through the mantle represents chemical and physical links between the interior and exterior. The mantle itself is flowing in a complicated pattern of convection. This pattern manifests itself at the surface as spreading centers and subduction zones, where vast slabs of lithosphere can be seismically traced along descending arms of the convecting system. The convection, which governs plate dynamics, may be limited to an outer layer of the mantle, possibly complementing another convection system delivering energy and material through an inner mantle. An alternative possibility suggests cells that convect through the whole of the mantle, directly linking the bottom boundary along the core to the surface characteristics of plate tectonics.

There are two layers to the Earth's core, recognizable from the distinctive behavior of seismic waves. The outer core is fluid. Only compressional waves propagate through it, while shear waves can be detected propagating within the inner core. The core is the nucleus of the internal domain, 2,900 km below the surface. Even from that remote depth it affects the crust and the atmosphere: the core is the source of the magnetic field.

Core, mantle, crust; lithosphere, hydrosphere, atmosphere, magnetosphere—every layer, every component of the earth system can be defined independently. But to understand the meaning of those definitions, the significance of the components, and the nature of the whole earth system requires consideration that transcends the specific. Exchanges between the innermost center and the outermost reaches of the earth system are ubiquitous and continuous. Earth scientists are discovering both explanations of the past and implications for the future by adopting this grand scale—the whole earth system—in their ongoing inquiries.

This expansive perspective solves old problems and presents new ones. For example, 25 years ago plate structure was recognized as a characteristic that specifically defined the lithosphere. Lately, motion of the lithospheric plates has gained a prominent position as a factor in processes that affect both mantle heterogeneity and global climate. Seismic studies have traced hot areas associated with spreading centers deep into the mantle and recently have detected slabs of cool lithospheric material descending deep beneath subduction zones. This cooler material persists over long periods; cool-temperature anomalies found in the mantle today are remnants of the breakup of the ancient continent of Gondwanaland about 150 million years ago.

The breakup of Gondwana also caused drastic changes in climate patterns. As Africa, India, and South America drew away from Antarctica and encountered the landmasses of the north, the equatorial currents of the Tethys Sea were interrupted and deflected. After India collided with Asia 45 million years ago, cold deep water began to accumulate off

Antarctica. Circulation around that continent became firmly established when the Drake Passage between South America and Antarctica was finally breached about 30 million years ago, and glaciation advanced over eastern Antarctica. After Arabia collided with Asia 15 million years ago, circulation of the warm saline waters southward out of the Indian Ocean ceased and the West Antarctic ice sheet grew. Finally, following the emergence of the Isthmus of Panama and complete interruption of extensive east-west equatorial ocean circulation, Arctic glaciation began and eventually grew to cover vast areas of North America and Eurasia. This modern glacial mode persists, despite the relatively small glaciated areas of the current interglacial period.

Climate changes result from ocean circulation changes. Ocean circulation in turn follows paths established by variations in plate distribution, according to the vagaries of plate tectonics. Evidence of plate tectonics is restricted to Earth, despite rigorous surveys of similar planets and planet-like bodies. The mantle convection that results in plate tectonics originated soon after establishment of the Earth-Moon system 4.5 billion years ago. Whatever the source of this remarkable process that renews much of the Earth's surface every few hundred million years, the result is a surface environment supporting the only known life in the universe. Understanding this dominant earth process is essential for maintaining the surface environment.

ORIGIN OF THE EARTH

Any theory of the Earth should consider its origin, and any well-rounded program in the earth sciences should consider it a component of the solar system. This consideration addresses scientific issues such as the formation of the solar system, the processes causing evolution of the planets, and the relationship of these processes to its structure and history. Knowledge about the origin of the solar system comes from studies of astronomical data on forming stars and planetary properties as well as from meteorites.

We can expect advances in the application of chemical physics to the interpretation of change in cosmic material in its journey from the stars and the interstellar medium to its eventual resting place in the Earth, the planets, and other solar system bodies. This will require a multidisciplinary approach. Laboratory work using innovative experimentation and analytical techniques, observational astronomy, and theoretical efforts is involved. Geochemical and cosmochemical studies of meteorites, comets, or their analogs can become more intimately related to the origin of the Earth. The Earth formed some 4.5 billion years ago. Its age is known from differences in its radiogenic lead isotope ratios from those of the Moon and meteorites.

Technical advances allow detailed computer simulations of early solar system evolution. The simulations apply the physical laws that control orbits, collision frequency, and accumulation mechanics as small objects grow to the size of planets. One conclusion of this work is that planet growth through the accumulation of small objects occurs relatively rapidly. To go from a multitude of dust grains in a solar nebula collapsing under its own gravitational attraction to a limited number of planets circling a central star requires only a few tens of millions of years. The most recently obtained isotopic dates from meteorites suggest that this process happened about 4.55 billion years ago in our solar system.

The simulations also indicate that as the size of colliding bodies grows, more kinetic energy from the incoming body is deposited deep in the planet. For a planet like the Earth, which grows to full size in only tens of millions of years, this heat cannot be transferred to its surface, and then into space, fast enough to keep the interior from reaching very high temperatures.

Recently, attention has focused on the hypothesis that a collision between the proto-Earth and an

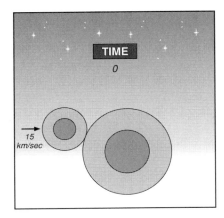
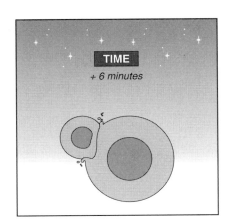
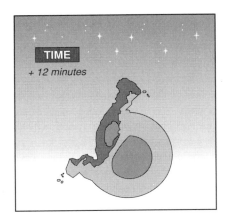
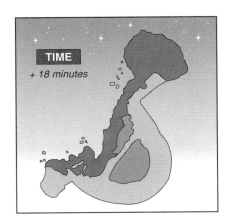

FIGURE 2.1 Model of the origin of the Earth's Moon by impact of a Mars-sized object on Earth.

object roughly the size of Mars, approximately one-tenth the mass of the Earth, may have been responsible for the formation of the Earth's moon (Figure 2.1). This impact would have been sufficiently energetic—comparable to a trillion 1-megaton atomic explosions—to propel material into Earth orbit, forming the Moon. The giant-impact origin for the Moon seems capable of explaining both the large relative size of the Moon and the geochemical similarities between it and the Earth that have been deduced from analyses of lunar samples returned by the Apollo missions.

Experiments on and calculations of the effects of impacts during accretion should lead to better understanding of the early history of the Earth, the role of a magma ocean in differentiation, and the history of outgassing and its relationship to the formation of the atmosphere and oceans.

Comparative Planetology

Many processes of the solid Earth must also occur on other planets but may frequently have rather different consequences. The proposal that the extinction of dinosaurs and many other species near the Cretaceous-Tertiary boundary was caused by the catastrophic impact of a 10-km asteroid or comet has sparked heated debate among geologists. But cosmochemists and solar system dynamicists argue that such events are probable, and researchers studying impact craters on the Moon, Mars, and Mercury say it is consistent with their data.

Study of lunar samples has led to new insights into the evolution of Earth. The earliest history of the Earth has been destroyed by plate tectonic processes. Early Earth history must be deduced from analyses of lunar and meteorite material and from exploration of other planets.

Detailed chemical analyses of lunar samples suggest that a largely molten Moon crystallized to form a thick crust composed of the low-density aluminum-rich mineral—plagioclase—overlying a dense magnesium and iron-rich interior. The lunar crust appears to have been formed by gravitational sorting according to the relative density of the minerals crystallizing from an original magma ocean. Age determinations show that much of the chemical differentiation of the Moon was complete by 4.35 billion years ago, or within 200 million years of lunar formation.

Debate continues about many issues concerning the evolution of the Earth. These issues include the degree of outgassing, the cooling of the planet, the heterogeneity of the mantle, the conditions necessary for plate tectonics, continental growth and crustal recycling, the creation and persistence of stable continental cratons, and the generation of the magnetic field. In the face of this array of questions, it would help greatly to have at least one other planet to use for comparison.

The planet Venus differs less than 20 percent from the Earth in mass; in mean density; and, as far as can be detected, in content of chemically active volatiles. Yet in secondary properties related to its interior, it is very different from the Earth. Radar imagery from the recent Magellan probe reveals a planet surface scarred by volcanic activity and bombardment from space, as well as deformation, but plate motions are not evident and the mountain-building processes contrast greatly with those of the Earth. Faults can be traced, and broad elliptical areas spanning hundreds of kilometers may represent mantle plumes rising beneath the Venerian lithosphere. With no apparent plate tectonics, heat may have to escape from deep within Venus in other ways. The comparative study has just begun, and it should be greatly enhanced by continuing analyses of data from the Magellan project.

Early Earth Evolution and Great Impacts

Obviously, any extraterrestrial impact capable of creating the Moon would have major consequences for the early evolution of the Earth. Computer simulations of giant impacts indicate that the energy released is sufficient to raise the temperature of the whole Earth by 3000 to 10000°C—more than enough to cause total melting of the planet.

Recent advances in high-pressure instrumentation have made possible studies of phase equilibria and element partitioning at pressures as high as 500 Gigapascals—5 million times atmospheric pressure—and temperatures up to 6000°C. These new limits allow experimental analyses over pressure and temperature ranges covering the entire interior of the Earth. In the next decade, integration of high-pressure experimental studies of earth materials and enhanced theoretical understanding of the dynamic behavior of the planet should provide dramatic advances in understanding these earliest stages of the Earth's evolution.

A hot early Earth accommodates speculations about formation of the Earth's large iron-metal core. Compared to the silicate minerals that make up the major portion of Earth, iron metal has a relatively low melting point. During the growth of the early Earth, the molten iron coalesced into masses of increasing size that eventually began to sink toward the center because of their high densities. The excess heat energy stored in the core is released slowly as the molten core crystallizes. At the present time, only about 5 percent of the core has crystallized, which indicates that continued crystallization of the liquid outer core may be a significant source of heat within the Earth. In addition, the heat released as the outer core crystallizes, and the transfer of that heat into the overlying mantle may provide the driving force for the convection in the outer core that is responsible for producing the magnetic field.

During their early histories, the Earth and Moon were subject to a high flux of relatively large impactors. Because the Earth is an active planet, no record of this flux is recorded, but evidence from the Moon suggests that the flux had died away by about the time of preservation of the oldest rocks exposed on Earth. The discovery of high iridium contents in some rocks a few hundred million years younger than the 3.8-billion-year age usually considered to mark a sharp drop in impactor flux indicates that this question may need reexamination.

Since that time in the earliest recorded history of the Earth, the flux of impactors has been slow, although the record of impacts is too poor to show whether it has been anything other than steady. Roughly 100 craters more than 1 km in diameter have been identified on the continents. The precise count depends on which criteria are regarded as strong evidence of impact. Some impacts are as old as 2 billion years, and the largest craters are 100 km or more across, perhaps indicating the impact of a 10-km-diameter object. Both asteroids and comets are likely to have been involved.

The distribution, characteristics, and possible consequences of impact in the more recent geological record are all active topics of research. The innovative suggestion about a decade ago that the great biological extinction 66 million years ago, which included extermination of the dinosaurs, resulted from impact has proved very stimulating. Evidence of impact at that time in the form of widespread high iridium concentrations, shocked quartz, and wildfire is persuasive. The giant crater at Chixulub in Yucatan is a strong candidate for the main impact. The possibility that other large craters, such as that at Manson in Iowa, are associated with the same event may require a cometary rather than an asteroidal encounter.

Current astronomical estimates of the flux of impactors make it clear that the number of impacts recognized on most continents is improbably low. This is also true of the number of impacts recognized within the continental stratigraphic record. Thus, much remains to be done in locating and studying ancient impact craters.

STRUCTURE AND DYNAMICS OF THE SOLID EARTH

One of the most significant advances in understanding the solid Earth took place within the past 30 years with the general acceptance that the solid interior is in motion and that movement of the surface plates is an expression of that motion. Everyday experience suggests that rocks are solid, but geological investigations reveal that earth materials behave very differently on long time scales and on human time scales. On a geological time scale, the solid mantle behaves like a fluid and convects. It is convection, not conduction, that is the main means of heat transfer within the Earth. This is the process by which heat is most effectively transported from the deep interiors of planets.

With this realization, it was no longer possible to view earthquakes, volcanic activity, mountain belts, sedimentary basins, or the general division between oceans and continents as isolated surficial phenomena. Temperature variations within the Earth control the convection that ultimately produces the magnetic field, surface topography, and active geology. Interactions between the rigid surface plates cause earthquakes and the majority of volcanic activity and provide the stresses leading to mountain building and basin formation. The plates are driven by the slow convective processes of the mantle. There is little question that subducted oceanic crustal plates penetrate at least a third of the way through the mantle to depths of 670 km. Some lines of evidence suggest that these plates may travel all the way through the mantle to form a layer around the core. Heat flowing out of the core may disturb the thermal boundary layer separating it from the convecting mantle to produce narrow plumes of uprising solid material that produce surface volcanism in settings like Hawaii and Iceland.

The exact style of convection is a subject of active research. Advances in understanding this process are being made through three-dimensional seismic imagery, increasingly sophisticated computer models, and laboratory simulations. Several avenues of research promise major breakthroughs in understanding the thermal state and evolution of the interior. It is now clear that the surface characteristics of the Earth originate in, and are being continually modified by, a complex interplay between the mobile surface plates and the dynamic interior of the planet.

Seismic Determinations of Earth Structure

The structure of the interior has been examined in detail for over 50 years, through the application of seismology—the study of natural and artificially generated vibrations traveling through earth material. The emphasis has been on determining the variations in physical properties, especially velocity, refraction, and reflection behavior, as a function of depth. These variations reveal major changes in composition with depth. For example, the core, an iron-rich alloy, is more than four times denser than the crust, which is made mainly of aluminosilicates. Because of advances in seismic instrumentation and analysis, the interior can be viewed in three dimensions from the surface to the center. Thus, the processes underlying near-surface geological phenomena can be mapped and understood.

A primary feature of present-day models of the interior (Figure 2.2) is the asthenosphere, a region of low seismic shear wave velocity in the upper few hundred kilometers of the mantle, where materials approach their melting point and where mantle flow may be concentrated. At a 400-km depth the first of the deep mantle discontinuities in seismic wave velocity occurs, followed by an even larger discontinuity at the 670-km depth. Current debate centers on the nature of these discontinuities.

One view is that they represent phase changes in a mantle of constant composition. In this scenario the increasing pressure causes silicate minerals to convert to more dense crystal structures (Figure 2.3) with depth. Other evidence suggests that these seismic discontinuities mark more than phase transitions and may be regions where the chemical composition of the mantle changes. If only the phase and not the composition changes, the discontinuities would not necessarily be barriers to convection. If, however, the discontinuities reflect compositionally induced density differences of sufficient magnitude, they would inhibit flow across them. In this case convection would be confined to a series of layers within the Earth. These two distinct types of convection have drastically different implications for the compositional and thermal evolution of the interior.

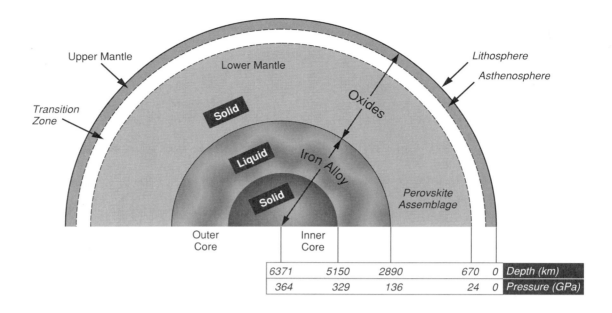

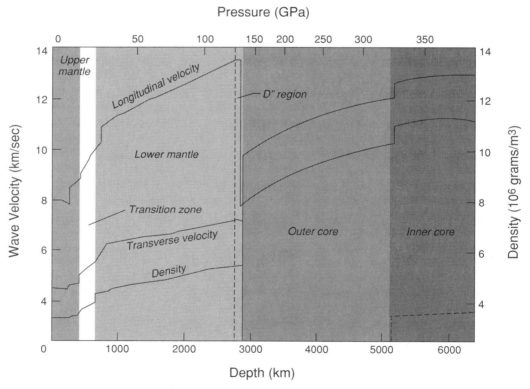

FIGURE 2.2   Cross sections of the Earth and its properties. The upper panel shows the seismologically determined regions and pressures as a function of depth (100 GPa = 1 Mbar = 1 million atmospheres). The bottom panel shows the average elastic parameters as a function of depth as determined by seismological analyses.

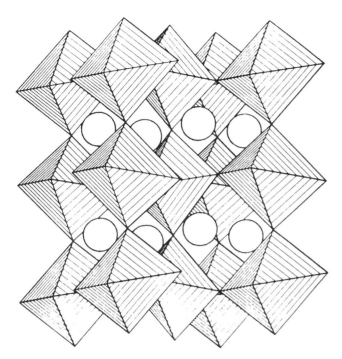

FIGURE 2.3 The bulk of the material making up the lower mantle is believed to have the perovskite structure.

At present, most physical property measurements have been carried out at room temperature and pressure. Extrapolation of these measurements to appropriate pressures and temperatures for seismic discontinuities carries sufficient uncertainties to allow either explanation for their origin. Only recently, with the development of several new approaches for high-pressure and high-temperature experimental apparatus, has it become possible to provide precise measurements of parameters such as density, compressional and shear wave velocities, and seismic wave attenuation. A new effort in mineral physics at high pressure and temperature, coordinated with interpretation of high-resolution seismic images of the mantle from advanced seismic instrumentation programs, such as those administered by the Incorporated Research Institutions for Seismology (IRIS), is enabling earth scientists to better understand the structure of the mantle.

Seismologists have used a wide variety of techniques to create three-dimensional images of the Earth's interior structure. One method is tomography, based on the same principle that is exploited in medical x-ray and ultrasonic imaging devices. It concentrates on small variations in the observed arrival times of seismic waves following earthquakes. A limitation of this technique is that it requires extremely dense arrays of seismic recording stations to provide high-resolution images.

This limitation has led to the development of sophisticated methods that utilize more of the complicated signals arriving at seismic stations during and after earthquakes. Large earthquakes excite free oscillations that are sensitive to the largest scales of heterogeneity and give direct constraints on lateral variations in density. Surface wave arrivals give good lateral resolution of upper-mantle structure but need to be augmented with other data to give good depth resolution. Body wave arrivals provide the best resolution in depth and can be used to map the topography of internal discontinuities.

Present-day images of the interior are of low resolution and uncertain accuracy. They reveal distinctive heterogeneities over horizontal distances extending thousands of kilometers. The heterogeneity is strongest near the top and bottom of the mantle, decreasing from 2 to 10 percent in the upper mantle to about 1 percent throughout the bulk of the lower mantle. The lowermost 100 to 300 km of the mantle, called the D'' (D double prime) layer, is also significantly variable, by 5 percent or more.

Thus, the strongest lateral variations in physical properties are associated with the major boundaries of the Earth: the surface and the interface between the mantle and the core. This result supports the assumption that the top and bottom of the mantle are two regions in which material moves horizontally, with little vertical motion. From a dynamic perspective, the seismically produced images suggest direct associations between these heterogeneities and the mantle's convective flow. The instability of these boundary layers ultimately produces crustal deformation through the forces of subduction and plumes. Thus, seismic tomography in principle can map out the underlying motions that drive plate tectonics at the surface.

One dramatic example of this type of mapping inside the Earth is the detection of cold slabs of lithosphere, the crust and uppermost mantle, sinking into the mantle beneath subduction zones. Because the slabs are cold, they appear as regions with anomalously fast seismic velocities relative to the velocities in the hot surrounding mantle. The presence of slabs in the mantle has been used to explain the existence of deep-focus earthquakes along the postulated extensions of near-surface subduction zones. Now actual images defining the dimensions of thermal anomalies—cold slabs—are being produced by tomography. Therefore, it is possible to detect subduction at depth, even in places where a slab may be seismically quiet. The present location of cold slabs at depth is an indication of past subduction because rock changes temperature very slowly. For the first time, observers can determine how far the slabs penetrate into the mantle, providing first-order information on the deeper aspects of the convection patterns associated with plate tectonics.

In addition, the shape of the slab is a direct reflection of the tectonic forces associated with convection. Observations of significant distortion of slabs in certain regions give evidence for variations in rock properties with depth and for background flows in the mantle convection pattern. This conclusion is supported by evidence of changes in the distribution and focal mechanisms of earthquakes with depth. All indications are that tectonic forces do vary along subduction zones, but the patterns are still indistinct.

Mantle Convection

Although many aspects of mantle convection are reasonably well understood, major scientific questions remain unresolved. These include the vertical structure and multiple scales of mantle convection and the efficiency of mantle mixing.

Plate tectonics is the surface manifestation of mantle convection. The rigid plates are the uppermost thermal boundary layers of mantle convection. These cool layers are rigid on geological time scales and behave as plates. But the plates become denser because of thermal contraction. Eventually, they become gravitationally unstable and founder into the mantle at subduction zones, defined by the ocean trenches. The weight of descending plates is a major force driving plate tectonics.

There is direct seismic evidence that the slabs of material subducted into the mantle at ocean trenches descend to depths of 670 km. An unresolved question is whether the downward limbs can penetrate this depth. The evidence is contradictory, and the views of experts are divided. If thermal convection penetrates this density barrier, whole-mantle convection occurs. If convection does not penetrate, then separate convection cells develop in the upper and lower mantle and mantle convection is layered. If mantle layers do not mix, significant variations in chemical composition and temperature could characterize the interior. There is also the possibility that both styles of convection can coexist. The amount of material transported across the entire mantle is currently the single largest uncertainty in understanding the Earth's thermal and chemical evolution.

Just as the plates are thermal boundary layers at the top of the convecting mantle, there should be boundary layers at the base of the convecting system. For layered mantle convection this boundary layer would be the result of heat transfer from the lower mantle; for whole mantle convection, it would be the result of heat transfer from the core. The gravitational instabilities in these lower boundary layers may generate ascending mantle plumes that are responsible for intraplate volcanism, such as that in Hawaii.

The fate of plates that founder into the mantle at ocean trenches has also been a subject of controversy. These plates are layered. The basaltic ocean crust extends to a mean thickness of about 6 km. Beneath that crust is a zone that has been depleted of basalt and is primarily composed of the refractory mineral olivine. This layer, which has a thickness of approximately 50 to 100 km, is gravitationally buoyant. Simple mass balance calculations show that ocean crust must be recycled through the mantle on a time scale of about 1.7 billion years or less. Therefore, present-day basaltic ocean crust has been processed through the plate tectonic cycle several times.

One hypothesis for the fate of the subducted ocean crust is that convection stirs it into the bulk of the mantle until it is nearly homogenous. Another hypothesis suggests that significant density differences between the basaltic ocean crust, which transforms to a dense phase called eclogite at depth, and the olivine-rich mantle result in gravitational segregation, with the depleted mantle rock overlying the crustal rock. The essential question that must be answered is whether convective mixing can homogenize the mantle before the buoyancy differences can cause layering.

The effect of an increase in temperature is to decrease the density of rock by a fractional amount—a few percent per 1000°C. Density variations caused by lateral temperature variations drive mantle convection. By the same token, the convective flows induce temperature differences. The flow field and temperature field are coupled.

Assuming that lateral variations of seismic velocities in the deep mantle can be ascribed to temperature variations, laboratory measurements of the changes in acoustic velocity with temperature can be used to infer the density variations in the mantle. In this way the buoyancy forces associated with mantle convection are obtained directly from seismic tomography.

Variations of the external gravity field can be calculated from the inferred density variations, taking into account not only the density distribution within the convecting mantle but dynamic topography as well. Dynamic topography is the deformation of the surface and internal boundaries, such as that between the mantle and core, that occurs because of convective flow in the interior.

By comparing the calculations with the observed variations in gravity determined from the analysis of satellite orbits, it has been found that the large-scale features of the Earth's external gravity are reproduced closely by the convection patterns deduced from seismic tomography. On a scale of 1,000 km, therefore, the seismic heterogeneity appears to be caused largely by temperature variations driving flow in the mantle. This analysis suggests that the large-scale pattern of convection and the associated buoyancy forces are being accurately imaged by seismic tomography.

Flow in the mantle is dominated by viscous forces, so inertially dominated turbulence is absent. The convective flow is amenable to both numerical and experimental modeling. Numerical models of mantle convection have improved with advances in the speed of computers and in computational techniques and with better understanding of the processes governing creeping flow in the Earth. Two-dimensional calculations now include temperature and pressure dependence as well as chemical and phase boundaries with realistic values for mantle parameters. Three-dimensional cartesian and spherical calculations with constant-property material have recently been completed. In the atmospheric sciences, general circulation models (GCMs) have become credible enough to be used for routine simulations by researchers in atmospheric chemistry and climatology—subjects far removed from solid-earth fluid dynamics. But within the next decade, mantle convection models should reach this level of acceptance, and applications should become widespread.

Core Dynamics and Geomagnetism

The outstanding geophysical problem involving the Earth's core is the generation of a magnetic field. The magnetic field is a product of dynamo action in the electrically conducting fluid outer core. A majority of planets in the solar system and an overwhelming majority of stars possess magnetism, and all of these magnetic fields are a consequence of dynamo action.

Despite this universality, our understanding of the Earth's dynamo remains rudimentary. Several key theoretical issues remain unexplained, including the physics of magnetic field equilibration and proper characterization of the energy sources in terms of fluid motions. Core convection may be driven by thermal buoyancy produced by heat loss to the mantle or by crystallization of the inner core. Core fluid dynamics is characterized by a wide spectrum of frequencies and spatial length scales. Filtering by the mantle allows only the lowest-frequency variations in the magnetic field to reach the surface. Thus, even the most informed theories are based on a heavily filtered, and therefore distorted, image of core processes.

At the present time it is not possible to develop consistent models of the geomagnetic dynamo. The relatively simple flows that are computed in numerical convection models generally do not work as dynamos. The successful theoretical dynamos strongly suggest that large-scale magnetic fields are produced by a broadband spectrum of fluid velocities that are chaotic and turbulent. In short, the difficulty is that simple flows do not produce simple dynamos—they produce no dynamo. Complex flows are required to produce dynamos; therefore, models of the dynamo process are necessarily complex.

The flow in the core required to produce the dynamo must be complex. An analogy with the winds in the atmosphere that determine weather patterns may help in comprehending core flow complexity. Scaling analyses show that 1 year of flow in the core corresponds to about 1 hour in the atmosphere; researchers are just now able to see the equivalent of 1 week's worth of weather in the core. To understand the magnetic field on a geological time scale requires the equivalent of understanding long-term climatic patterns in the flow of the outer core.

One outstanding question is why the magnetic field varies with time. A suitable solution must explain the reversals in polarity that have occurred frequently—at roughly 1-million-year intervals—throughout geological history. These variations have proved useful for applications in geology and geophysics, most notably in paleomagnetic documentation of tectonic motions of the crust. Mapping magnetic orientations of rocks has illustrated continental growth by the assembly of preexisting blocks.

Among the most exciting results of the past few years are the first reliable observations of changes in the magnetic field during a polarity reversal. Both the pattern and the strength of the field appear to change rapidly over 10,000 years, the duration of a single reversal. In fact, observations of the same reversal, as recorded in rocks from widely separated locations around the globe, are just becoming available. The data tentatively suggest a much more complex magnetic field configuration during a reversal than had been expected.

Studies now in progress concentrate on the nature of the field at the beginning and end of a transition, the idea being to document exactly how reversals are triggered. Such data may become important in light of a recent theoretical breakthrough in understanding the origin of reversals. This breakthrough postulates that polarity reversals result from the interaction between two separate time-varying components of the magnetic field. Detailed configurations of the entire field are now known for the past several hundred years, based on historical records. This configuration is extremely important because it reveals the flow pattern in the fluid outer core, where the geomagnetic field is created by magnetic and hydrodynamic processes.

Geomagnetism offers one of the only tools for exploring the nature of the deep interior far into the geological past. Effective documentation of long-range trends in field intensity and in reversal frequency could provide important constraints on the geological evolution of the core and core-mantle system. It should be emphasized that these advances depend on developments in the study of mineral and rock magnetism. Only with a thorough understanding of how magnetic remanence is acquired, and how it can subsequently be altered, can reliable determinations of the paleomagnetic field be made.

Core-Mantle Boundary

The boundary (see Figure 1.12) between the Earth's mantle and core is the most significant interface within the planet in terms of the contrast in materials and properties. The changes in density and seismic wave velocities across this boundary are larger by far than those across the boundary between the mantle and crust. The heterogeneity of the lowermost mantle rivals the geological heterogeneity observed in the crust. The time scales on which the heterogeneities in the core-mantle boundary layer are disrupted and at least partially mixed back into the overlying mantle are a matter of great interest.

The development of a new broadband digital network of seismometers and increased computational capabilities to interpret such seismic data markedly improves prospects for deciphering the current nature of the core-mantle boundary. High-resolution, three-dimensional images of this region, combined with the results of geodynamic modeling and ultra-high-pressure laboratory simulations,

should provide revelations about the dominant structures and processes of this dynamic region. Researchers anticipate that geomagnetic anomalies may be associated with seismological heterogeneities found at the base of the mantle. If so, this would open up the exciting possibility of documenting changes in the core-mantle boundary structure through the geological past. Documenting such changes, through paleomagnetism or other approaches, is important because thermal, mechanical, and electromagnetic coupling across this boundary cause the geological evolution of both the core and the mantle and, ultimately, of the crust.

EARTHQUAKES: CONSEQUENCES OF A DYNAMIC MANTLE

Earthquakes are recurrent demonstrations that the Earth is indeed an active planet. Earthquakes occur over a large range of areas and vary by more than 20 orders of magnitude in energy release. Close study of their geographic distributions, their depths and associated geological settings, and their various magnitudes has provided some of the most basic clues to plate tectonic theory and subsequent insight about the dynamics of the solid earth. Earthquakes are among the most destructive of natural hazards, and as such their nature and predictability are considered at length in Chapter 5 on hazards.

Geographic Distribution

As the surface jigsaw puzzle of major plates shifts in association with convective motions of the interior, the relative plate motions are accommodated by episodic slips along major faults, discontinuities in the Earth's crust. A familiar example is the San Andreas Fault in California (Figure 2.4), which separates the Pacific and North American plates. The two plates are moving horizontally relative to one another at a rate of about 5 cm/year. The motions in the deep Earth proceed as slow and continuous flow because the rocks there are hot. Near the surface, however, the rocks are cold and brittle, and here the faults respond to the plate motions in one of two possible ways. The less violent is a continuous creep, without earthquakes. The other response is the building of elastic strain that is released abruptly by frictional sliding of the rock along the fault—an earthquake.

The San Andreas Fault is called a strike-slip fault because here the plates are sliding past each other horizontally. Along other major plate boundaries the crust is either spreading apart, as in ocean ridges

FIGURE 2.4   The San Andreas Fault on east edge of the Carrizo Plain. Rugged, dissected terrain west (right) of the fault trace is the Elkhorn scarp. Photograph courtesy of R.E. Wallace, U.S. Geological Survey.

or continental rifts, or is converging, as in subduction zones and continental collisions. At these boundaries the faults range in angle from horizontal to vertical. Understanding the diverse types of fault motion in each of these three environments—strike-slip, spreading, or converging—was critical to reconciling the global distribution of earthquakes to large-scale plate tectonic processes.

North of the San Andreas Fault, in the Pacific Northwest, the major plate boundary is conver-

gent, and the ocean crust plunges under the continent. Associated with this subduction zone is a linear chain of active volcanoes extending from British Columbia to Mount Lassen in northern California. The largest earthquakes that have occurred around the world in this century have been located in subduction zones, including the 1964 Alaskan event that devastated Anchorage. Events of comparable size may well occur in the Pacific Northwest.

Seismologists routinely determine the kinds of faulting associated with all large earthquakes almost as they happen. This composite information gives a picture of the ongoing tectonic motions, which is critical to understanding the dynamic Earth system. Earthquakes occur as deep as 670 km in regions where cold ocean plates sink into the mantle. Study of the faulting involved in deep earthquakes sheds light on the process of subduction and on the fundamental nature of mantle convection.

Understanding Earthquakes

While seismologists have made progress in characterizing the global distribution of earthquake activity and the types of faulting involved, a basic understanding of the physics of earthquake rupture is still lacking. Without this it is difficult to assess seismic hazard, and for this reason little progress has been made on the short-term prediction of earthquake occurrence. A concerted interdisciplinary effort is under way in an attempt to achieve a better fundamental understanding of the fault mechanisms that produce earthquakes. Established ideas are being reexamined, and new questions such as "Why do earthquake fault surfaces appear to have so little friction?" are being posed.

Although characterizing the physical behavior of an earthquake fault remains a challenge, local studies of earthquake distributions in space and time have permitted improved estimates of the probability of earthquake occurrence in specific regions. As a result, administrators have been able to develop response plans to reduce the disastrous effects of earthquakes. Operational prediction of earthquakes is at present a distant goal, but it could eventually become one of the primary means of earthquake hazard mitigation.

Seismological approaches to understanding earthquake rupture processes are expanding dramatically. Creating high-resolution images of active faults is one approach. This is accomplished by determination of fault orientations for earthquakes of all sizes. It is now recognized that faults are remarkably complex, with many undulations and intersecting branches. One of the most important advances in the past decade has been the expansion of the record of large earthquakes beyond the limited historical record to tens of thousands of years of prehistoric time, through the new subdiscipline of "paleoseismology." By geological means, including the evaluation of stratigraphic records in excavations made across suspected active faults, and by quantitative geomorphic analysis of the ages of fault

scarps, many previously unknown large earthquakes have been identified around the world, and knowledge of the global patterns and timing of large earthquakes has been greatly enhanced. Another of the dramatic advances in the past decade involves the ability to map a rupture in space and time as it expands during an earthquake. This information provides insight into the physical processes, such as stress accumulation, that govern initiation and termination of earthquake rupture. The degree of ground shaking produced by a given earthquake is influenced by the detailed nature of the rupture process. The field of strong ground motion seismology involves quantification of the shaking induced by an earthquake as rupture spreads over the fault. Predicting strong ground motion also requires an understanding of the interaction of seismic radiation with complex crustal structure, and a major effort is under way to develop the necessary three-dimensional wave propagation capabilities.

Laboratory studies of rock physics and rock mechanics help in understanding the nature of earthquakes. Rocks are three-dimensional aggregates of mineral grains containing a complex assemblage of defects such as dislocations, grain boundaries, and fractures, which often contain impurities or fluids. Researchers in rock physics strive to understand how the properties, proportions, and arrangements of the component phases interact to determine the overall properties of the rock bodies.

Theoretical modeling and laboratory measurements of friction have been used in studies of earthquake instabilities, orientation, and distribution of faults and rupture mechanisms. The effect of fluid pressure in promoting brittle fracture and frictional sliding of faults is being addressed by a variety of experiments. An essential question in understanding the behavior of strike-slip faults like the San Andreas involves the absolute stress levels on the fault. The absence of a high heat flow anomaly, along with low average seismic stress drops, suggests that the mean stress level is low. But laboratory friction studies indicate that high stresses are needed to cause the fault to slide. High fluid pressures on the fault may reconcile these observations, or perhaps a better understanding of stress levels during rupture will be needed.

Earth Deformation

Many technologies are developing for measuring regional strain on both local and global scales. Two color laser-ranging techniques, small trilateration nets, level lines, and stretched-wire creepmeters are

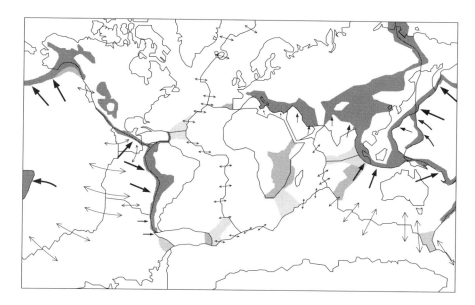

FIGURE 2.5 Map showing idealized plate boundaries, velocities between plates, and regions of deforming lithosphere. Plate velocities are shown by arrows; the length of the arrows shows what the displacement would be if the plates were to maintain their present relative angular velocity for 25 million years. Plate separation rates are shown by symmetrical diverging arrows; plate convergence rates are shown by asymmetrical bold arrows. Stippled areas are those areas of deformation.

used to measure localized deformation along faults. For example, the pattern of tectonic strain throughout California and the western United States has been delineated by laser-ranging surveys during the past 20 years. Tiltmeters, linear strain meters, and volumetric strain meters are among instruments currently used for studying tectonic strains on both the continents and seafloor. Recently developed techniques using very-long-baseline radio interferometry (VLBI), satellite laser ranging (SLR), and the global positioning system (GPS) provide the capability to measure larger-scale motions. In fact, these measurements have directly confirmed the theory of plate tectonics, by resolving relative plate motions, and these measurements offer the potential of detecting departures from the motions predicted by rigid plate models (Figure 2.5).

As geodetic measurement precision and global coverage increase, it will become possible to directly monitor surface deformation, which is an important component of the dynamic Earth system. For example, in many subduction zones the cumulative slip that can be accounted for seismically is a small fraction of the convergence indicated by plate tectonic rates. Quantifying the nonseismic component, and determining whether it proceeds episodically or continually, will contribute to our understanding of earthquake occurrence and the nature of the large-scale dynamic system. Already, new seismological techniques are being developed to search for silent earthquakes, which involve nonsteady creep events that do not excite seismic waves. With this development, the spectrum of Earth motion measurements from seismology and geodesy will become continuous.

VOLCANIC ACTIVITY: CONSEQUENCE OF CONVECTING MANTLE

Volcanism is another expression of the interaction between the dynamic interior and the surface of the planet. About 1,000 potentially active volcanoes dot the Earth's surface. For the past several years, an average of 50 eruptions per year have occurred at some 85 volcanoes. This activity not only reflects the continuing activity of the inner Earth, it also furnishes an intimate physical and chemical probe into Earth's magmatic life cycle. Molten rock—magma—migrates to the surface because it is hot, fluid-like, and of low density. The magmas record the temperature at depth, the chemical composition of deep-seated source rocks, the physical and chemical nature of the magmatic plumbing system, and the dynamic conditions that result in volcanoes.

Recently, manned and remotely controlled submersibles and new high-resolution sonar images of the seafloor have enhanced our understanding of the most active volcanic system on Earth, the ocean ridges. Also, satellite and aircraft remote sensors can monitor eruptions and track volcanic gases and aerosols around the globe. Synoptic satellite-borne instruments provide the means of obtaining regional views of the distribution of volcanoes and volcanic activity and for making quantitative measurements of thermal and volatile emissions from volcanoes.

Unraveling the mystery of the causes and consequences of volcanism depends on understanding the details of the physics and chemistry of the origin and evolution of magma. Research on volcanism is discovering more about the physics of magma production, accumulation, and eruption. This research asks seemingly simple questions: How and why does rock melt? How does melt escape from partially molten rock? How does magma rise? How long does it take? Is there a chamber of magma a short distance below every volcano? Do the magmas in these chambers flow turbulently or are they stagnant? Will the eruption be a catastrophic explosion or a quiescent extrusion? What is the cause of huge ash flows from some volcanoes?

Volcanism at the surface is caused by flow in the interior as it tries to rid itself of the heat of Earth formation and the heat generated by radioactive decay. The patterns of flow in the mantle, however, may be controlled or determined by events outside the mantle, such as enhanced heat flow from the core or changes in the motion of the rigid surface plates. The realization that the core, mantle, and crust interact to determine the surface characteristics is an outgrowth of the plate tectonics paradigm.

Geographic Distribution, Style, and Scale of Eruptions

Volcanoes are not randomly distributed about the Earth's surface. Like earthquakes, volcanic activity is concentrated primarily along the edges of tectonic plates. Along the ocean ridges, the most active volcanic setting on Earth, mantle material rises slowly to fill the gap created by plate separation. As the mantle material ascends to the surface, it melts. When it reaches the surface it creates the ocean crust—a thin veneer of layered lava flows about 6 km thick, which separates the oceans from the mantle. On a worldwide basis, volcanism at ocean ridges produces approximately 20 km³/year of lava.

Another concentration of volcanism is found above places where the plates are being returned to the interior at subduction zones. The arcuate shape of these tectonic features is well displayed by the ring of fire around the Pacific Ocean margin. This ring traces patterns of volcanic activity along all of western South America, the Cascades, Alaska, the Aleutians, Kuriles-Kamchatka, and Japan, as well as extending into New Zealand, Fiji-Tonga, and the New Hebrides in the southwest Pacific.

The volume of lavas erupted at arcs is less than at ocean ridges, with estimates of total arc magmatism, including intrusive magmas, of about 2 km³/year. The composition of the rock produced in arc volcanism is distinct from that of the ocean ridges. It is somewhat closer to the estimated average composition of the continental crust. This chemical similarity, coupled with the survivability of thick buoyant arc crust, has led to the suggestion that the continents are constructed from a patchwork quilt of assembled arc fragments.

A large fraction of volcanism is concentrated near plate boundaries. However, there are examples of volcanic activity well removed from the boundaries of the surface plates. Hawaii, in the middle of the Pacific plate, is one example of intraplate volcanism. The origin of this type of volcanism was poorly understood until the late 1960s, when it was proposed that major isolated centers of intraplate volcanism, such as Hawaii, originate over narrow ascending plumes of deep mantle material. This plume, or hot-spot, model successfully explains many features of intraplate volcanism, from their age progression to their geochemistry.

Most continental volcanic activity can be related to either plume, rift, or convergent margin processes. Melts generated in the mantle pass through the continental crust, sometimes exchanging material with continental wall rocks to cause changes in the ultimate composition of the erupted product. Continental rift valleys produce unusual lavas with high alkali content, although basalts and rhyolites also occur in the same tectonic environments. These lavas are formed by partial melting of the mantle beneath the continents, with the involvement of high concentrations of volatile components, water and carbon dioxide. A melt of related composition is kimberlite, which is noteworthy because its explosive eruption brings diamonds into the crust from depths greater than 150 km, as well as fragments of the mantle through which it came. Study of these mantle fragments confirms that the upper mantle has been modified chemically by the passage of fluids in the form of melts or dense vapors.

Flow and Storage of Magma

An interesting problem in volcanology addresses how magma, once separated from its source, travels through solid rock to erupt at the surface. A number of external parameters will influence the basic process. These parameters include the composition of magma relative to wall rock, the stress state of the area of magma intrusion, the mechanical behavior of wall rock—especially its permeability and its ease of fracture—and the rate of magma supply relative to eruption.

Because rocks are a mixture of mineral components, melting occurs over a broad temperature range, in some cases over several hundred degrees centigrade. Consequently, the melting process must not only provide the heat necessary to induce melting but also that necessary to continue raising the temperature of the partially molten assembly to melt the more stable components. As a result, total melting seldom, if ever, occurs within the Earth. Instead, magma is produced as interconnected tubules, or as thin skins of liquid along grain boundaries. Even at high degrees of melting, the mixture of solid and liquid rock probably never exceeds a crystal-dominated mush.

The relatively low density of the liquid rock results in separation, with the liquid rising toward the surface. The magma may ascend along grain boundaries like liquid flowing through a porous solid. At the confining pressures within the Earth, a liquid cannot move unless replaced by some other material; gaps do not remain within rocks of the interior. Liquid must be kneaded, or squeezed, out of the solid component of magma-producing rock. This occurs through recrystallization that fills gaps left behind as the liquid travels upward along grain boundaries. This process allows the liquid to coalesce into accumulations of increasing size. At some stage the pressure from accumulating magma forms veins and cracks in the overlying rock. The flow of magma through rock shares many similarities with the movement of other fluids, such as groundwater and petroleum. A better understanding of the physical processes of fluid flow through the Earth could provide answers to a wide variety of problems in the earth sciences.

Extensive geophysical observations of Hawaiian eruptions carried out by the Volcano Observatory of the U.S. Geological Survey have produced tremendous improvements in our understanding of magma transport and storage (Figure 2.6). This work is a major advance in volcanology because it has established a predictive framework for eruptive activity in a volcanic system such as Hawaii's. The nearly continuous volcanism occurring in Hawaii offers a rare opportunity to study geological processes operating on human time scales.

Hawaii represents only one of the many types of volcanic systems active on Earth. Many evolutionary steps in Hawaiian volcanism are similar to those that occur in other settings, but the distinct end products erupted at the surface testify to different evolutionary paths for magmas produced in different tectonic settings. Continental volcanism almost invariably involves longer storage times in larger intermediate-depth magma chambers, resulting in the production of more chemically mature volcanics.

Volcanic Eruptions

Once a magma has been produced within the Earth's interior, the likelihood that it will erupt on the surface depends strongly on its chemical composition. Silica-rich magmas are more viscous than those low in silica. In other words, the lower the silica content, the more easily the magma flows. Consequently, basalt—the common silica-poor lava—erupts in great abundance on the surface to form the ocean floor, islands like Hawaii and Iceland, and large flood basalt provinces on the continents, such as those of the Columbia River Plateau and the Deccan Traps of India. Basaltic magmas also tend to have relatively low concentrations of volatiles, which, along with their low viscosities, generally allow basalts to be erupted quiescently. As a result, basaltic eruptions rarely cause loss of life because the flow paths are easily predicted and the flow velocities generally are slow enough to allow effective evacuations. Destruction of immovable objects, however, can be significant because relatively low viscosities allow even small-volume flows to cover substantial areas. The largest basaltic provinces can be devastating in this respect. For example, the Columbia River basalts cover an area of about 200,000 km^2, an area roughly equal to Virginia, Maryland, West Virginia, and Delaware combined.

As the silica content of the magma increases, eruption becomes less likely. This is because the high viscosity of silica-rich magma makes it susceptible to heat loss and crystallization during its slow ascent through the crust. Consequently, high-silica subsurface intrusions, which cool to become granite, are abundant, while low-silica intrusions are rarer. However, granitic magmas that do reach the surface erupt violently and may cause widespread destruction.

Eruptions of silica-rich magmas devastate because they are driven by the explosive exsolution of volatiles, particularly water. Even small eruptions, such as that at Mount St. Helens, can be accompanied by tremendous explosions that destroy life and landscape over a wide area. The 1883 eruption of Rakata, on the Indonesian island of Krakatoa, was a moderate-sized event that exploded with the energy equivalent of 20,000 Hiroshima-sized atomic bombs. The explosion ejected approximately 80 km^3 of rock into the atmosphere. Some of the finer-grained dust, ash, and vapor circled the globe and stayed in the atmosphere for several years. The explosion left a crater 6 km in diameter and induced tsunamis that killed more than 36,000 people.

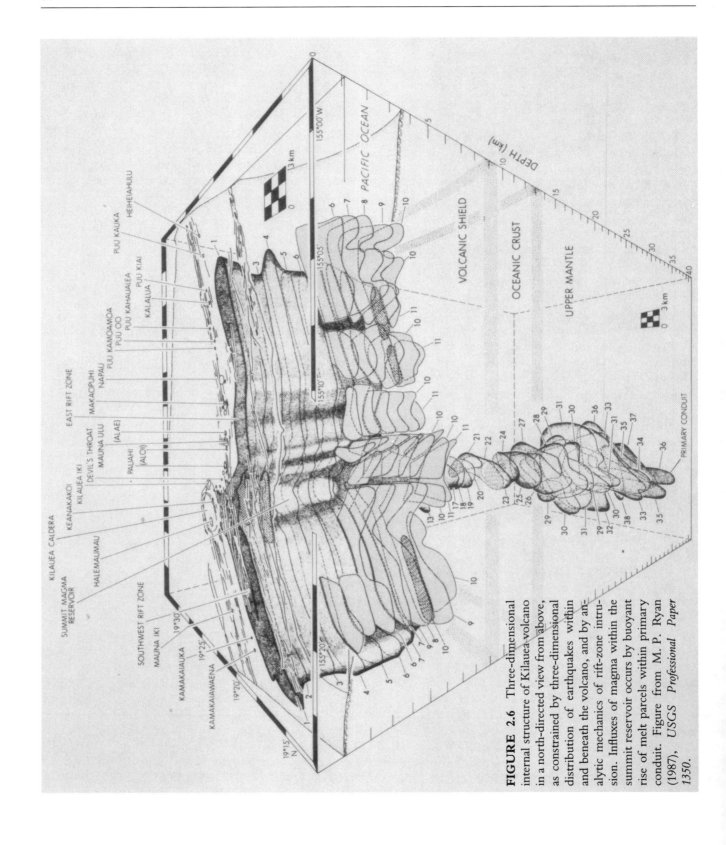

FIGURE 2.6 Three-dimensional internal structure of Kilauea volcano in a north-directed view from above, as constrained by three-dimensional distribution of earthquakes within and beneath the volcano, and by analytic mechanics of rift-zone intrusion. Influxes of magma within the summit reservoir occurs by buoyant rise of melt parcels within primary conduit. Figure from M. P. Ryan (1987), *USGS Professional Paper 1350.*

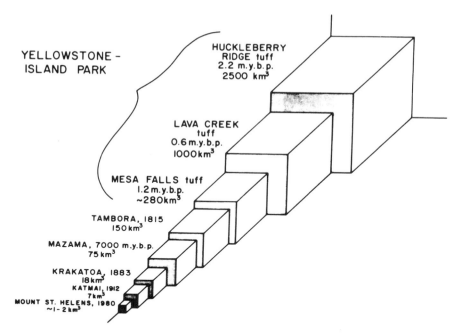

YELLOWSTONE –
ISLAND PARK

HUCKLEBERRY
RIDGE tuff
2.2 m.y.b.p.
2500 km³

LAVA CREEK
tuff
0.6 m.y.b.p.
1000 km³

MESA FALLS tuff
1.2 m.y.b.p.
~280 km³

TAMBORA, 1815
150 km³

MAZAMA, 7000 m.y.b.p.
75 km³

KRAKATOA, 1883
18 km³
KATMAI, 1912
7 km³
MOUNT ST. HELENS, 1980
~1–2 km³

FIGURE 2.7 Relative volumes of some well-known volcanic eruptions. Figure from R. B. Smith and L. W. Braile (1984), in *Explosive Volcanism,* National Academy Press.

The largest explosive eruptions have left craters 10 to 50 km in diameter. Fortunately, such eruptions predate human experience, but their occurrence in the future is a certainty. For example, large caldera-forming eruptions occurred in the Yellowstone volcanic area of Wyoming about 2.2, 1.2, and 0.6 million years ago (Figure 2.7), leading to the not unreasonable expectation of another major eruption within the next few hundred thousand years. The latest Yellowstone eruption created sizable deposits of ash as far away as Kansas. The first caldera explosion at Yellowstone, 2.2 million years ago, has an identifiable volume of ash of about 2,500 km³. It is estimated that the total volume erupted was about twice this value and represented only 10 percent of the magma chamber. Therefore, a magma chamber of between 20,000 and 40,000 km³ was involved.

Direct observations of volcanic eruptions generate estimates and measurements of ascending columns of steam and ash, duration of eruptions, and rates of dome growth, of magma production within the dome, and of lava flow. Space-based observations are capable of locating eruptions that otherwise would not be detected. Satellite-borne sensing devices also provide a clear image of interaction between volcanic emanations and the atmosphere by tracing the global dispersal of gas and dust following an eruption. A new appreciation of volcanic activity's effect on climate has been realized. Clear evidence from historic and contemporary eruptions is now available to show that significant global decreases in temperature accompany the injection of large volumes of volcanic aerosols into the upper atmosphere.

A better understanding of the interactions of volcanic emissions and the atmosphere and hydrosphere is critical. The Krakatoa eruption is only one of the historic events that have dramatically modified the Earth's climate for several years by introducing large volumes of dust and gas into the atmosphere. Recent speculation suggests that large submarine eruptions along ocean ridges may alter ocean temperatures, establishing the El Niño condition with its subsequent implications for climatic variations. Further understanding will come from global monitoring of volcanism, most likely through satellite-based remote sensing and a better theoretical grasp of the relationships dictating climatic response to heat and mass transfer from the Earth's interior to the hydrosphere and atmosphere.

The large-scale effects of volcanism on the atmosphere, climate, and ecosphere have been recognized as a component of the earth system and of cooperative activities such as the International Geosphere-Biosphere Program (IGBP) and the International Decade for Natural Disaster Reduction (IDNDR). Precursor activity of Lascar Volcano, Chile, was first recognized from thermal measurements taken by Landsat, while several stratospheric volcanic plumes containing sulfur dioxide have been discovered and measured by means of the total ozone mapping spectrometer (TOMS) instrument onboard the Nimbus 7 spacecraft. In addition to the use of satellites to assess volcanic hazards, such programs will include several interdisciplinary investigations that incorporate the effects of volcanism within climate modeling, the analysis of different ecosystems, and the investigation of sea-surface/atmosphere interactions.

The energy of a major volcanic eruption is well beyond what can reasonably be expected to be controlled by engineering. Consequently, for the foreseeable future, humanity can best deal with volcanic phenomena by supporting programs aimed at predicting the occurrence and understanding the likely consequences of an eruption.

OCEAN BASIN PROCESSES

During the middle and late 1960s, widespread recognition of the lithosphere's plate structure crowned 20 years of postwar research in the oceans. Since then, research efforts have driven the investigation of the dynamic Earth in two directions: downward, to discover how plate activity relates to the deep mantle, and laterally outward onto the continents, to establish how the continents are involved in today's plate activity and how earth dynamics operated in the past. There have been surprises, such as the discovery of the black smokers, and there have been breakthroughs, such as the establishment of the age of the ocean floor. There have also been new puzzles, such as the significance of the huge oceanic plateaus representing the eruption of vast volumes of basalt over short intervals. The intellectual momentum provided by plate tectonic theory continues to affect the research directions of many earth scientists, including those who study the interior-driven oceanic processes that appear particularly important to the overall system.

Ocean Spreading Centers

Ocean basins open along axes where the crust is torn by the force of plates moving away from each other. Basins spread along centers where partial melting of the shallow mantle generates basalt that ascends toward the surface and is solidified as ocean crust (Figure 2.8). New plates are continuously created. On average, spreading centers lie about 2.8 km below the ocean surface, where they can be detected as topographically high, elongated areas, often with valleys at their crest. A wide range of methods has been applied to the study of spreading centers. They include surface and ocean-bottom geophysical techniques, manned and remotely operated submersible observation, and direct sampling by dredging and drilling. Samples of rocks and

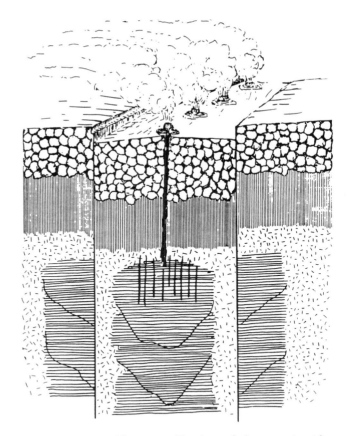

FIGURE 2.8 After crystallization of the oceanic ridge magma chamber, tectonics or late magmatic activity breaks open the plated barrier layer at the top, and water actively penetrates into the hot cumulates. This is the stage of high-flux geothermal activity, where high temperatures can be attained even in very permeable rocks. Figure from *The Mid-Oceanic Ridge: A Dynamic Global System* (1988), National Academy Press.

fluids have been analyzed physically, chemically, and isotopically. The highly specialized biota that characterize ocean spreading centers have also been studied in detail.

Detailed sampling along ocean spreading centers shows that the intensity of melting is inversely proportional to depth below sea level. The depth of a spreading center, and any accompanying ridge, can be taken as an indication of the average mantle temperature beneath it. High-temperature material will produce a higher degree of partial melting and thus a thicker oceanic crust. The relatively narrow range found in the composition of spreading-center volcanic products worldwide strongly suggests that melting beneath an ocean spreading center occurs through a straightforward process of decompression melting. This conclusion is based on a relatively small sample—ocean centers and their ridges are not easily accessible for direct study.

Recent studies of the ocean volcanic system show that an assumption of uniform magma production along the axis of a spreading center is too simplistic. Close examination of ocean spreading centers and ridges shows them to be segmented at finer scales. Within each segment is a topographic high that is thought to reflect the center of an isolated magma chamber. These chambers feed magma to the remainder of the segment but not beyond the segment boundaries. The spacing of these magma chambers is quite regular—each is on the order of 50 to 70 km long.

The regular spacing of ocean ridge segments is one case in which theoretical understanding of a phenomenon preceded its observation. Fluid mechanical models for melt separation from a large partially molten zone indicate that the melts are drawn from the host rock to coalesce into local concentrations of magma. These local concentrations begin to rise toward the surface because, compared to the surrounding rock, they are low in density and are buoyant. The rising magma bodies are fed by melt extracted from a much wider area than that represented by the bodies themselves. The regular spacing of the globules, and of the volcanoes they eventually create, can be thought of as the optimal size of the melt feeding zone for the globule. Globules formed too close together would not have enough magma available for them to grow large enough to rise efficiently through the overlying rock. Globules formed too far apart are not capable of extracting magma from the midpoint between the globules. Widely separated globules leave behind a source of magma that eventually will create another globule between the original two.

The observed segmentation of ocean spreading centers conforms to this model and suggests that there is a continuous melt zone at depth feeding the isolated volcanoes of each segment. Researchers are attempting to identify and understand the effects of segmentation. They are especially interested in the cooling and chemical fractionation history of magmas erupted along the spreading centers.

The discovery of black and white smokers—vents of very hot, mineral-laden seawater—in the late 1970s clearly showed that the high heat of volcanically active spreading centers drives hydrothermal circulation of ocean water. This circulation allows extensive chemical exchange between water and newly formed crust, leading to concentration of rare metals, significant modification of seawater composition, and maintenance of the giant clam and

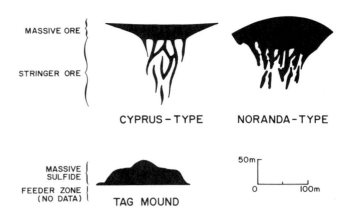

FIGURE 2.9 Cross sections showing the shapes of undeformed Cyprus-type and Noranda-type massive sulfide deposits and the massive sulfide mound at the TAG hydrothermal field in the rift valley of the mid-Atlantic ridge. The age of the Cyprus deposits are about 70 million years and of the Noranda deposits about 2.6 billion years. Figure courtesy of Peter Rona, NOAA.

tube-worm populations around the vents of hot water (see Plate 6).

Remnants of submarine black smokers were initially observed in submarine massive-sulfide deposits found on the continents. These ancient representatives had been tilted on their sides and planed off by nature, which permitted detailed examination of these deposits in cross section (Figure 2.9). The active black smokers have two features rarely observed in the ancient ones. One is the peculiar biota able to grow in the dark by feeding on sulfur-metabolizing bacteria, and the other is the chimney form, which has since been identified in some ancient examples.

Researchers now model aspects of spreading centers that incorporate the variables of rift propagation, magma chamber size, spreading rate, and rock flow. They can also run models experimenting with variables of (1) chemical composition, (2) temperature of partial melting, (3) depth of equilibration, and (4) isotopically distinct sources. All of these computer-aided techniques should contribute to successful explanations of variation along the 40,000-km length of the ocean spreading centers.

Intraplate Volcanism: Hot Spots and Oceanic Plateaus

Enormous amounts of volcanic material erupt at hot spots such as Hawaii. Unlike that at spreading centers, however, the rise of mantle beneath hot spots is not a passive response to movement of the overlying plate. Rather, hot-spot volcanism appears

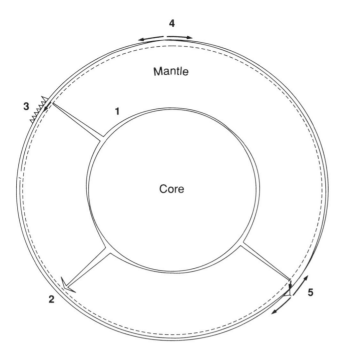

FIGURE 2.10 Various types of mantle plumes are shown. A thermal boundary layer (1) is produced by heating from the core (plumes arise where the boundary layer is thick); the conduit of the plume tapers upward because viscosity decreases upward; new plumes (2) start with a large head; hot-spot chains (3) are generated as a plate moves over a plume and hot material is entrained from the asthenosphere by a plate; normal spreading centers (4) tap the adiabatic interior; and plume material is entrained into hot-spot ridges (5).

to be actively triggered by plumes of mantle material that penetrate the general convective circulation patterns of the Earth's interior. The amount and composition of mantle melt relate closely to the temperature and pressure of melting. This correlation has focused new interest on Hawaii and on other hot spots, including one at an ocean spreading center—Iceland.

Researchers have been able to model the fluid dynamics of plumes that generate hot spots. Integrated models that incorporate trace element and isotopic data, such as the high proportion of ^{3}He compared to ^{4}He at hot spots, address previously intractable questions. Numerical modeling and experimental simulations of mantle convection suggest that plumes arise from unstable boundary zones between convective layers. The most likely source for the largest plumes, such as those under Hawaii and Iceland, is the boundary separating the core and mantle (Figure 2.10). Another possibility places the source at a hypothetical boundary sepa-

rating the upper from the lower mantle, which would result in layers with distinct convective systems. Hot spots are so diverse that some could come from each boundary. Surface occurrences of hot-spot volcanism correlate with several large-scale geophysical features of the deep Earth. These features include gravity field anomalies and zones of low seismic velocity, suggesting that hot spots may be a direct link between surface activity and events occurring in the deep interior, perhaps even within the core.

The lazy L-shaped track of the Hawaiian hot spot (Figure 2.11) shows that the Pacific plate changed direction about 45 million years ago as it moved over the plume. The track's concentricity with other hot-spot tracks on the Pacific plate confirms plate rigidity and indicates as well that hot spots do not move very fast with respect to each other. Hot spots and their tracks are very useful in interpreting plate rotations and in attempts to search for a component of true polar wander in the relative motions between continents and geographic poles.

Plateaus are extensive areas of above-average

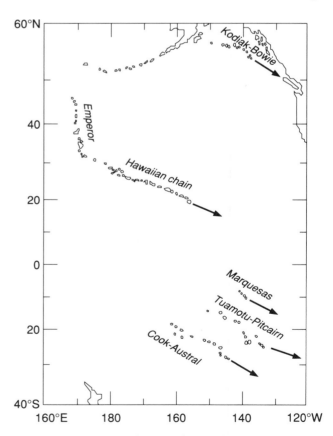

FIGURE 2.11 Some linear volcanic chains of the Pacific with arrows showing directions of propagation about a rotational pole.

elevation, and oceanic plateaus are areas largely or wholly flooded by the waters of the ocean. Some oceanic plateaus are known to be fragments of continental crust such as the Seychelles plateau in the Indian Ocean, the Rockall plateau in the Atlantic, and the Campbell plateau southeast of New Zealand. But most appear to be giant analogs of Iceland or Hawaii, up to 1 million km^2 in area and tens of kilometers thick, representing an enormous amount of partial melt from the mantle. The largest group of these oceanic plateaus, including the Shatsky, the Manihiki, and the Ontong Java plateaus, lies in the western Pacific Ocean. Most appear to have been erupted about 110 million years ago, in some cases over intervals as brief as 20 million years or less. Magnetic anomaly patterns show that some of these oceanic plateaus were formed at ocean spreading centers, indicating that complementary features—plateau twins—may have formed on symmetrically growing plates that have since been subducted. There are no representatives of the western Pacific type of oceanic plateau within the continents, so it can be inferred that they, like the plates they formed on, subduct into the mantle. Distinctive slivers in rocks marking ancient convergent boundaries within the continents suggest that this has indeed been the case. Oceanic plateaus may record sporadic and episodic events in mantle convection, so closer scrutiny could help to answer some questions about time dependence of mantle processes.

Plate Kinematics

The creation of new ocean crust at spreading centers is an essential feature of the plate tectonic theory. The general details of the formation, evolution, and destruction of ocean basins are now relatively well understood. Ocean basins form with the splitting of an arc or of a continent in the process of rifting. The basin increases in size by additions of volcanic crust on each side of the spreading center at rates ranging from 1 to 10 cm/year along the entire spreading center's length, increasing the width of the ocean basin by up to 200 km every million years. At the same time the ocean deepens away from the ridge crest, the ocean crust and lithosphere become more dense and less buoyant.

Development of plate tectonic theory led to determination of the rates and directions of motion on the largest plates (see Figure 2.5). The motions of plates are calculated with respect to each other and with respect to various slowly moving reference frames. Using earthquake mechanisms, plate

boundary geometry, and the youngest ocean-floor magnetic anomalies, researchers can estimate plate velocities in the general range of centimeters per year. These velocities, which represent averages for the past 3 million years, have been steadily refined. Now, with a small number of local exceptions, plate velocities are well established. Space geodesy shows that over intervals of years plates move at about the same speeds as they are estimated to average over millions of years. Over longer intervals, as much as 100 million years, estimates of ancient plate motions have been steadily refined with mapping and remapping of magnetic anomalies and fracture zones. The challenge of the next decade is to resolve the motion in small plates and in complex plate boundary zones and to track irregularities in plate motion that appear over short intervals and locally within plates.

The plate tectonic cycle and seafloor spreading can explain the general evolution of ocean basins, so attention has focused recently on the specifics of the spreading process. These specific problems include propagation of spreading centers, reorganization of plate structure, formation of new plates, and amalgamation of older plates. An important result has been to show how changes recorded in rocks and continental structures relate to changes in plate structure recorded in the ocean. The crowning success in this field was the recognition that the strike-slip motion in California, which has produced the complexity of the San Andreas Fault system, began when the East Pacific Rise spreading center reached the convergent margin of North America. Refinements in understanding continental tectonics will continually refer to the structural history of the oceans. The most obvious needs are in less-well-known areas such as the Arctic, the South Atlantic, and the Indian oceans.

Ocean Convergent Plate Boundaries: Island Arcs

The plate tectonic cycle is completed when the ocean floor subducts at an ocean trench. The ocean floor material returns to the mantle along subduction zones, which are often associated with island groups, such as the Marianas and the Lesser Antilles and in the Central Aleutian, Scotia, Vanuatu, and Tonga arcs. Early plate tectonic researchers recognized that volcanoes in oceanic island arcs erupted some rocks that are similar in chemical composition to the average outcrop of continental crust. So, theorists suggested, a plausible origin for the continents might have been assembly by the sweeping together of island arcs. Island arc volcanism is accompanied by igneous intrusions. If these arcs swept together, the underlying intrusions could eventually form the huge batholiths that form the roots of cordilleran mountain ranges characterizing ocean-continent convergent plate boundaries. This is the process going on today in Southeast Asia, especially the Philippines and Indonesia.

Detailed study of oceanic island arcs, and especially the Central Aleutians, has cast doubt on this simple picture. The average composition of the arcs differs significantly from that of the continents. For example, silica contents average 50 percent rather than the 60 percent typical of continental crust. A likely explanation appears to be that the upper parts of the oceanic arcs, which are richer in silica, are concentrated in the continents during the process of continental assembly. If this suggestion proves valid, it has far-reaching implications for the evolution of both crust and mantle. Earth scientists have generally considered that the bulk of the buoyant material of island arcs and continents remains at the Earth's surface once it has formed. It is a new and challenging concept that some fraction of this material is being returned to the mantle.

Volcanism along convergent margins is a perplexing subject of debate. The problem is compounded by lack of theoretical and observational information on the temperature and flow structure in the mantle beneath convergent margins. The premise is that the subduction process injects a cold surface plate into the mantle. Intuitively, this should cool the mantle that surrounds the subducting plate, but somehow the subducting plate instigates volcanism.

Two distinctive features of volcanism in the environment are the involvement of volatile components and the eruption of andesite in addition to basalt. Melting seems to be aided by the transport to depths greater than 90 km of water and carbon dioxide held within the subducting plate. Water- and carbon-dioxide-rich fluids may be released from the subducting plate as it becomes heated. These fluids then rise into the overlying mantle to act as fluxes that trigger melting. Trace element and isotopic analyses of arc lavas have shown that the sediment coating of the subducted oceanic crust indeed is transported to the sources of the volcanoes to play a role in initiating magma genesis.

Experimental studies of magma source materials at appropriate temperatures, pressures, and varying fluid contents have outlined possible melting conditions under a variety of thermal conditions. Geochemical studies of the volcanic products have described their variability in volcanic products and have identified the physical processes that contribute to determining magma composition. But these studies have not identified a unique origin or a differentiation path for arc magmas.

The current initiatives to establish a global network of broad-dynamic-range seismometers and abundant portable seismometers will produce detailed seismic images of the mantle beneath arcs. Some work already has been done on this topic, particularly in Japan, which has provided new information on the temperature distribution beneath arcs. Additional information on the character of mantle flow beneath arcs may derive from studies of anisotropic seismic wave propagation in these areas. Seismic waves passing through olivine, the dominant mineral in the upper mantle, travel at different speeds, depending on their orientation with respect to the crystal growth axes of the olivine. The crystals can be preferentially oriented by the stresses associated with mantle flow. This orientation causes detectably different transit times for seismic waves passing parallel and perpendicular to the flow direction. Because of this anisotropic behavior of olivine, seismic data may possibly be used to determine the direction of mantle flow beneath arcs. Detailed seismic studies of the subarc mantle, especially with the improved resolution expected from the next generation of digital seismic sensors, may provide much clearer images of these inaccessible regions.

Volcanism at convergent margins may be unique to the Earth, and understanding the process is particularly important for several reasons. Arc volcanism may be the primary means by which the continents are formed. If that is true, the continents ultimately owe their origins to subduction and the volcanism it instigates. Subduction returns surface materials to the mantle, which keeps the interior fertile for continued volcanism. If it were not for subduction, continued crustal formation would have removed most of the easily meltable components from the mantle, leaving a residue immune to melting and hence incapable of driving the plate tectonic cycle.

Arc magmas often are very rich in dissolved volatile components that interact with hydrothermal circulations, resulting in economic concentrations of certain elements, particularly copper, but also gold, silver, and molybdenum. This same characteristic makes arc magmas likely to result in dangerously explosive eruptions when they near the surface—explosions driven by the violent boiling of the gases that were dissolved in the magmas at greater depths. Arc volcanism, consequently, represents one of the prime volcanic hazards to human populations.

Important and difficult questions challenge investigators of subduction zones. These questions address the controls over subduction and magma genesis; the fate of subducted slabs; and the proportions of converging plates that remain near the surface, that descend only to reemerge as volcanic material, and that sink far into the mantle to affect the convecting material at great depth. Because of the complex nature of subduction zones, answers to these questions are likely to come slowly and only by integration of a wide variety of data. As with many areas of the earth sciences, understanding the causes and consequences of subduction zone volcanism and the history of the assembly of the continents from volcanic arcs will result from integration of the increasingly detailed data provided by geological, geochemical, and geophysical observations with powerful numerical models.

CONTINENTAL STRUCTURE AND EVOLUTION

The other terrestrial planets lack the surface division into continental and oceanic crust that is a distinctive feature of the Earth. The continents may owe their existence to another distinctive characteristic: the Earth's abundance of free water. As described in preceding sections, oceanic water forms submarine hydrothermal vents at oceanic ridges, where associated hydration of the juvenile ocean-floor basalt forms minerals such as chlorite, serpentine, and amphibole with water bound into their structures. The subsequent dehydration of these minerals when they are subducted provides a flux of solutions into the overlying mantle wedge that initiates arc magmatism and causes the distinctive geochemistry of the magmas that build the island arcs. Parts of these in turn provide the raw material that is built into continents.

Studies of the continental crust and its margins will be a prime focus of geological research in the twenty-first century. Deciphering the complex interplay between tectonism, volcanism, climate change, sedimentary deposition, and geomorphic processes is vital for understanding the nature of global change. The development of this field will have a major influence on the intellectual development of the solid-earth sciences as well as on exploration for material resources. A large variety of investigatory methods will be applied.

Seismic Imaging of the Crust

In the early 1960s an active program of seismic investigation gathered data about the continental crust. As a result, the general characteristics of the continental crust in North America and the relationship between crustal structure and tectonic features were recognized. However, the usefulness of the seismic data was restricted by the limited capabilities of existing instrumentation and the absence of sophisticated processing and modeling techniques for interpretation. A rejuvenation of seismic crustal studies occurred in the 1980s with the availability of modern digital recording instruments and powerful computers.

Seismic information on the structure of the continental crust has been provided by three primary types of experiments—deep seismic reflection profiling, refraction and wide-angle reflection methods, and analyses of earthquake network and array data. A stimulus in understanding the present structure of the continental crust has come from the work of the Consortium for Continental Reflection Profiling (COCORP) and other deep reflection profiling efforts, both in the United States and abroad. Their mission was to extend the methods of seismic reflection profiling, which had been developed with great sophistication for shallow depths in oil exploration, to depths of tens of kilometers. In this way the structure of the continental crust, which is normally between 30 and 50 km thick, could be analyzed. This technique produces images detailing intrusions, shear zones, complex configurations of faults, boundaries between rocks of different composition, and reflectivity associated with variations in physical properties such as porosity. Deep reflection has mapped areas of ancient continental collision and located many dormant rifts, both within continents and along Atlantic-type margins.

The long-range capabilities of refraction and wide-angle reflection profiling techniques make these methods particularly suitable for studies of the deep structure of the continental crust and upper mantle (Figure 2.12). These capabilities have led to the discovery that crustal velocity structure, thickness of the crust, depth to the crust-mantle boundary, and the configuration of reflective interfaces within the crust are most strongly related to the nature and timing of the most recent major tectonic event to have affected the continental crust. Scientists had assumed that these properties were determined by the characteristics of the crust at the time of its origin. For example, profound changes in the seismic structure of the crust are produced by

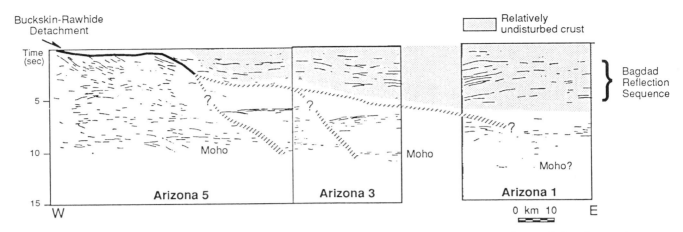

FIGURE 2.12 Interpretive line drawing from seismic reflection data. The upper crust where the Bagdad Reflection Sequence is observed has remained relatively intact throughout Tertiary deformation, limiting the location of significant deformation associated with detachments. From R. K. Litak and E. C. Hauser (1992), *Bull. Geol. Soc. Am. 104*, 1315–1325.

rifting, extensional events, thermal events, and volcanic activity. Therefore, detailed seismic studies of these features provide information not only on present characteristics but also on the geological and tectonic evolution of the continental crust.

Substantial amounts of new information are generated nearly continuously by the recording of earthquake-generated seismic signals on network and array stations. These data are being used to locate and study earthquake sources and to map crustal structure variations beneath the network of stations.

Higher-than-average velocities for seismic shear waves are observed in the upper mantle beneath continents to depths of at least 150 km. Though diminished in amplitude, the high velocities locally appear to extend down as far as 400 km, as in Canada's 2.7-billion-year-old Superior geological province. This evidence for continental roots suggests that the mineralogical constitution, and hence bulk chemical composition, of the upper mantle beneath ancient continental crust differs from that of the surrounding mantle. One explanation for the origin of this distinct mantle beneath continents is that it represents the residue left behind when partial melts were removed to form overlying crust. Melt removal leaves a residue that is less dense than the original material. Therefore, a melt-depleted mantle root could be buoyantly stable beneath a continent even though it might eventually cool to lower temperatures than the surrounding mantle. The presence of the anomalous mantle material may help to protect the overlying continental block from the effects of convection at greater depth. Indeed, old continental crust may owe its long survival in an otherwise very active and changing surface envi-

ronment to the distinctive composition of its underlying mantle.

Direct samples of the subcontinental mantle reach the surface as fragments torn off conduit walls of certain types of explosive volcanic eruptions. Pressure- and temperature-dependent changes during eruption have left their signature in differences between certain minerals in these rocks. Recognition of the signatures allows the depth of origin and original temperature of these materials to be determined. Based on this type of information, a well-catalogued sample suite spanning a depth range to 200 km is available in many areas, particularly southern Africa. Results of the analyses of coexisting minerals in these xenoliths have provided estimates of temperature as a function of depth down to 150 to 200 km, giving a fossil geotherm for the date of the eruption, tens of hundreds of millions of years ago. This remarkable geophysical result from mineral analyses is a good example of the interdependence of different approaches to the earth sciences. Field and petrological studies identified the rocks as samples derived from the mantle, experimental calibrations and thermodynamic calculations defined the mineral compositions in terms of pressure and temperature, and refinement of the electron microprobe facilitated analyses of sufficient accuracy that the calibration could be applied to the rocks. These samples of subcontinental mantle are depleted of their easily meltable component but, curiously, are enriched in a number of trace elements that also would be expected to be depleted by melt removal. The pattern of trace element enrichment of subcontinental mantle mirrors that of the continental crust. This suggests that the subcontinental mantle may serve both as the ultimate source of the magmas that form the continents and as a filter that selectively passes incompatible trace element-rich fluids rising from below.

Mountain Building: Metamorphism and Deformation of Continents

Convergence of tectonic plates causes contraction in the crust, resulting in mountain ranges with many folds and faults. The formation of mountain ranges transports rock masses through changes of pressure and temperature. The rock masses respond by changing texture, structure, composition, and mineralogy as they approach equilibrium with the new conditions. During the complex changes, dissociation reactions release volatile components and solutions migrate through the rock masses. Study of these metamorphic changes enhances information about the thermal structure of the continental lithosphere and of mountain ranges, the time scales for mountain building, and the mechanisms and scales of fluid flow through the crust.

A long-term approach of metamorphic petrologists to understanding mountain-building processes has been to calibrate naturally occurring mineral assemblages in terms of the depth, a pressure equivalent, and the temperature at which they last equilibrated. This approach is complemented by the forward approach, in which the thermal response of the rocks to tectonism is determined by computer modeling of the transient temperature distribution in a rock mass having specified properties, as it is depressed into warmer regions or uplifted toward the surface at specified rates. Combining these two approaches promises greater understanding of processes. Experimental determination of mineral reactions as a function of pressure, temperature, and different volatile components—such as H_2O, CO_2, and O_2—provides a depth-temperature framework, or grid, of reaction boundaries for the location of many common mineral assemblages. Reactions in various rock types were initially modeled in terms of phase diagrams, using observations of natural mineral assemblages. A second generation of grid models was derived through the combination of petrological observations and experimental results on selected mineral reactions. Sufficient thermodynamic data are now becoming available for the calculation of a third generation, allowing the whole family of grids to be calculated once the thermodynamic parameters have been chosen. Predictions of mineral reactions and compositions, based entirely on thermodynamic data, agree well with petrolog-

ical observations, but additional experimental work is required before further refinements can be made.

The application of temperature and pressure estimates—called geothermometry and geobarometry—to certain rocks in the European Alps yielded the surprising result that these continental rocks had been buried to a depth of at least 100 km, at a temperature of 800°C or more. It had been assumed that light, relatively buoyant continental rocks had not been buried very deeply. The explanation of how these rocks returned to the surface from such depths without completely recrystallizing is another challenge for structural geologists. In recent years investigations in mountain ranges on other continents have suggested that similar occurrences are not uncommon.

The principles of geothermometry and geobarometry, when applied to zoned minerals or to incompletely reacted mineral assemblages, help to define the paths of depth, or pressure, and temperature (P-T paths) followed by the individual rocks. These paths represent part of the tectonic history of the whole rock mass, or mountain range, and provide important insight into geological processes. Interpretation involves unraveling details of crustal thickening, folding, uplift, and local heating by igneous intrusions. Chemical data on the zonation of minerals can provide a wealth of information on the thermal processes that took place during mountain building as well as on details such as the growth history of minerals. Future work coupling inverse theory, experimental diffusion, and crystal growth data with new high-technology measurements of chemical zonation in minerals will provide an exciting revelation of the kinetic history chemically preserved in rocks that have undergone a wide variety of tectonic excursions in the Earth. This information will open a new window into the internal workings of the crust and upper mantle and will help to test our current views on the workings of plate tectonics. Improvements in analytical instruments have recently produced a highly significant advance. Ion-microprobe measurements of ages within a single zoned garnet in a rock provide, in addition to the P-T path, information about time (t). The newly developed laser probe dating technique produces results accurate enough to discriminate between several episodes of mountain building by measurements made on a single crystal.

Rates of mountain building can be estimated as well. For example, when measurements of radiometric ages are matched with the geobarometric and geothermometric evidence, P-T-t paths are derived that yield rates of geological processes associated with mountain building. The focus of metamorphic petrology today is shifting from a static mode, which reports the mineral assemblages found in the field, to a much more dynamic mode, aimed at working out the processes involved in metamorphism. This shift emphasizes not only the thermodynamic variables but also nonequilibrium aspects and the kinetics of metamorphic processes. In particular, kinetic study has been extended to the generation, motion, and characterization of metamorphic fluids. There are several processes that must be quantified in describing the results of metamorphic reactions between rocks and fluids so as to provide answers to questions about heat transport processes, both conductive and convective; fluid mass transfer processes; solute mass transfer processes—convective and diffusive; mineral surface reaction processes, both dissolution and growth; and nucleation of new minerals.

The P-T-t histories of rocks commonly include extensive expulsion and migration of fluids. These fluids usually move near the surface, but problems emerge with fluid movement much deeper in the Earth. For example, at depths near 6 km the fluid pressure approaches lithostatic pressure. It is difficult to understand convective fluid flow, potential for channelization, or flow rates and volumes at such high temperatures and pressures.

The chemical reactions taking place involve moving fluids. Progress has been made in obtaining thermodynamic data for minerals and fluids, but the behavior of fluids as a function of composition over a wide range of pressure and temperature conditions and the equilibrium description of solid solutions needs more data. Models need to be developed that account for the dissolution, transport, and precipitation of chemical components during the flow of fluids through a series of rock units. The quantitative treatment of isotope exchange between minerals and fluids needs to be elaborated further to make use of the increasing isotopic data base on reactions. An understanding of the chemical evolution of the crust must be based on knowing the transport properties of the fluid, such as whether flow is diffusive or convective, as well as the complex heterogeneous kinetics taking place at every mineral surface in contact with the fluid.

Extensional Deformation of Continental Lithosphere

Deformation in mountain belts is dominated by contraction. It is possible to simulate the great folds of the Alps, Himalayas, and Rockies by pressing on the leaves of a book or by pushing a napkin along a table. Although the rheology is very different, there are resemblances to what happens in nature, both in the flat layering of the original material and in the detachment at the base of the deforming layers from the material beneath. Analyses of rock deformation in a contractional regime on mega, macro, and micro scales have become very sophisticated, and environments can be satisfactorily modeled, especially where complementary data on metamorphic state are available.

The concept of converging tectonic plates that contract the crust and produce mountains fits well with our understanding of plate tectonic theory. Within the past decade researchers have emphasized another process that actively deforms vast tracts of continental crust. That force is extensional deformation, and its widespread effects have come as a surprise. Contraction and extension are both characterized by detachment of the crust from the underlying mantle; in the case of extensional deformation the detachment occurs in the form of normal faults.

Surface expression of extension commonly occurs in the form of rifts, and active rift zones can be seen in many continental areas. The character of the modern rifts, however, shows considerable variation. The rift of East Africa spans almost the entire continent from north to south. At its northern end, extension and volcanism have been considerable, leading to the formation of the Red Sea basin that now separates the once-connected Africa and Arabian peninsula. Through central Africa, rifting has been less successful in splitting the continent. Volcanism in this area is rarer, and the rift also contains a higher proportion of peculiar alkali-rich lava types than the northern rift. In the United States the Rio Grande rift that splits the Colorado Plateau from eastern New Mexico is only tens of kilometers wide, whereas the Basin and Range province exhibits continental extension and rifting over a zone more than 600 km wide. Volcanism is widespread in each of these rifts but generally consists of isolated, relatively small volume eruptions.

The rifting of continental material can evolve in one of two ways: either the extension in the rift can develop until new ocean floor forms at the rift site (such a rift is called a "successful" rift), or extension can cease before a new ocean forms and the rift can become inactive within the continent (such a rift is called a "failed" rift). The former course represents a progressive step in the cycle of the opening and closing of the ocean basins, while the latter represents only one more episode within the evolution of the continental lithosphere. Whether a rift system "succeeds" or "fails" is presumed to be controlled by the disposition and sum of the plate-driving forces at some critical time in rift evolution.

The "successful" course of rifting is typically represented around the shores of the Atlantic Ocean and other similar margins attributable to continental rupture. A considerable research effort is developing on what has happened in the earliest stages of ocean formation in these areas. Stimuli include the lessons learned in the Basin and Range province and the recognition that the typical dog-leg shape of the shorelines of the Atlantic resembles the shape of the East African rifts. As in East Africa, nodal (hotspot) volcanic areas are concentrated at the dog-leg bends in the Atlantic shore. Marine geophysics calibrated by the drilling of research holes of the Ocean Drilling Project is revealing that enormous amounts of basalt were erupted onto the young ocean floor at the dog-leg bends. In some cases much smaller volumes of lava continued to be erupted as the ocean continued to grow, forming hot-spot tracks that lead to active hot spots such as Iceland, Jan Mayen, Reunion, and Tristan da Cunha.

More than 100 ancient rift systems have been recognized within the continents. The oldest go back to nearly 3 billion years ago, indicating that ancient continents behaved rather like their modern counterparts. Regional extension of ancient continents is also indicated by the preservation of parallel swarms of hundreds of narrow vertical dikes occupied by basalt, all emplaced at the same time, that extend for distances of 1,000 km or more across the ancient continents.

The cause of the variations in volcanic output and degree of crustal extension in continental rifts is not understood at present. Significant advances can be made on this topic through detailed field studies employing modern techniques of remote sensing, chemical and chronological analyses, and paleomagnetism, in addition to traditional field mapping. Scientifically oriented drilling into continental rifts can do much to illuminate their structure and the sequence of volcanism and sedimentation that records their evolution. Combined with the increasing resolution of geophysical techniques capable of providing images of subsurface structure, computer modeling and laboratory simulation of the response of rigid crust to extensional stress may illuminate the underlying causes of continental rifting. The Basin and Range province is proving a marvelous area in which to study extension. Detachment systems in that region are the subject of integrated geological, geophysical, and geochemical study.

The detachment systems accommodate extension by the separation of relatively undeformed crustal blocks, with individual faults transecting the upper 15 km or so of the crust and having displacements typically measured in tens of kilometers. The end products of extension are often exposures of wide tracts of middle crustal rocks veneered with patches of upper crustal rocks and subsequently eroded material. Such extended domains have been identified throughout the Basin and Range province. They are on the order of 100 to 200 km wide and are interspersed with relatively unextended upper crustal blocks (Figure 2.13).

Paradoxically, the topography of these tracts, as well as gravity anomalies and reflection seismic data, indicate that despite the vast area and substantial heterogeneity of upper crustal strain—large enough in some cases to form an ocean basin—the crust maintains a near-constant thickness across the boundaries. These observations have led to the recent suggestion that the upper crust may be floating on a deeper layer within the crust. This differs from the usual assumption that the thickening and thinning observed in the upper crust are accommodated at depth by flow in the upper mantle. It appears that the upper crust and upper mantle are probably strong, while the lower crust is generally weak and ductile in this environment. Multidisciplinary studies of the lithosphere in these regions promise major advances in understanding deformational processes in the continental lithosphere over the next decade. Particular advances are anticipated in understanding the bulk rheology, strain field, and deformation mechanisms of the deep crust. These advances will construct a framework with the potential to unify a broad range of disciplines, including petrologic and isotopic thermobarometry, geophysical imaging, geochemical studies, physical modeling, and field geology.

Evolution of the Continents

The plate structure of the lithosphere was accepted in the 1960s. Soon afterward came the realization that plate tectonic processes represented the dominant way in which the Earth dissipates its internal heat. The next question addressed the length of time that plate tectonics has played a major role in earth processes.

There is no strong evidence, even in the oldest rocks, of any processes that were radically different from those of modern plate tectonics. There is

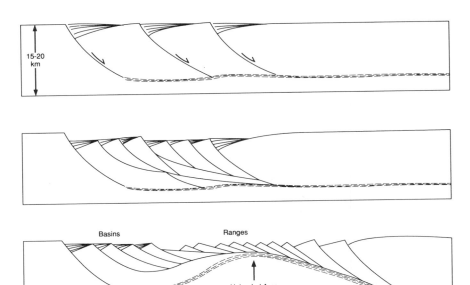

FIGURE 2.13 A conceptual model of a migrating zone of isostatic uplift during extension; although the detachment surface is originally subhorizontal, its ascent through the brittle crust occurs along a relatively steep zone.

persuasive evidence, however, that some processes were not exactly the same. For example, because the Earth was hotter when it was young, a larger proportion of the mantle melted at higher temperatures. This produced magnesia-rich lavas that are abundant in rocks formed between 3.8 billion and 2.5 billion years ago but are very rare in the more recent rock record.

Direct evidence of rigid plate rotation across the surface of the Earth extends no farther back than the formation of the oldest preserved ocean floor—only about 170 million years ago. From this evidence, the dominance of plate tectonics during that interval—5 percent of earth history—is essentially proven. For earlier times indirect inference becomes necessary. The record of older earth history must come from the continents because ocean floor is subducted back into the mantle. Geologists establish what has happened to the continents during the past 170 million years of earth history and then project those assumptions to earlier intervals.

The world map indicates that plate tectonic processes have led, over the past 170 million years, to the operation of a global cycle in continental evolution. Continents rupture, as in East Africa, and open into new ocean basins such as the Red Sea. Ocean spreading centers expand the basins until they reach at least the size of the Atlantic. Eventually the basins shrink when subduction eats away at their edges. Finally, oceans close, as the Mediterranean is closing. The ocean basins disappear, their location marked by sites of continental collision, as in the Himalaya and Tibet.

Within this broad cycle, complexity develops.

Arcs break away from continents, as the Japanese arc rifted from Asia about 30 million years ago. Elsewhere, arcs collide with continents, as have Taiwan, Panama-Colombia, and Timor. Along the Andean cordillera, mountains and volcanoes were built at the convergent boundary, were rifted into the Pacific ocean, and were swept back into the South American continent. Further complexity results when spreading centers slip beneath continents—slivers of material slide along the edge of the continent, and associated elevated areas form as continental rifting develops. This is the origin of the San Andreas Fault system in California and the neighboring Basin and Range province (Figure 2.14).

Projections of this complex cycle to earlier intervals suggest that similar tectonic forces have dominated since Earth's early history. The geological record shows evidence of suturing where ancient ocean basins closed and continents collided. It also contains the arc and continental fragments that collided and contracted and have been assembled into the continents.

No known continental lithosphere is older than 3.9 billion years, and the oldest sedimentary rocks date to about 3.8 billion years. But detrital grains as old as 4.2 billion years have been discovered in western Australia in rocks that date to 3.6 billion years ago. This discovery supports the assumption that continental crust of even greater age once existed but has been destroyed.

All the continents contain fragments of material that is older than 3.0 billion years. The best examples are preserved in extensive areas of Greenstone belt terranes in the Canadian, southern African, and

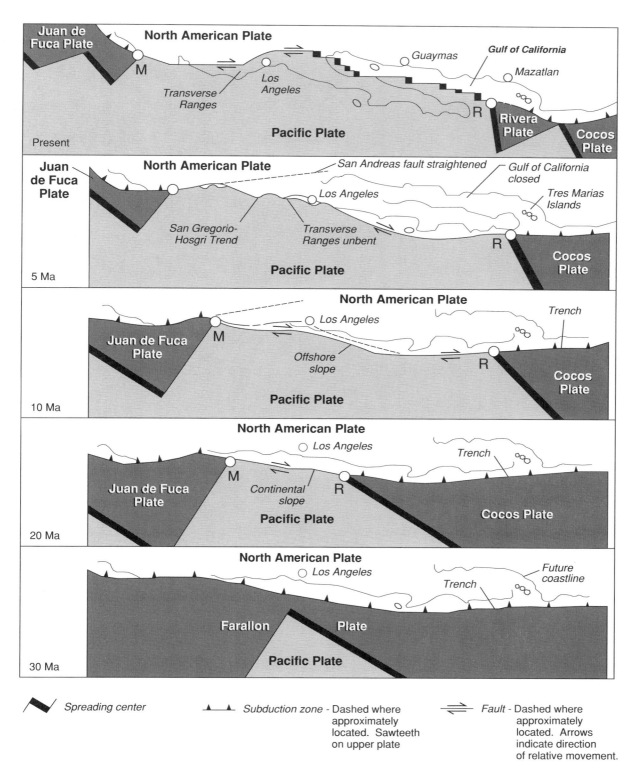

FIGURE 2.14 Sequential diagrams showing plate-tectonic evolution of the San Andreas transform fault system. Early transform faulting was west of the present-day San Andreas Fault and presumably separated young oceanic rocks of the Pacific plate from rocks of the North American plate. Over time, the transform faulting has stepped eastward, and virtually all of the present fault is now in rocks of the North American plate. Figure from *USGS Professional Paper 1515.*

West Australian shields. This old crust represents material assembled from island arcs in ancient ocean basins. Those primeval island arcs were surprisingly similar to the modern Marianas, Central Aleutians, or Lesser Antilles.

The mantle lithosphere beneath these greenstone belt terranes is unusually cold, and the lithosphere is twice as thick as elsewhere. This may be explained if ocean spreading centers of the early Earth operated at higher temperatures than they do today. Higher temperatures would have extracted more material from the mantle to form a thicker ocean-basin floor. During subduction the thicker ocean floor would produce a larger proportion of light material that would accumulate beneath the old continents as they were constructed by the assembly of colliding island arcs.

Sedimentary Basins

Rocks above sea level, particularly those forming the mountain belts of the continents, weather from reaction with the atmosphere, biosphere, and hydrosphere. Streams carry the weathered material downslope until it is deposited in low-lying areas forming the world's sedimentary basins. Sedimentary basins form by long-term subsidence of the crust through sediment or tectonic loading and by thermal subsidence. Many sedimentary basins have a subsidence history that covers about 100 million years and characteristic horizontal dimensions of several hundred kilometers. They are of particular interest because they are the source of fossil fuels and many mineral deposits. (Sedimentary basins are covered in more detail in Chapter 4.)

Over long time scales, sedimentary basins (see Figure 1.13) provide a record of the geological processes active in the crust. Their structure is often attributable to flexure of the crust. Other sedimentary basins are attributable to displacements on major faults. Examples of this latter type include both rift and foreland basins.

Reconstructing time-temperature histories for sedimentary basins is of direct economic application to the petroleum and mineral industries but also is of interest to the general geological community. Seismic stratigraphy of the sedimentary deposits in basins has provided a wealth of data on the variations of sea level with time. Models that use variations in the volume of mid-ocean ridges to displace water appear to explain long-term (~100-million-year) variations in sea level. Some short-term variations can be explained by episodic glaciation. However, many observed variations in sea level

remain unexplained. Models that include sea-level change, subsidence, and sediment supply can be used to obtain synthetic stratigraphic sections, and inverse modeling has been used to determine the role of these factors from observed seismic sections. Changes in the Earth's orbit influence sedimentary deposition. Evidence of this can be discerned in sediments, particularly during glacial epochs.

The broadest sedimentary basins are the oceans themselves, and thick deposits of sediment collect along continental slopes and on continental shelves. When rivers flow directly into open oceanic currents, their sediment loads disperse over wide areas. When they flow into protected gulfs and small seas, rivers deposit their sediment loads as soon as they reach relatively quiescent waters and the sediment accumulates into deltas, such as the Nile, the Mississippi, and the Ganges. When a river carries very large quantities of sediment, extensive deltas develop even if they flow directly into the open ocean as in the Amazon system.

During certain periods of earth history, when ocean spreading centers were very numerous and broad, water was displaced from ocean basins, and high seas flooded extensive tracts of low-lying continental crust, creating inland seas. Both inland seas and deltas may subside beneath the weight of accumulating sediments until thicknesses reach thousands of meters and lower layers consolidate under the pressure. Long after the inland seas have disappeared and the rivers have changed course, these buried porous sandstones, siltstones, and limestones act as reservoirs for groundwater and petroleum resources. Smaller expanses of lake beds and river valleys also serve as sedimentary basins that accumulate deposits and later serve as reservoirs for fluid concentration. Finally, when continental crust rifts or collides with other plates, the thick deposits in sedimentary basins are subjected to forces of extension, compression, and intrusion. They are deformed and metamorphosed. Sometimes they are carried down into subduction zones at convergent plate boundaries and recycled into the overlying crust. When the rocks of sedimentary basins are thrust to the surface during mountain building, they are exposed once again to weathering processes and to the sedimentary cycle.

Continental Collision

Because the volume of the Earth has varied little since its earliest history, the operation of plate tectonic processes on the global scale must have always involved complementary plate making at spreading centers in the oceans and plate destruction at convergent plate boundaries. The cordilleran process that characterizes convergence between oceanic and continental plates involves the rifting of fragments hundreds of kilometers long from the continent and the collision and lateral motion of objects of comparable size. At times in earth history, plate motion has inevitably led to collisions between objects of continental dimensions.

The huge Alpine-Himalayan mountain chain represents the site of an ongoing continental collision between Eurasia and three fragments of the former continent of Gondwanaland: Africa, Arabia, and India. An earlier large-scale continental collision involved Gondwanaland itself, which collided with Laurasia 300 million years ago to form the short-lived supercontinent of Pangea. Comparable collisions may be vaguely discerned in the older rock record—for example, in the original assembly of Gondwanaland about 600 million years ago.

In the Mediterranean area the Alpine-Himalayan chain shows some of the complexities of continental collisions. Continental fragments collided tens of millions of years ago to form the Alps, the best studied of all mountain ranges. Crustal rocks were locally carried to depths of 100 km during this mountain-forming process, and they have been intensely deformed during several episodes. Widespread collapse of the elevated mountains is evident, a process also encountered at present in western North America. Collapse of mountain ranges by extension may be as general and as significant a part of their history as their construction by shortening. The kinds of integrated geological and geophysical approaches applied in western North America suggest that the mountain belts of the Alps and the Mediterranean region are uncoupled from the underlying mantle and are moving independently. Lateral movement, perhaps in response to the collision of the Arabian continental protuberance, is accommodated by the earthquake-generating faults in Turkey. This lateral motion may play a major part in the collision that is closing the Mediterranean. For example, over the past 25 million years west-to-east movement of most of the Italian peninsula has pushed up the Apennine and Dalmatian mountains and, in its wake, has opened a small basin now filled by the Tyrrhenian Sea.

Perhaps the most distinctive feature of the collision of Arabia with Asia, in the sector of the Alpine-Himalayan belt immediately east of the Mediterranean, is the role it has played in forming the enormous accumulation of oil and gas in Saudi Arabia, the Gulf states, Kuwait, Iran, Iraq, Syria,

and Turkey. This remarkable resource accumulation has depended on the association of the collision along with a number of other special circumstances; these are considered in Chapter 4, Resources of the Solid Earth. As a first approximation, oil generated on the old continental margin that was the northern edge of Arabia appears to have been driven up-slope at the collision by shortening in front of the collisional Zagros Mountains.

The collision between India and Asia represents the most advanced stage of continental collision on Earth at this time. Phenomena associated with this collision include elevation of the highest mountains, the Himalaya; formation of the world's largest sedimentary body, the Bengal fan; and generation of violent earthquakes, including the most deadly earthquake in recent decades at Tangshan, near Beijing, China. The collisional zone in Pakistan and India has yielded far less oil and gas accumulations than that in Arabia. This is, at least in part, because the Indian continental margin formed at high latitudes where the accumulation of organic material was less abundant.

The study of continental convergence between Eurasia and fragments of Gondwana serves to unite scientists as well as science. Researchers from India, Nepal, Pakistan, the Commonwealth of Independent States, China, and the countries of Southeast Asia, in cooperation with each other and with solid-earth scientists from Europe, Australasia, and North America, are suggesting solutions to questions concerning the ongoing collision. They conclude that India is dipping under Asia, but how fast and by how much are unknown. Recent investigations considering earthquake mechanisms and Landsat imagery propose that China is bulging eastward into the Pacific to escape the squeeze of the collision. And innovative interpretations speculate that the Tibetan Plateau, unable to escape the squeeze, has reached a limit of thickness and is now beginning to collapse.

The Himalayan/Tibetan area stands 5,000 meters above sea level over an area of more than 1,000,000 km^2, which is as large as the lower-standing mountainous area of western North America. Researchers propose that this mass of continental crust has become detached from its underlying lithospheric mantle, which has foundered into the convecting mantle layer. If recycling of continental crustal material and its roots is indeed happening on such a grand scale beneath the Himalaya, the Alps, and perhaps the American cordillera, current ideas about continental crustal growth and processes of recycling should be reexamined.

The extreme elevation of Tibet has affected atmospheric circulation over much of Asia, perhaps influencing not only the development of the Asian monsoon but also the formation of the Sahara and even the onset of Northern Hemisphere glaciations. The extensive high ground in western North America may have played a similar role, but other tectonic events, such as the closing of the Panama Isthmus and the Zagros Ocean, have modified oceanic circulation and may have been more significant. Both kinds of major barriers to warm equatorial circulation would have generated profound climatic and environmental changes within the past 15 million years. Recognition of tectonic influences on the evolution of the Earth's climate demonstrates the potential for intellectual breakthroughs resulting from the study of the earth system.

Continental collision is a major research interest in the solid-earth sciences. Although most attention focuses on the active Alpine/Himalayan belt, older collisions, especially those recorded in the ancient Precambrian rocks, yield complementary information. Rocks from high-temperature and high-pressure environments buried deep within the modern mountain chains are preserved and well exposed for study at the surface in the old belts. Traditionally, these provocative exposures provide both information and inspiration to earth scientists.

Growth of the Continents Through Time

Much research remains to be undertaken to test and modify the simple picture of continents assembled by the process of arc collision and modified by the cordilleran, continental collisional, impact, hotspot, rifting, flooding, and erosional processes. Establishment of the history of the continents through time will provide an important test of how they have evolved.

Additions to the continental crust are known to have occurred throughout recorded geological time, from about 4 billion years ago to the present. With the average age of continental surface rocks around 2 billion years, and with a wide variety of ages spanning most of recorded time, it seems possible that the continental crust has grown in volume through the history of the Earth. However, the details of this growth are still uncertain, and determination of a crustal volume versus age curve is of considerable importance for understanding the Earth's evolution. There is geological and geochemical evidence suggestive of periods of enhanced crustal growth, although this picture may be clouded by the geographically patchy distribution of the data. Large volumes of continental crust give isotopic signatures indicating that the material from which they formed became fractionated from the mantle (by the processes of partial melting that take place at divergent plate boundaries and beneath volcanic arcs) relatively recently. For example, much of the crystalline basement of Arabia, Egypt, and eastern Sudan formed from the mantle some 600 million to 900 million years ago. Estimates of the present-day rate of crustal addition in island arcs fall short of the rate of average growth of the continents (about 2 km^3/year, a figure obtained by dividing the present volume of the continents by the age of the Earth). If island arc addition has been the main way of making continents, rates must have been far higher in the past.

We have a clear picture of the present distribution of continental material today in the large bodies of Eurasia, Africa, North and South America, Australia, and Antarctica and in smaller objects like Greenland, New Zealand, Madagascar, Japan, and the Seychelles. The motions of these fragments over the past 200 million years since Pangea began to break up are reasonably well known from the history of the intervening ocean basins. Some idea of how continental fragments were assembled into Pangea, between the assembly of Gondwanaland about 600 million years ago and its final collision with Laurasia about 290 million years ago, has also emerged, but no clear picture has yet been obtained of how continental material was distributed about the surface in earlier times. Ancient latitudinal indicators for these older times have provided a confused picture.

Determining when and how all the pieces of all the continents were formed and how they came to be in their present positions is proving a substantial research exercise. It looks as though the greater part of North America was assembled into one piece by 1.7 billion years ago, but Asia is a continent put together only within the past 600 million years. We do not yet know whether these temporal and geographic differences record stochastic operation of plate tectonic processes or whether they reflect systematic changes in the Earth's behavior in space and time.

The central part of North America—with perhaps 75 percent of the present continental area—was put together long ago. Establishing a subsequent history dominated by the peripheral addition of relatively small exotic blocks and fragments is an active research frontier. Questions such as where the blocks came from and how they were incorporated into North America are rendered more challenging by the violent changes wrought on these objects at and after collision. Only the western side of North America has wholly escaped episodes of collision by major continents over the past billion years. These huge collisions and subsequent ruptures have made the story even harder to unravel in such areas as the Canadian Arctic, the Appalachians, and the Ouachitas.

One intriguing idea is that the cycles of ocean opening and closing that have operated throughout earth history have caused the continents to have episodically come together to form a huge single landmass. After a few hundred million years, these supercontinents have been torn apart into fragments that drifted away from each other, much as the modern continents were dispersed from Pangea. One hypothesis for why this occurs is that the cold thick continental fragments are pushed away from areas of active upwelling of the underlying mantle, much as the skin on a pan of heated milk moves away from violently boiling areas. Eventually, this retreat from hot mantle brings the continental fragments together where they collide and weld again. At a certain size, however, the thick continent hinders heat transport out of the mantle that underlies it. Eventually, the underlying mantle heats up to a point at which it forms uprising plumes, which, if of sufficient size, break through the overlying continent. If continued, this plume activity evolves into a general upwelling of the mantle, which again causes the breakup of the supercontinent and the formation of an ocean basin.

Determining long-term feedback relations between the elements of mantle convection, including plate formation and subduction as well as mantle plumes, and the elements of surface geology, including sea-level change, continental assemblies, and continental rupture, is a major interdisciplinary challenge. Regional geology, mantle geophysics, and geochemistry need to be used together in a new way. The identification of cold regions in the mantle, interpreted as subducted slabs over 100 million years old, marks an important step in this new direction.

GEOCHEMICAL CYCLES

Only some 20 to 25 years ago was it realized that surface materials could be returned to the Earth's interior through the process of plate subduction. The importance of recycled crustal materials in modifying mantle composition has been revealed only in the past 5 to 10 years. The consequences of crustal recycling and the involvement of crustal

material in mantle convection are still being worked out (Figure 2.15). The crustal material includes water, carbon dioxide, and elements that are significant in the biogeochemical cycles occurring at and near the surface, and the interchange between biogeochemical cycles and the deep interior needs to be established. Cycling of material from the reservoirs of the deep interior through the Earth's surface systems typically occurs on time scales ranging from tens of millions to billions of years, but cycles in the near-surface systems driven by solar energy can be as short as a year or as long as hundreds of millions of years. Interchanges have to be recognized as happening on a large range of time scales. With realization of the significance of the interchange, it has become clear that the Earth as a whole—its interior, crust, oceans, and atmosphere—all interact through chemical cycling. To understand the characteristics of one reservoir, the nature of the exchange with the other components of the earth system must be understood and taken into account. The near-surface cycles are dealt with in Chapters 3 and 4. Here, emphasis is on geochemical cycles in the mantle and crust with reference to their connections with the hydrosphere and atmosphere.

Extraction of the continents from the mantle has left an identifiable chemical imprint on the interior, causing the mantle, as sampled by the basalts erupted along the worldwide ocean ridge system, to be depleted of the same elements as those by which the continental crust is enriched. Isotopic data for oceanic basalts erupted throughout earth history show that the composition of their mantle source was modified early in earth history. The original modification could have been caused by the extraction of continental crust prior to 4.0 billion years ago. Significant volumes of continental crust of this age, however, are not preserved. If the early chemical differentiation of the mantle was caused instead by removal of a crust more like oceanic than continental crust, it is possible that this ancient crust was subducted into the mantle and is no longer visible at the surface. An additional increment of chemical modification of the mantle would then have occurred when the majority of continental crust was formed, roughly 2.5 billion to 3 billion years ago. This second step in chemical differentiation of the mantle, in fact, is observed in the isotopic record of mantle-derived rocks and crustal sediments.

One of the major recent achievements of mantle geochemistry has been the identification of recycled surface materials as an intrinsic part of the mantle. The first major consequence of crustal cycling is its

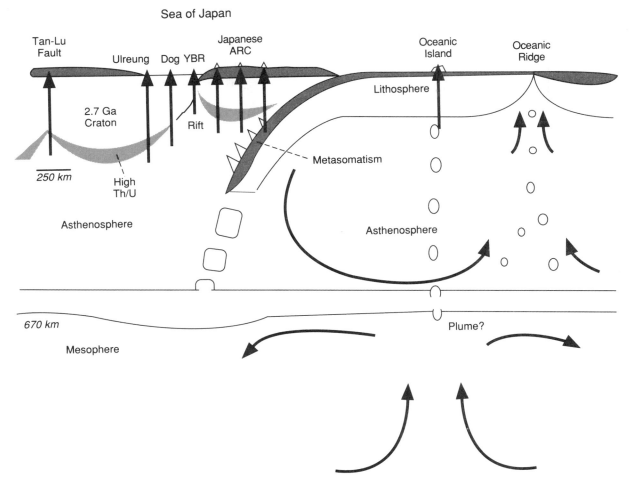

FIGURE 2.15 Schematic diagram illustrating plate recycling.

instigation of volcanism landward of subduction zones. Arc volcanic rocks display a chemical signature best explained by the contributions of subducted oceanic sediments and altered oceanic crust to their source regions in the mantle. The recent discovery of an isotope of beryllium in arc volcanoes, but in no other volcanic system on Earth, provides clear evidence that even the upper few meters of sediment on the ocean floor are transported to a depth of at least 150 km in subduction zones, the depths at which the processes of dehydration and perhaps partial melting initiate the generation of arc magmas. There is a direct link with cycles in the atmosphere and hydrosphere. The radioactive isotope, [10]Be, is transferred from the upper atmosphere by rain and becomes concentrated in ocean sediments. The fact that it decays relatively rapidly places constraints on the time interval between raining from the atmosphere, sub-

duction in sediments, and emergence at the surface in volcanic lavas.

A distinctive feature of arc lavas is their relatively high water content, which is largely derived from the rocks of the subducted oceanic plate. The combination of experimental-phase equilibrium studies on source rocks and lavas with geophysical modeling of temperatures at depth confirms that most of the subducted water is removed by dehydration or melting by a depth of 150 km or so, whence it returns to the surface for involvement in the shallow cycles. However, under some conditions there is opportunity for small amounts of water to escape the magmatic processes and to be transported to greater depths for long-term storage within the mantle. Recent theoretical and experimental studies at very high pressures have led to proposals that the lower mantle, deeper than 700 km, may contain water in significant quantities: at least 0.3 percent by weight, the equivalent of about two masses of ocean water. The recent synthesis of iron hydride at high pressures corresponding to conditions within the Earth's core has raised the question of how much hydrogen and oxygen may be present within the core. Thus, water may be involved in earth processes all the way from the core to the atmosphere.

Another example of the depth of penetration of subducted material is provided by diamonds. Diamonds crystallize only at depths greater than about 150 km in the mantle. A significant fraction of diamonds have distinct carbon isotopic compositions, like that found for organic carbon-rich sediments at the surface. Thus, some diamonds may once have been living matter at the surface that was subducted with an oceanic plate to depths where the elevated pressure and temperature turned the dispersed carbon into diamond.

Small, but observable, heterogeneities in the chemical and isotopic composition of the mantle have been documented by detailed studies of the basalts erupted in the ocean basins. The important feature of this heterogeneity is that it appears to reflect mixing between a depleted residual mantle, formed by extraction of continental and oceanic crust, and two to three other more "enriched" components. Two of the enriched components are believed to be basaltic ocean crust and continentally derived sediments. These materials are injected into the mantle in subduction zones; are carried along by the general mantle circulation, during which time they partially mix with surrounding mantle; and eventually return to the surface beneath ocean ridges and in mantle plumes. Estimates of the time necessary to travel from subduction zone to ridge are on the order of 1.7 billion years.

The complete implications of this process represent an area of active research that links geochemical and seismic observations of the mantle with theoretical treatments of the nature of mantle convection and the efficiency with which convective "stirring" can rehomogenize the injected components. What was once viewed as a one-way transfer of material from the mantle to the crust must now be seen as a continuation of global geochemical cycles, in this case including all of the planet. The ability to recycle surface components back into the mantle may be responsible for keeping the Earth's interior unexpectedly close to its original composition. Recycling of crust may be unique to the Earth as a consequence of its active plate tectonic system and may explain its continuing vigor.

Some of the key questions in chemical geodynamics or deep geochemical cycling include:

1. What is the mass balance of the cycling? Has all of Earth's interior once been at the surface, or is some material reaching the surface for the first time?
2. Can the age and geographical distribution of returning recycled oceanic crust be used to trace flow patterns and velocities in the mantle and provide clues for the locations of past subduction?
3. What is the balance of fluid transport? Are the oceans growing with time, or is water being subducted into the mantle faster than it is emitted at ocean ridges? What fraction of the water in arc volcanism is from the mantle, and what fraction is recycled from the oceans?
4. Is subduction responsible for continent formation? Are an active plate tectonic system and a liquid ocean necessary prerequisites for the development of Earth-like continents on a planet?
5. How has recycling of chemically differentiated material from the Earth's surface affected the composition of the mantle? Is the Earth still an active planet because its attempts at internal chemical differentiation were reversed by the rehomogenization accompanying recycling and convective mixing?

Key areas for understanding deep geochemical cycles, and their connections with the near-surface biogeochemical cycles, are the two tectonic environments where most material enters the interior and where most material emerges from the interior: subduction zones and oceanic ridges. Detailed multidisciplinary investigations of the fluxes of energy and matter in these and related environments (e.g., continental rifts) are of fundamental importance.

Major parts of the biogeochemical cycles involve material transferred relatively rapidly in and out of the biosphere and material transferred between rocks and the fluid envelopes during weathering processes.

INTERACTION BETWEEN THE SOLID EARTH AND ITS FLUID ENVELOPES

The surface water of the Earth and to a lesser extent the atmosphere play a critical part in the recycling of material that is one of the planet's most distinctive features. The outer layer of solid Earth is in contact with the fluid envelopes of the hydrosphere and the atmosphere. Seventy-one percent of the solid surface is innundated by ocean water, and the remaining 29 percent, the continental surface, is flushed by water at intervals with a frequency governed by climatic zone. The crust can be thought of as a catalytic bed of minerals and fluids of great diversity consisting of several trillion square kilometers of surface area. Most of this surface consists of the interface between minerals and water. The reactions that occur at these interfaces, some of which are catalytic in nature, directly affect the planet we live on and the way we live. They play critical roles in determining the quality of our fresh water supply, the development of soils and the distribution of plant nutrients within them, the genesis of certain types of ore and hydrocarbon deposits, and the geochemical cycling of elements.

Mineral-water interface geochemistry is not a new field. For example, observations relevant to silicate mineral dissolution were made more than 150 years ago. However, it was not until the past 10 or 20 years that the instrumental means were developed to directly study mineral surfaces and the reactions that occur at mineral-water interfaces at the molecular level. Much of the technology needed has been provided by developments in the field of surface science, which has traditionally been in the domains of chemistry (heterogeneous catalysis, electrochemistry) and applied physics (semiconductors and integrated circuits). The transfer of this technology by geochemists is relatively recent (within the past few years), but it is already leading to new fundamental knowledge about how minerals dissolve and undergo reduction or oxidation, how chemical species partition from fluids to mineral surfaces, and how the hydrosphere interacts with crustal rocks.

An immediate application of this knowledge to a problem of societal relevance is the development of more robust and accurate models for predicting the transport of contaminants in groundwater. Certain contaminants can be strongly chemisorbed on mineral surfaces under certain conditions, thus removing them from the fluid phase. However, we must understand the molecular-level mechanisms for such reactions to model them properly; incorrect assumptions about the stoichiometry of sorption reactions can lead to errors of several orders of magnitude in predictions of the partitioning behavior of chemical species. Errors of this magnitude simply cannot be tolerated in predicting how a contaminant plume disperses.

Although the field of interface geochemistry is in its infancy, it is already leading to changes in the way we think about many geochemical and mineralogical processes. Continuing studies in this rapidly growing field will undoubtedly lead to a more fundamental understanding of the chemistry and physics of mineral-water interfacial phenomena and how chemical species are partitioned between minerals and aqueous fluids in the crust.

RESEARCH OPPORTUNITIES

The Research Framework (Table 2.1) summarizes the research opportunities identified in this chapter and in the relevant panel reports, with reference also to other disciplinary reports and recommendations. These topics, representing significant selection and thus prioritization from a large array of research projects, are described briefly in the following section. Processes operating near the surface are, for the most part, reserved for Chapter 3, although there is no sharp boundary between the deep-seated processes and surficial geology.

The research areas are interrelated. There is continuity between the processes occurring at the core-mantle boundary, mantle convection, the physical deformation of the lithosphere and mountain building in the crust, and the geochemical transfer of material from mantle to crust. In the crust the material is exposed to physical and chemical interaction with the atmosphere, the oceans, and the hydrosphere, which transgresses the two fluid envelopes and the crust. Research opportunities arise when methods are developed to explore new regions, such as the inaccessible ocean floor, continental lower crust, and the even more inaccessible deep interior.

Research Area II: Global Geochemical and Biogeochemical Cycles

Evolution of the Crust and Its Relationship to the Mantle

Examination of the Earth's crust through the use of isotopes, trace elements, and rare gases continues to be a fundamental frontier topic. Study of the growth rates of the continents can provide a long-term view of the Earth's evolution, and these rates should be investigated by a concerted effort that couples detailed field studies with high-precision geochronology and isotopic studies to determine mantle separation ages for different continental terrains. Within the continental rocks lies an historical record of mantle convection, changes in

TABLE 2.1 Research Opportunities

Objectives

Research Areas	A. Understand Processes	B	C	D
I.				
II. Geochemical and Biogeochemical Cycles	■ Evolution of the crust and its relationship to the mantle ■ Fluxes along the global rift system ■ Fluxes at convergent plate margins ■ Mathematical modeling in geochemistry			
III. Fluids in and on the Earth	■ Mineral-water interface geochemistry ■ Role of pore fluids in active tectonic processes ■ Magma generation and migration			
IV. Dynamics of the Crust and Lithosphere	■ Oceanic lithosphere generation and accretion ■ Architecture and history of continental rift valleys ■ Sedimentary basins and continental margins ■ Continental-scale modeling ■ Recrystallization and metasomatism of the lithospheric mantle and lower crustal metamorphism ■ Thermal structure, physical nature, and thickness of the continental crust ■ The lithosphere at convergent plate boundaries ■ Tectonic and metamorphic history of mountain ranges ■ Quantitative understanding of earthquake rupture ■ Rates of recent geological processes ■ Real-time plate movements and near-surface deformations ■ Geological predictions ■ Modern geological maps			
V. Dynamics of the Core and Mantle	■ Origin of the magnetic field ■ Core-mantle boundary ■ Imaging the Earth's interior ■ Experimental determination of phase equilibria and the physical properties of Earth ■ Chemical geodynamics ■ Geodynamic modeling			

Facilities - Equipment - Data Bases
■ Remote sensing of the continental unit from satellites
■ Global digital seismic array, with broadband instruments
■ Portable seismic arrays
■ Ion microprobes
■ Accelerator mass spectrometers
■ Ultrahigh pressure temperature instrumentation
■ Microscale, in situ analytical instrumentation
■ Synchrotron radiation facilities
■ Ocean-bottom seismometers and other geophysical instruments
■ Instruments for micromagnetic measurements
■ Deep continental drilling
■ Parallel-processing computers
■ Data storage and distribution facilities

crustal growth and destruction by recycling, temporal variations in the temperatures in the interior, variability in the nature of the plate tectonic process, and the compositional evolution of the mantle.

Fluxes Along the Global Rift System

The rift system transports energy and material from the interior to the lithosphere, hydrosphere, and biosphere. An important goal of research in this area is understanding the geophysical, geochemical, and geobiological causes and consequences of this transport. The flow of mantle material, the generation of melt, its emplacement along spreading centers, and its transformation into crystalline oceanic crust are primary problems. Aspects relevant to magmas at hot spots beneath oceans and continents are a significant related concern.

Fluxes at Convergent Plate Margins

Convergent plate boundaries are the sites of material transport from the surface to its interior. It is necessary to determine the relevant mass and heat fluxes, treating the generation and flow of vapors and melts as part of a dynamical system.

Mathematical Modeling in Geochemistry

Practically all theoretical treatments of geochemical cycles use overly simple models, with each box representing a chemically homogeneous reservoir and the transfer of material being represented by fluxes between boxes. Intense efforts are required to improve theoretical modeling. There is an urgent need to determine the connections between the relatively near-surface short-term biogeochemical cycles covered in Chapter 3 and the longer-term geochemical cycles extending deep into the Earth's mantle.

Research Area III: Fluids in and on the Earth

Mineral-Water Interface Geochemistry

This rapidly growing field uses technology transferred from the domains of surface chemistry and physics (catalysis and semiconductors). Continuing studies in this field will lead to a more fundamental understanding of the chemistry and physics of mineral-water interfacial phenomena, how chemical species are partitioned between minerals and aqueous fluids, and how the hydrosphere interacts with crustal rocks.

Role of Pore Fluids in Active Tectonic Processes

The fluid state may be most important in understanding the mechanics of deformation, including earthquakes, and insight can be gained from laboratory experiments of rock fracture and rock friction with fluids present. A program of continental

drilling would greatly add to our empirical knowledge concerning pore fluids.

Magma Generation and Migration

Magma generation and migration are fundamental processes. Their analysis needs the coupling of phase equilibria and trace element partitioning data to fluid dynamical equations, with deformable solid matrices. Magma study provides research opportunities ranging from the thermodynamics and geochemistry of rocks in the molten state, through their role in chemical differentiation of the Earth, to their manner of emplacement in and on the crust. The physical processes associated with upward migration of the magma and the mechanisms of eruption are only partially understood. The transition from micro- to macropermeability is relevant. Magma-driven fractures are also likely to play an important role. Fractures are amenable to being characterized as fractal, and fracture geometrics is a promising area of active research.

Research Area IV: Dynamics of the Crust and Lithosphere

Oceanic Lithosphere Generation and Accretion

Continued interdisciplinary studies of ocean ridges, with their associated magmatic processes and interaction with the hydrosphere, promise excellent scientific returns. The global correlation of ocean ridge basalt geochemistry with axial depth and crustal thickness resulting from temperature variation in the mantle is an example of recent interdisciplinary successes. Patching together diverse results from three decades of geophysical surveys and geochemical analyses has led to the formulation of a model for the detailed structural and magmatic segmentation of mid-ocean ridge spreading centers.

Architecture and History of Continental Rift Valleys

Better comprehension of continental separation and the formation of continental margins, where natural resources are commonly concentrated is needed. There has been a revolution in our understanding of extensional tectonics with the discovery of basement-penetrating normal fault systems. Determination of the rheology of the lithosphere requires deep seismic profiling and laboratory experiments on rock strength. Isotope and trace element analysis of rift valley magmas and xenoliths will test rift models deduced from independent data sets.

Sedimentary Basins and Continental Margins

Gravity data for many sedimentary basins, including both foreland and cratonic basins, indicates the presence of buried loads. The origin of these loads and their relation to thermal subsidence remain unexplained challenges. We need a better way of estimating the original masses of ancient sediments. Also, to improve modeling we need a much better inventory of sediments beneath present-day continental margins.

Continental-Scale Modeling

This area of computational tectonophysics has great potential. Its benefits would include the testing of qualitative hypotheses for orogenies, the discovery of new orogenic mechanisms (e.g., delamination, Moho buckling), and in situ determinations of the rheology of the lithosphere. Dynamical systems approaches (fractals, chaos) offer much promise for attacking major unsolved problems in crustal dynamics.

Recrystallization and Metasomatism of the Lithospheric Mantle and Lower Crustal Metamorphism

Intensive study should integrate the full complement of isotopic and trace element tools integrated with field work, geophysics, the study of deep crustal/upper mantle xenoliths—"meteorites from the mantle"—and the calibrations of experimental petrology, in terms of depth, temperature, and fluid compositions.

Thermal Structure, Physical Nature, and Thickness of the Continental Crust

These are key variables governing the state of stress and attenuation of seismic energy in the crust. Study of the ancient crust may help to establish the history of the thermal structure of the Earth. A fundamental area of research in active tectonics is the source, distribution, and propagation of the energy that drives tectonic changes. Heat flow studies, in situ stress measurements, and measurements of strain are among the elements in need of study. The opportunities range from field measurement of these elements to theoretical modeling of how they interact.

The Lithosphere at Convergent Plate Boundaries

Multidisciplinary studies of this environment are essential for understanding mantle dynamics and geochemistry as well as the chemical differentiation of the Earth.

Tectonic and Metamorphic History of Mountain Ranges

A new dimension, time, has recently been added to the depth-temperature framework provided by experimental petrology. Isotopic analyses yield ages for different stages of mineral growth, and another approach to "time" involves geophysical calculations on the thermal history of rocks.

Quantitative Understanding of Earthquake Rupture

The physical processes applicable to earthquake rupture can be approached by integrated seismological, geophysical, and geological studies of earthquakes and the faults on which they occur. Direct measurement of stress by drilling techniques for accessible faults is now feasible. Improvements in three-dimensional imaging capabilities are needed to map subsurface faults that are often nonplanar.

Rates of Recent Geological Processes

New insights into the rates at which processes take place can be expected, especially for those occurring over the less-than-1-million-year time scale. New and improved techniques for dating materials and events offer rewarding opportunities. There should be extensive dating of Quaternary landforms and sediments.

Real-Time Plate Movements and Near-Surface Deformations

The Global Positioning System (GPS), which is rapidly evolving into a powerful tool, is used to monitor these motions. Geodetic techniques, based on land and in space, provide some of the best direct evidence about short-term changes in the crust.

Geological Predictions

The revolutionary field of geological prediction in geological evolution within 1-million-year time frames requires increasing emphasis on surficial geology and neotectonics. The documentation and understanding of rapid geological processes are an entirely new dimension for earth scientists to explore, with computers now making it possible.

Modern Geological Maps

Maps constitute an important data base in the solid-earth sciences. Field relations must be continually reexamined in the light of new theoretical concepts, and no substitute exists for continuing geological mapping and analysis of relations in the field. Maps should include three-dimensional data on geophysics and geochemistry and data from satellite-based remote sensing. Digitizing the different data sets that are to be used together is essential.

Research Area V: Dynamics of the Core and Mantle

Origin of the Magnetic Field

Magnetic field generation is one of the universal processes in the cosmos, and the outstanding unsolved geophysical problem involving the core is generation of the geomagnetic field. Important advances in the near future will concentrate on more limited problems such as (a) the origin of the dipole inclination, secular variation, and the westward drift; (b) the role of the mantle in influencing magnetic field structure; and (c) a determination of the power source driving the dynamo.

Core-Mantle Boundary

With the deployment of a new broadband digital network of seismometers, the likelihood of deciphering the nature of the core-mantle boundary is excellent. There is the prospect that geomagnetic anomalies can be associated with seismological heterogeneities found at the base of the mantle, leading to the possibility of documenting changes in the core-mantle boundary through the geological past.

Imaging the Earth's Interior

The new digital recording seismometers with broad wavelength sensitivity and large dynamic range include both portable varieties and permanent stations that will be applicable to global studies. The three-dimensional distribution of velocity anomalies in the mantle obtained through these data can then be used to infer relative temperatures and compositions within the mantle.

Experimental Determinations of Phase Equilibria and the Physical Properties of Earth Materials at High Pressure and Temperature

New high-pressure apparatus extends the range of experimentation. Properties of materials (e.g., density, seismic velocity, melting temperature) can be measured in situ using, for example, high-intensity x-rays produced by synchrotron sources. Comparison of the high-resolution seismic images of the interior with direct experimental determinations of the physical properties of earth materials at high pressure and temperature will advance the understanding of the interior's temperature and compositional structure to an unprecedented degree. This can yield better insight into how the high temperatures drive internal motions that in turn determine the geological history of the Earth's surface.

Chemical Geodynamics

The combination of geochemistry and geophysics continues to reveal the scale of heterogeneities within the mantle. Isotopic variations in mantle-derived rocks provide time-dependent information about the creation of mantle heterogeneities by partial melting or lithosphere subduction and about the efficiency of convection in remixing the mantle components. The nature and source of the mantle plumes responsible for generation of at least some volcanic hot spots remain a tantalizing problem, one that may link phenomena at the core-mantle boundary to massive volcanic eruptions.

Geodynamic Modeling

The recent success in quantitatively relating geoid anomalies and the results of seismic tomography, at least on scales of thousands of kilometers, prompts similar investigations over smaller distances. The amount of material transported across the entire mantle by convection is currently the single largest uncertainty in our understanding of the Earth's thermal and geological evolution. This transport is part of the major geochemical cycles of the Earth.

FACILITIES, EQUIPMENT, AND DATA BASES

The earth sciences offer special opportunities for the development and application of new technologies, as exemplified by the instrumentation that has recently been created for use in the field and in the laboratory. Technological developments in the earth sciences are at the forefront of research.

The emphasis in geochemical instrumentation continues to be on attaining higher spatial resolution while maintaining high sensitivity and accuracy in isotopic and trace element analyses. One of the greatest needs in the field of isotope geochronology is to bring more ion microprobe instruments into operation. New generations of instrumentation include the super high resolution ion microprobe (SHRIMP) and the x-ray microprobe.

For cosmogenic nuclide geochronology the major goal is a better understanding of variations in production rates. Accelerator mass spectrometer measurements are especially well suited to studies of the interrelationships among the solid earth, the atmosphere, and the hydrosphere.

The development of new ultra-high-pressure/temperature instrumentation for simulating the deep interior remains a major goal. This high-pressure research requires additional development of techniques for microscale in situ analysis of small samples or of physical properties (e.g., elastic constants).

Continued access to synchrotron radiation facilities is important for precise crystallographic information as well as compositional data.

Ocean bottom deployment of geophysical instruments is an area generally in need of technological development. Until this becomes routine, achieving global coverage from satellite measurements will continue to be a major problem.

High-sensitivity, high-spatial-resolution analytical techniques are in demand for carrying out micromagnetic measurements on minerals and rocks as functions of temperature and field.

The increasing availability of massive computational capabilities provides opportunities for attacking complex problems. Parallel-processing computers hold much promise for a wide variety of simulations in the earth sciences. Less powerful special-task computers are seeing increasing applications in the field and the laboratory. There is an urgent need to improve the capabilities for communications and storage of data. Networking between computers will be necessary, particularly for large-scale data transfer and high-speed interactive computing.

BIBLIOGRAPHY

National Research Council Reports

NRC (1980). *Studies in Geophysics: Continental Tectonics*, Geophysics Study Committee, Board on Earth Sciences and Resources, National Research Council, National Academy Press, Washington, D.C., 197 pp.

NRC (1982). *Studies in Geophysics: Climate in Earth History*, Geophysics Study Committee, Geophysics Research Board, National Research Council, National Academy Press, Washington, D.C., 198 pp.

NRC (1983). *Opportunities for Research in the Geological Sciences*, Committee on Opportunities for Research in the Geological Sciences, Board on Earth Sciences, National Research Council, National Academy Press, Washington, D.C., 95 pp.

NRC (1983). *The Lithosphere: Report of a Workshop*, U.S. Geodynamics Committee, Board on Earth Sciences, National Research Council, National Academy Press, Washington, D.C., 84 pp.

NRC (1984). *Studies in Geophysics: Explosive Volcanism: Inception, Evolution, and Hazards*, Geophysics Study Committee, National Research Council, National Academy Press, Washington, D.C., 176 pp.

NRC (1986). *Studies in Geophysics—Active Tectonics*, Geophysics Study Committee, Board on Earth Sciences and Resources, National Research Council, National Academy Press, Washington, D.C., 266 pp.

NRC (1987). *Earth Materials Research: Report of a Workshop on Physics and Chemistry of Earth Materials*, Committee on Physics and Chemicstry of Earth Materials, Baord on Earth Sciences, National Research Council, National Academy Press, Washington, D.C., 122 pp.

NRC (1987). *Geologic Mapping in the U.S. Geological Survey*, Committee Advisory to the U.S. Geological Survey, Board on Earth Sciences, National Research Council, National Academy Press, Washington, D.C., 22 pp.

NRC (1988). *The Mid-Oceanic Ridge: A Dynamic Global System*, Ocean Studies Board, National Research Council, National Academy Press, Washington, D.C., 351 pp.

NRC (1989). *Margins: A Research Initiative for Interdisciplinary Studies of Processes Attending Lithospheric Extension and Convergence*, Ocean Studies Board, National Research Council, National Academy Press, Washington, D.C., 285 pp.

NRC (1990). *Studies in Geophysics: The Role of Fluids in Crustal Processes*, Geophysics Study Committee, Board on Earth Sciences and Resources, National Research Council, National Academy Press, Washington, D.C., 170 pp.

NRC (1990). *Studies in Geophysics: Sea-Level Change*, Geophysics Study Committee, Board on Earth Sciences and Resources, National Research Council, National Academy Press, Washington, D.C., 234 pp.

NRC (1991). *International Global Network of Fiducial Stations: Scientific and Implementation Issues*, Committee on Geodesy, Board on Earth Sciences and Resources, National Research Council, National Academy Press, Washington, D.C., 129 pp.

NRC (1991). *Opportunities in the Hydrologic Sciences*, Water Science and Technology Board, National Academy Press, Washington, D.C., 348 pp.

Other Reports

Basaltic Volcanism Study Project (1981). *Basaltic Volcanism on the Terrestrial Planets*, Pergamon Press, Inc., New York, 1,286 pp.

Interagency Coordinating Group for Continental Scientific Drilling (1988). *The Role of Continental Scientific Drilling in Modern Earth Sciences Scientific Rationale and Plan for the 1990's*, 151 pp.

Interagency Coordinating Group for Continental Scientific Drilling (1991). *The United States Continental Scientific Drill-*

ing Program, Third Annual Report to Congress, 35 pp. + appendices.

Inter-Union Commission of the Lithosphere (1990). *International Lithosphere Program*, International Council of Scientific Unions, 119 pp.

NASA (1991). *Solid Earth Science in the 1990s, Vol. 1*. Program Plan, NASA Office of Space Science and Applications, Washington, D.C., 61 pp.

USGS (1987). *Volcanism in Hawaii*, U.S. Geological Survey Professional Paper 1350, R. W. Decker, T. L. Wright, and P. H. Stauffer, eds., U.S. Government Printing Office, Washington, D.C., 1,667 pp.

USGS (1990). *The San Andreas Fault System, California*, U.S. Geological Survey Professional Paper 1515, R. E. Wallace, ed., U.S. Government Printing Office, Washington, D.C., 283 pp.

3

The Global Environment and Its Evolution

ESSAY

A familiar image shows the Earth hanging against the black marbled coldness of deep space. The blues sparkle, barely dulled by patches of brown. Swirling white whorls veil the bright sphere only slightly. Twenty-five years ago that image had not yet been seen. Furthermore, the idea of the Earth as an integral unit was not a prevalent one. Study of the planet proceeded at local or regional scales. Then plate tectonics began to weave regional studies into one planetwide dynamic model.

The past two decades have also seen the emergence of a new perspective in the earth sciences—or to use a more recent term, earth system science—emphasizing changes in the global environment that occur over spans of geological time. The changing environments leave geological evidence that permits investigation of a wide range of geographic, oceanographic, climatic, and biotic transitions. Such evidence includes information about environmental changes that cannot be directly observed today. The record of the rocks reveals that certain factors force changes in the global environment and that some ecosystems are more sensitive than others to those changes.

The surface has been changing for over 4.5 billion years. Many of these changes are fluctuations within definable extremes. A familiar example is that of water. Sediments record changes in the hydrologic cycle during which rising seas engulfed extensive continental tracts and then drained away. Some of these changes resulted from ice ages that left wide swaths of continental shelf exposed to the air during glacial accumulation and sent torrents to the oceans during intervals of melting. Other cyclic fluctuations recorded in the geological record include geochemical exchanges through reservoirs: atmosphere, ocean, biomass, sediment, crust, and mantle. Ocean basins rise and sink and expand and contract in cycles, and mountain ranges thrust upward and then waste away.

While cyclic changes leave recognizable patterns in the geological record, they continually alter the components that are recycled. Inevitably, new conditions are created by these natural movements. The original state can never be exactly regained. The new conditions are the result of what is referred to as secular change—change with time.

The most obvious secular change that has taken place during earth history is the early transformation of its surface from a landscape of naked rock, barren seas, and toxic atmosphere to a landscape seething with life, with organisms that exist on a variety of scales and in a medley of forms. As the cycles have churned away and new secular changes have occurred, sporadic catastrophic events have thrown the whole dynamic system into chaos.

Geoscientists use an assortment of techniques and instruments to investigate the complex interactive systems that have created the surface environment. A 200-year-old tradition of field mapping and detailed description offers a solid foundation—called ground truth—for new technologies like remote sensing and for new conceptual models such as ones that explain how biological evolution has altered the chemistry of the atmosphere.

Remote-sensing technology is not limited to satellite imagery and geodesic laser measurements; remotely sensed magnetic and gravitational anomalies help trace the vertical and horizontal movements of the crust.

Fine-tuned seismic reflection research has also provided a new way to "see" inside the Earth. Seismic reflection can now produce subsurface images that rival, in an areal extent, drilled cores for locating geological boundaries; the cores, however, are often needed to provide ground truth. Tomography combines sets of seismic reflectance data to create cross sections through various planes.

Mapping also reveals patterns of past environments, ranging from tropical seas to jungles and deserts. These environments are identified by a great variety of evidence, including fossil occurrences, key sedimentary rocks, and the isotopic composition of shells. Maps of fossil occurrences show how organisms were distributed in space and where they lived bears directly on how they evolved.

Once geologists have determined what the patterns are, they can study how those patterns changed; on a planetary scale, this dual effort is at the heart of studies of global change. Geochronology provides the framework for arranging temporal sequences in the geological record. It also provides estimates for rates of chemical, physical, and biological change. Paleontology and stratigraphy led to the first arrangement of geological time scales. Intervals of the Earth's past gained names like Precambrian, Devonian, Cretaceous, and Pleistocene. Today, new quantitative methods for analyzing fossil occurrences are improving our ability to compare ages of strata throughout the world.

Techniques that use naturally occurring radioactive isotopes can now provide dates within a 2-million-year range of accuracy for events that affected particular materials 2.5 billion years ago. At the other extreme of recency and resolution, close documentation of tree-ring patterns yields chronologies accurate to less than 1 year.

Reconstruction of ancient seas, climates, and continental geography for key intervals of the past provides data for constructing and testing numerical and conceptual models that portray global circulation patterns for ancient oceans and atmospheres. Such models are shedding light on the agents that triggered ice ages and other shifts to new environmental states on a global scale.

Ancient life has tracked changes in habitats—sometimes migrating, sometimes evolving, and sometimes disappearing from the Earth. In fact, fossils provide the only direct record of large-scale evolution and extinction, and this record can be understood only in the context of past global change. Evolutionary theorists depend on the paleontological record to test their hypotheses, and the application of innovative quantitative techniques is providing new insights into rates, patterns, and modes of evolution.

The unifying theory of a dynamic Earth and the vast scale of coverage provided by satellite imagery complement recent advances in international cooperation at all levels. Geoscientists have organized the International Geological Correlation Program (IGCP) and the Global Sedimentary Geology Program (GSGP) and have helped set up the International Geosphere-Biosphere Program (IGBP). Over the past 25 years, internationally based ocean drilling programs have concentrated on obtaining cores from the ocean floor—nearly three-quarters of the surface—that have improved patchy data and supported stratigraphic correlations with an unexpected degree of success. The drill ship has been to geologists what the telescope is to astronomers, allowing geologists to study the most remote parts of their domain. Such systematic examinations of the Earth are filling in gaps in knowledge as though they were pieces of a giant spherical jigsaw puzzle.

Investigators who correlate the geological information from remote regions like the deep ocean find signs of environmental change that affected the whole planet. Some of these sweeping changes periodically caused widespread extinction of species. The study of such extensive occurrences is called global event stratigraphy. The most familiar example of global event stratigraphy is the evaluation of the iridium anomaly found in widely distributed strata that date to 66 million years ago. The iridium anomaly may signify a meteorite impact that resulted in the extinction of as much as 66 percent of the species, including all the dinosaurs. The most pervasive extinction occurred 225 million years ago when perhaps as much as 95 percent of all species died off.

More recently, 75 percent of the giant mammal species that roamed the spacious North American plains south of the Laurentide ice sheet disappeared 11,000 years ago as the ice sheet melted away. This extinction may have had other causes besides the climatic fluctuations that occurred in the wake of the retreating ice front. Additional threats were posed by the hunting skills of human beings.

Anthropogenic factors—those caused by the activities of humans—affect the surface environment on a vast scale. As modern technology offers striking images of a bright blue globe against a black abyss, it also provides evidence of environmental crises, some caused by humankind.

While efforts continue to monitor global change and to attribute the causes to humans or to nature, no consequences can be predicted without an understanding of the records of past global change. Through the record in the rocks, the past has become the key to the future.

THE GLOBAL ENVIRONMENT: A GEOLOGICAL PERSPECTIVE

The Changing Land Surface

The processes of weathering, erosion, and soil formation that together degrade upland areas have operated throughout earth history. Variations in the way the processes operate have generally been dominated by climate. Glaciers form in polar or subpolar regions or at high altitudes; deserts develop around 20° from the equator; and rain forests, with their great rivers, grow in equatorial and temperate latitudes.

Against this image it has been surprising to learn from new methods of dating and from quantitative studies of material fluxes that the times in which we live are unusual—not for one but for two reasons. The first is that since Northern Hemisphere glaciation developed about 2.5 million years ago, fluctuations in ice volume have been large enough to repeatedly change world sea level by as much as 100 m over time scales of tens to hundreds of thousands of years. Despite recent fluctuations, Antarctic ice caps have remained stable for as long as 40 million years. For this reason the dominant major cause of sea-level change during the past 2.5 million years has been the accumulation and ablation of Northern Hemisphere ice sheets. As an example of fast change, consider that 18,000 years ago, when sea level was about 100 m lower, rivers reached the sea at the edge of the continents. Since then they have retreated so much in response to rising sea level that some river mouths (e.g., the Susquehanna) now lie far back from the continental shelves in estuaries like Chesapeake Bay. Such short-term changes prevent the establishment of an equilibrium in weathering, erosion, and deposition, and during the past 2.5 million years they have rendered the surface an unusually dynamic place.

We ourselves provide the other reason for unusual conditions. The condition of the soil and processes of erosion have been substantially changed since agriculture began. These changes have increased in recent decades because of such processes as dam building, forest destruction, widespread irrigation, and flood control. For instance, sediments that formerly were carried to the Mississippi delta are now being impounded behind dams; this is contributing to the encroachment of the Gulf of Mexico upon the delta. The problems of marine transgression along the Gulf Coast are discussed throughout this report; here, attention is drawn to progress in understanding the major processes that dominate the change in land surface above sea level, with emphasis on the peculiar problems and opportunities that result from the apparently exceptional time in which we live.

Landforms

Landforms are continually changing, but most changes are subtle and generally escape notice. Although great attention is rightly given to catastrophic events such as earthquakes, volcanic eruptions, and landslides, the time scales of importance in geomorphology—the study of landforms and the processes that shape them—range from seconds to millions of years, and the space scales range from single hillsides to global dimensions. The challenge is to characterize the ways in which landforms respond to both common and uncommon events. The geomorphic record contains information about the ways in which present and past environmental changes have modified the processes operating at the surface, both in intensity and duration. A long-term view is essential because the time spans of contemporary monitoring are too short to represent the range of possible conditions. Long-range perspectives from the geomorphic record permit testing of models of environmental change, whether they apply at global, regional, or local scales.

A landscape can provide information about the magnitudes and return frequencies of natural processes. This information can lead to identification of geomorphological thresholds that precede disastrous events. Some responses are immediate—floods, landslides, and debris flows—but others can be spread over years or decades—upland soil erosion, glacial to interglacial cycles, river-channel change, and sea-level change. The rates of some of these processes have been greatly accelerated by human activities.

Geomorphic events of the past, which are recorded in landforms and stratigraphy, can provide usable analogs of anticipated environmental change.

For example, the rate of sea-level rise in the next century could become 1 cm per year, a rate not unlike that at which sea level rose at the end of the last ice age. Study of the coasts drowned at that time will provide information relevant both to biotic response to a future sea-level rise and to the anticipated acceleration of sedimentation in the lower reaches of river systems.

Geomorphologists and geochemists have been using particular isotopes, produced in the atmosphere and in rocks by cosmic radiation, to determine ages of landforms and to date events such as floods, landslides, fault movements, lava and debris flows, and the onset of glaciation. The application of accelerator mass spectrometry to carbon-14 (^{14}C) dating provides a means of dating samples both older and smaller—by a thousand times—than the type of sample conventionally used.

Various new methods are being developed to determine the age of a landform, measuring the time elapsed either since the rocks forming it were deposited or since they were exposed by erosion. These methods represent a breakthrough in quantifying landform dynamics, permitting more precise dating of key events in the evolutionary development of landforms and yielding erosional rates for the various features of a landscape.

The Himalaya, for example, have long been considered not only the highest but also the fastest rising mountain range in the world (Figure 3.1). Just how fast they are rising is something we are only beginning to understand. Uplift rates may be as great as 5 mm per year— 5 km per million years—in places along the front of the mountain belt during the past 20 million years. Crystals of microcline were eroded from the surface and deposited in the sediments of the Bengal deep-sea fan south of Sri Lanka only a few million years after they became cool enough to stop losing radiogenic argon by diffusion, which occurred at a depth of 5 km inside the mountain belt.

Weathering and Soil Formation

The interaction of the atmosphere and hydrosphere—in the form of groundwater—with the rocks at the surface is very complex, partly because organisms ranging in size from bacteria to trees are involved but largely because the relevant time scales vary so widely. Water has a residence time as short as a single storm and typically cycles on an annual scale; trees have lives of decades to centuries; and the minerals formed in the weathering processes can have residence times in the soil of as high as thousands to millions of years. Varieties of soil are strongly controlled by local climate, and with the growing interest in global change soils are being looked at anew to learn what they record about past climatic changes on time scales from decades to millions of years.

Soils lie at the interface between the geosphere, biosphere, and atmosphere. They have unique properties that derive from the intimate mixing of partly weathered geological substrata, dead organic matter, live roots, microorganisms, and an atmosphere high in carbon dioxide, nitrogenous gases, and moisture. Ions of potassium, sodium, calcium, magnesium, sulfur, and phosphorus are released from minerals by hydrolytic weathering. Many of these ions, as well as carbon and nitrogen, are also released from dead organic matter by microbial digestion. The geological composition of the soil passively influences soil biota through ionic deficiencies or release of toxic elements. The soil biota actively influence weathering rates through respiration and production of organic acids.

Soils are open systems that gain and lose energy and matter as they evolve through time (Figure 3.2). Gains, losses, translocations, and transformations occur continuously in the soil column; the relative magnitudes of these processes determine the types of horizons formed along the column. If the vectors of these processes remain relatively constant, the intensity of their expression in horizons is time dependent. Time-dependent soils are useful for correlating geographically separated geomorphic surfaces. This property of soils has been widely used in studies concerned with the rates and timing of tectonically and climatically driven geomorphological processes.

In the past, most research on rates and pathways of soil development centered on careful chemical, physical, and microscopic dissection of soil horizons developed during known time intervals. Future research will emphasize collection of in situ measurements to document fluxes of gases and solutes, rates of mineral weathering, and biological interactions. Already, these studies reveal the physiology of soil to be modulated by complex feedback mechanisms that are nonetheless subject to catastrophic breakdown under external influences such as fire, erosion, or dramatic climatic change. If we are to understand the contribution of soils to ecological stability, we must also study the effect of anthropogenic disruptions, such as toxic chemical discharges, on soil processes. Soils are great purifiers, but we do not know what level of disruption constitutes a lethal dose.

FIGURE 3.1 View of the Himalaya.

Rivers and Material in Transit from Mountains to Sedimentary Basins

The surface is shaped locally by weather. The temperature and moisture regimes, with the weather intensities and rates of change, represent climate. The effects of climate change are becoming more clearly understood through the use of general circulation models. Climatologists, who develop general circulation models to predict the consequences of global change, test their models by comparing results obtained with variables that define past conditions against the evidence found in the geological record. That evidence includes the landforms themselves, sedimentary sequences, and fossil biotas. Closer study of the response generated on the surface by climatic changes that last for short intervals, up to hundreds of years, or long intervals,

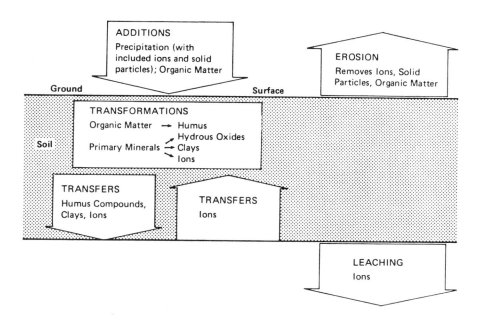

FIGURE 3.2 Flow of the major processes in soil development.

hundreds of thousands of years, allows prediction of how future climatic fluctuations can alter the landscape. For example, the sizes of river channels formed in the past 10,000 years and the character of their contained sediments have been used to reconstruct a history of long-term change in the magnitudes of high-frequency floods in the Upper Mississippi valley (Figure 3.3). That record helps to test general circulation models of climatic fluctuations over the past ten centuries.

Stratigraphic evidence also extends the existing record of river behavior beyond the limits of data collected through observation over the past 100 years. Advances have recently been made in estimating the size and frequency of ancient floods,

effectively extending hydrologic records for up to thousands of years, by combining interpretations of river and sediment behavior with results from geological dating methods.

The intensity and sequence of climatic events are crucial factors in molding the landscape into a distinctive form and producing a recognizable geological pattern. The lone occurrence of a moderate flood might accomplish little erosion and deposition, but the clustered occurrence of two or three moderate floods can destabilize entire channel systems and cause them to become sensitive to the erosional effects of even small floods.

Studies of the erosion, transportation, and deposition of sediments in contemporary drainage sys-

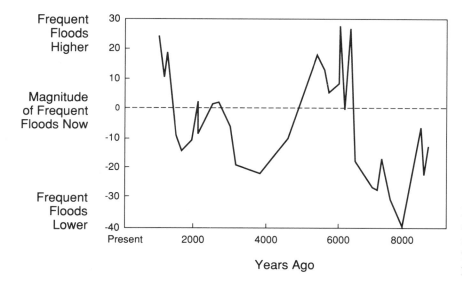

FIGURE 3.3 Variations in the size of annual and biennial floods in the upper Mississippi Valley over the past 9,000 years; the size is expressed as a percentage of the present flood size.

tems are showing departures from the steady state. Modern watersheds are accumulating more sediment than they pass on. In many rivers a large proportion of the sediment is transported by flood events that prevail only a few days of the year. How the climatic conditions that favor these events relate to the magnitudes and frequencies of erosional and transportation episodes is emerging as an area of great scientific interest.

Erosion by other agents, such as glaciation and wind, plays some part in the degradation of topography, but most eroded material is carried to the sea by rivers either in solution or as detrital sediment. River transport varies enormously with both climate and source area; the Huang He (Yellow River) alone, for example, accounts for 6 percent of the world's total suspended matter river load, because this river drains the readily eroded windblown material of the loess plateau in the Chinese interior. In addition, the beginnings of agriculture more than 4,000 years ago appears to have greatly increased the rate at which sediment is moving into the river.

Winds and Glaciers: More Material in Transit

Although most of the material carried into sedimentary basins is transported by rivers, appreciable amounts are carried by wind and glaciers. Deserts, where wind erosion and deposition in the form of sand dunes are most important, have become the subject of increased research activity. This activity has been sparked by such diverse influences as the rapid extension of the Sahara into the Sahel within the past 30 years, the discovery of wind-made landforms on Mars, and the availability of satellite and radar images of the Earth's remote deserts (Figure 3.4). Again, new techniques of age determination have yielded exciting results.

Recent measurements of the thermoluminesence of rocks from surfaces buried beneath dunes in the high plains of the American West have shown that the dunes were moving in response to desert winds much more recently than had previously been realized. Windblown dust mixed into the deep-sea sediments of the North Pacific helps to show how continental climatic fluctuations in China relate to the orbitally induced climatic changes that are well known from the deep-sea record. On a longer time scale, windblown dust in sediments from the deep Atlantic indicates that the Sahara first became a huge desert about 10 million years ago, possibly as a result of changes in atmospheric circulation related to the uplift of the Tibetan plateau.

Glaciers and glacial deposits reflect the tremendous fluctuations in the climate of the current ice age. Only 20,000 years ago glaciers extended as far south as New York City, and there may have been as many as a dozen comparable advances and retreats of Northern Hemisphere ice during the past 2.5 million years. A new research effort to integrate continental and oceanic data from the past 2.5 million years should produce a picture of how the surface environment adapts to rapid climate change. Glacier ice provides a unique record of short-term change, which is discussed in the part of this chapter dealing with cyclical change.

Lakes: Interruptions in Transit

Lakes represent a peculiar part of the earth system. If the solar-driven heat engine entails erosion of mountains, transit of eroded material from the mountains to the sea, and deposition of sediments at the edges of the continents and on the ocean floor, then lakes represent an interruption to the smooth flow of the system. As such, lakes are usually quite short lived because they fill with sediments. The familiar outlines of North America's Great Lakes are less than 20,000 years old and are unlikely to last more than another 20,000 years. Only a few of the world's existing lakes are more than a million years old, and the oldest, Lake Baikal in Siberia and the Great Rift Valley lakes of East Africa, are found in places where the continents are being ripped apart by tectonic forces.

Lakes and lake sediments provide crucial information about climatic change during the past 2.5 million years. In recently glaciated areas, some lakes have produced annual cycles in sedimentary layers called varves. Some of these varve-layer exposures indicate thousands of years of continuous deposition. The most productive method of retrieving environmental information from lake deposits is the analysis of fossil pollen, which provides a record of changes in nearby vegetation. In tropical areas, lakes expand during intense monsoonal episodes, while in temperate latitudes high lake levels may indicate more rain or less evaporation. Fluctuations in shoreline elevations are used with the pollen record to reconstruct recent environmental changes.

The long-term rock record contains scattered evidence of ancient lakes; the oldest of these is more than 2 billion years old, half as old as Earth itself. The study of old lake deposits has become a major research activity. Interest has been stimulated not only by studies of the geologically recent lakes but also by the discovery that lake beds are commonly

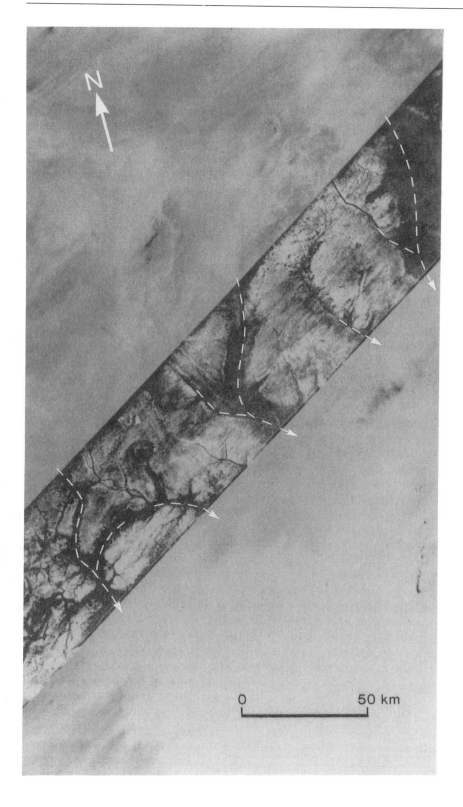

FIGURE 3.4 SIR-A radar scan (diagonal band) reveals aggraded valley segments that were barely perceptible on Landsat images of eastern Sahara in northwest Sudan (19.7°N, 25.2°E).

a source of petroleum—an idea promoted by Chinese geologists and now broadly accepted. Some of the finest remains of early humans have been found around the shores of old lake beds, and often the richest sites for dinosaur fossils are associated with the rocks along ancient shores.

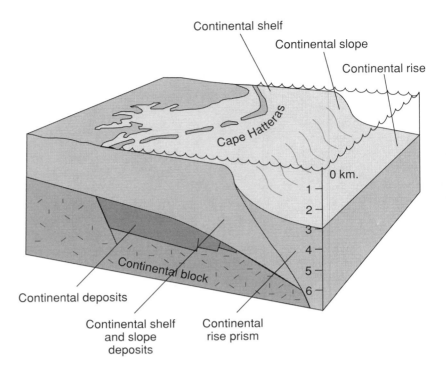

FIGURE 3.5 Atlantic-type (passive) continental margin along the eastern United States. Continental deposits accumulated in fault-bounded basins as the early stages of rifting separated North America from Africa. When the continents separated enough so that seawater could enter to form the Atlantic Ocean, marine deposition commenced.

Deltas and Estuaries

Rivers reach the sea in deltas and estuaries. Deltas form at river mouths when the prevailing current becomes too slow to carry detritus, so that enormous amounts of sediment are dumped close to the continental margin. Where deltaic deposition has continued for tens of millions of years, deltas have extended onto the deep ocean floor. Huge petroleum resources in the Mississippi, Niger, Orinoco, and other deltas are currently being developed, and related understanding has contributed to the way in which deltas are perceived from the complementary resource, environment, and hazard viewpoints.

Estuaries, in contrast to deltas, are broad embayed river mouths that have been flooded since the end of the most recent glaciation when sea level rose and the river's sediment load diminished. Water circulation in estuaries is much more restricted than in the open sea, and, as a result, sensitivity to environmental modification is very high. For this reason estuaries are important targets for interdisciplinary research in biogeochemical dynamics.

Beneath the Sea

The large-scale structure of the ocean basins has been established by the operation of the Earth's internal heat engine, which causes rupture and drift of continents and island arcs, formation of new

ocean floor at spreading centers, and establishment of new arc systems where plates converge. The operation of these processes leads inexorably to the opening and closing of oceans, to island-arc and continental collisions, to the assembly of continents, to the addition of new arc material to existing continents, and to recycling of both ocean-floor rocks and continental material into the mantle. Solar heat modifies the ocean floor mainly by deposition of detrital sediment eroded from the land and by precipitation of calcium carbonate and silica from oceanic waters—partly by marine organisms—to form limestone and chert. Together, these processes degrade the thermally generated submarine relief, not so much by erosion, the process dominant above sea level, as by deposition that smooths the topography.

Deposition at Atlantic-Type Margins

Sedimentary deposition below sea level is controlled primarily by the tectonic framework of the ocean basins. The largest volumes of sediments accumulate at the rifted continental margins, called Atlantic-type margins because they are best developed around the Atlantic Ocean (Figure 3.5). The most rapid additions to these types of margins at present come from rivers draining the world's largest areas of high elevation—from the Mississippi and Mackenzie rivers that drain western North

America and from the Ganges and Indus rivers that drain the Himalaya and Tibet. Accumulations of rock as much as 15 km thick can develop along Atlantic-type margins. These, the greatest thicknesses of sediments in the world, are accommodated by subsidence as the lithosphere—thin and hot at continental rupture—cools and thickens. The load of accumulating sediments depresses the lithosphere farther and amplifies this subsidence.

Broad continental shelves extending for many tens of kilometers from the edge of the ocean at depths of only a few tens of meters below sea level are characteristic of Atlantic-type oceanic margins, because of deposition by powerful river systems and perhaps because of extensive lithospheric thinning at the time of continental rupture. Those off the coast of New England and the Alaskan coast of the Bering Sea are typical. The recent sea-level changes that mark responses to glaciation and deglaciation have led to repeated erosive episodes of the unconsolidated sediments of continental shelves. Large masses of such sediments have been off-loaded through submarine canyons onto deep-sea fans and abyssal plains.

Limestone shelves develop where there is little sediment eroded from the land and in areas of abundant biological activity. Sediments originate mainly from the calcareous skeletons of shallow-water, bottom-dwelling marine organisms. These limestone shelves respond to sea-level change in a way very different from sand and mud shelves. When sea level falls, exposure to fresh water as rain or runoff produces cementation that binds the loose skeletal sediment. As a result, carbonate-dominated banks and shelves reflect conditions of deposition during high stands of the sea and of subsequent cementation during low stands. Much of Florida and all of the Bahama banks were produced by this process.

Submarine canyons carry sediments from the continental shelves to the deep oceans and the submarine fans, where that sediment settles. When they were first recognized about 50 years ago, the prime question asked was how such enormous features could form. Computation of the huge volume of sediments in the fans showed that turbid sediment-laden flows pouring from the continental shelves in times of glacially controlled low stands of the sea could have readily carved even the greatest of submarine canyons, many of which are much larger than the Grand Canyon. Modern research on both submarine fans and canyons is accelerating because of the availability of new instrumental capabilities. Deep-sea drilling, multibeam echo

sounders, side-look scanning sonars, and manned and remotely controlled submersibles are providing a much more detailed picture than was formerly obtainable.

Research on the sedimentary development of the Atlantic-type margins has expanded enormously during the past decade, largely in response to two stimuli: an appreciation, following the plate tectonic revolution, of how continental rupture happens and an understanding of how the sediment wedge at the continental margin evolves through time. The latter owes much to oil exploration, which led to the development of the technique of sequence stratigraphy, where coherent packages of distinctive strata in reflection seismic data—calibrated against the record of local oil wells—can be used to establish a detailed history of the transgression and regression of the sea. Lively controversy persists as to exactly how and whether the seismic stratigraphic records can be linked to global sea-level fluctuations.

Deposition at Convergent Plate Boundaries

The greatest variations in topographic relief are produced at convergent plate boundaries. The giant peaks of the Himalaya, 8 km high, result from the collision of India with Asia, and the 11-km extreme of oceanic depth is found in the Marianas Trench, where subduction is carrying the Pacific plate into the mantle. Two dominant sedimentary depositional environments are associated with these immense topographic contrasts: trenches and foreland basins.

Subduction-zone trenches contain substantial sediment accumulations only where the supply of sediments is large enough to exceed the rate of removal by subduction (Figure 3.6). This type of accumulation is occurring, for example, at the eastern end of the Aleutian Trench close to sediment sources in the North American continent and at the southern end of the Lesser Antilles Trench close to the South American continent at the mouth of the Orinoco River. Sediments accumulate at the front of the related arc system to form a thick wedge, or accretionary prism, that extends along the trench. An exciting challenge—currently being addressed by deep-sea drilling, multibeam echo sounding, and other new techniques—is to establish exactly how unconsolidated sediment entering a deep-sea trench becomes solid rock in the accretionary prism, a process that involves both intense deformation and the expulsion of vast quantities of water.

The other sites of substantial sediment accumulation in convergent plate boundary zones are the

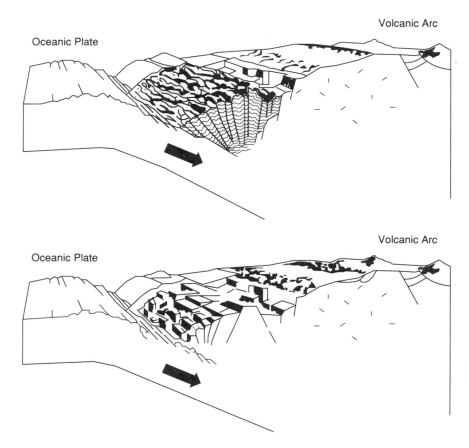

FIGURE 3.6 Accretionary (top) and erosional (bottom) end-members of convergent margins. Accretionary prisms form only where sufficient sediment is supplied to a trench.

foreland basins that develop next to continental margin mountain belts in areas where the load of the mountains depresses the lithosphere profoundly. Sedimentary thicknesses in the foreland basin depressions are huge and may rival those of the Atlantic-type margins, especially in places where the mountains are being actively thrust over the basin. At present, many of the best-developed foreland basins are accumulating sediments above sea level. This is the case in the finest examples of all, the foreland basins lying to the east of the 5,000-km-long Andean chain, which has been rising for the past 3 million years.

During most of the geological past, when sea level was higher, foreland basins accumulated marine sediments. Recent research in foreland basins emphasizes an integrated approach that models how episodes of uplift in the mountain belts modify sediment supply and interact with sea-level changes and with thrusting of the mountain load over the basin. The methods of sequence stratigraphy address these problems. Further pursuits in oil exploration have led to studies of how fluids migrate through the foreland basins for distances of up to hundreds of kilometers.

Deposition in the Deep Ocean

Very little of the material eroded from the continents reaches the central areas of the oceans, and much of what does is in the form of windblown dust. In these remote regions, far from the continents, accumulation of the skeletons of microorganisms that live in the oceanic waters dominates the depositional process. While calcareous skeletons are most important at shallower depths, they dissolve in the deepest and coldest waters faster than they can accumulate. Beneath the deepest waters siliceous skeletons form a significant part of the sediment pile.

The calcareous sediments located around the ocean's abyssal plains are proving to contain an astonishingly informative record of the history of the water masses of the world ocean. The oxygen-isotopic compositions of the skeletons of foraminifera, which make up most of the calcareous oozes, reflect the isotopic composition of the water in which they lived. The ratios between different isotopes reflect the size of the world's ice sheets and the temperature and salinity of the water in which the organisms grew. Cores from the deep seafloor show

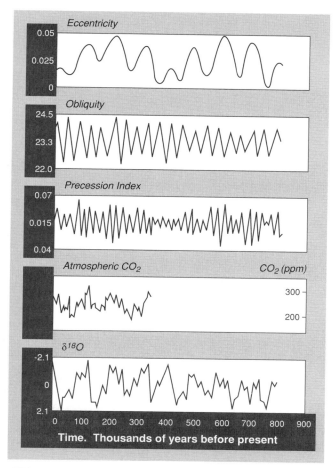

FIGURE 3.7 Variations in insolation caused by changes in the Earth's orbital parameters (eccentricity, obliquity, precession) can be correlated with variations in fossil atmospheric CO_2 contents (from ice sheets) and global ice volume (as determined from $\delta^{18}O$ of ocean sediments).

isotopic variations that record the history of the ice ages and yield persuasive evidence of relations between ice ages and the Earth's orbital parameters (Figure 3.7).

The deep-sea sedimentary record reveals long-term alterations in oceanic circulation that reflect changes in the way the Sun's energy has been dissipated in the oceanic waters. Cycles of change, secular variations, and catastrophic perturbations of the system can all be discerned in this remarkable record.

CHANGE IN THE GLOBAL ENVIRONMENT

Interest in global changes focuses especially on time scales of decades to centuries and in attempts to characterize human-induced change and to discriminate between that change and natural variability.

The geological record provides a unique perspective on change because it extends back for more than 3.5 billion years. This is the extent of the cumulative record built up from shorter individual sequences of rock, each of which rarely spans more than a few tens of millions of years. The overall record is integrated from numerous and disparate sources by correlation. In ancient rocks the precision of correlation is poor, with deviations exceeding a million years, but in younger rocks it is more reliable. Because of new dating techniques, all anchored to time scales based on the decay of natural radioactive isotopes, better temporal resolution in both young and old rocks can now be attained.

The most recent past preserves the best record, so study of the youngest parts of earth history is intense. But high-resolution evidence of annual and more frequent events has been preserved in very old rocks. There are annual layers in ancient lake beds dating to more than 2 billion years, and 14-day tidal cycles have been discerned in rocks 700 million years old. But even global correlation cannot patch together a complete record spanning the more than 4 billion years of active geological processes. Many of the processes acting in the system are as likely to destroy the record as to preserve it. Comparisons of sediment accumulation rates in modern environments with the isotopically dated rates for ancient intervals in similar environments have always shown that the preserved sediments could have been deposited in a fraction of the measured interval. The apparent rates are much lower than modern ones because the duration represented by a preserved section includes times of nondeposition and erosion as well as deposition.

Herein lies a unique challenge of the earth sciences—the lack of conspicuous evidence can be a clue, as well as the less subtle direct testimony. When expected evidence is missing from the record, that absence poses new questions and suggests new mysteries about the geological processes at work.

Cyclical Change in the Global Environment

One way of looking at the geological record is to consider it as preserving evidence of cyclical processes. The rise and fall of sea level take place on time scales ranging from days to tens of millions of years; superimposed on these cyclical changes are noncyclic or secular changes, such as biological evolution, that have been occurring throughout earth history. The way in which the cyclical processes operate has been modified to some extent by secular changes, but it is one of the more exciting recent developments in the study of earth history that so many of the cyclical processes have changed very little since the formation of the earliest preserved rocks.

A cyclical change can be chaotic but fluctuates between two or more distinct extremes. Time elapses during all cyclical processes, so when the cycle is complete conditions can never be identical to those at the start, but the repetition of similar phenomena is usually considered a cycle. The rock record contains evidence of cycles operating on daily scales, such as tidally controlled sedimentation; annual scales, such as varved lake deposits; and tens to hundreds of thousands of years, such as cycles controlled by variations in the Earth's orbit around the Sun. Sea levels change over millions of years due to variations in the average age of the ocean floor; ocean basins open and close over tens to hundreds of millions of years; and the isolation of major compositional reservoirs in the mantle may be the result of cycles that take billions of years.

These cycles involve transfers of energy and material among reservoirs of different sizes. Materials can be concentrated and even isolated in individual reservoirs for varied intervals. Usually the longer the interval the larger the reservoir. The largest parts of the Earth constitute the largest reservoirs: the core, mantle, crust, asthenosphere, lithosphere, hydrosphere, atmosphere, and biosphere.

The idea of cycles in earth activity is at least 200 years old. However, the strength of the cyclical concept has been fully realized only since we have come to appreciate the Earth as a complex system consisting of a multitude of interacting subsystems. Its recently achieved strength stems from new analytical and observational capabilities that allow characterization of the discrete reservoirs within the earth system in a way that was not formerly possible. Modern data-handling and especially computational capabilities are essential for the interpretations now being made.

The Rock Cycle

The processes operating at the surface constitute part of an enormous cycle. Mountains form and material is eroded from them, which is eventually deposited in basins. The sediments in the basins become caught up in new mountain building and are then eroded themselves to repeat the cycle. This is the rock cycle, and it involves interaction among all the outer parts of the Earth—atmosphere, ocean, biosphere, crust, and upper mantle.

Arbitrarily, we can start the rock cycle at the surface with weathering: liberation of a chemical element to solution and its inclusion within secondary minerals. During weathering some elements are also exchanged with the atmosphere, a process involving the active participation of plants and soil biota. The liberated element is carried in solution, together with eroded particles traveling in suspension, by streams that eventually empty into the ocean. Here it is deposited on the seafloor, either as eroded and transported particles or as part of a newly formed precipitate from seawater, which is often biological in origin, such as a calcium carbonate seashell.

Materials accumulate on the seafloor as sediments. Upon burial these become sedimentary rocks, which may themselves become involved in mountain-building processes. They may be transformed into metamorphic rocks by increased temperature and pressure or into igneous rocks by partial melting. Igneous material erupts onto the surface of the continents or onto the seafloor to form volcanic rocks, which can react with seawater to change the chemical composition of both. Other igneous material may crystallize slowly at depth to form plutonic rocks. During the formation of igneous and metamorphic rocks, volcanic gases migrate upward to the surface to enter the atmosphere. Sedimentary, metamorphic, and igneous rocks are uplifted by tectonism into the zone of weathering to begin the cycle anew. The process entails weathering, erosion, river transport, deposition, burial, metamorphism, melting, volcanism, degassing, uplift, and weathering again—a rock cycle that has been repeated many times during the history of the Earth.

Plate tectonics and hydrologic processes combine to drive the rock cycle at rates that vary over a wide range. This results in complex interconnections between very different earth processes. For example, at various times in the geological past—millions to billions of years ago—changes in the level of atmospheric carbon dioxide (CO_2) may have resulted from increases in the rate of degassing of CO_2 by volcanic activity and from decreases in the uptake of CO_2 by rock weathering. An increase in atmospheric CO_2 would have tended to elevate global temperatures through the greenhouse effect, which in turn might have influenced the evolution of life. There is some evidence that about 100 million years ago just this situation developed. As interest grows in long-term changes in atmospheric CO_2 content, a parallel research effort is focusing on short-term cycles on the order of tens of thousands of years. These changes are well correlated with variations in the Earth's orbit around the Sun.

Orbital Cycles

It was recognized in the nineteenth century that variations in the Earth's orbit would cause changes in incoming solar radiation that could be important in controlling ice ages. Theoreticians first calculated how these variations would interact, and the study of deep-sea sediments has yielded persuasive evidence that the recurring ice ages of the past million years are indeed closely associated with orbital cycles. These cycles cause subtle changes, particularly in high latitudes, in the seasonal variation of the incoming solar radiation, called insolation, and may be reliably calculated from celestial mechanics. The ice ages themselves are recorded in the ratio of oxygen isotopes in deep-sea sediments. This is because the elevated fraction of light isotopes in fresh water evaporated from the ocean surface and stored in ice sheets is reflected by an increased fraction of heavy isotopes in the precipitated carbonate skeletons of microorganisms living in the remaining ocean water. Variations in the oxygen isotope ($^{18}O/^{16}O$) ratio with depth in a sediment core are widely interpreted as indicating total land-ice volume as a function of time.

The glacial record, as revealed in ice cores, sedimentary sequences, landforms, and other related phenomena, is especially useful for understanding past changes and anticipating the characteristics of future changes. In some instances, fossil pollen and other specific environmental indicators are also present in stratigraphic records.

As better cores are examined and dating procedures are refined, it has become apparent that the changes in insolation correlate with subsequent changes in ice volume. However, a correspondence between the two records requires allowance for the slow buildup of ice sheets over several tens of thousands of years in contrast with their relatively rapid decay, which introduces a degree of nonlinearity into the system response. Presumably the periodic changes in insolation are the ultimate cause, but the precise mechanism remains obscure. The presence of ice, however, does not appear essential for a cyclical response. Recently compiled geological records of 200-million-year-old lake sediments in the eastern United States show a sequence of cycles of approximately the same intervals as the present orbital cycles, spanning a period of 40 million years. These lakes were then in the tropics. No evidence of continental glaciers or sea ice exists

for this period, but local climate and lake levels were apparently influenced by a strong stable external control.

The relationship of the orbital cycles to climatic variation is a fertile research field. The deep-sea sediment record indicates that about 900,000 years ago the governing periodicity of cycles switched from 40,000 years to 100,000 years. This sudden change has not yet been explained. High-resolution records of the glacial cycles come from fast-sedimentation-rate deep-sea cores, from cores in the Greenland and Antarctic ice sheets, and from cores of mountain glaciers at low and mid-latitudes. Cores from the Vostok station, high on the Antarctic ice cap, have extended the record back to about 160,000 years ago, so a remarkably complete record is now available of how temperature varied through the whole of the last glacial cycle. Analyses of air bubbles in ice cores show that temperature and atmospheric CO_2 content generally varied sympathetically. Geologists are currently investigating whether orbital variations drove the system and whether changes in oceanic circulation and biology affected the atmosphere's trace gas content, amplifying the climatic oscillations.

Abrupt changes in environmental conditions are recorded in the Greenland record, where the Dye-3 core indicates a switch from glacial to interglacial conditions within one century. Some researchers suggest that such sudden changes during the last glacial period could have been triggered by major diversions of meltwater draining from the Laurentide ice sheet. In a more recent time frame, dust and oxygen isotopic records associated with an ice record from the Peruvian Andes indicate that local transition from the Little Ice Age to current conditions could have occurred about 100 years ago and in as short a time span as a few years.

These examples show that large climatic changes can occur on many time scales, including those of critical relevance to modern society. Climate change may be initiated by variations in atmospheric and oceanic circulation patterns driven by feedback connections to other terrestrial environmental factors, such as changes in vegetation cover or in physical composition of the atmosphere influenced by volcanic or human activity. Therefore, the geomorphic history, geographic distribution, and rates of glacial advances and retreats need to be documented to permit understanding of the interconnected global associations of environmental change and to seek causal connections. These data can provide very important independent tests of the atmospheric general circulation models (GCMs) that are used to estimate consequences of future environmental change.

The most recent million years of earth history offer exceptional opportunities to reconstruct global environments. For the latter part of this interval, geological and human time scales overlap, and under certain circumstances radiocarbon dating can be pushed back to several tens of thousands of years before the present. Radiocarbon dating applies directly to fossil material, providing remarkably accurate dates for certain biotic events of the recent geological past. The vast majority of species that have lived during this interval are alive today, so that detailed knowledge of their ecological traits in the modern world can be used to reconstruct ancient environments. Such details help to improve the resolution of research on environmental change.

Deep-sea cores provide an extensive record for this interval, with radiocarbon dates scaling a chronology for the most recent few tens of thousands of years. Studies of tree rings, pollen sequences in lake and estuarine sediments, fossils of terrestrial insects, and cave deposits supply even greater detail, offering high resolution for events that have occurred within the past 10,000 years. For the most recent past, ice cores from arctic and alpine glaciers exhibit yearly bands that allow events to be traced back from the present with a level of resolution that approximates 1 year. The record of this historic and barely prehistoric past reveals global events that produce very rapid change, not only when scaled against traditional geological chronologies but even when scaled against a human lifetime.

Remote sensing, the set of processes by which we observe large areas of the Earth from outer space, also provides valuable information for assessing environmental changes of the recent past. Combinations of different perspectives and frequent coverage inform us about rates of tectonism, volcanism, and other processes that have altered landforms. The images display the effects of past and ongoing climatic changes. For example, radar and laser altimeter data indicate the degree to which alluvial fans in arid regions have weathered and become mantled with windblown sediments, and visible, near-infrared, and thermal infrared imaging spectrometers reveal the degree to which exposed rocks and sediments have weathered to become clay minerals. Remote sensing also detects recent migration of sand dunes and changes in the levels of ancient lakes, both of which testify to climatic change. Satellite technology helps in distinguishing how much current deforestation, beach erosion, and desertification is a product of human activities and how much the result of other natural causes.

Cycles of Sea-Level Change

The level of the world's oceans has oscillated up and down over a range of more than 300 m in the course of earth history (for at least the past 100 million years or so). Because continents generally slope gradually toward their margins, high stands of sea level sometimes have spread shallow marine waters rapidly over broad continental areas. In fact, for most of the time during the past 600 million years, sea level stood higher than its present position, so that continents were more extensively inundated. For this reason, large volumes of ancient marine sediments and sedimentary rocks are currently exposed on modern continents.

Several factors control sea level. The growth or shrinkage of ice sheets cause seas to rise or fall as much as 100 to 150 m at rates that might have exceeded 10 m per 1,000 years. The desiccation of isolated ocean basins may generate sea-level fluctuations at the same rate, but only within a range of about 15 m. Spreading rates of mid-ocean ridges produce much larger changes (in the range of 300 m) but more slowly—at rates of only 1 cm per 1,000 years or less.

Even changes in sea level of only a few meters leave their imprint in the rock record. In tropical areas of limestone deposition, for example, calcareous algae and other sediment producers tend to maintain the floors of shallow marine carbonate platforms within a few meters of the sea surface. A small lowering of sea level exposing the platform will terminate sediment deposition, subjecting these deposits to subaerial weathering that will be recognizable hundreds of millions of years later. Conversely, a rapid rise in sea level by several meters can virtually halt shallow-water carbonate production because such a sea-level rise will exceed the rate at which carbonate-producing algae can produce sediment. Thus, minor fluctuations in sea level can leave imprints in the stratigraphic record. In ancient rocks deposited in lakes and on carbonate platforms, geologists are now recognizing cycles that seem to match distinctive orbital periodicity. It may be appropriate to invoke climatic forcing from orbital variations, even without evidence of glacial activity.

Deciphering sea-level change in the rock record is currently one of the most active areas of geological research. Here a primary new tool is sequence stratigraphy, which uses seismic reflection data to study the spatial relationships and, indirectly, the temporal relationships of sedimentary strata lying below the surface. Petroleum geologists are the pioneers in this field and use seismic data obtained from modern continental margins to interpret the history of global sea-level change. The focus has been on the Atlantic-type margins, which form when continents rift apart. Once ocean spreading has carried a continental margin far from the uplifted center, tectonic subsidence becomes very slow. The structurally quiescent margin serves as a yardstick against which global sea-level change can be measured.

Sequence stratigraphy makes use of the discovery that buried surfaces, appearing as lines in reflection profiles, have been formed simultaneously. Many appear to mark brief interruptions in deposition when compaction or lithification took place. The configuration of these surfaces, which can represent both subaerial and submarine topography, depicts the fluctuation of ancient shorelines through time. When adjustments are made for subsequent compaction and subsidence, positions of ancient shorelines reveal past positions of sea level. The chronology is established primarily by the study of microfossils recovered from well cores drilled in the area of the seismic data.

Because modern ocean basins and continental margins are no older than about 200 million years, sea-level analysis for older rocks using sequence stratigraphy focuses on rocks exposed on continents. Here other methods help. Deposits whose sedimentary characteristics and fossils represent very shallow marine conditions provide indications of past sea levels, as do unconformities that represent intervals of subaerial exposure.

Some indices of ancient sea levels in the rock record are ambiguous, so that the sea-level curve for the past 600 million years is under constant revision. The configuration of this curve has great significance, in part because the total area occupied by shallow seas has a controlling influence over certain climatic conditions. Shallow epicontinental seas supply moisture to surrounding terrestrial regions. They also absorb sunlight and conserve heat because they do not mix extensively with cool deep waters and because the heat capacity of water is so high that they do not lose heat rapidly to the atmosphere. Because of their thermal stability, shallow seas—like large lakes—moderate temperatures in nearby terrestrial regions.

Cycles of Opening and Closing of Oceans

Once earth scientists became aware of the plate structure of the lithosphere—the outer rigid layer of the Earth—it was a short step to recognizing that some of the world's oceans are opening and others are closing. Opening oceans, such as the Atlantic, have active spreading centers but lack major convergent plate boundaries. Closing oceans, such as the Pacific and the Mediterranean, are bounded largely by convergent plate boundaries. Researchers concluded that perhaps earth history operates within a framework of complex cycles of opening and closing ocean basins. The evidence for these ocean cycles is found in the sediments deposited in the oceans and along the ocean's margins. The depositional evidence of ocean development depends on the solar-atmosphere-ocean system and its associated latitudinal patterns of temperature and rainfall, which profoundly affect erosion and its consequences.

A continent drifting latitudinally records its transit in the character of its sediments. Use of these paleolatitudinal indicators—as well as magnetic evidence that can track paleolongitudinal drift—has allowed the construction of maps (Figure 3.8) that show the distribution of continents and oceans in the past. The record farther back than about 500 million years ago is too patchy at present to justify the construction of world maps, but evidence from the better-known areas, such as North America, is consistent with the idea that similar ocean-opening and -closing processes were operating in the remote past.

Geochemical Cycles

Ingenious sampling methods often assist in estimating the distribution of individual elements, compounds, and nuclides among the various terrestrial reservoirs. Patterns are readily discernible: iron and nickel are concentrated in the core and volatile elements in the atmosphere and ocean. Many of the distribution patterns appear to be basically simple. But the study of how concentrations within discrete reservoirs change with time promises valuable information about how the Earth's chemical systems behave.

Isotopes of the same element, particularly radioactive isotopes, that travel by different paths through the reservoirs to a common end, can be especially informative about geochemical cycling. The change in the strontium isotopic ratio ($^{87}Sr/^{86}Sr$) of seawater as recorded in marine shells is an excellent example.

^{87}Sr is a stable nuclide continuously formed by the decay of radioactive rubidium (^{87}Rb). Like the other alkali metals, rubidium has become concentrated in

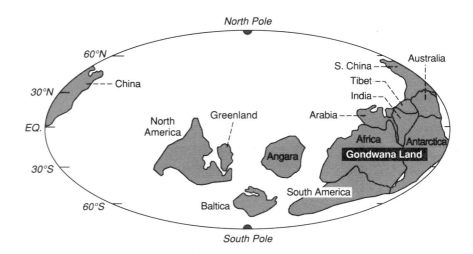

FIGURE 3.8 Map showing how the continental blocks were distributed on the Earth's surface about 530 million years ago. The ability to construct high-resolution maps of this kind allows integration of geological information and the construction of models embodying a range of paleoenvironmental data. Numerous, now widely dispersed continental blocks were assembled in the great continent of Gondwanaland at this time.

the continents through the repeated partial melting processes that fractionate the mantle. ^{86}Sr is a stable nuclide that remains more strongly linked to the mantle. The $^{87}Sr/^{86}Sr$ ratio of seawater has varied with time (Figure 3.9) over the past 500 million years. The variation can be attributed to changes in the relative influence of erosion from the continents, which promotes concentration of the rubidium daughter ^{87}Sr in seawater, and of volcanism beneath the sea, which samples the mantle and promotes the concentration of ^{86}Sr. With the fall of sea level over the past 100 million years, the continental contribution has generally been rising. A serendipitous application of the change in $^{87}Sr/^{86}Sr$ of marine shells is that stratigraphers are using the ratio to date sedimentary rocks.

A more familiar example involves the cycling of carbon and oxygen. The concentration of CO_2 in the atmospheric reservoir has risen rapidly in recent decades. These concentrations are usually in chemical equilibrium with dissolved CO_2 and bicarbonate ions in the ocean waters and with calcium carbonate in the oceanic sediment reservoir. The cyclical transfer of CO_2 through these three reservoirs appears to be considerably perturbed by the rapid rise in the atmospheric component; model simulations indicate that it could take hundreds of years to restore equilibrium to this subsystem.

The geological record offers information about a past scenario that involved disequilibrium among these same three reservoirs. About 100 million years ago there was about twice as much underwater volcanism as there is at present because seafloor spreading was more rapid and the total volume of the oceanic ridges was about twice what it is now. This condition tended to perturb the CO_2 cycle in two ways. The extra volcanism added CO_2 to the ocean and the extra volume of young hot rock on the ocean floor displaced the oceanic waters so that they flooded the continents to an exceptional extent, thus reducing the area of rock available for the weathering that extracts CO_2 from the atmosphere. There is a strong likelihood that the CO_2 content of the atmospheric reservoir rose in response to these perturbations. The evidence indicates that climatic conditions were warmer than today, as the greenhouse principle would suggest.

A comforting general observation is that feedback mechanisms will come into play to ameliorate any extreme consequences of perturbations to cyclical processes. The shallow waters of the flooded continents (some 100 million years ago) were an ideal environment for the deposition of limestone, and the process of limestone deposition pulls CO_2 out of the atmosphere and processes it through the oceanic waters into the rock reservoir. This feedback system would have brought the cycle back to a more normal state.

The geochemistry of carbon is uniquely exciting, primarily because of carbon's role in life. Carbon also forms economically important resources, in-

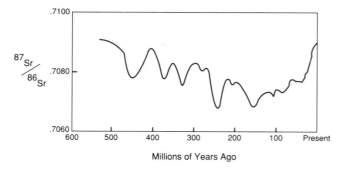

FIGURE 3.9 Variations in strontium isotopes of seawater.

cluding oil, coal, and diamonds. And ^{14}C, a radio-active isotope with a half-life of 5,000 years, is the most important means for timing environmental events over the past 40,000 years. The existence of two readily fractionated stable isotopes and a single short-lived radioactive isotope, along with the preservation of carbon from a wide range of environments throughout the geological record, means that interpretation of the geochemical cycles of carbon is particularly informative. The resulting understanding of the rates of past changes allows researchers to assess ongoing changes.

The carbon cycle has played a major role in the development of the global environment. In any body of water, dead organic matter settles to the bottom where animals and bacteria have a chance to oxidize the contained carbon. But mud and the minerals produced by organisms also settle to the bottom. Their accumulation may be rapid enough to trap and bury organic material before it can be consumed by oxidation, sometimes preserving it to become fossil fuel. The abundance of fossil fuels and other organic debris in the sedimentary shell is considerable, and every atom of that organic carbon, as it was buried, left behind a molecule of O_2 that was released into the surface environment.

Balancing the accounts should be possible. From the inventory of elements in sedimentary rocks, we should be able to calculate the amount of oxidizing power that the buried accumulation of organic carbon left behind at the surface—and also the timing, or history, of the accumulation. There are two problems. We cannot collect samples of all buried rocks for analysis of carbon content, and even if we could take a perfect inventory of all existing sedimentary rocks we could not account for rocks destroyed by erosion, altered by metamorphism, or subducted into the mantle.

There is another way to approach the accounting. The problem can be restated usefully by asking: What fraction of the carbon passing through the system has been buried in the form of organic material? This turns out to be a question that we can answer with the help of the two stable carbon isotopes, ^{12}C and ^{13}C. The ^{12}C isotope is more abundant, amounting to 98.895 percent of all terrestrial carbon; most of the remainder is the ^{13}C isotope. Because both isotopes are stable, their abundances have not changed throughout earth history. At any time, the isotopic composition of the carbon entering the surface part of the system—the atmosphere, biosphere, and hydrosphere—is given by the terrestrial average, but the two processes of biomass synthesis and carbonate precipita-

tion tend to slightly separate the carbon isotopes. At present, for example, carbonate forming in the ocean contains 1.113 percent ^{13}C, and, on average, organic material being buried in sediments contains 1.086 percent ^{13}C.

Measuring isotopic abundances at that level of precision is not simple, but it is incomparably easier than constructing a global inventory of carbonates and organic material, and it provides a way to monitor the behavior of the carbon cycle. By calculating the abundances of buried organic and inorganic carbon from the total carbon and ^{13}C mass balances, indications are that at present about 30 percent of the carbon passing through the hydrosphere, atmosphere, and biosphere is being buried. Characteristics of ancient carbon cycles can be similarly determined. For every interval it is necessary only to obtain globally representative carbon isotopic abundances for carbonate sediments and organic carbon.

Paleoceanography: Cycles in the History of Oceanic Waters

Recognition of changes in variables such as the chemistry of the oceans, the global sea level, the configuration of ocean basins, the three-dimensional thermal structure of the ocean, and the history of marine organisms permits the description of ancient conditions, which even during the past 18,000 years have undergone remarkable transformations. On a broader scale of time, changes have been even more profound.

About 70 million years ago, shortly before mammals inherited the Earth from dinosaurs, the oceans supported a huge population of calcareous nannoplankton. They were so abundant that their minute skeletal remains rained down on the seafloor to produce thick deposits of chalk that stand now as the White Cliffs of Dover in England and the cliffs of the Selma Chalk in Alabama. Today, photosynthesizing calcareous nannoplankton survive as very important producers in the marine food chain but have never again generated such widespread deposits of chalk; they suffered severe losses at the same time the dinosaurs met their end. Probably part of the explanation is that they never rediversified fully because other taxa took their place. Certainly another important factor is that relatively cool climatic regimes, which do not favor calcareous nannoplankton, have prevailed during the past 60 million years. On the other hand, diatoms—silica-precipitating organisms that thrive in cold water—expanded greatly during that time. The deep sea, which today is close to freezing, was warm between 100 million and 70 million years ago. This is inferred from the isotopic composition of foraminifera that then lived on the deep seafloor. Paleoceanographic research suggests mechanisms that may have caused the refrigeration of the deep sea since that time. Geologists are investigating the effects of those thermal changes as well as related aspects of ocean evolution on time scales that range from thousands to billions of years.

Although the history of seawater is an important subject in its own right, it also serves as an indicator of processes that have shaped the outer parts of the Earth through time. Limits for the composition of ancient oceans are determined from the mineralogy and chemistry of marine evaporites, the sediments formed by the evaporation of seawater, but even these indicators leave a wide range of uncertainty. The most useful technique currently available to define the major compositional variation of seawater over the past 600 million years requires extraction and analysis of brines trapped in the rock salt found within marine evaporite deposits. In many instances these brines appear to have suffered little, if any, alteration. Their composition is not that of seawater, but the mass compositional parameters of the parent seawater can be reconstructed from the brine chemistry by correcting for the effects of evaporation and for the precipitation of limestone, gypsum, and rock salt. The results of the analysis of more than 100 inclusion fluids from marine evaporites covering the past 550 million years of earth history suggest that the chemical composition of seawater has not changed greatly. This observation has come as something of a surprise, because the isotopic compositions of sulfur, strontium, and carbon in seawater have varied significantly. During the next few years the chemical evolution of seawater should be defined much more precisely, and we anticipate gaining a clearer understanding of the mechanisms that have controlled the composition of seawater.

We know less about patterns of circulation for modern oceans than about those of the modern atmosphere because of the logistical difficulty of gathering oceanographic data. This deficiency limits the accuracy of paleoceanographic modeling. However, conditions within ancient oceans can be reconstructed by using patterns of modern oceanic circulation to reassemble the thermal structure and dominant currents in ancient oceans and by selecting especially important physical, chemical, and biological indicators in the geological record to plot distributions. Nowhere has this approach been un-

dertaken more effectively than in the Climate: Long-Range Investigation, Mapping, and Prediction (CLIMAP) project and its successors, broad international initiatives inaugurated in 1971 to recreate the ice age world of the past million years.

Although CLIMAP's broad goal was to investigate global climates for the past million years, its crowning achievement was the production of a climatic map of the world as it existed 18,000 years ago. This was the time of the most recent glacial maximum. In the overall strategy of CLIMAP, the most important element was reconstruction of sea-surface temperatures for the time frame of 18,000 years ago. Fossil occurrences of living marine species were used to chart the geographic distribution of ancient temperatures.

The most general conclusion drawn from the CLIMAP model was that 18,000 years ago the average sea-surface temperature was 2.3°C cooler than it is today. The high spatial resolution of the analysis permitted many more specific results. The equatorial Atlantic and Pacific oceans did not cool as much. Waters near the sea surface were generally cooler and more mixed than they are today, with a less pronounced thermal contrast between surface and deep waters. Ice floes extended to much lower latitudes in the North Atlantic—the Gulf Stream flowed eastward toward Spain, not Great Britain. And in the North Pacific, radiolarian species that today are restricted to cool waters from northern California to Washington ranged at least 1,000 km farther south. From other evidence we know that glacial expansion took place primarily in the north, with ice caps centered in Hudson Bay, Greenland, and Scandinavia, but marked climatic changes occurred in the Southern Hemisphere as well.

Reconstructing ocean temperatures and current patterns for earlier times is more difficult. Nevertheless, certain striking oceanographic changes that occurred tens and even hundreds of millions of years ago are clear. Deep-sea conditions changed drastically over geological time in response to profound global changes in shallow marine thermal regimes and in terrestrial climates. Today, throughout the globe the deep sea remains only slightly above freezing because its waters are derived from polar regions. At those high latitudes, surface currents cool so severely that they become much more dense than the underlying water. The chilled water sinks to the bottom and spreads along the deep seafloor to equatorial latitudes, forming a cold basal layer in all the oceans.

Fifty-five million years ago, many regions of the Earth were much warmer than they are today. At that time, unlikely as it may seem, southeastern England was cloaked by tropical jungles like those of modern Malaysia. Fossils of deep-sea ostracodes, minute crustaceans that are distant cousins of crabs and lobsters, reveal that a major change took place in the deep sea about 40 million years ago. The types of ostracodes that occupy the oceanic abyssal plain today, having adapted to frigid conditions, made their first appearance at that time. Oxygen isotopes in foraminifera confirm the observation of this trend toward frigidity.

Before that cooling began, the deep sea may have reached temperatures as warm as 15°C. It did not cool to its modern temperature immediately, of course, but gradually lost heat until 30 million years ago when the cold basal currents of the modern oceans became firmly entrenched.

Plate tectonics offers a possible explanation for cooling in the Southern Hemisphere 40 million years ago. In the vicinity of Antarctica, microfossils preserved in deep-sea sediments testify to a drastic change in thermal conditions. Millions of years earlier the supercontinent of Pangea had rifted apart to form many of the fragments that constitute the continents of the modern world. South America and Australia remained attached to what is now Antarctica, which was positioned on the South Pole close to its present location. While these connections remained, cool water was deflected equatorward and warm water poleward along the coasts. Microfossil and other data indicate that about 40 million years ago South America began to drift away from Antarctica, allowing a continuous current to flow around Antarctica—the Circumantarctic Current. This current isolated the continent thermally, and the change in circulation marked the origin of the refrigeration system for the deep sea that operates in this region today, trapping water and allowing it to cool and sink. About the same time this refrigeration system was supplemented by another. The Arctic Ocean became connected to the Atlantic over the Iceland sill, allowing cold Arctic surface waters to descend into the deep sea.

Throughout earth history a cold basal layer of ocean water must have formed each time at least one of the poles became frigid. Fossil data verify the occurrence of such an event 450 million years ago, when the supercontinent of Gondwana encroached on the South Pole and accumulated massive ice sheets that left extensive glacial deposits in what is now the Sahara Desert in Africa. Careful stratigraphic research into the period has shown that brachiopods and other creatures that had colonized the seafloor at cooler high latitudes progressively shifted into deep-water habitats at all latitudes, apparently tracking the movement of cool waters into the deep sea.

Even after the origin of the modern cold basal layer and before the start of the modern ice age, the oceans experienced major thermal changes. Substantial alterations occurred between about 22 million and 5 million years ago. Important clues have come from carbon isotopes in fossil foraminifera. Gradients of $^{13}C/^{12}C$ ratio, detected by the study of fossil foraminifera, reveal that prior to about 14 million years ago water flowed from the Mediterranean Sea into the Indian Ocean and southward toward Antarctica. It traveled at intermediate depths, apparently having sunk below the surface because it was more saline than normal seawater. The intense salinity resulted from a high evaporation rate in the Mediterranean region. This water and others that it entrained apparently joined the Circumantarctic Current at depth. The outflow of this saline plume ended about 14 million years ago, probably when collision of Arabia with Asia closed the eastern end of the Mediterranean. The cutoff of warm-water flow toward Antarctica may have resulted in the buildup of the West Antarctic ice cap, which has been documented to have occurred at this time period on the basis of other geological evidence. Apparently the tectonic movements that pinched off the flow from the Mediterranean had profound climatic repercussions in regions thousands of kilometers away.

Problems in distinguishing among the effects of glacial expansion, temperature change, and variation in salinity frustrate detailed investigations of paleoclimate that use the oxygen isotopes preserved within fossil skeletons. A partial remedy is now in sight, and it comes from an unexpected source. A particular family of calcareous phytoplankton includes several living species that produce lipids called alkenones. The degree of hydrogen saturation in these fatty compounds varies markedly with the temperature at the time of production. They retain their original chemical composition over millions of years, even after bacterial decay releases them and they end up in deep-sea sediment. Their changing patterns of chemical composition, as displayed in ice age cores, correlate closely with those of oxygen isotope ratios, but the alkenone composition can be scaled to approximate absolute temperature.

The analysis of fossil alkenones promises a large volume of ocean temperature data extending back tens of millions of years. One important controversy inviting resolution relates to the warm interval that preceded the origin of the modern cold basal ocean waters. Climatic modeling and analysis of oxygen isotopes cannot yet produce a consensus as to whether the tropics were also warmer at that time or whether they were cooler than today, generating gentler latitudinal temperature gradients. In the near future, fossil alkenones may yield a general temperature map for the 50-million-year-old global ocean.

At any time in earth history, deep ocean water is generated from the densest water masses that develop at shallower levels and have access to the major ocean basins. The high density of these waters may result from either low temperature or high salinity. So when deep waters are relatively warm, they probably are saline, after descending from marginal basins with high evaporation rates. When relatively warm waters occupy the deep ocean basins, thermal gradients are weak, deep-sea circulation is sluggish, and bottom waters become depleted of oxygen. In contrast, at times such as the present the dynamic descent of cool polar waters constantly replenishes the oxygen, which maintains biological respiration and the oxidation of organic and inorganic compounds. In addition, the spread of cold water masses into low latitudes scours out areas of the deep seafloor. Cores of deep-sea sediment display features that reflect these conditions. At some levels, cores contain a substantial amount of red oxidized sediment or evidence of depositional hiatuses. These particular anomalies represent intervals when cold currents from polar regions plowed through the deep seafloor, supplying oxygen or eroding sediments respectively.

A particularly interesting interval extended from about 110 million to 90 million years ago, when huge concentrations of hydrocarbons accumulated. The organic matter that finds its way into marine sediments, and is the source of most petroleum, derives ultimately from phytoplankton. As primary producers, phytoplankton utilize solar energy to synthesize inorganic carbon sources into organic material. Their biomass fuels the marine food web, in which energy required by other organisms is produced by the metabolic oxidation of organic material. If the food web operated with perfect efficiency, all organic material would be oxidized and the underlying sediments would contain no organic carbon. Where high amounts of organic carbon are found in sediments, the food web must have operated at low efficiency, allowing organic material to escape oxidation. There are two possible explanations for the vast reservoirs of organic carbon remaining in the 100-million-year-old deposits. Either the conditions of the water column or sediments were not favorable for the growth of efficient

recyclers, or the production rates were so high that the food web was unable to utilize all of the supply. Close study of the sediments from that interval favors the first alternative, although the second might have been important in places. Even though large amounts of organic carbon were preserved, production rates were low in comparison to present values.

Because the organisms most efficient at reoxidizing organic carbon require free oxygen, this combination of high preservation despite low production could characterize a deep ocean containing little or no dissolved oxygen. Sediments from this period show fine laminations that are completely undisturbed by trails or burrows, indicating that oxygen-dependent animals were unable to colonize the deep environments despite an abundance of available food. On the basis of this interpretation, episodes of global deposition of organic-carbon-rich sediments have been termed oceanic anoxic events.

The climate prevailing 100 million and 90 million years ago was warmer than today, and seawater would thus have held less oxygen. Moreover, the positions of the continents were different at that time. North and South America were in the early stages of separation from Europe and Africa; the Atlantic Ocean already had some deep narrow basins but widened as the Mid-Atlantic Ridge created new seafloor and moved the continents apart. At the Mid-Atlantic Ridge and similar active spreading centers, free oxygen reacted with sulfides and other oxidizable materials that were being introduced at unusually high rates.

That warm world may have been completely free of ice caps; the north polar region is known to have been covered with lush forests. The inventory of water was almost entirely in liquid form, so sea level was elevated. Two additional factors made the high stand even more pronounced. First, water expands when it is warmed, and the average temperature of seawater was much higher than at present. Second, the volume of the ocean basins decreased because the rapid extrusion of new hot crust at spreading centers meant that large areas of ocean floor rose upward. The resulting elevation of the sea level produced maximal flooding of the continents. In some localities even the oxygen-poor deeper waters extended onto continents, depositing marine sediments rich in organic carbon. When such sediments became deeply buried and were heated to temperatures of 90° to 120°C, they became important sources of petroleum. In summary, a powerful combination of global conditions discouraged the availability of oxygen in deep ocean waters, which in turn prevented the efficient consumption of organic carbon and eventually resulted in the accumulation and preservation of material now used as a nonrenewable resource.

Such episodes of enhanced burial of organic carbon represented globally significant perturbations of the carbon cycle. This change is reflected in the carbon isotope record, which indicates that the fraction of carbon buried in organic form was 50 percent higher during anoxic events than during the intervals immediately preceding and following them. As a result, atmospheric levels of oxygen must have risen even as the deep seas became oxygen depleted. Atmospheric levels of CO_2 declined, contributing to global cooling and the consequent termination of the conditions that had prevailed. Though many factors related to anoxic events can be identified and discussed, the details are not yet fully understood.

Paleoceanographers are investigating the changes over the past 100 million years in the vertical thermal structure of the oceans at increasingly finer scales of resolution. The heart of this research, which addresses biogeographic patterns as well, concentrates on diagnostic elemental isotopes. The isotopes distinguish plankton living at particular water depths characterized by unique thermal conditions. Areas of upwelling leave their own kinds of evidence, including phosphate deposits along ancient continental margins. Many of these deposits have considerable economic value, and paleoceanographic models contribute to their discovery.

Paleoclimatology: Cycles in Past Climates

A number of important climatic indicators help to establish how climate has changed. Most of them are applicable to the record of the past few hundred million years, and some portray conditions that dominated billions of years ago. These indicators are coals, soils, evaporites and sand dunes, glacial deposits, marine reefs and bedded carbonate deposits, and land plants.

■ *Coals*: Present in the modern world as peat, coals form from organic accumulations in areas combining high rain and poor drainage. The optimal conditions are located in equatorial rainforests or in moist areas at higher latitudes best represented today in zones about 55° north and south of the equator.

■ *Soils*: Soils rich in the clay kaolinite develop in warm humid climates. Associated with these soils are laterites, in which iron and aluminum are con-

centrated. Bauxite, the main ore of aluminum, also characterizes very hot moist conditions. All of these climatic indicators are end products of protracted weathering processes and resist subsequent alteration. Despite later cooling or aridity, such evidence of tropical climate can be preserved in the rock record for vast stretches of geological time.

■ *Evaporites and Sand Dunes*: These sedimentary features reflect arid conditions. Thick evaporite deposits require continuing replenishment from seas or lakes and extremely dry air. Dunes imply strong winds, a source of sand, and a lack of stable vegetation. They are best developed in the subtropical dry climatic zones but can form at mid-latitudes in landlocked regions such as the Gobi Desert or in rainshadows of mountains such as the Sierra Nevada.

■ *Glacial Deposits*: Glaciers leave boulders in erratic deposits and produce icebergs that release dropstones to lake bottoms and seafloors. Their most characteristic traces, however, are striations—the scratch marks found on pebbles or boulders that glaciers transported or on bedrock that glaciers scoured. Only continental glaciers have broad climatic significance because mountain glaciers commonly form at high altitudes, even near the equator.

■ *Marine Reefs and Bedded Carbonate Deposits*: Today, limestones accumulate mainly within about 40° of the equator. Those formed primarily by calcareous algae are probably restricted to this zone in part by sunlight requirements. Others, including massive organic reefs built by organisms specific to different time periods, are limited by thermal requirements. Modern coral reefs are confined to within about 30° of the equator.

■ *Land Plants*: Terrestrial floras are excellent indices of paleoclimates. Flowering plants are especially useful because of their conspicuous fossil record, which extends for about 100 million years. Climatic conditions are reflected in the basic leaf morphology (Figure 3.10) of flowering plants. Perhaps most valuable is leaf outline—a strong, positive, linear relationship exists between the percentage of species in fossil floras with smooth leaf margins and the mean annual temperature of the habitat. While the slope of the curve may have varied with time, the kind of gradient that we observe today has almost certainly characterized flowering plants since early in their history. The visible characteristics and the composition of fossil vegetation, as determined from pollen, spores, and seeds as well as leaves, give a general picture of climatic changes in North

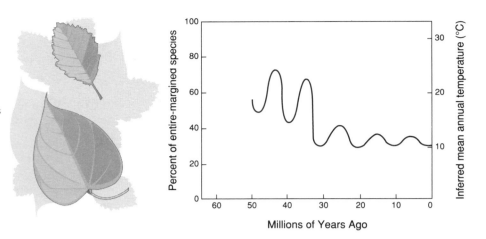

FIGURE 3.10 Entire-margined leaves became less common in rocks (30 million years and younger) of the Pacific Northwest. This has been used as an indication of mean annual temperature and shows a cooling. Entire-margined leaves are common in hot, humid climates and those with incised margins in cool climates.

America for the past million centuries. Efforts are under way to extend this technique to vegetation that existed before the rise of flowering plants.

Paleoclimatic models for intervals of time tens or hundreds of millions of years before the present have been determined by these criteria. For more recent intervals, more detailed information is available. For example, in the 1980s the Cooperative Holocene Mapping Project (COHMAP) produced climatic simulations for time frames that ranged from 18,000 years ago, the most recent glacial maximum, to the present. Empirical input came from estimates of numerous variables. These included insolation controlled by orbital variations, aridity produced by mountains and glaciers, trace gas concentrations in the atmosphere, distribution of sea ice and snow cover, albedo fluctuations, and effective soil moisture. The simulations were based on a model constructed from data concerning those variables in the modern world. The simulations were then tested against other geological indices of terrestrial climates, including the distribution of

fossil pollen, levels of ancient lakes, and the distributions of fossil plankton in nearby oceans that had been analyzed by the earlier CLIMAP project. For the most part, the empirical data supported the simulations. There were discrepancies that suggest imperfections in the models. Simulated July temperatures for the southeastern United States for the interval from 18,000 to 12,000 years ago were substantially lower than temperatures indicated by data from the stratigraphic record, and ancient dunes indicate that wind directions may not have been properly modeled. Discrepancies such as these call for further research.

Testing the models for earlier periods of earth history becomes even more difficult because climatic indicators are less precisely documented. But general climatic models developed for periods approaching half a billion years offer provocative results that must be taken seriously. One example is a model for 250 million years ago (Figure 3.11). A linear version of a two-dimensional seasonal energy-balance model shows a temperature response that depends on differences in the heat capacities of

FIGURE 3.11 An energy-balance model based on this distribution of the continents about 250 million years ago generates the temperature ranges indicated. The interior of Gondwana had an extreme range.

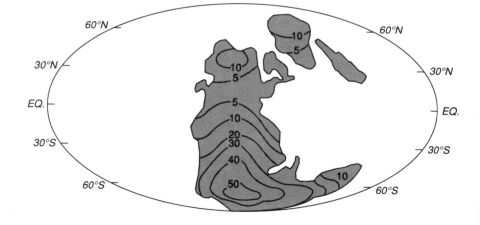

land and water. Other components of the model include the effective heat capacity of the earth-atmosphere column, the isotropic heat diffusion coefficient, the solar constant, the regional distribution of solar radiation, and the albedo of the earth-atmosphere system.

These models yield results for the annual climatic cycle that generally agree with observations. For 250 million years ago, the model indicates remarkably pronounced seasonality, owing to the accumulation of continental crust into a single giant continent. Simulated subtropical mean temperatures for the Southern Hemisphere range from about 10°C in the winter to about 40°C in the summer, with summer temperatures occasionally reaching the value of 45°C (113°F). These results have profound implications for terrestrial life. A large group of mammal-like reptiles called therapsids—whose fossil record ranges from South Africa and Virginia to Russia—occupied the subtropics of the Southern Hemisphere, where it appears that temperatures underwent large great seasonal oscillations. This occurrence is inconsistent with the idea that therapsids, which have no living representatives, were cold-blooded like their reptilian relatives. The implication is that, like their mammalian descendants, they employed thermoregulation. In fact, fossil trackways of these creatures suggest that they were warm-blooded. Their footsteps were usually far apart, like those of fast-moving mammals and unlike those of cold-blooded reptiles.

A more general conclusion of this energy-balance model applied to the world 250 million years ago is that continental glaciation in polar regions was favored by positioning of small continental areas near poles. Summer temperatures remained low because of the high heat capacity of neighboring seas, which caused abundant winter snow accumulation to persist throughout the year. If a very large dry continent were situated more centrally over a pole, pronounced seasonality would produce warm summers that could prevent the snow from persisting and glaciers from expanding.

Secular Change in the Global System

The numerous cyclical processes that interact to make up the entire earth system are superimposed on unidirectional, mostly gradual, processes of which radioactive decay and biological evolution are the most fundamental. Catastrophic events that affect large parts of the earth system are capable of interrupting the operation of familiar cycles and can greatly modify the established direction of gradual

secular change. During the past decade, in one of the most exciting developments in the study of the Earth, there occurred a deviation from traditional assumptions of general gradualism to consideration of a possible role for catastrophes.

The history of the Earth can be conveniently divided into three major intervals distinguished by important secular changes in the geological record. Many surface processes have changed somewhat throughout the three phases, but the biggest changes are seen in the sedimentary and fossil evidence. The record from the oldest rocks used to be extremely sporadic in quantity and quality, but gradually it has been improved by the accumulation of superior data.

The Precambrian Record

The interval of time that preceded the development of a rich fossil record of invertebrates with skeletons, nearly 600 million years ago, is called the Precambrian. Representing nearly 4 billion years of earth history, its unique features have led geologists to use techniques that are given less emphasis in the evaluation of later intervals. One limitation of this most ancient record is that rocks older than about 1.4 billion years lack any fossils that can be used to correlate strata and establish synchroneity from place to place. Limited correlation using fossils is possible among rocks between 1.4 billion and 600 million years old. Nonetheless, global developments of environments and life forms during the Precambrian were so significant that many general features have been deciphered and assigned at least approximate dates on the basis of isotopic dating. Among these was a major radiation in phenotypic diversity in eukaryotes, organisms with a cell with a true nucleus, at about 1.6 billion years ago or possibly even earlier—an event that may have been linked to changes in the proportion of atmospheric carbon dioxide and oxygen.

Both prokaryotic, or prenucleus, and eukaryotic single-celled organisms are remarkably well preserved in rocks of this period. Beginning around 550 million years ago, preskeletal multicellular animals left an abundant fossil record consisting of tracks, trails, and body imprints.

Study of early environments, and the organisms that evolved in them, is intertwined with the study of secular trends. These earliest secular trends were more profound and influential than any of the changes characterizing the most recent 600 million years. Not only was the Sun weaker and the greenhouse effect stronger, but calculations suggest that the Moon was circling the Earth in a smaller orbit, which would produce stronger tides in the ocean. Fossils provide clues about less ancient Earth-Moon relationships. Corals preserved from 360 million years ago have about 400 growth bands in each annual interval, revealing that the Earth rotated more rapidly and that days were shorter. Rotation must have been more rapid still during Precambrian time.

The oldest preserved sedimentary rocks exposed at the surface are about 3.8 billion years old, although there has been a discovery in Australia of zircon grains, preserved in somewhat younger sediments, that were formed 4.2 billion years ago. The oldest sediments do not look very different from later ones, which suggests that surface environments have not changed all that much. A notable exception is a recent discovery of a very high iridium content in rocks about 3.6 billion years old from South Africa and Australia. This evidence has been interpreted as indicating that the flux of meteorites to the surface of the early Earth was much higher than had previously been considered likely. Meteorite impacts had a substantial role in the evolution of the earliest surface. But, by analogy with evidence from the Moon, it has generally been considered that the flux rapidly decreased at about the time the oldest preserved rocks formed.

Speculation about conditions at the surface before the oldest rocks formed has changed as ideas about the Earth's origin and earliest development have evolved. The possibility of an impact of a Mars-sized body on the Earth suggests a mechanism for Moon formation. Such a scenario would have occurred within a few tens of millions of years after the solar system's origin 4.56 billion years ago. After the impact the Earth was probably wholly molten, although mantle and core would have remained separate with no great amount of chemical interaction. Volatiles, those elements and compounds that tend to vaporize, would have been lost to space or dissolved in the mantle silicates. In the former case the materials making up the subsequent atmosphere and ocean may have reached the surface in further meteoritic flux, originating farther out in the solar system, within a few more tens of millions of years. This material was probably dominated by water, nitrogen, and carbon dioxide, like the early atmospheres of Venus and Mars.

By about 4.46 billion years ago, when the solar system was 100 million years old, the Earth was approaching its present state. It had a core, a mantle, some kind of basaltic crust, and an atmosphere and ocean. Conspicuous differences from later times might have included the absence of a discrete inner core, much more heat dissipation from a hotter mantle and more vigorous convection, a much higher meteorite flux involving the generation and dissipation of more heat, and the repeated rupturing of the lithosphere. Unless some form of organism had been delivered by a meteorite or cometary dust, the Earth would have been still barren of life at this early stage.

The earliest erosion and deposition would have operated in much the way they do today, with the former dominant above sea level and the latter below. Rapid convection in the mantle would have led to rapid operation of cycles that open and close ocean basins, although there is no way of knowing when rigid extensive plates of lithosphere first characterized the surface. An intermediate arrangement could have involved a less organized convective system dominated by numerous Hawaii-like hot spots.

These speculations suggest that the surficial processes operating when the oldest rocks formed were similar to processes operating today. The geologist studying the oldest rocks is less like a playgoer seeing the curtain rise on the first act of a drama than one walking in on a performance that has been in progress for some time.

The oldest preserved sediments, at Isua in Greenland, still portray the drama in progress at the time of their formation, despite having been heated to 800°C and deformed intensely during subsequent mountain building. Detrital sediments indicate that both weathering and erosion proceeded nearby and that deposition was under water. The presence of small amounts of limestone may indicate that organisms were active in the depositional basin. The oldest structures acknowledged as exhibiting bacterially controlled limestone deposition are about 3.5 billion years old, 300 million years younger than the rocks at Isua. But the presence of limestones among the oldest preserved sediments strongly suggests that life had already originated by 3.8 billion years ago. Evidence involving the carbon cycle further implies that even then, as now, organisms played a substantial role in modifying the environment.

Solar energy reaching the Earth has probably increased over time because solar luminosity has risen as a consequence of the conversion of the Sun's hydrogen to helium. That conversion increases mean solar atomic weight and hence the temperature necessary to maintain thermal pressure against gravitational collapse. When the oldest rocks formed, the Sun's energy output was about 30 percent less than it is today. The surface would have been warmed so gently that much of the ocean would have frozen. But the early atmosphere was rich in carbon dioxide, a strong absorbent of the infrared wavelengths that conduct heat reflected from the surface. The abundant carbon dioxide would have created the effect of an enhanced greenhouse, sheltering the fragile surface of the planet from a complete freeze that would have been difficult to reverse.

Development of the modern atmosphere from those primeval conditions accompanied biological evolution. The oldest carbon compound deposits, stored in limestones of the rock reservoir, were produced by organic precipitation. Those early organisms consumed carbon dioxide in photosynthesis, released free oxygen to the atmosphere, and left their carbonaceous skeletons to accumulate as limestone on the ocean floor.

The record of change in the oxidation state of the atmosphere-ocean system is fragmentary, and interpretation is necessarily speculative. Even so, a case can be built on a collection of evidence that suggests a gradual buildup toward saturation. There are remnants of detrital grains that could not have resisted oxidation in an oxygen-rich atmosphere. These grains are preserved in sediments that are at least 2.0 billion years old, and the grains themselves may be 2.5 billion years old. Also, abundant iron ore deposits chemically precipitated occur in sediments deposited between 2.8 billion and 1.8 billion years ago. These ores are richly oxidized compounds that when formed would have acted as an oxygen sink; that means newly freed oxygen would have had to react with such exposed minerals before any atmospheric buildup could have become available for biological functions. Furthermore, the oldest uncontestable evidence indicating biological use of oxygen dates to 2.8 billion years ago. Finally, the first evidence of persistent oxygenation of surface environments occurs in rocks that formed 2.2 billion years ago.

The interpretation of much of this evidence is disputable, and active research, including a search for stable carbon isotope variations, is attempting to clarify the picture. Recent isotopic investigations emphasize the role of the carbon cycle in modifying the environment. These investigations focus on another episode that witnessed the burial of large quantities of organic carbon in sediments between 900 million and 550 million years ago—a period that preceded an explosion of multicellular organisms throughout the world's oceans. Complementary release of oxygen from carbon dioxide would have driven atmospheric oxygen concentrations up to present levels or higher. This interpretation supports those evolutionary biologists who speculate that multicellular life forms could not develop until the atmospheric partial pressure of oxygen became high enough to diffuse the element across multiple cell layers. Paleontological evidence of prolific multicellular life fills the stratigraphic record immediately after the isotopic evidence of increasing oxygen.

Projecting the carbon isotope record forward to 550 million years ago produces evidence of numerous smaller cyclical changes. For example, significant variations in ^{13}C abundances have been linked to complementary shifts in the abundances of sulfates and sulfides, the oxidized and reduced forms of sulfur. It seems likely that, when larger-than-average quantities of organic material have been buried and the carbon cycle frees excess oxygen, the excess is consumed by the oxidation of sulfur minerals exposed at the surface. The oxygen release during the carbon burial event 900 million to 550 million years ago was apparently so large that oxidation of sulfur could, at best, attenuate it.

Record of Change Between 600 Million and 150 Million Years Ago

During this period organisms colonized vastly differing environments through a variety of physical adaptations. Among the plants, ferns and conifers evolved, and animal life developed from marine invertebrates to fish, insects, amphibians, reptiles, and mammals. Interpretation of change between 600 million and 150 million years ago profits from the richness of the fossil record and the relatively accurate reconstructions of former continental positions based in part on paleomagnetic data (see Figure 3.8). Sedimentary particles containing iron tend to settle in magnetic alignment with the Earth's magnetic field, and, similarly, iron-bearing lavas become magnetized as they crystalize while cooling. A rock magnetized in one of these ways is, in effect, a paleocompass that reveals its own orientation to the magnetic pole at the time when it formed. Reliably preserved data come from the magnetic declination, which shows directional orientation, and from the magnetic inclination, which increases at higher latitudes. These relative magnetic deviations are set into a rock body when it forms, allowing geologists to determine the orientations of ancient continents as well as their latitudinal positions.

Rock and fossil distributions can also indicate possible paleogeographies, and maps of ancient continental positions are under constant revision. As might be expected, the paleogeography of older time intervals is the least certain. It has even been suggested that as long as 600 million years ago nearly all continental lithosphere formed a single supercontinent—an earlier Pangea—which then separated into Gondwana and smaller fragments. Subsequent reassembly in a different configuration formed Pangea about 250 million years ago.

Many forms of life that existed between 600 million and 150 million years ago belonged to genera, or even families, that are now extinct, so only broad environmental inferences can be drawn from their geographic distributions. As a result, analysis of tectonic patterns remains the focal point of global environmental reconstructions for the entire interval. Since 150 million years ago the dominant process has been the breakup of Pangea to form the modern continents and the oceans that separate them. The record is much better preserved, and paleoenviromental analysis is correspondingly more refined.

Record of Change from 150 Million Years Ago to the Current Ice Age

For the past 150 million years, the breakup of Pangea into the modern continents and their subsequent distribution have dominated the processes recorded in the rocks. The changing plate positions can be projected for that interval from information preserved in the ocean-floor record. The high quality of the youngest segment of the geological record allows earth scientists to use additional techniques in evaluating global change over the past 150 million years. Many sediments of this age remain soft, so they readily yield their fossils for study. Furthermore, the fossils are commonly quite well preserved; some marine skeletons faithfully retain even the original ratio of stable oxygen isotopes, shedding light on temperature change during the recent past.

The oldest dated floor of modern ocean basins covers only a small area and is about 180 million years old. Progressively larger areas of younger ocean floor are preserved within the ocean basins. The Ocean Drilling Program, like its predecessors, utilizes the relatively complete sedimentary record beneath the deep ocean where erosion is much less pervasive than on the land. Close examination of the microplankton fossils found in deep-sea cores, partly through the study of oxygen and carbon isotopes in microfossils and partly through analysis of microfossil paleoecology, has proved rewarding.

Thanks to this work, paleoceanographic research that is focused on the past 150 million years reveals dramatic changes in the thermal structure of the oceans, which have culminated in present patterns.

The method of correlation known as magnetic stratigraphy is used for sediments deposited during the growth of an ocean basin. The magnetic field, which is generated by fluid motions within the outer core, reverses itself episodically. These reversals of polarity are recorded in magnetization of sediments and volcanic rocks, as described previously, and they serve as valuable events for global correlation. Magnetic reversal sequences in sediments can match marine with nonmarine strata, which have few fossils in common. The magnetic reversal pattern recorded in igneous rocks formed at spreading centers is laid out as stripes on the floor of the world's oceans. Matching the characteristics of these patterns as they appear on either side of an ocean rift can be used to reconstruct the size and shape of the oceans and the distribution of the continents. For much of the past 150 million years, the magnetic time scale has been calibrated by isotopic dating, so the scale is not simply a means of correlating strata but also a source of information about ages of sediments and fossils.

Fossils are not only important in correlating strata separated in space, but also they are invaluable for characterization of ancient environments. For example, flowering plants, which include not only plants with conspicuous flowers but also hardwood trees and grasses, have been called the thermometers of the past because of their diagnostic value in assessing ancient thermal regimes on the land since they became abundant forms of life 100 million years ago.

Recognizing Environments from the Geological Record

Our picture of global environmental conditions is a composite of local studies. Many of these studies are rooted in detailed analyses of depositional environments—environments in which sediments, and often fossils, have accumulated.

During the past 25 years, sedimentologists have developed a set of tools for distinguishing particular environments of deposition and for unraveling the histories of these environments from clues left in sequences of sedimentary strata. Some of the diagnostic features are small in scale. Examples are accumulations of lens-like inclined sand beds draped with mud, produced by ripples migrating across a tidal flat that left a record of ancient oscillating currents. Other examples are beds in which scours are filled by coarse sands at the base, which are covered by finer sands, then by silt, and finally by clay. This sequence, or package, indicates that the entire bed formed when a turbid flow charged with particles of many sizes—a turbidity current—swept down onto the deep seafloor at the margin of a continent. Upon deceleration it dropped its coarse debris first and its slower-settling fine debris later.

Key diagnostic features involve the associations between individual sediment units in the vertical sequence. For example, meandering rivers, as they migrate back and forth across a valley floor, produce repetitions of deposits that suggest cycles of sediment accumulation. Coarse sandy or even gravelly deposits with inclined bedding indicate an active stream channel. These pass upward into finer deposits that culminate in muds recording the lateral migration of the channel with deposition limited to flood events. The cycle is complete when evidence shows an abrupt cut through the sequence that is filled with another coarse deposit and subsequent fining upward.

By analyzing sedimentary sequences, geologists can recognize a wide range of ancient environments, ranging from alluvial fans that form along the bottom slopes of mountains to mountain belts that have incorporated sediments squeezed up when deep floors of ancient oceans converged with continents along subduction zones. Depositional environments bear witness to most ongoing earth processes—they only need accurate interpretation. And there are immediate applications. Reconstruction of sedimentary environments plays a major role in the search for petroleum and gas, indicating locations of natural traps for these fluids.

Accurate dating of rocks is critical to paleoenvironmental reconstruction on all spatial scales. When isotopic dates are not known, tight temporal correlations between areas must be dependable. Improvement of existing techniques continues, as does the invention of new ones. Only recently, for example, geologists have applied and improved quantitative techniques for correlating strata on the basis of the earliest and latest appearances of fossil species. Quantitative analysis of populations of fossil conodont—minute tooth-like structures of an extinct group of marine vertebrates—can correlate rocks more than 400 million years old. Calculations indicate that these correlations are accurate to within just a few hundred thousand years.

Breakthroughs have come in other areas, too. Stratigraphers have lately discovered that many limestones exhibit sufficient paleomagnetism to display reversals in the magnetic field. This character-istic permits geologists to assign strata to positions in the global paleomagnetic time scale. Surprisingly, the magnetism results from forms of bacteria that colonize tropical seafloors and produce minerals containing iron. Using this approach, stratigraphers have dated limestone cores taken during drilling operations in the Bahama Banks. From these dates they can calculate rates of subsidence for this huge limestone platform and determine the times when global lowering of sea level left it standing far above marine waters.

New means of isotopic dating are also continually being developed and refined. Single grains of zircon from Precambrian rocks more than a billion years old can now be dated with a precision of just a few million years, as discussed previously. Other exciting techniques, still in the early stages of development, should soon permit dating of terrestrial sediments that are just a few tens or hundreds of thousands of years old—too old for radiocarbon dating, or lacking any carbon that might be dated, and too young for other dating methods.

Paleogeography

Researchers have proposed that not only the positions of the continents but also the uplift of mountains exert control over global climate patterns. Mechanisms that physically alter environments on a regional scale have traditionally been accepted, but theories that suggest that tectonic forces may cause climatic changes on a global scale still inspire controversy. For example, the uplift of the Sierra Nevada in eastern California exemplifies regional, and direct, effects. Today, the Sierra Nevada is an imposing structure, a block of granitic crust heaved upward to form a towering eastward-facing scarp that was a formidable barrier to pioneers attempting to reach California. Fossil plants dating from 10 million to 15 million years ago that could not have lived as much as 1 km above sea level are found today on the crest of the range, nearly 3 km high. Only in the past 5 million years has the Sierra Nevada approached its present height; the consequences of this elevation are enormous for the Basin and Range Province to the east, in Nevada and southern California. This area, which had been covered by broadleaf evergreen forests, came to lie in the rain shadow of the Sierra Nevada in the past 5 million years and developed a savannah vegetation. During the past 1.8 million years climates became drier on a global scale. The Basin and Range Province became the desert that we know today, although during glacial maxima it received substantial rainfall.

The environmental changes produced on a global scale by uplift involve more complicated processes. Important simulations have been produced by researchers who used the National Center for Atmospheric Research's Community Climate Model with variables adjusted for three different orographic conditions. The three variations are plotted for no elevation, half elevation, and full mountain elevation in two separate areas: the extensive high ground that spreads from the Rockies to the Sierra Nevada and the Cascades in the western United States, and the Tibetan Plateau bordered to the south by the lofty Himalaya. The three variations of uplift correspond respectively to conditions that dominated 40 million years ago, 8 million years ago, and during modern times. The results suggest that the uplift of those two areas during the past 50 million years modified the Northern Hemisphere's atmospheric circulation, in a domino-like series of events that could have promoted the glaciations of the current ice age. The uplifted plateaus would have rerouted dominant air circulation, establishing violent monsoonal weather systems. Those weather systems, in turn, would have accelerated chemical weathering from the mountains—weathering that draws carbon dioxide from the atmosphere. An atmosphere depleted in carbon dioxide would have a global cooling effect and would encourage ice buildup in the higher latitudes.

A differing hypothesis points out that increased weathering, induced by frost, could remove material from an already high plateau. Traditionally, geologists have considered coarse massive deposits as evidence of uplift. They have tended to assume that uplift exposes more rock to weathering and produces piles of boulders and cobbles downstream. The new hypothesis attributes the piles of rubble to climate change—unloading the upstream regions generates uplift because of isostatic compensation.

Another tectonic mechanism that may have contributed to the ice age in the Northern Hemisphere is the closing of the gap between North and South America by the uplift of the Isthmus of Panama. Changes in plankton preserved in deep-sea sediments date the uplift of the Isthmus to 3 million years ago. Blockage of the westward-flowing equatorial currents from the Atlantic to the Pacific would deflect warm equatorial waters out of the Gulf of Mexico. The Gulf Stream delivers large masses of warm water to the northern Atlantic, where it evaporates and increases humidity. The humidity precipitates as snow at high latitudes. The hypothesis that a strong Gulf Stream engendered the ice age attributes the accumulation of continental ice sheets to that increase in precipitation (Figure 3.12).

The cause of the current ice age probably incorporated facets of all these changes and more. The implication is quite astounding. Small tectonic events plunged the entire Northern Hemisphere into a persistent ice age. This current ice age has persisted for the past 2.5 million years. All human civilizations have occurred within one geologically brief interglacial interval, and there is no indication when this ice age will come to an end.

Throughout earth history, lateral plate movements have caused profound environmental changes. Paleolatitudinal data indicate that between about 450 million and 400 million years ago, the great supercontinent of Gondwana encroached on the south pole and underwent major climatic changes that appear to have influenced all latitudes. The polar region of Gondwana, which now constitutes northern Africa, was the center of a vast glaciation. Today, in the Sahara Desert we find massive deposits 450 million years old that bear striations produced by grinding along the base of glaciers and that contain boulders. Many species failed to adapt to the new thermal regimes or to migrate successfully to hospitable ones. This was the time of one of the most severe mass extinctions to have taken place in the marine realm during the past 600 million years.

Anomalous distributions of fossils have been related to plate movement in quite a different way. The distribution of some fossil species simply does not make sense. During the 1970s such abnormal distributions provided the first clues in both eastern and western North America to the existence of exotic terranes—large blocks and fragments of lithosphere that have been sutured to the edge of the continent (Figure 3.13). The indication that these terranes had been rafted into place from far away came from the observation that their fossil marine faunas represented paleobiogeographic provinces quite different from those of neighboring continental regions. In this way a major mechanism of continental growth was demonstrated.

History of Life

Life on Earth develops in an ever-changing environment, and paleobiologists are reaching out to other disciplines for help in assessing the effects of environmental changes on ancient life. Prominent questions are those about the kinds of environmental change that influence the evolution and extinction of life and about the time scales on which the

30 Million Years Ago

20 Million Years Ago

10 Million Years Ago

Present

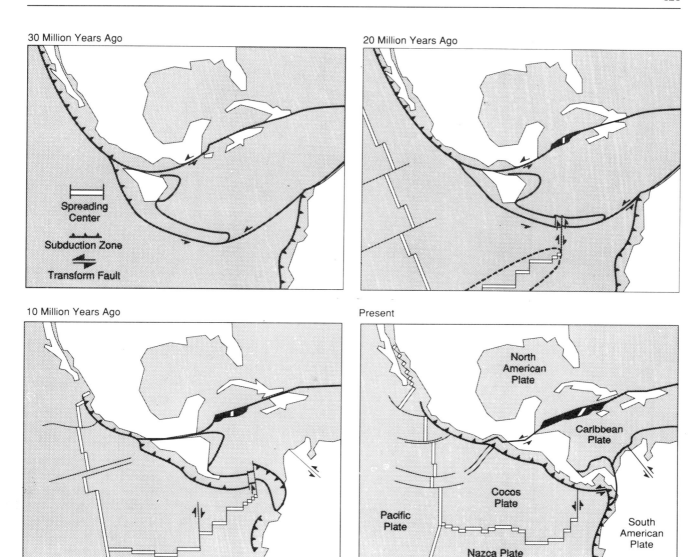

Spreading Center

Subduction Zone

Transform Fault

North American Plate

Caribbean Plate

Cocos Plate

Pacific Plate

Nazca Plate

South American Plate

FIGURE 3.12 Evolution of Caribbean land bridge through plate tectonic processes over the past 30 million years. This changed ocean circulation and may have led to the accumulation of continental ice sheets.

changes operate. These questions not only bear on past rates and patterns of evolution and extinction, as revealed in the fossil record, but also illuminate impending extinctions that humans may be able to prevent. Some predictions suggest that half of all living species may disappear within the next 50 to 100 years. This rapid disappearance is supposed to be biased toward loss of terrestrial species, especially those of the tropics. However, there has been little work on present rain forest diversity or on natural rates of extinction in rain forest areas during the past 50,000 years. The little research that has been done produced conflicting results. Conservation programs can only be improved by learning more about the environments and biology of these areas during the recent past.

Specifically, we need to find out the climatic and vegetational histories of the Amazon Basin and of similar rain forests in western Africa and Asia. The Amazon rain forest shrank and broke up into pockets of verdant growth separated by stretches of savannah as the building northern glaciers repeatedly removed much of the oceanic and atmospheric water from free circulation. Some researchers claim that there was a 90 percent reduction in rain forest area during this period of Northern Hemisphere glaciation, with the inevitable accompanying extinctions.

The modern world is impoverished in terrestrial species of mammals, the class to which humans belong. Only 12,000 years ago the American landmasses were populated by beavers the size of black

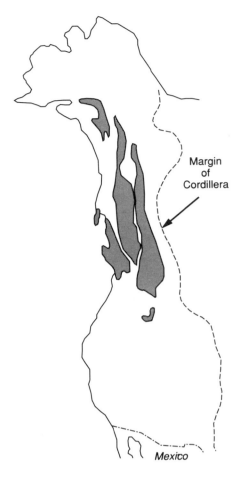

Margin
of
Cordillera

Mexico

FIGURE 3.13 The shaded areas are among those terranes in western North America that have been added to the continent during the past 250 million years. This process has been important in building continents throughout earth history.

bears, ground sloths the size of elephants, and lions much larger than their living relatives. Close to 11,000 years ago these giant mammals and many others disappeared. Large animals such as these tend to be especially vulnerable to extinction because they have small populations. Perhaps the appetites of humans accounted for the demise of the behemoths; one scenario for the extinctions implicates human hunters armed with advanced new weaponry. The other possible culprit is climatic change, which caused heavy extinction in the oceans earlier in the current ice age. In the North Atlantic and Caribbean, where climates became cooler and more seasonal beginning about 3 million years ago, thousands of shallow-water marine species died out.

Whatever the causes of the various extinctions during the current ice age, the fossil record gives some indication that the modern world is biotically

deprived. This paleontological perspective makes the faunas and floras that survive today especially precious. The magnificent mammals of the Serengeti Plain provide an unusual glimpse of a rich savannah fauna, but even this environment is depleted in diversity compared to the fauna that inhabited the same area 2 million or 3 million years ago.

Rates of evolution and extinction show how the modern ecosystem came into being. The number of species in any higher biological category is a result of the positive process of the multiplication of species, called speciation, and the negative process of extinction. But speciation represents only one mode of evolution—another is phyletic evolution, the transformation of already existing species. Both multiplication and transformation of species result primarily from natural selection. Evolutionary processes became more fully elucidated in the twentieth century through progress in modern genetics. Still, the relative importance of phyletic evolution and speciation within segments of the tree of life for particular intervals of time remains controversial. Some scientists question whether speciation has contributed more to the total amount of evolutionary change—the punctuational model—or whether phyletic evolution has played the larger role—the gradualistic model. If phyletic evolution is sluggish, accounting for a relatively small fraction of all change, the traditional idea that many species respond to gradual environmental change by undergoing adaptive evolution should be reexamined. Instead, the punctuational model should be considered. This model, which emerged during the 1970s from interpretation of the fossil record, has engendered much research in both paleontology and biology. While insight has and will come from the field of biology, the fossil record will be the ultimate testing ground. On a scale of millions of years, evidence comes uniquely from the fossil record.

Traditionally, studies of rates, trends, and patterns of multiplication and extinction use the more general biological categories, genera and families, as units of analysis. During the past few years, however, numerous studies have been undertaken at the level of species, improving the fidelity of macroevolutionary research. Analogy with demography, which normally monitors individual organisms rather than species, contributes to assessments of increase and decline. Multivariate statistics and computers have made it relatively easy to handle morphological data and to test alternative models of evolutionary processes. Paleontologists have learned to use more than sophisticated mathematical tools to analyze the fossil record. Interpretation of trends and patterns involves familiarity with functional morphology, which brings extinct plants and animals to life; with local paleoecology, which places them in their environmental context; and with paleobiogeography, which reconstructs their broader distribution.

Speciation and Extinction

A variety of demographic techniques is available to assist in analyzing the waxing and waning of biological groups through analogy with the birth and death processes of individual organisms. One example is logistic growth. A population of cells in a petri dish that expands from a single original cell will grow at a rate of cell division that initially exceeds the rate of death. In a similar way, new groups of animals and plants have diversified because their rate of speciation exceeded their rate of extinction. The number of species expands almost exponentially at first and then suffers a decline in its rate of increase, so the diversity curve levels off. Growth rate in the petri dish declines because of crowding and perhaps because of waste accumulation. For a group in nature, ecological crowding may eventually reduce the rate of speciation to a level that approximates the rate of extinction. Alternatively, highly efficient predators or competitors may evolve or the habitat may deteriorate, increasing the rate of extinction. The demographic analogy extends to groups that arise during brief geological intervals. They are comparable to cohorts of individuals born more or less simultaneously within a population and are amenable to various methods of survivorship analysis.

The factors that govern rates of speciation and extinction for any group of organisms may change greatly from place to place or time to time. More fundamentally, there are also intrinsic differences in the rates characteristic of different taxa. Mammals, for example, have experienced higher rates of both speciation and extinction than many groups of fossil marine invertebrates. Sets of rates such as these reveal patterns, and patterns in turn suggest the nature of the factors that regulate the rates. We now recognize that at times of biotic crisis some normal patterns of extinction have been altered. Statistically significant differences between the biological traits of victims and survivors serve to test extinction hypotheses. For example, in the mass extinction that swept away the dinosaurs, marine phytoplankton that could form biologically inactive resting spores were relatively unaffected.

Though not definitive, this pattern is consistent with the hypothesis that during the crisis oceanic photosynthesis was drastically diminished, a condition that might have initiated a cascading trend of food chain collapse.

The basic nature of speciation and extinction must be understood in order to interpret the rates at which they occur. Species are separated from one another by reproductive barriers. Although biologists debate the importance of population divergence to form new species where there is no spatial separation from parent species, it seems clear that most speciation entails at least partial isolation of the diverging population. This isolation may occur by fragmentation of the parent species' population or by migration or transport of a subpopulation to a separate location. Simple arithmetic reveals that speciation is quite infrequent. If the average time span for a species has been reasonably estimated at 5 million years, and if global diversity has not changed markedly during a 5-million-year period, then during this long interval an average species will have spawned only about one descendant species.

Extinction has been the fate of the vast majority of species that have inhabited the Earth. Usually it is caused by an accentuation of the limiting factors that naturally restrict the distribution and abundance of the species. Extinction amounts to an ultimate decline in both areal extent and population size. These limiting factors include competition with other species, efficiency of predators, availability of resources, conditions of the physico-chemical environment, and chance factors. Often two or more limiting factors conspire to cause extinction, with chance playing its most important role when species are rare and spatially confined.

To explore more fully how patterns suggest controls over probable speciation and extinction, controlling factors can be assigned to three categories: intrinsic traits of the species being considered, limiting factors of the biotic environment, and limiting factors of the physical environment.

Intrinsic Biological Traits

The fossil record reveals that rates of speciation and extinction correlate in predictable ways with geographic dispersal, characteristic size and stability of populations, and behavioral complexity. For example, weak dispersal can favor speciation by fostering the frequent isolation of small populations and can favor extinction by restricting geographic distribution. These relationships cannot be examined easily in the modern world, which represents but a brief slice of geological time. However, some segments of the fossil record represent useful reality checks for the mathematical models that examine such relationships.

Biotic Interactions

Ecological relationships between groups can be positive or negative. On the positive side, diversification in one group can promote speciation in another. Furthermore, ecologic interactions between two groups can promote the diversification of both through a kind of evolutionary synergism—combined actions that produce enhanced results. For example, during the past 150 million years the diversification of flowering plants went hand in hand with an expansion of insect pollinators. New varieties of plants offered new food resources for insects, while at the same time new varieties of pollinators promoted reproductive isolation, and hence speciation, in flowering plants. Unfortunately, this relationship has rendered both groups more vulnerable today. The two groups are so interdependent that extinction of species within either will often lead to extinction of species within the other.

Negative interactions between groups have included both competition and predation. As one example, during the period dominated by the dinosaurs, the progressive decline of several major groups of seafloor life is attributed to the expansion of predatory groups that remain prominent in modern seas: crabs, bony fishes, and carnivorous snails. The declining groups include certain kinds of bivalves, snails, and calcareous algae that were especially vulnerable to attack according to studies of functional morphology. The long-term result was a wholesale transformation of seafloor life.

Changes in the Physico-Chemical Environment

Nonbiological aspects of the environment also change in ways that can have positive or negative effects on particular forms of life. Some of these changes influence evolution and extinction by removing barriers to migration, thus allowing species to move to new regions. Tectonic events and changes in global sea level have had this effect by connecting landmasses or oceans that had previously been separated. The tectonic origin of the Isthmus of Panama, for example, served as a natural experiment for testing faunal equilibrium. Migration of mammals at first gave the appearance of maintaining equilibrium, but this has broken down

in the past few hundred thousand years. Numerous northern forms infiltrated the South American system without having drastic effects on existing elements, until South American carnivores experienced heavy extinction. Apparently, the northern immigrants had some ability to migrate not possessed by their southern counterparts.

Environmental changes can also result in new habitats that tend to produce diversification of the groups that first gain access to them. The origins of islands and lakes epitomize this phenomenon. Just as modern mouth-breeding fishes proliferate rampantly in the lakes that have formed recently in rift valleys of Africa, primitive fish groups underwent spectacular diversifications in the rift valley lakes that developed as North America separated from Europe and Africa. Today, their 200-million-year-old fossils are preserved in lake sediments of eastern North America.

Other changes in the physical environment have had global effects. During the past 35 million years or so, a decrease in mean annual temperature at the surface and an increase in seasonality and aridity promoted evolutionary changes in plants that propagated up the food chain. A proliferation of new species of seed-producing herbs and grasses contributed to a rampant diversification of seed-eating rodents and song birds, which in turn fostered a great increase in the diversity of predatory snakes (Figure 3.14).

Ironically, deterioration of a species' habitat can also promote evolutionary diversification if its effect is to fragment the habitat, producing isolation and eventual speciation. This occurred when the fragmentation of tropical rain forests in Africa and South America during dry intervals of the current ice age led to the origin of many new species in the small remnants of forest that survived until the return of better times.

Adaptive Radiation

Rapid proliferation of species within a group constitutes adaptive radiation. This process accounts for most evolutionary change. Typically, numerous distinctive new taxa arise during relatively brief intervals in the early stages of adaptive radiation. Fossil discoveries that date from 600 million years ago, when organisms first developed hard skeletons, record the initial explosive radiation of animal life. But more recent, and more modest, radiations present special research opportunities when they can be studied with high-quality analytical tools that test diversification as well as spatial and temporal distri-

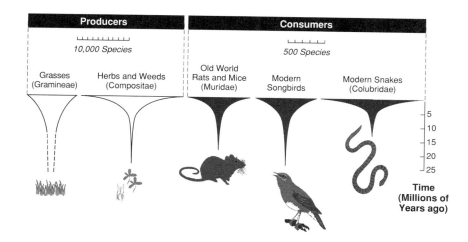

FIGURE 3.14 A proliferation of new species of seed-eating rodents and song birds fostered a great increase in the diversity of predatory snakes.

bution. Most adaptive radiations have resulted from unusually high rates of speciation, but low rates of extinction, compared to those of closely related groups, have been important in some cases.

The adaptive shifts associated with radiations must be understood before the evolution of the biosphere can be adequately explained. Radiations often followed the evolution of adaptive innovations that greatly expanded the inhabitable domain of life. A few examples among many are the evolution of oxygen-mediating enzymes in Precambrian eubacteria, the protective roll-up reaction in trilobites, and advanced limb mechanics in mammal-like reptiles that were ancestral to dinosaurs.

When judiciously focused, functional morphological analysis—the study of anatomical forms that suit particular functions—underpins most identifications of adaptive breakthroughs. Adaptive radiations of some groups have been triggered by the extinction of competing groups or, in some cases, predatory groups. The adaptive radiation of mammals following the mass extinction of the dinosaurs may have resulted from decreases in both competition and predation. Lately, detailed studies of this adaptive radiation have revealed that a single mammal genus (*Protangulum*), after surviving that mass extinction, gave rise to some three dozen genera during the first 2 million or 3 million years of the ensuing Age of Mammals.

Pulses of Extinction

Pulses of extinction are, in a sense, the negative equivalents of adaptive radiations. Normally, extinctions occur on local scales at a regular rate, but mass extinctions affect many groups throughout broad regions of the globe. The record of mass extinctions shows more sudden and drastic change than has ever been achieved by global radiations. No simple formula can be established for the optimal analysis of these events. During the 1980s, research into this exciting branch of paleobiology has found that the most fruitful approaches fall into two categories.

One approach investigates nonbiological evidence for causation at certain levels. This approach calls attention to iridium, an element generally very rare in the crust, and to shocked minerals, particularly quartz grains with distinctive sets of intersecting planar structures. Both are found worldwide in anomalously high concentrations at the level of strata that dates to about 66 million years ago—the period of the dinosaur's extinction. These anomalies have been construed to be evidence of an extraterrestrial impact.

The second approach studies patterns of extinction—selective survival—in a search for clues to causation. This research investigates the characteristics of those species that remained after a mass extinction, like the marine phytoplankton with resting spore stages that survived the crisis of 66 million years ago. This research indicates a tendency for mass extinction to strike most heavily in the tropics, which suggests an important causal role for climatic cooling. During climatic cooling, tropical organisms have no thermal refuge.

The dominant theme pervading both approaches to mass extinctions is the need for interdisciplinary research. Geochemical, sedimentological, tectonic, and even astronomical data have come into play. The endeavor to obtain the best possible stratigraphic resolution has helped to establish a reliable sequence of events. Paleoclimatic and paleoceanographic data provide frameworks that can support dependable interpretations.

Efforts directed at explaining the extinctions that

mark the geological record beginning 34 million years ago exemplify these points. This crisis is especially amenable to study because it happened relatively recently. The stratigraphic record is of high quality, and the well-established pattern of magnetic reversals provides an excellent temporal framework. Fossil records of planktonic foraminifera, calcareous nannoplankton, and terrestrial mammals show that the crisis was protracted, suggesting multiple pulses of extinction. As we have seen, changes in terrestrial paleofloras indicate that climates cooled in many areas; shifts in oxygen isotopes of foraminiferan shells reveal that water masses underwent major changes; and plate reconstructions suggest that the Antarctic cooling system for the deep sea originated at this time. Nonetheless, the ultimate cause of this crisis remains controversial.

Pulses of extinction, when they have removed large numbers of species by imposing unusual and lethal conditions, can reset biological systems in ways that may have had nothing to do with the victim's ability to adapt successfully to ordinary conditions. This circumstance is epitomized by the extinction of the dinosaurs, which emptied ecospace to the benefit of the mammals. In recent years, dinosaurs have been recast as active, ecologically adept creatures—not backward lumbering forms that were inherently inferior to mammals.

A continuing question about global biotic crises in general is whether, through geological time, they form the tip of one tail in a unimodal distribution of extinction rates or whether they represent a statistical outlier; a sparse data base currently frustrates valid statistical testing. The second condition would imply unique causation rather than simply an accentuation of normal agents of extinction. On the other hand, periodic spacing of mass extinctions may suggest a highly abnormal cause. The possible regularity of these crises stimulates current debates.

At the other end of the distribution of extinction rates lies another question: When the record shows extinctions occurring at minimal rates, do species still disappear in groups or do they die off independently in piecemeal fashion? The former condition would require an expansion of catastrophic ideology to embrace even relatively calm intervals of biotic history. Finer resolution of the stratigraphic record may solve this basic problem of normal, or background, extinction levels.

Macroevolutionary Trends

Macroevolution transcends species boundaries, involving changes in the more generalized taxo-nomic levels, such as genus or family. Macroevolutionary changes include those trends in the marine realm that have been driven by the expansion of sophisticated predators on the seafloor over the past 100 million years. Phyletic evolution can produce macroevolutionary trends, as can differences in rates of extinction and speciation among groups of species. An important question here relates to the relative importance of phyletic evolution in producing trends in the history of life. If it has been of secondary importance, as asserted in the punctuational model, differential rates of extinction and speciation gain importance. A further question concerning phyletic evolution addresses the degree to which it has been gradual and the degree to which it has followed a stepwise course, in which established species have undergone most of their changes during very brief intervals of geological time.

Trends can be documented only on the basis of careful taxonomic and biostratigraphic studies, and they can be interpreted only by considering the functional morphology and ecology of component species.

Origins of Major Biological Categories

Fossil data provide estimates of the times when higher biological groups evolved. For groups with excellent fossil records, early estimates remain unchanged even though large volumes of new data have accumulated. For those with poor fossil records, the estimates are vulnerable to new discoveries. Thus, although *Archaeopteryx* is remarkable for its intermediate morphological position between dinosaurs and birds, the early fossil record of birds is so poor that a recent claim for a much earlier bird cannot be rejected out of hand. The fossil record offers a unique opportunity to assess the origins of taxonomic groups at general levels—the origins of kingdoms in Precambrian time and then of phyla.

Times of origin for the more general groups, which are estimated from molecular data, do not always agree with those from fossil data. Molecular data clearly have great potential for the assessment of relative times of origin, but only well-dated fossils will estimate actual rates of divergence between forms. The environmental sites of major evolutionary breakthroughs are difficult to predict. These events may be concentrated in relatively unstable habitats, such as those of high latitudes or nearshore marine habitats, or in more stable habitats, such as those of the tropics or offshore marine habitats at middle or low latitudes. Hostile habitats generally offer more vacant ecospace, but more hospitable habitats sustain more varieties of life and harbor more biological groups, which means that they can support greater total rates of speciation.

Origins of higher groups frequently reflect the evolution of adaptive innovations. Studies of ancestors and descendants in the fossil record can reveal the morphogenetic mechanisms that gave rise to innovations. Among these mechanisms are changes in the relative timing for development of different morphological features. This area of research offers great potential for fruitful interaction between paleobiologists, developmental biologists, and geneticists.

Phylogenetic Reconstruction

During the past two decades, cladistic analysis—the study of similarities resulting from common origins—has emerged as a powerful quantitative method for reconstructing the genealogical relationships called phylogenies. It is based on assessment of how traits that have evolved only once are distributed among taxonomic groups. Some proponents of cladistic analysis disregard the stratigraphic distribution of groups, grounding reconstruction of phylogenies on judgment as to which characteristics are primitive and which are derived. An alternative approach reconstructs phylogenies by evaluating stratigraphic and morphological distances between groups. Comparisons have shown that the two approaches sometimes yield identical results. Such comparisons are of great value, and methods that combine the two approaches warrant further examination. Molecular data also can reconstruct phylogenies; however, some of the techniques, including DNA hybridization, remain controversial, and their results must be compared to those achieved with the other approaches.

Catastrophes in Earth History

Geologists call sudden violent changes catastrophes, and they contrast catastrophes with the changes in the rock record attributed to constant gradual processes. There is obviously a continuum between frequent events, moderate events, and the occasional violent happening, and this simple relationship is readily expressed in empirical laws. Such relationships can, for example, link earthquake frequency with earthquake magnitude.

Nearly two centuries ago polarized positions were assigned to geologists—they were either catastrophists or uniformitarians. Since then geological interpretation has accommodated the occurrence of occasional violent events, of the kinds experienced within our own lifetimes—hurricanes, earthquakes, and volcanic eruptions. A uniformitarian approach has dominated because it is very effective in analyzing the rock record and in making correlations.

The possibility that the four or five great biological extinctions of the past few hundred million years marked catastrophic events never received serious attention because no evidence of a promising catastrophic mechanism had been recognized. That changed a decade ago when researchers, studying a few sites scattered over the globe, reported that strata that marked the mass extinction of about 66 million years ago contained anomalously high concentrations of iridium and related platinum-group metals. Similarly, high concentrations have now been reported from dozens of localities worldwide of the same age. They are considered strong evidence of a catastrophic event close to the time of extinction of many groups of animals, of which dinosaurs are the most notable. This latter extinction event is one of the five largest extinction events to have occurred since fossils became abundant—the largest, 250 million years ago, is represented by a less detailed record. But many things were happening 66 million years ago; evidence indicates that the atmosphere and ocean were cooling and that the sea level was rising again after having fallen. Because the last vestiges of some life forms ceased to appear in strata that date to before the time of the iridium anomaly, it is argued that a catastrophic event may have been the final blow at a time of general environmental deterioration.

Two kinds of nonbiological catastrophic perturbations that have been suggested are a global volcanic episode and the impact of a meteorite. A high platinum-metal concentration could indicate origins from within the mantle, though such a volcanic episode would have to be extremely intense, and one of the largest lava eruption events of the past few hundred million years formed the Deccan basalts in India at just the right time. Catastrophic perturbation by extraterrestrial impact seems more probable, not only because meteorites exhibit high concentrations of platinum-group metals in the right proportion but also because of an association of the iridium-rich horizon with quartz grains showing the effects of intense shock. Shock features are commonly found with meteorite impacts but are unknown in volcanos. And perhaps even more significant is the existence of small beads of glassy material of the type produced by the heat released by meteorite impacts—microtectites. Several mechanisms have been proposed that show how meteorite impact could result in sufficient environmental modification to cause catastrophic mortality. One computer simulation portrays the ejection of a dust layer into the high atmosphere that would cut out sunlight for months at a time, ending photosynthesis and sinking all latitudes into a deep cold. After the dust settled, high levels of atmospheric carbon dioxide would warm the Earth by producing a greenhouse effect. Another model suggests that the fast-traveling ejecta from the impact site could have been at temperatures high enough to cause atmospheric oxygen and nitrogen to combine, forming clouds that would precipitate into nitric acid rain. A third hypothesis, which is gaining increased acceptance, is that an impact of the proper age, formed the 180-km-diameter Chicxulub crater in the Yucatan of Mexico. Chicxulub strata is composed of thick sulfate-rich evaporite and carbonate deposits; an impact into such deposits could eject huge amounts of sulfate aerosols into the stratosphere, resulting in large-scale global cooling and several years of acid-rich rains as the aerosols settled out of the atmosphere—both the cooling and acid rains could have devastated the food chain. Several issues complicate the impact scenario—one strong argument cites evidence of animal groups that suffered heavy extinction before the primary iridium anomaly settled into place.

Regardless of what the eventual consensus on its cause may be, the iridium anomaly has stimulated intense geological research. Not only have the contemporary rocks been studied closely—the anomaly has been documented at about 100 sites throughout the world—but the deep ocean record has been scrutinized, revealing a perturbation of the earth system that lasted half a million years. The most significant result of this intense geological research may be that serious consideration of sudden global catastrophes is now acceptable.

All meteorite impacts are now receiving great attention. The roughly 100 known impact craters on Earth are being looked at anew, their ages are being reassessed, and attempts are being made to see whether their incidence could be periodic. An impact site in Sweden, 50 km in diameter, has been drilled to a depth of more than 6 km, allowing observations about the effects of large-body impact at depth. The discovery of iridium anomalies in some of the oldest rocks strongly suggests a flurry of meteorite impacts as late as 3.6 million years ago. And the most drastic of catastrophes, collision between the Earth and a Mars-sized body before 4 billion years ago, is now considered a possible explanation for the Moon's formation.

Observations of asteroids show that the current meteorite flux is likely to be higher than had been considered, and in 1989 one carefully observed asteroid passed closer to us than any other in the past 50 years. There is a calculable, if remote, chance that the Earth will be struck by an object more than a few kilometers in diameter in the foreseeable future. Such an event would cause massive destruction, and a case can be made for assessing the possibility of diverting an incoming object to avert such a potential collision.

No other global iridium anomalies have been recognized, although a locally defined anomaly dated at 34 million years ago corresponds to a substantial extinction. No iridium or shocked quartz horizons have been found for the greatest extinction ever, which was 240 million years ago; however, the absence of evidence, especially in older rocks, should not lead to a positive conclusion. There is certainly evidence that other ancient mass extinctions were complex events, extending over intervals of several million years.

The mainstream of the earth sciences has shifted away from the extreme uniformitarian position, which attempts to explain all phenomena in terms of directly observed processes. Now researchers, confident in the soundness of their inductive methodology, can consider the possibility of exceptional events. There is, of course, an appropriate reluctance to invoke exceptional events as causal mechanisms, and intense skepticism deservedly awaits all such suggestions. The study of catastrophic effects on earth systems requires further work. Promising directions for general research include their continuing consequences, such as how perturbations propagate through the earth system and how secular change is altered permanently by catastrophe.

MODELING THE EARTH SYSTEM

An Incomplete Record

Solid-earth scientists, like all other scientists, have long used conceptual models as an aid in understanding. For example, it was recognized in ancient times that the occurrence of seashells in rocks on the continents required a model in which the disposition of sea and land changed with time. By the nineteenth century this idea had developed further with the recognition that there had been widespread episodes of continental flooding that were correlatable over great distances. In the mid-twentieth century the repeated flooding and emergence of North America over the past 550 million years were found to provide a strong framework for understanding historical geology, and, during the past 20 years, flooding and emergence have been found to be approximately simultaneous for at least three continents. This coincidence has led to the suggestion that the main driver of sea-level change is within the world ocean. The amount of seafloor spreading going on at a particular time seems a strong candidate for the dominant control. Young seafloor is hot and shallow, and aged seafloor is cold and lies deep. The average rate of spreading and the total length of spreading center at any one time can be quantitatively associated with the extent of flooding of the continents.

Conceptual models that are progressively refined may be considered successful for parts of the earth system but are not really comprehensive. The question of continental flooding helps to show why this is so. The amount of water that is locked up in ice has decreased since about 20,000 years ago with the melting of North American and Eurasian ice sheets. This decrease has accounted for about 100 m of sea-level rise over the same interval. This is a very large change in sea level compared with the range of continental elevation. About 80 percent of the continental area lies within 1 km of the present sea level. Although we can estimate the change in the average age of the ocean floor for the past 100 million years we cannot do so well for the amount of water locked up in ice sheets. Evidence that there were no major ice sheets 100 million years ago is strong, but it is not clear how long ago the great Antarctic ice sheet formed, nor can we tell whether it has fluctuated much or even disappeared and been renewed. This example illustrates a general situation. Comprehensive modeling of earth systems is very difficult, because although some parts of the system are understood, fluctuations in critical variables for other parts of the system are too poorly known to justify the construction of elaborate models. The challenge represented by the fact that some parts of the earth system can be readily modeled whereas others cannot has been met by concentrating on modeling those subsystems that can be handled well and by seeking innovative ways of quantifying information about subsystems that cannot. A further problem is linking the models of system parts that operate at different rates, such as the atmosphere, which changes on a very short time scale, and the ocean, which changes more slowly. The most successful and productive models reconstruct cyclic changes involving limited components of the earth system. Only a few components, or variables, can be tested at one time, but every successful simulation provides further understanding.

Some aspects of paleoceanography and paleoclimatology lend themselves to quantitative modeling in isolation from the whole. However, most are not, and, because of the interdependencies, the models should be global in scale, link adjacent water masses, and couple the atmosphere with the oceans, so great is the interdependence of these two great fluid envelopes. Atmosphere and ocean are linked especially by exchanges of heat, momentum, carbon dioxide, and, of course, water. The surface temperatures of the ocean influence these fluxes. The salinity of waters that reach the ocean bottom, a variable related to climate, also affects oceanic circulation. And changes in global sea level affect the levels of atmospheric carbon dioxide, which in turn influence global climate through the greenhouse effect. Outstanding successes include modeling of the glacial cycles of the past 0.8 million years, as controlled by variations in orbital parameters, and modeling of sedimentary basin fill, as controlled by loads applied to the lithosphere.

Modeling of atmospheric systems, which mainly use energy balance models and general circulation models (GCMs), has grown to be extremely sophisticated in recent years, and solid-earth scientists have found some community of interest with atmospheric modelers. For example, at the time of the most recent glacial maximum, the solar energy input at the surface was less than it is today. A GCM of that time also indicates wind and temperature distributions that are very different from those of today. These indications have been compared with the geological record of the most recent glacial maximum with some success. Simplified energy balance models for the atmosphere for 100 million and 375 million years ago, when continental distributions were very different and much of the surficial climate was warmer than it is now, have also met with some qualified success. There are some problems, too—for example, existing GCMs for the glacial maximum will not predict conditions that would grow the large ice sheets that existed then.

Intellectual Frontiers

Understanding the processes that are active today in establishing the surface environment and understanding how those processes have operated throughout geological time are the basic challenges addressed in this chapter. Intellectual frontiers relate to questions about the surface environment that have not been asked in the past, either because it was not possible to ask them or because it was not recognized that they had meaningful answers. A distinctive new development is the perspective that views the environment as a complex of interactive systems. Now specific problems are posed, and solved, as part of a broader framework of global understanding. Remotely sensed imagery from space and organized international cooperation have done much to stimulate the global approach. In the next decade the operation of higher-resolution instruments on advanced space platforms, such as those envisaged for the Earth Observing System, will enhance the global perspective. But perhaps the most important efforts toward global understanding are made through programs that depend on international cooperation among scientists. The Ocean Drilling Program exemplifies this trend, as does the innovative International Geological Correlation Program and the International Geosphere-Biosphere Program.

Dating Past Events

Because geochronology scales physical, chemical, and biological events against time, it plays a fundamental role in the earth sciences. To appreciate what has happened, we need to know the sequence of events and the rates of change.

Quantitative biostratigraphic techniques yield correlations with accuracies approaching a few hundred thousand years for bodies of rock that are hundreds of millions of years old and lie thousands of kilometers apart. These methods of correlation are integrated with others, including paleomagnetic methods and radiometric dating of marker beds such as volcanic ashes. Together they encourage the search for high-frequency events and for regular patterns in such events. Correlation techniques and isotopic dating serve as checks on each other. Isotopic dating methods can focus on scales from billions to mere thousands of years, but when possible they should be integrated with other dating methods.

Global event stratigraphy correlates the worldwide expression of certain events. It provides a framework of additional instantaneous markers against which intervening events can be calibrated. The stratigraphic evidence of rapid global sea-level change falls within this category, as does the identification of global chemical signals. The chemical signals include both narrow stratigraphic markers that formed during brief moments of geological time and long-term secular trends that trace continuing developments. The iridium anomaly, which marks the mass extinction of 66 million years ago

and may signal the impact of a huge meteorite, is the most famous geochemical marker in the stratigraphic record. But others have been, and will continue to be, discovered.

All of the challenging areas of research described in the following paragraphs—historical studies of oceans and atmospheres, terrestrial environments, and life on Earth—depend on advances in geochronology.

Atmospheric and Oceanic Chemistry

On the longest time scale, the geological record indicates a gross change from a reducing to an oxidizing state of the linked atmosphere-ocean system. The details of timing and the reason for this secular change still provide topics for lively debate. On shorter time scales, the record of the rocks preserves evidence of cyclical changes. Geologists can trace variable concentrations of carbon dioxide in the atmosphere and ocean on time scales ranging up to hundreds of millions of years. They have also distinguished episodes during which much of the deeper ocean was anoxic. Intervals when widespread anoxia in deep waters expanded to flood broad areas of the continents are especially interesting, because they resulted in the massive accumulation of valuable hydrocarbons from sources in black anoxic sediments. Some researchers think that the storage of so much organic carbon implies the possibility of an increased oxygen content in the atmosphere at such times in the past. Others consider that an inflammatory concept.

Past oceanic composition, recorded within ancient sediments, reflects many aspects of the global environment. These include the mantle contributions through volcanism and continental input through erosion, the global climate, the presence or absence of ice caps, and the level and kinds of biological activity. The history of ocean chemistry can be established from the rock record—an endeavor that is rewarding because it has been so successful. For the past 150 million years, the interval when the sediment now carpeting the deep seafloor has been accumulating, ocean chemists can study changes among the individual water masses that together make up the world ocean.

The most fundamental variable controlling atmospheric and oceanic chemistry has been the temperature of the deep sea. But patterns of upwelling and shallower currents and high biological productivity have undergone dramatic shifts at frequent intervals. Changes in oceanic conditions, especially sea level, have exerted a strong control over evolution and extinction and also over the formation of valuable resources.

Dynamics of the Global Environment

The evolution of the environment is an important area of research in modern earth sciences. The study of ancient conditions, when climates were warmer than today, offers insight into potentials of modern greenhouse warming. In addition, studies of different global ecosystems of the past may reveal peculiarities of the present world that render it especially vulnerable to certain forcing factors. The most recent segment of the geological record provides the temporal resolution and geographic control needed to identify very sudden environmental changes.

During the first 4 billion years of earth history, life arose and evolved through many intermediate stages to a point at which a variety of multicellular plants and animals existed. Evolution and extinction during this interval were tightly interwoven with profound changes in the physical nature of our planet, especially its atmospheric chemistry. This intimate relationship between life and environment serves to underscore an important point about future directions of research. Different intervals of earth history require distinct scientific strategies, because the intervals were characterized by different kinds and degrees of environmental change and are represented by different kinds of geological records.

The prospects are exciting. But they require a prodigious amount of research to chronicle global environmental change for a variety of past intervals with the resolution required. Once models can accurately simulate environmental conditions of key intervals during earth history, they can be used for predicting future conditions. Collaborative modeling projects that unite workers in the geological sciences with meteorologists and oceanographers are beginning to yield results.

Life Through Time

Numerous advances have breathed new life into paleontology. The contributions of the fossil record to the study of evolution and extinction uniquely document myriad forms unknown in the modern world. A cumulative chronology indicates the rates of evolution and extinction, and the timing of major events in the context of environmental change. Such a chronology establishes fluctuations and patterns that can point to new questions about the history of life. The development and application of quantitative techniques will continue to play a prominent role in future studies of life through time. Key methods will assess morphological change; patterns of evolution and extinction; and theoretical modeling of taxonomic, stratigraphic, and environmental data obtained from the fossil record.

The fossil record also provides evidence of the timing of evolutionary branching. The molecular clocks used to estimate the times when certain extant groups branched from others must be calibrated against fossil data, and conclusions must be tested against macroscopic evidence. Of special importance are fossil data that reveal the occurrence of adaptive breakthrough—evolutionary innovations that ushered in new modes of life that transformed the ecosystem. These can be recognized in the fossil record only by inferring function from morphology, an activity that merits increased support. Many adaptive breakthroughs have triggered adaptive radiations—diversification from a single life form that results in the invasion of a variety of ecological niches. Such radiations have special significance because they account for most of the major evolutionary changes in the history of life. Understanding the general diversification of life on Earth requires an understanding of adaptive radiations.

Sudden extinctions—the negative equivalent of adaptive radiations—have repeatedly reordered the global ecosystem and opened the way for new evolutionary directions. Mass extinctions can be understood only in the context of global environmental change. The taxonomic, temporal, and geographic patterns that have characterized these events are especially significant; compilation of new data demands the continued generation of high-quality biostratigraphic and taxonomic information.

Whether global biotic catastrophes have occurred in combination with background extinction or instead have been quantitatively or qualitatively distinct is a question that can be answered only by understanding the patterns and causes of noncrisis extinctions.

The Most Recent Past

Geologists know that the record of the most recent past is exceptional because its complex history can be better established than that of any earlier period. Powerful new techniques are determining the ages, isotopic compositions, and temperatures of materials deposited during the past 2.5 million years. But the most significant recent development is the worldwide awareness of changes in the global environment caused by humanity. This new awareness has stimulated a need to better understand the

TABLE 3.1 Research Opportunity Framework

Research Opportunities

Research Areas	A. Understand Processes	B	C	D
I. Paleoenvironment and Biological Evolution	■ Soil development, history, and contamination ■ Glacier ice and its inclusions ■ Quaternary record ■ Recent global changes ■ Paleogeography and paleoclimatology ■ Paleoceanography ■ Forcing factors in environmental change ■ History of life ■ Discovery and curation of fossils ■ Abrupt and catastrophic changes ■ Organic geochemistry			
II. Global Geochemical and Biogeochemical Cycles	■ Geochemical cycles: atmospheres and oceans			
III. Fluids in and on the Earth	■ Analysis of drainage basins ■ Mineral-water interface geochemistry			
IV. Crustal Dynamics: Ocean and Continent	■ Landform response to change ■ Quantification of thresholds, response rates and feedback mechanisms for landforms ■ Mathematical and computer modeling of landform changes ■ Sedimentary basins ■ Sequence stratigraphy			
V.				

Facilities, Equipment, Data Bases
■ Exploit new tools and techniques (e.g., isotopes, trace compounds, DNA sequencing and hybridization, digitizing techniques)
■ Exploit new dating techniques (e.g., radiometric methods, trends in isotope ratios, biostratigraphic correlation, chemical markers in stratigraphy)
■ Acquire high-quality data bases and establish information systems

immediate past that is forging closer links between geologists and other earth scientists. The history of human evolution is beginning to unfold in sufficient detail to reveal the kind of environmental influences that affected human ancestors. All the challenges that have been identified here as intellectual frontiers have special possibilities for resolution when addressed in light of our understanding of the ongoing global changes during this most recent geological period.

RESEARCH OPPORTUNITIES

The Research Framework (Table 3.1) summarizes the research opportunities identified in this chapter, with reference also to other disciplinary reports and recommendations. These topics, representing sig-

nificant selection and thus prioritization from a large array of research projects, are described briefly in the following section. The relevant processes operate near the surface for the most part, although there is no sharp boundary between surficial geology and the deep-seated processes covered in Chapter 2. There are excellent prospects for generating better models of the earth system, both present and past, including global tectonic models, coupled ocean-atmosphere biogeochemical models of the fluid envelope, and paleoecological models of the biosphere. The fossil record of the biosphere provides information about evolution as well as evidence of environmental changes and migration of continents.

The relevant Research Areas for this chapter are interrelated, and many of the research topics relate to more than one area. The carbon cycle (Area II), for example, links to life and biological evolution (Area I) and has influenced paleoatmospheres and paleoceans theme (Area I), as well as playing a major role in continental weathering theme (Area III). It also has connections to the mantle through the igneous and metamorphic processes in the lithosphere and crust (Area IV), as discussed in Chapter 2. Some of the many applications of these Research Areas are outlined in Chapters 4 and 5.

Research Area I. Paleoenvironment and Biological Evolution

Soil Development, History, and Contamination

If extended to soils, the new age-determination capabilities that are being applied to landforms and drainage basins could quantify temporal aspects of soil development that have been the subject of much speculation. More complete understanding of soil processes as aspects of environmental, especially climatic, evolution will improve both paleosoil and soil-contamination studies.

Glacier Ice and Its Inclusions

Given the important role of glaciers in controlling sea level and climate, and considering the evident instability of glacial volume, it is essential that we gain a clearer picture of the history of glaciation since the onset of the recent ice age about 2.5 million years ago. Especially important too is further study of the annual layers in ice cores, including analysis of oxygen isotopes, carbon dioxide, and dust to obtain a record of the past few thousand years. We need a much better understanding of glacial expansion and contraction and the role of environmental thresholds in governing these processes.

Quaternary Record

The record of change in the Quaternary (the past 1.6 million years) is important because it is the most complete available record for any part of the past and thus provides the best picture of environmental change. Synthetic and snyoptic studies of Quaternary history (e.g., CLIMAP and COHMAP) have led the way. Extending studies of this kind with better spatially distributed data and higher temporal resolution will help to show how phenomena occurring simultaneously in different areas were related to each other, and will indicate sequences of events. This kind of information is needed to test atmospheric and oceanic models that are being used

to assess what might happen in future global environments.

Recent Global Changes

The evolution of the ocean, atmosphere, and life during the first 4 billion years of the planet's history is of great intellectual appeal and deserves continued emphasis. Nonetheless, of greatest practical value are studies of global change during the most recent interval of geological time. The geological record of the past 10,000 years has great potential for shedding light on the ways in which climatic changes affect life. In light of the global warming anticipated for the coming decade, warm intervals of the past should be scrutinized for possible lessons for the future. In addition, intervals exhibiting large-scale environmental change or mass extinction of life warrant detailed scrutiny for patterns and causes.

Paleogeography and Paleoclimatology

The past few years have seen major advances in paleoclimatology and terrestrial paleogeography. Sedimentary indicators and fossil biotas, especially floras, continue to shed light on climates of the past, but there is also a need to refine the existing kaleidoscopic picture of changing continental configurations, especially for the long pre-Mesozoic (older than 66 million years) interval of earth history. Global paleogeographic reconstructions must continue to be developed and refined in all possible ways, including paleogeographic interpretations, global paleobathymetric reconstructions, identification of accreted terranes and microcontinents both through anomalous fossil distributions and paleomagnetic determinations. Determination of the rates of geochemical cycling requires knowledge of the sizes, elevations, and positions of continents in the past.

Paleoceanography

Changing climatic regimes, plate positions, and levels of land and sea have had powerful effects on the thermal structure and current patterns of the world ocean. At the forefront of paleoceanography today are studies of the three-dimensional oceanic structure, including thermohaline circulation, conditions in the deep sea, upwelling, and vertical zonation of plankton. The Ocean Drilling Program should continue to play a key role in this kind of research for the most recent 150 million years of earth history, as should the use of isotopic data. The global influence of events in polar regions is in special need of further study.

Forcing Factors in Environmental Change

We are only beginning to understand the interrelationships between continental configurations, the dynamics of the ocean and atmosphere, and the distributions of life on Earth. In modeling global environmental change, sensitivity experiments that suggest what forcing factors have pushed environmental conditions across thresholds to new states are often more successful than detailed global simulations. At present, models are outstripping the data needed to constrain and test them, and research that will provide additional data is badly needed.

History of Life

Inasmuch as the fossil record represents a unique store of information on rates, trends, and patterns of evolution and extinction, we must continue to exploit it to understand these aspects of the history of life. There is still no consensus on such issues as the incidence or cause of evolutionary stasis, the degree to which extinction occurs in pulses, or what environmental changes trigger rapid evolution, including bursts of speciation. New quantitative techniques, including morphometric methodologies, must play an important role in research here. There is also a need to develop methods of phylogenetic analysis that integrate morphological and molecular approaches with stratigraphic data. Interactions between life and the environment—for example, how much the changes in atmospheric composition have been responses to evolutionary change and how atmospheric change has influenced evolution—must be established.

Discovery and Curation of Fossils

Our understanding of the processes of biological evolution continues to be refined by the discovery and description of fossils that fill gaps in the record. Examples of great steps forward made within the past decade are the discovery of new localities and material indicative of the diversity of marine life 550 million years ago, identification of the conodont animal and appreciation that it was a vertebrate, critical new specimens of the first terrestrial vertebrates, and (stepping back into the ocean?) a whale with vestigial limbs. The kind of relatively unglam-

orous fieldwork that leads to these successes requires ongoing commitment.

Abrupt and Catastrophic Changes

Sudden events have a lot to teach earth scientists, particularly where their record extends over the whole world or at least very large areas. What are the causes of these events? Are they of impact, volcanic, or other origins? Does the rock record indicate precursory phenomena? Was there subsequent environmental change? If there is evidence of change, how long did it last? The behavior of the environment under stressed or extreme conditions is likely to be informative and rapid climatic changes are of special interest in the field of global change research. It is worth noting that had this report been written a dozen years ago little emphasis would have been given to catastrophes.

Organic Geochemistry

There are diverse ways in which organic geochemistry is yielding new information. New techniques for isotopic analysis of specific organic compounds, "chemical fossils," provide opportunities for reconstructing the temperatures, compositions, and oxidation states of the ancient ocean. Working out the role of ancient microbial communities in sediments and illuminating the thermal histories of sedimentary basins are some of the challenges.

Research Area II. Global Geochemical and Biogeochemical Cycles

Geochemical Cycles: Atmospheres and Oceans

It is crucial that we improve our understanding of geochemical cycles to learn how the atmosphere and oceans have changed in the past and how they may change in the future. Controls of atmospheric carbon dioxide, and the resulting greenhouse effect, are of especially great significance. Geochemical cycles are complex dynamic systems that entail geological, biological, and extraterrestrial processes. Changes in fluxes and in sizes of chemical reservoirs must be more accurately quantified for the geological past. Among the controls needing further study are the compositions and abundances of sedimentary rocks, magmatic and metamorphic degassing, biological uptakes and emissions, sea-level change, rates of weathering, and isotopic shifts for key elements. Even fluxes to and from the modern ocean are poorly known, as are the contributions of relevant organic biogeochemical processes. Mathematical techniques for modeling biogeochemical cycles can be greatly improved.

Research Area III. Fluids in and on the Earth

Analysis of Drainage Basins

To understand past changes in our terrestrial habitat and to anticipate future changes and their consequences, it is imperative that geomorphologists undertake quantitative analysis of drainage basins. This analysis should quantify linkages and pathways of weathering sediment erosion, storage, transportation, and contamination. Links to sequence stratigraphy (see Chapter 4) will come from identifying the impact of base-level changes on fluvial systems. One important question is how the history of river flooding for the past few thousand years has related to climatic change. On long time scales there is a need to relate the rates of surficial processes to tectonic activity.

Mineral-Water Interface Geochemistry

The rapid growth in this field will lead to a more fundamental understanding of weathering, the chemistry and physics of mineral-water interfacial phenomena, how chemical species partition between minerals and aqueous fluids in crust, and how the hydrosphere interacts with crustal rocks.

Research Area IV. Crustal Dynamics: Ocean and Continent

Landform Response to Change

Landforms represent the products of tectonic, climatic, and hydrologic processes. The ability to date surfaces and shallow sediments using cosmogenically generated nuclides and other direct and indirect methods provides the opportunity for improving understanding of how these processes generate landforms. The current strong interest in how tectonic, climatic, and hydrologic processes change with time makes this opportunity particularly timely.

Quantification of Thresholds, Response Rates, and Feedback Mechanisms for Landforms

Further definition is needed for stability/instability thresholds, response and recovery times for landforms following destabilizing events, and the significance of return frequencies of key climatic events for landform stability.

Mathematical and Computer Modeling of Landform Changes

Modeling will elucidate landform changes over longer periods of time than it is possible to observe them. Process-based models are essential for understanding mechanisms and for predicting future changes. Exciting progress has been made with, for example, hillslopes and river channel patterns.

Sedimentary Basins

Depositional basins are likely to reward detailed study because they record so many diverse kinds of information. The integrated approach to basin studies—which embraces structural, erosional, and depositional evolution as well as thermal, chemical, and fluid-flow history and uses field, seismic, well-log, geochemical, and isotopic data—is proving extremely powerful. Foreland basins and rifted continental margin basins, including great deltas, harbor the world's largest volumes of hydrocarbons. We are learning about how the sediment accumulation in these environments relates to tectonic and sea-level changes and apparently to orbital parameters.

Sequence Stratigraphy

The identification of sediment packages that are separated by what appear to be abrupt temporal boundaries is an exciting new tool because it permits characterization of environmental changes that can be linked to episodes outside the area of deposition. Some researchers correlate boundaries in sequence stratigraphy globally, but this practice is questioned by others. Improved access to well-calibrated sequence data may help to resolve this issue.

Facilities, Equipment, and Data Bases

Data Bases in Well-Managed Computer-Based Information Systems

The major problem facing paleogeographers, paleoclimatologists, and paleoceanographers is that the ability to predict and understand the earth system at high resolution is outstripping the availability of the data required to test the methods and models. Vigorous collection of and rigorous quality control on data from the ancient record is urgently needed. Data and samples taken on continental margins and from wells drilled by industry represent an invaluable resource, as do many kinds of fossil data. The data base available to scientists could be expanded by several orders of magnitude.

New Tools and Techniques

New equipment offers much potential that has yet to be exploited in paleontology in the areas of isotopic analysis; identification of trace organic and inorganic compounds, DNA sequencing and hybridization, new multivariate techniques, digitizing techniques for three-dimensional quantification of morphology, application of geographic information systems (GIS) to quantitative historical biogeography, and computer-interfaced digitizers (some three-dimensional and some linked to light microscopes or scanning electron microscope systems) for morphometric studies.

New Dating Techniques

During the past few years, new dating techniques have greatly improved resolution in the temporal correlation of geological and biological events and in the measurement of rates for a wide variety of processes. These techniques include identification of key chemical markers in stratigraphic sequences, quantitative biostratigraphic correlation, application of new radiometric methods, and utilization of secular trends in stable isotope ratios. Not only the development of new dating techniques, but also the refinement and ingenious application of existing methodologies, will benefit numerous areas of research in the decades to come.

BIBLIOGRAPHY

National Research Council Reports

NRC (1982). *Studies in Geophysics: Climate in Earth History*, Geophysics Study Committee, Geophysics Research Board, National Research Council, National Academy Press, Washington, D.C., 198 pp.

NRC (1983). *Opportunities for Research in the Geological Sciences*, Committee on Opportunities for Research in the Geological Sciences, Board on Earth Sciences, National Research Council, National Academy Press, Washington, D.C., 95 pp.

NRC (1986). *Global Change in the Geosphere-Biosphere: Initial Priorities for an IGBP*, U.S. Committee for an International Geosphere-Biosphere Program, Commission on Physical Sciences, Mathematics, and Resources, National Research Council, National Academy Press, Washington, D.C., 91 pp.

NRC (1989). *Technology and Environment*, Advisory Committee on Technology and Society, National Academy of Engineering, National Academy Press, Washington, D.C., 221 pp.

NRC (1989). *Margins: A Research Initiative for Interdisciplinary Studies of Processes Attending Lithospheric Extension and Convergence*, Ocean Studies Board, National Research Council, National Academy Press, Washington, D.C., 285 pp.

NRC (1990). *Research Strategies for the U.S. Global Change Research Program*, Committee on Global Change, U.S. National Committee for the IGBP, National Research Council, National Academy Press, Washington, D.C., 291 pp.

NRC (1990). *Studies in Geophysics: Sea-Level Change*, Geophysics Study Committee, Board on Earth Sciences and Resources, National Research Council, National Academy Press, Washington, D.C., 234 pp.

NRC (1991). *Toward Sustainability: Soil and Water Research Priorities for Developing Countries*, Committee on International Soil and Water Research and Development, Water Science and Technology Board, National Research Council, National Academy Press, Washington, D.C., 65 pp.

Other Reports

The International Geosphere-Biosphere Programme: A Study of Global Change (IGBP) of the International Council of Scientific Unions (1992). *The Pages Project: Proposed Implementation Plans for Research Activities*, Pages Scientific Steering Committee, 110 pp.

NASA (1988). *Earth System Science: A Closer View*, Earth System Sciences Committee, NASA Advisory Council, Washington, D.C., 208 pp.

Office of Science and Technology Policy (1992). *Our Changing Planet: The FY 1992 U.S. Global Change Research Program*, Committee on Earth and Environmental Sciences, 90 pp.

4

Resources of the Solid Earth

ESSAY: NATURAL EXPLOITATION

All organisms exploit the surrounding environment to sustain their metabolism, and in the process they produce waste. Prokaryotes, single-celled bacteria, have colonized every niche on the Earth's surface, absorbing local chemical compounds and excreting unused or altered compounds. Cities with vast human populations do the same thing—they use natural resources and produce waste. Whether a single-celled organism or a complex society, the efficient location and exploitation of resources and the production and appropriate disposal of neutral or recyclable wastes determine success or failure, life or death.

The resources required for a complex civilization are more diverse than those needed by simpler organisms. Besides the obvious needs for primary materials—chemical and energy sources—shared by all living beings, humans depend on physical and biological systems that process primary resources. The ability to invade potentially hostile environments depends on the use of many of the Earth's subsystems. We require large amounts of water and landscapes that provide the means to feed and shelter us.

Water may well be the Earth's most vital resource. While regions subject to drought are usually concerned with the quantity of available water, all communities must now care for the quality of water. In the United States, irrigation methods and personal habits deplete ancient aquifers and drain mountain runoff. At the same time, toxic chemicals and dangerous pathogens can seep into water supplies. Only with conservation and careful planning will water become a renewable resource again; education programs are needed to deter both sophisticated cultures and simple communities from fouling their own supplies.

All organisms run on energy. In plants and most animals energy is supplied by proteins and sugars. Human societies require fuel to operate the mechanisms that supply food, warmth, and services. Less complex communities depend on wood for energy, but since the Industrial

Revolution energy sources have expanded to include gas, coal, oil, steam, and fissionable materials.

Mineral resources produce the tools that enable society to supply water, food, fuel, and manufactured goods to growing populations. Exploration and extraction respond quickly to market values of minerals, and market values vary according to technological innovation. Certain minerals will always be needed, but there is a point at which the price of supplying a product to market is more than the price of developing techniques that may render that product obsolete. Because technological developments cannot be predicted in the long run, the need for some minerals cannot be predicted.

Curiously enough, research into the formation of all these resources involves an understanding of fluid movement within solid materials. Separate investigations of the solid, fluid, and gaseous envelopes of the Earth are no longer possible. Researchers now are reorienting their goals to address an earth system that integrates geology, hydrology, and atmospheric sciences with biochemical elements. The breadth of the earth system allows for full appreciation of the consequences of concentration and redistribution of material by natural causes and human intervention.

Often the search for resources is frustrated, and even when it is successful their extraction can be difficult and dangerous. To warrant the effort there must be a degree of profit. As market values change, managers must decide whether a particular resource is *economic*. If it is economic, it is worth the effort to discover and extract deposits. There are unlimited amounts of all the compounds and elements that humankind needs, but they are not usually concentrated into deposits that are economically worthwhile. If a deposit is worth the effort, or if it can be reasonably assumed that it will become worthwhile, it is classified as a *reserve*. New discoveries and new extraction techniques contribute to the increasing quantity of what can be considered reserves.

But the rising population continually puts more demand on available resources. As accumulations of waste material grow and concentrations become more hazardous, a new factor affects assessments of a resource's economic value: the cost of landscape reclamation and by-product neutralization. Communities have come to understand that they can no longer dump sewage into streams and let solid wastes accumulate indefinitely. Mining enterprises have come to understand that they can no longer leave mountainsides gouged and stripped, with piles of tailings leaching toxic compounds into the watershed.

As understanding of the earth system grows, resource scientists can moderate the effects of society on nature. One exciting research area is the ability of wetlands to clean contaminated fluids. Wetlands and marshes were once considered useless wasted tracts. If not left to hunters and trappers, these regions were often drained. Researchers have now concluded that the chemical and biological processes that prevail in marshy environments can remove toxic compounds from contaminated water. Constructed wetlands are now being introduced as systems that cleanse wastewater and return acid levels to natural levels. Specific bacteria can clean up specific compounds through bioleaching, biosorption, and bioreduction.

Researchers have developed ways to reclaim mining sites by conservation methods and recycling. Techniques used for extraction of minerals from low-grade ore can be adapted to detoxify contaminated soils as well as water. And mining practices in use now, such as surface landscaping and sealing mine walls to reduce seepage through the mine into the water table, have successfully minimized environmental modification. Incorporating the traditional practice of backfilling spent caverns with tailings to prevent subsidence, modern reclamation efforts strive to recreate the original scene as closely as possible.

Recovery of petroleum resources also can threaten the environment. Oil spills come to mind—the Exxon Valdez spill and the oil spill released into the Persian Gulf during the fighting to reinstate the status quo of the Kuwaiti Emirate. But petroleum geologists are proud of their efforts to reduce the dangers of spillage during the production phase. Today, oil lost to the environment during production is much less than the amount released naturally through seepage. The greatest dangers from oil spillage are found in the transportation phase—and tanker transportation is the most dangerous of all.

The burning of coal produces emissions that accelerate global warming trends. Nuclear energy creates radioactive by-products that can threaten the environment in a variety of ways—as emission into the atmosphere or invasion into the groundwater. Even burning wood causes smog. But nature does have the ability to absorb some of this abuse—it has a carrying capacity. The carrying capacity is the upper limit of a system's ability to support all components within the bounds of available resources. When the carrying capacity is exceeded, a threshold is crossed, and new equilibria establish themselves. Often a new equilibrium spells disaster for the components supported by the former system. For living beings crossing such a threshold can mean extinction.

The carrying capacity of a natural system may be threatened by various means. Natural climatic changes can devastate landscapes and destroy soils just as efficiently as humans can in their roles of gatherer, farmer, or skyscraper builder. Thresholds of stress are crossed whenever a natural disaster hits—earthquakes, volcanoes, and floods all result from natural systems thrown into a different order. Seismologists, volcanologists, and hydrologists experiment and observe, in search of greater understanding of the natural systems, their carrying capacities, and their dangerous thresholds. That understanding should result in the ability to predict the potential for problems, and such predictions might permit mitigation of the intensity of an event—or at least provide time to warn populations to evacuate endangered regions and thus save lives.

The humans who occupy this planet are profoundly dependent on the ordered operation of its natural systems. Human populations exploit the minerals, fuels, soils, and waters, straining against capacities and adapting to limits. That straining and adapting is what characterize life: it's completely natural—as extinction will be, if the human species strains too hard or adapts too slowly.

ROCK-FLUID INTERACTIONS

Traditionally, the science of geology concerned rocks and minerals—the solid part of the natural environment. But over the past two centuries, geologists have become increasingly aware that divisions between solid, liquid, and gaseous environments can be unnecessarily restrictive. The phases are interactive and interdependent; they are not distinct. The nature of the crust cannot be understood separately from the atmospheric and oceanic systems. Most recently, earth scientists have been surprised to detect evidence of organic forms thriving in the depths of the crust and in the heights of the atmosphere. Rock, water, air, and biota interchange molecules throughout a layer of skin covering the planet. The list of interdependent processes lengthens every day as new discoveries are made, and it reflects much geological research—as it is currently defined.

Understanding of rock-fluid interactions began with the study of groundwater flow. Empirical laws were developed in the 1800s to define the rate and intensity of the underground flow feeding springs and wells. Further studies have revealed characteristics of fluid systems that are determined by permeability of the rock, composition and volume of the fluid, and pressure and temperature gradients within the system. These characteristics can be determined close to the surface through wells, but rock-fluid interaction occurring at great depth must be deduced from examination of outcropping rocks that formerly lay far below the surface. Outcrops generally provide evidence of the effects left by deeply circulating fluids, only rarely offering samples of the fluid itself.

Indications from wells and outcrops suggest that fluids may be present in significant amounts at most crustal levels, although they are locally sporadic, and that the surrounding rocks undergo chemical modification as a result of contact with migrating fluids. These conclusions are consistent with observations of the pivotal role played by fluids in the genesis of both mineral and fossil fuel concentrations. Recent geophysical studies have added to the insights about fluids within the crust. Electrical conductivity inferred for great depths suggests a persistent aqueous fluid phase, while seismic reflections from possible low-velocity zones imply abnormally high fluid pressures. This information supports the opinion that extensive fractures and large volumes of aqueous fluids permeate the deep crust.

Researchers testing the behavior of crustal rock

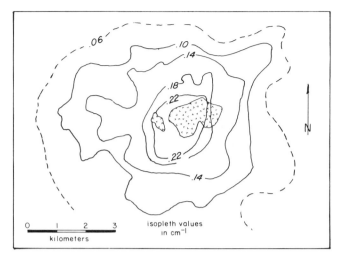

FIGURE 4.1 Vein/joint abundance map of the Sierrita system, Arizona. The total number of fractures per centimeter increase as the pluton (patterned area) is approached. Figure from S. R. Titley, NRC, 1990, *The Role of Fluids in Crustal Processes.*

have concluded that fluid proportion decreases with depth, but they do not understand the transition between the extremes of saturation and scant traces. At every level water acts to redistribute heat in hydrothermal systems. Because an increase in temperature raises fluid pressure, any substantial introduction of magma into the crust will initiate a convecting system of groundwater around the magma body (Figure 4.1). The convecting water transports heat from the magma, controlling its cooling rate, which affects the crystallization rate of the rock and thus some of its subsequent identifying characteristics as a solid.

The convecting water also redistributes chemical components from the intrusive body, and from the surrounding host rock, to more remote locations. Even if the surrounding rocks are initially impermeable, the heat generates enough pore-fluid pressure to open hydraulic fractures, creating permeability. Subsequent episodes repeat the whole process until the fractures fill with minerals, permeability decreases, and fluid flow abates. At the same time, the magma must release its influence along a different front to shut down the intricately integrated feedback system.

Horizontal and vertical movement—convection and advection—of fluid through rock disperses chemical components and supports chemical reactions between the minerals and fluids. Dissolution, precipitation, ion exchange, and sorption continue as the fluids migrate through the matrix. Material that does not dissolve may be forced into migration along the front of the fluid plume; this is what happens to oil and gas. The pressure of the aqueous fluid drives hydrocarbons into reservoir rocks where they collect and remain, if a seal successfully traps them. Otherwise, the pressure of groundwater and their own buoyancy will force them to the surface.

Understanding the role of fluids in the crust necessitates analysis of the interacting thermal, chemical, mechanical, and hydrological processes. Researchers believe that role may extend to influencing, perhaps even initiating, tectonic events. Extremely high pore-fluid pressures, which are characteristic of actively tectonic regions, may facilitate major crustal movements. Frictional resistance to slippage and faulting becomes negligible in certain cases of high pore-fluid pressure.

While seismologists analyze the influence of fluids on fault susceptibility, research continues to reveal how aqueous fluids move in subduction zones and return to the surface through volcanic eruptions. Aqueous fluids also cycle through crustal material at ocean spreading centers, spewing from vents loaded with particulate matter as "black smokers." On continents such fluids bubble to the surface as geothermal springs and geysers.

Ores, those materials that contain valuable metals or other materials, can form by many concentration processes involving chemical reaction with water. Water can seep through soil horizons, leaching solutes away and leaving residual materials such as the bauxite deposits that form aluminum ore. When it reaches solid bedrock, water sustains weathering of minerals and carries away the residue. Deep within the crust, water percolates through the metamorphic zones where igneous intrusions shoulder into the native rocks and contributes to the process of change. And where hot igneous rock and cool saline water make contact along the 40,000-km length of the oceans' spreading centers, researchers can watch minerals precipitate. These observations support analogies that help describe processes characterizing other areas of igneous activity, such as volcanic arcs and continental rifts. Eventually, even the oldest water returns to the surface through uplift, exposure, and erosion; then it quickens again and churns through a shorter episode along the surface.

But once the fluids reach the surface, they do not cease their interaction with the rocks. Oceans continue to pound against the shores, as waves and rocks break each other. Rivers erode the mountains and carry them away. Rain pelts against outcrops, dislodging a grain at a time. Rainwater seeps into fractures and pores, expands on freezing, and thus weathers the rocks mechanically. Water flowing along the surface also dissolves rocks through chemical weathering, forming sinkholes, caverns, caves.

Geologists have always had an appreciation of the links between the solid earth and its fluid envelopes, but they are now realizing that those envelopes permeate the patches of soil that clothe the continents and the ooze that shrouds the ocean floor. Hydrogen, oxygen, and carbon compose a major proportion of the elements that circulate through the air, ocean, and crust in a variety of fluid forms—coupling into compounds, migrating through pores, dissolving, and precipitating. That circulation of hydrous fluids, in different phases and forms, through the few kilometers between the mantle and space makes this planet what it is.

WATER RESOURCES

The Earth is the water planet. This claim is not based on a mere fluid veneer. The distinctive features that set the Earth apart from other solid planets—the deep, wide oceans; the abundance of living beings; even the buoyant, mutable, silica-rich continents—can be attributed to circulation, and concentration, of water. In the narrow view, water is the most critical resource required for human survival. In the wider view, it is a necessary component of many earth subsystems (Figure 4.2).

Most water exists, rich with salt, in the oceans, and many theorists agree that this has been the rule for nearly all of earth history. Theories proposing a hydrous mantle that slowly generated the ocean waters by gradual degassing do not resolve problems associated with retention of large amounts of water-bearing minerals within the host mantle. Water is continuously cycling through the shallow mantle (Figure 4.3). Lithospheric slabs plunge down into the mantle at subduction zones, notably along the deep oceanic trenches. These slabs, detected by instruments that sense density anomalies, contain water in the form of hydrous minerals—inorganic chemical compounds that incorporate the components of water. The volatile water returns to the surface in complex processes associated with subduction zones, resulting in volcanic arcs that arise where one ocean plate subducts beneath another ocean plate or in volcanic spines that run along the edge of a continent where an oceanic plate subducts beneath it. Certain ancient volcanoes, such as those found in the South African interior, generated material containing hydrous minerals that came from

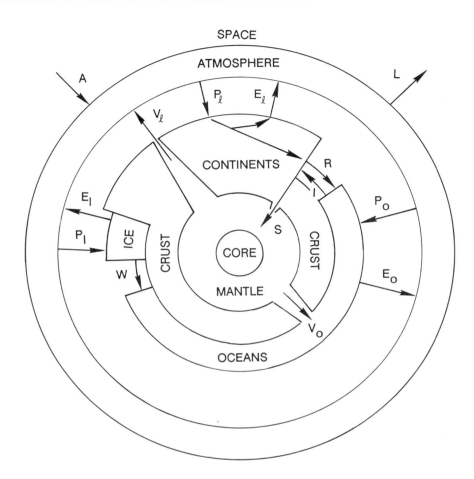

FIGURE 4.2 The hydrologic cycle as a global geophysical process. Enclosed areas represent storage reservoirs for the Earth's water, and the arrows designate the transfer fluxes between them. Figure from NRC, 1991, *Opportunities in the Hydrological Sciences*.

depths exceeding 150 km. These data suggest that some surface water has been recycled to great depths. Hydrous minerals have even been found imbedded in diamonds that form only at temperatures and pressures found at depths greater than 100 km.

All but a fraction of the water remains at or near the surface in six interacting reservoirs that can be listed according to decreasing volume (Figure 4.4). The largest volume of water—1.35 billion km³—circulates freely throughout the oceans. There is a substantial amount of water frozen in glacial ice sheets and ice caps, but it is still less than 2 percent of the oceans' volume. Permeating the pores and cracks of the crust, groundwater forms a reservoir with 0.7 percent of the oceanic volume. For all the vast reaches of fresh water we swim in and sail on, only 130,000 km³—not even 0.01 percent of all the oceans—fills the lakes and rivers. And only trace amounts, relative to the oceans' vast capacity, cycle through the atmosphere as vapor and through the crust as aqueous fluids.

Water circulates among these reservoirs in a system known as the hydrologic cycle, and it abides

in the disparate reservoirs for varying times. In an average year about 60,000 km³ of water is carried over the United States in the atmosphere. This represents an amount sufficient to cover all the land areas to a depth of 30 cm. Of this amount, one-tenth falls as meteoric water—rain or snow. About two-thirds of that precipitation returns immediately as evaporation or moves up through plant roots, carrying nutrients from the soil, and enters the atmosphere as transpiration through leaves. The remaining one-third runs along the surface, accumulating in arroyos, brooks, and creeks. They flow together, gathering as streams and rivers, and pause to form lakes. Eventually, after carving or molding the land and sustaining or sapping the groundwater, rivers run down to sea level, building deltas and mixing with the salty ocean. Ocean, runoff, rain, and vapor in the atmosphere constitute the overwhelming bulk in the hydrologic cycle. The other stores of water involve much slower cycling, although the volumes they contain are substantial. Of particular significance—economically, scientifically, and socially—are the residual products resulting from reactions between subsurface water and the

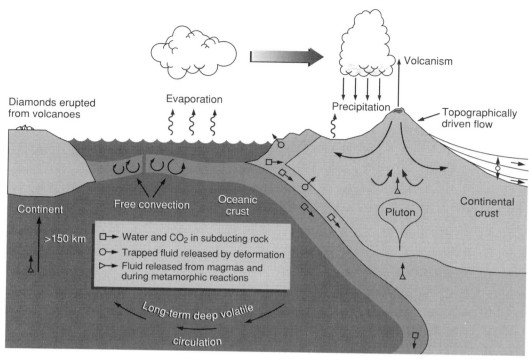

FIGURE 4.3 Circulation of water in a variety of reservoirs. Water and other volatiles return to the surface in complex processes associated with expulsion in subduction zones and related volcanism.

minerals it encounters. The small proportion of rainfall that enters the groundwater reservoir may return to the surface through natural springs or feed dwindling stream flows. Or it may move through rocks of various composition, dissolving minerals and carrying them away. Or it may displace lighter fluids, forcing them into other rocks.

Water Quality

Water circulates through the hydrologic cycle, dissolving elements, sometimes carrying them over great distances, and eventually depositing them into sinks. Earth systems depend on that circulation to constantly supply their moisture requirements. Wa-

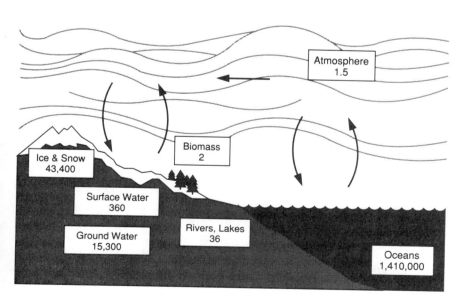

FIGURE 4.4 The volumes of water in the near-surface reservoirs of the Earth (in thousands of cubic kilometers). The total volume is about 1.5 billion km³; a comparable volume may be dispersed within the Earth's interior.

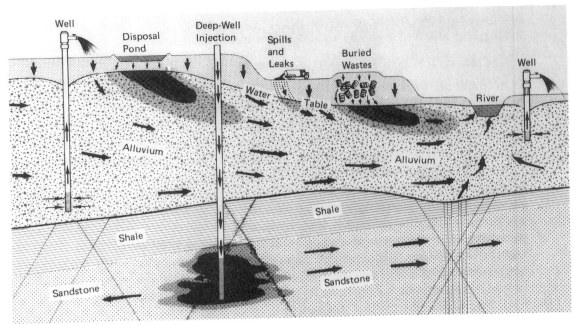

FIGURE 4.5 Schematic representation of contaminant plumes (arrows indicate flow direction) possibly associated with various types of waste disposal. From NRC, 1984, *Groundwater Contamination*.

ter dissolves salts from soil, carries them to the ocean, and then evaporates, leaving the collected residue to accumulate. Moving through the atmosphere, water arrives again over the land surface to replenish the puddles, rivulets, lakes with fresh water. But within the past few decades Americans have realized that increased concentrations of toxic residues are contaminating fresh water prematurely—long before it can naturally process the material—largely because of the actions of humans.

Traditionally, waste products have been allowed to flow into the nearest body of water. As a consequence, streams, estuaries, and large parts of the coastal environment have become contaminated. Coastal zones, including estuaries, are areas of active sedimentation, and deposited materials often contain toxic wastes. These wastes may seep to the surface or landward into aquifers. They may be exhumed—either naturally or through channel dredging and widening—and redistributed by erosion. Solid wastes placed in landfill sites and other terrestrial depositories are gradually dissolved by moving groundwater; in many cases they contaminate the groundwater supply (Figure 4.5). Because streamflow is often nourished by groundwater, dissolved contaminants can eventually make their way into streams and other water courses.

Groundwater contamination studies have produced many surprises. Volatile organic compounds are mobilized through the atmosphere and may

become components of the soil, where they remain for long periods of time. Transport through the soil and groundwater is generally slow, and long periods are required to move contaminants significant distances. Hydrologists agree that once contaminated, groundwater will remain contaminated unless remedial action is taken. Much of the water contamination identified today is the result of waste disposal practices 20, 30, 40, or more years ago.

There is a significant difference between pollution by toxic *compounds* and pollution by toxic *elements*. Compounds may spontaneously decompose, through inherent instability, or may be decomposed by heat, biological action, or catalytic properties of earth materials. Either course may result in the loss of toxic character. Heavy metals are elements and remain toxic unless they are immobilized into an insoluble state or detoxified by chelation. Chelation constrains metal atoms within an innocuous but stable chemical species; the heavy metals are tamed or caged by an organic compound that establishes bonds with the metallic ions and eliminates the potential for the metal to react with any other compound.

It is possible that ill-conceived detoxification tactics might accidentally exacerbate the predicament. For example, the native element mercury is toxic but only very slightly soluble. Conversion by industry produces the compound methyl mercury $(Hg(CH_3)_2)$, which is both soluble and toxic.

Methyl mercury accumulates in short-lived species throughout the food chain with no effect, but when contaminated fish provide the major source food for a human population, mercury concentrations reach malevolent levels. Mercury poisoning results in disorders of the nervous system, mental impairment, and irrational behavior; the mercury compounds used in the production of felt hats during the nineteenth century also produced the stereotypical Mad Hatter.

Plants acquire necessary nutrients by absorbing enriched water from the soil. They return almost pure water to the atmosphere through transpiration and leave unwanted minerals, such as residual salts, concentrated in the soil. Salt buildup is an agricultural problem as ancient as the practice of irrigation. On one hand, it is desirable to apply only the amount of water that the plants actually require. On the other hand, sufficient water must be applied to transport residual salts and minerals beyond the root zone. An effective conservation program in irrigated agriculture requires carefully calculated amounts of water to address both plant requirements and residue removal. Further complications arise when the residual salts or minerals collect in sinks before they reach the ocean. For example, in the Central Valley of California, selenium occurs as a trace element in soils and is mobilized by irrigation water. It moves into the drainage canals and to holding reservoirs where it is concentrated. Eventually, aquatic birds high in the food chain receive large doses of this selenium. Currently, there is no obvious solution to the problem other than to retire a substantial area from irrigation or to flush the reservoir continually into the ocean. Either alternative is costly and temporary.

One major deficiency in our hydrologic understanding and management has been a lack of systematic measurements of water quality. The National Water Quality Assessment (NAWQA) programs are beginning to address this problem. But for all the millions of cubic kilometers of flowing water, the U.S. Geological Survey (USGS) maintains only a small long-term network of approximately 500 gauging stations where water quality is routinely measured. Water samples from these stations are chemically analyzed for the major anions and cations, while some are tested irregularly for traces of metals and organic materials.

To assess the state of the nation's water quality, the Environmental Protection Agency (EPA) relies largely on reports from water quality agencies in the 50 states. The EPA has questioned the adequacy of this information because the states do not share a standardized data base. The agency has begun nationwide sampling to assess the concentrations of particular constituents in surface water, but the appraisal of groundwater is even more limited than that of surface water. Contaminated groundwater has been located at many sites, but there is uncertainty as to the extent of the problem. The contamination is detected in areas devoted to specific land-use patterns, leading to conclusions that are then extended nationwide. This line of reasoning estimates that 0.5 to 2.0 percent of the usable groundwater in the United States is contaminated. Even the higher percentage may be an underestimate: pollutants are generally produced in highly populated areas, while investigations providing the data base tend to concentrate in areas where groundwater use is important, and the two kinds of areas do not necessarily coincide. For instance, there is little information on groundwater contamination in the northeastern urban corridor of the United States, a region dependent on surface water supplies, but common sense suggests that groundwater in that area would be contaminated.

Systematic collection of water quality information is difficult and expensive. The problem is one of complexity. There are several hundred, perhaps thousands, of potentially toxic chemicals for which every water sample could be analyzed. Lakes, rivers, and groundwater all occupy three spatial dimensions that make representative samples expensive to obtain. Water quality also varies through time, so that recurrent sampling is required. And, finally, sampling in an extensive geographic area will probably reveal unsuspected links between various reservoirs.

Three priorities have been established by the USGS: detecting major contaminated reaches of rivers; determining baselines in uncontaminated areas as benchmarks; and detecting groundwater contamination for those parts of the country where rivers or aquifers, or both, provide major water supplies. The cost for this minimal assessment phase is on the order of $100 million annually. This impressive sum is only 1 percent of the approximate $10 billion spent building new sewage treatment facilities each and every year.

Water Supply and Use

Every human being requires about 2 liters of fresh water every day to maintain the minimal physiological functions. But many people use more than that amount; the added quantity varies according to an individual's personal habits and standard of living. In North America every individual consumes or uses material that requires an average of 1,500 liters of water per day. This total includes the water that cools the turbines of power plants and the water that irrigates cotton fields, as well as the moisture that plumps the artichokes and provides the bulk of milk or muscatel. It is the fluid that showers bodies and washes cars. It drips from faucets and drowns the roots of suburban lawns. When it rains, this fluid runs in silken sheets along the slopes of parking lots and collects in open foundations at construction sites. It gets pumped into the sewer system and mixed with organic waste and then runs out—to someplace downstream. As the population expands, and as material expectations rise, the need for water increases at an exponential rate.

Over the past 300 years, the amount of water removed from freshwater reservoirs by humans has increased more than 35-fold. Most of the water used in North America does return to the hydrologic cycle after only a short interruption. The problem is that it frequently returns in a very different state and often to a very different reservoir. If pure water is taken from a river and returns to the river contaminated by fecal matter or toxic materials, it may take long reaches of that river out of the supply side. And if vast quantities of water are mined from a 100,000-year-old aquifer and run through an irrigation system, resulting in extensive evapotranspiration, the aquifer may not be replenished for another few hundred thousand years.

Because the hydrologic cycle transfers water from one reservoir to another at various rates and because it does not always transfer to the location most convenient for the schemes of civilized minds, humans must use available water resources very carefully. Water cannot be manufactured economically from its component elements and ocean water cannot be desalinated without incurring large, usually unacceptable, expenses. So humans must adapt to the natural limitations on available fresh water imposed by the hydrologic cycle.

In the United States the opportunity exists to deliberately formulate water distribution and wastewater treatment policies according to principles of conservation, wisdom, and justice. The United States uses 770 km^3 of fresh water every year. About 340 km^3 is consumed—exposed for evapotranspiration or consigned to uses sequestered from the hydrologic cycle. The rest is recycled into the system as wastewater. Irrigation uses 330 km^3—41 percent of the total—and accounts for 215 km^3 of the consumption. Domestic uses amount to 66 km^3, consuming 20 km^3; and industry uses 294 km^3, but consumes only 29 km^3.

Most of the river and lake water becomes available in the spring and early summer when the snow melts, ice breaks up, and rain falls to flush out—perhaps to flood—systems that run low in late summer. Mitigation of droughts and prevention of floods traditionally require water control projects. Management of water supplies—circulating 740 km^3 through U.S. cities, suburbs, forests, and fields—necessitates planning that modifies seasonal supply to match independently fluctuating demands.

In the half of the United States west of the 100th meridian—which slices through Texas, Oklahoma, Kansas, Nebraska, South and North Dakota—stream runoff is less than 1 inch in an average year. Throughout that region runoff comes in early spring, largely as snowmelt from mountains hundreds of kilometers away. The water flows through the region months before the optimal time for watering crops. Storage of water—either in artificial surface reservoirs, lakes, or aquifers—dampens the lag in timing between supply and demand. But no reservoirs create water; they only allow a delay of the transfer within the hydrologic cycle. Artificial reservoirs and lakes can be refilled when the runoff returns. Underground aquifers—the groundwater—serve as vast and wonderful reservoirs; but they often cannot be refilled in the next season.

Several western states now depend on mining underground aquifers—taking water out faster than the rate of recharge. In the past, official policy toward groundwater use endorsed the idea of safe yield. Safe yield was a concept accepted by hydrologists as a maxim—an aquifer should not be pumped faster than it is naturally recharged. In the early 1960s this idea was replaced by one that treated underground water as a nonrenewable resource: depletion of groundwater is justifiable if it creates an economy that can afford to buy more expensive water when the well runs dry. Adoption of this maxim reflects the development of irrigated agriculture in the High Plains as well as population migration to the Sunbelt. Now, urban and energy developments are coveting the water available to agriculture—especially in the Southwest.

This competition will undoubtedly intensify, posing two major issues for society: how local, state, and regional communities can manage increased competition for water and to what extent the country can wean itself from irrigated agriculture in the West. The present domestic and industrial water requirements can be effectively met without serious impact on agriculture. Diverting 10 percent of current agricultural water consumption

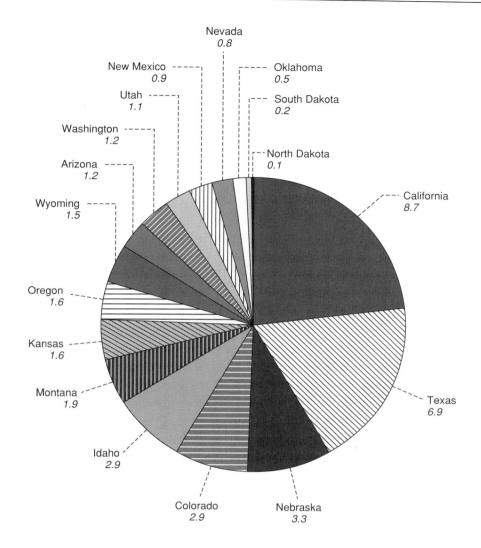

Nevada
0.8

New Mexico
0.9

Oklahoma
0.5

Utah
1.1

South Dakota
0.2

Washington
1.2

North Dakota
0.1

Arizona
1.2

California
8.7

Wyoming
1.5

Oregon
1.6

Kansas
1.6

Montana
1.9

Texas
6.9

Idaho
2.9

Colorado
2.9

Nebraska
3.3

FIGURE 4.6 Amount of irrigated acreage (in millions of acres) in the 17 western states.

would permit a doubling of water use by others. But even without the competitive pressure from population centers, irrigated agriculture is in trouble.

In the western states, irrigation accounts for more than 90 percent of consumptive use, and the surge in groundwater withdrawal over the past 25 years—since the philosophy of safe yield was rejected—has been due almost exclusively to irrigation. In the conterminous United States, the 17 western states consume 84 percent of the country's fresh water, mostly for agriculture. The acreage irrigated in the 17 western states is shown in Figure 4.6. In Texas and California, which account for 42 percent of the total, almost one-half of this water is pumped from the ground. Irrigation is also being increased in the more humid areas of the country to raise crop yield.

Hydrologists calculate the volume of freshwater supply in terms of relative depletion. Relative depletion is calculated from total consumption plus the total water exported from each drainage basin, divided by the total input of rain and snow. The ratio is expressed as a percentage. Groundwater is not considered part of the total supply because it is not replaced by precipitation. The exclusion of groundwater from the calculation results in relative depletion that can exceed 100 percent in some areas. Several regions in the United States are considered to be critically depleted; in most of the lower Colorado River basin, in Southern California, and in Nevada, depletion exceeds 100 percent. In south-central California, which includes the San Joaquin and Owens valleys, in the High Plains of Colorado and West Texas, and in most of New Mexico, depletion exceeds 75 percent.

These relative depletion figures are incomplete and probably low, because instream flow requirements are not included. The instream flow requirement represents the minimum streamflow necessary to preserve aquatic and associated ecosystems.

Maintenance of the instream flow is becoming an important constraint on water availability as the sensitivity of individual ecosystems is recognized. A familiar example is the case of Mono Lake in eastern California. Mono Lake is the last puddle left of an inland sea that has occupied the basin for more than 1 million years. During the most recent glaciation 12,000 years ago, the lake reached depths of 250 m and extended over 870 km². In the course of postglacial warming, the lake level fell, but it has managed to maintain a rich variety of biotic systems thriving in its salty waters. In 1941 the city of Los Angeles—540 km to the south—began to divert the water that feeds Mono Lake. The lake's area decreased nearly 25 percent and its surface fell almost 20 m. If the trend continues, there will be no ecosystem left in the lake: its instream flow requirement will not be met, and the lake will be too saline and too alkaline to support any life. A 1991 court decision has forced Los Angeles to stop diverting much of the water that normally would flow to Mono Lake, which may be sufficient to save the lake. If not, it will share the fate of Owens Lake, which once glistened 150 km to the south. All Los Angeles has left is the lake bed—250 km² of alkaline salt flats.

The withdrawals from Owens and Mono lakes are only a fraction of the water exported to urban centers. In 1990 approximately 50 percent of the U.S. population depended on groundwater for domestic use. Efforts to eliminate the overdrafts of groundwater in Arizona have led to increasing dependence on surface supplies from remote locations. The Central Arizona Project provides surface water to the Phoenix area from the Colorado River, over 200 km to the west. Arizona has also implemented a new law to eliminate groundwater mining by the year 2020.

California has eliminated relative depletion in the Central Valley by increasing surface water supplies. Southern California continues to import 6.0 billion m³ of water every year. The city of Los Angeles alone extends its water delivery system over hundreds of miles throughout the Southwest and draws the power for the system from projects reaching across six states. It has been involved in disputes over water with the state of Arizona and most recently with Mexico. But in California the cities use only 15 percent of all the water consumed. The rest goes to irrigation. In 1990, 16 percent of the state's water supply irrigated fields of alfalfa, which is used to feed horses and cattle. That adds up to more water than is needed to supply 30 million people for a year. Difficult choices lie ahead for urban and rural Californians as droughts continue.

The southern High Plains represent an area of the United States that will return, sooner or later, to dry-land farming. The transition will come sooner and with fewer ecological and economic crises if the agricultural industry is weaned gradually from its dependence on groundwater irrigation. If nothing is done until all the accessible water in the Ogallala aquifer has been removed, the transition will be ecologically dangerous and economically dreadful. Approximately two-thirds of this water is in Nebraska, which has an enormous reserve of groundwater and is nearest to the most productive recharge area. In Texas and New Mexico more than 10 percent of the water initially in storage has been depleted. However, the depletion statistics for an aquifer such as the Ogallala are deceptively optimistic simply because it has not been economically feasible to remove all the water in underground storage because of the high cost of pumping from great depths. It is estimated that 50 to 60 percent of the water might be removed under favorable economic conditions. Some areas in Texas and New Mexico have already reached the point at which irrigation from the Ogallala aquifer is no longer practical. In such areas of severe groundwater depletion, episodic surface runoff might serve for aquifer recharge if appropriate capture strategies were devised.

Influencing the Water Cycle

Humanity plays an integral role in the hydrologic cycle. While consuming less than 3 percent of the rainfall and returning large portions of that small percentage to the system, humans still profoundly disrupt the natural cycle. Human influence on our water resources is disproportionate to actual use because of the tendency to concentrate efforts in areas where the natural water supply is sparse. The earliest irrigation was practiced in the deserts of the Middle East and was characterized by the same aims as today: to redistribute water geographically and seasonally—in space and time. Mesopotamians built canals and dams that would divert and delay the water flowing through the Tigris and Euphrates rivers. Early Egyptians worked to maintain the ponds and sluices that would trap the flood waters of the Nile and its rich silt. Both the water and the silt would be redistributed over fields at the proper time.

Water may have been a truly renewable resource in early history. Although monsoons and even the annual Nile flood might occasionally fail to take place, they always returned in a year or in seven.

But today hydrologic practices in some areas use water resources faster than they can be replaced. A lake impounded behind a dam may fill with sediment, although the chains of impoundment on many major rivers effectively reduce this risk. And ecological systems downstream of a dam may depend on the natural flood to flush pathogens. Rivers that flow into arctic waters maintain a cap of fresh warm water over the cold saline water that sinks to the bottom. If stream flow were sapped from these systems, it could affect global ocean-atmosphere-climate links.

Awareness of these and similar influences modifying the environment is growing at an astonishing rate, not just among scientists and engineers but in the world at large and especially among politicians. Global change is a concept that has captured attention. Interest has focused on the changes that are likely to follow most directly from the burning of fossil fuel and the related increase in atmospheric carbon dioxide.

Human involvement in modifying the hydrologic cycle has received limited attention as an aspect of global change, but the obvious can no longer be ignored. In the past 100 years human endeavor has destroyed natural streamflow patterns by damming up most of the major rivers of the Northern Hemisphere. Aquifers in arid and semiarid areas have been mined to dangerous levels. Water quality has deteriorated because of the disposal of wastes in streams, lakes, and aquifers, and water salinity has increased in critical areas as a result of irrigation practices. The dissolved salt and the organism components of groundwater are receiving more attention with the recent recognition that irreversible changes are occurring in important aquifers. They are being contaminated by sewage and also by seawater. When fresh water is removed too quickly from groundwater reservoirs along coastal plains, the spaces fill with saltwater. Because of such saltwater intrusions, the cities of Miami, Pensacola, Daytona Beach, Brooklyn, and Los Angeles have had to abandon important aquifers.

Impoundment of a major reservoir behind an artificial dam illustrates how human-induced change affects the hydrologic environment which in turn propagates through the earth system. Impoundment of Lake Nasser by the Aswan Dam was expected to change agricultural practices downstream that had responded to the annual Nile flood for thousands of years. However, the generation of earthquakes caused by the load of the lake on the lithosphere was not considered, nor was the higher elevation of water levels in the Nubian sandstone

beneath the desert on either side of the lake. Neither did anyone expect an epidemic spread by snails that had formerly been flushed from the marshy river banks with the annual flood or depletion of the soil and incursion of the Mediterranean (and loss of fisheries) into the delta region because sediment supply had been cut off. On the other hand, rising water levels in the Nubian sandstone have made water wells more productive over a large area. The drought and famine in Ethiopia and neighboring countries upstream from Aswan have not affected Egypt because of the availability of irrigation water from Aswan. The dam has also prevented several dangerous floods.

Humanity can learn from its mistakes. If so, one of its major goals will be that of understanding the complex earth systems well enough to predict the possible consequences of ongoing changes. In the field of hydrogeology, a variety of crises have stimulated accelerated studies and actions, especially an improved understanding of groundwater flow. But many aspects of our water resources remain undefined or even unrecognized.

The geoscientist's role is to understand and synthesize the multiple factors involved in the maintenance of adequate water resources so that policy makers—regionally, nationally, and internationally—can reach informed decisions. The demands of that role represent a remarkable opportunity for newly integrated studies of the hydrologic cycle and particularly for the application of new observational, analytical, and modeling approaches to understanding the groundwater system.

MINERAL RESOURCES

The word *mineral* can be defined in a variety of ways. One definition includes any naturally occurring substance that is not a vegetable or an animal. The geoscientist defines minerals as all the naturally occurring solids—plus a few liquids—that display distinctive chemical composition and crystalline structure and that are the components of rocks. To the economist, minerals are materials extracted from the Earth that have current or potential economic worth, and any site from which such material can be recovered is considered a mineral deposit.

For this section the most appropriate definition includes economically valuable material that is not considered mineral in the strict geological sense, such as the synthetic open-framework silicates—the types of zeolites that are used as molecular sieves. Mineral resources include the naturally occurring combinations of elements, the ores that contain metals such as gold and iron. Understanding the location of metallic ores and how to mine them is one of the most ancient fields of geoscientific endeavor. Research involving mineral deposits is similar in some ways to research in oil and gas exploration. Both are strongly affected by economic fluctuations and international politics. When the industries thrive, the larger companies manage extensive research programs that foster close links between exploration projects and scientists from government and universities. At the same time, smaller enterprises can justify ventures that would be considered too risky in a sluggish market. Advances in geophysics, geochemistry, and data-gathering and data-handling capabilities have revolutionized mineral exploration and development, as they have the petroleum industry. Both industries are especially stimulated by applications of new theoretical concepts.

Mineral resources are generally concentrations, sometimes exceptionally high concentrations, of one or more of the materials that constitute the solid earth. Ninety-nine percent of the crust is made up of only nine chemical elements: oxygen, silicon, aluminum, magnesium, iron, calcium, sodium, potassium, and titanium. It is the other 1 percent of the crust that captures the interest of the mineral exploration geoscientist who wants to know how, when, and especially where concentrations of minerals occur. Interest centers on understanding the processes that form the mineral deposits, the environments in which those processes operate, and the distribution of deposits through space and time. From this understanding, new deposits can be predicted, discovered, and developed, and existing ones can be exploited efficiently.

Mineral deposits form through a wide range of geological processes that may operate at uncommon levels of effectiveness or in unusual associations. In some situations, rocks such as sandstone and limestone qualify as mineral deposits because they are of economic value in a particular place. Sands and gravel deposited in the bed of a mountain stream provide a simple example of such a situation. Where a stream erodes crustal rocks that contain gold, the downstream gravels may contain anomalously high concentrations of the precious metal. This is because the weathered gold particles are denser than the gravels and sands. The stream deposits the gold particles when it loses the capacity to carry particles of that density and size, so the gold accumulates within a particular reach of the stream. These are the *placer* deposits that have been the sites of gold mining throughout history, from the mythical deposits of Colchis where Jason found the Golden Fleece to those of Sutter's Mill in California that inspired the forty-niners of a century and a half ago.

Less than one part in 10,000 of the metals present in the upper kilometer of the continental crust is concentrated in known mineral deposits. The remainder is widely dispersed at low concentration and, for that reason, is unsuited to economical recovery. Ore deposits are rare indeed! A simple illustrative plot for the metal lead shows how concentration of a typical element varies in the crust (Figure 4.7). The range of concentrations is so large that it is necessary to represent it logarithmically. Common crustal rocks contain about one part in 100,000 of lead by weight; this is represented by the crest of the bell-shaped curve. There are virtually no rocks in which the amount of lead is too small to be measured. The smooth curve represents the gradual increase in lead abundance, with the maximum representing the most probable value for a randomly selected sample. The fall in abundance from this value is not symmetrical. Economically workable lead concentrations—rocks with a lead context of more than a few percent—occupy a place at the far right end of the curve.

Deposits with lower lead content are expected to become economically profitable at some time in the future when all larger deposits are exhausted. A threshold identified as the *mineralogical limit* separates a deposit in which rocks contain lead-dominant minerals from a deposit of lead so evenly distributed through the rock that it does not form discrete mineral concentrations. Although by definition all lead ores fall within the first category, there is a far greater total amount of lead distributed thinly throughout crustal rocks.

Lead is just one example of crustal element occurrence that can generate a bell-shaped distribution curve. Such distribution patterns indicate the challenge of mineral exploration, which is to discover the small amounts of concentrated materials constituting ore bodies. These concentrations not only are rare but can be extraordinarily localized. For example, seven gold fields covering an area of about 5,000 km^2—no larger than the county of Los Angeles—within the Witwatersrand basin of South Africa have produced more gold than has been discovered over the remainder of the surface. The mercury deposit at Almaden, Spain, has yielded more mercury than any other source and still retains the bulk of the world's reserves. The Bushveld intrusion in South Africa contains 98 percent of the world's chromium reserves, most of them in a single layer.

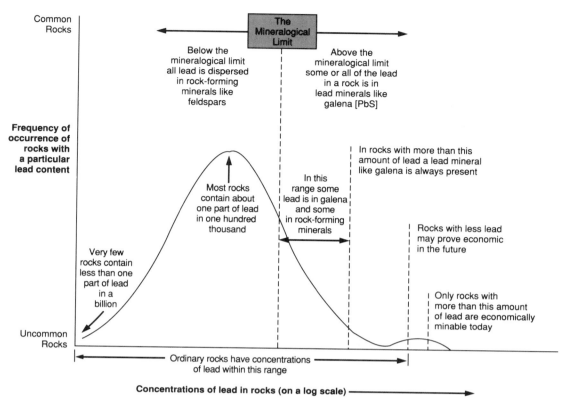

Common Rocks

Frequency of occurrence of rocks with a particular lead content

The Mineralogical Limit

Below the mineralogical limit all lead is dispersed in rock-forming minerals like feldspars

Above the mineralogical limit some or all of the lead in a rock is in lead minerals like galena [PbS]

Most rocks contain about one part of lead in one hundred thousand

In this range some lead is in galena and some in rock-forming minerals

In rocks with more than this amount of lead a lead mineral like galena is always present

Rocks with less lead may prove economic in the future

Very few rocks contain less than one part of lead in a billion

Only rocks with more than this amount of lead are economically minable today

Uncommon Rocks

Ordinary rocks have concentrations of lead within this range

Concentrations of lead in rocks (on a log scale)

FIGURE 4.7 Schematic grade-tonnage curve for scarce elements; although this curve is for lead, similarly shaped curves also apply to copper, tungsten, chromium, and other metallic elements.

These are unusual concentrations of gold, mercury, and chromium, but researchers question whether they are unique. Such rich deposits represent only minuscule proportions of otherwise ordinary parts of the crust. The possibility exists that similar deposits are hidden in the unexposed basement rocks of the North American mid-continent, beneath the jungles of South America or Southeast Asia, or under thin cloaks of glacial drift in Canada. While exploration has the potential to be a very prosperous undertaking, success in a venture characterized by cryptic clues and expensive tests requires steadily increasing levels of understanding about deposits and their environments.

To illustrate the fact that new sources of strategic minerals continue to be discovered, consider two commodity sources that once were thought to be unique but have proven otherwise: the Climax molybdenum deposit and Southeast Asian tin. The Climax molybdenum deposit at Fremont Pass, Colorado, was for half a century the overwhelmingly predominant source of molybdenum—used in the steel industry—although by-product molybdenum was also recovered in significant amounts from some copper deposits. Despite a ready market,

other major molybdenum deposits were not identified until the 1960s, when the essential geological characteristics of the Climax deposit were recognized in numerous other localities. The result was that a series of major molybdenum deposits was discovered and brought into production. Similarly, the marine placer deposits of Southeast Asia dominated the world tin market for years, because they could provide tin more cheaply than any other source. However, in the 1980s the world tin market collapsed, and many Southeast Asian operations were sharply curtailed when abundant low-cost tin emerged from the Rondonia district of Brazil. Thus, molybdenum and tin are examples of mineral resources in which suppliers with no rivals suddenly found themselves downgraded or displaced by more economically competitive sources.

But new deposits are not the only reasons for change. Other factors may include price fluctuations in response to demands or cartels; environmental and other concerns that restrict mining, milling, or smelting; or national policies that encourage or discourage mineral exploration and development. Changes also result from new or improved technologies. Within the past two decades the steel industry has developed a new technological approach called the minimill. Rather than producing bulk standardized steel, with profit gained through tonnage alone, minimills produce customized types of steel in smaller batches. This means that the steel product does not have to be milled again by the final manufacturer. Minimills are even more profitable when they can be located in regions where the iron and alloy minerals necessary for a specific product occur together or in proximity to each other. Predicting suitable locations for particular minimill sites is a new task assigned to the mineral geologist.

The locations of undiscovered major mineral deposits are among nature's best-kept secrets. Except for bulk materials such as sand or limestone, most deposits constitute minute targets in complex but valueless rocks. Ore concentrations are rare and difficult to find, but their values can be immense. An example is the recently developed Red Dog zinc-lead-silver deposit in the arctic expanse of northwest Alaska, which contains metals worth $15 billion to $20 billion within 1 square mile. The Red Dog deposit is remarkable because high-grade zinc ore was exposed at the surface—and visible to a curious pilot who talked a geologist into flying over some strange rocks. Such obvious occurrences are rare. Most exposures of significant concentrations appearing directly at the surface have long since been discovered by prospectors throughout the world. The tops of shallow deposits commonly are worn by weathering, and their bulk is concealed by soils, gravels, and lavas or is submerged beneath swamps, lakes, and rivers. Unexposed deposits usually indicate their presence only by meager evidence. Thus, earth scientists have the challenging task of detecting mineral caches to maintain society with its basic resource requirements. Just as necessary is the identification of deposits with economic potential, to select optimum strategies for land use. This latter task could help to avoid future conflict involving exploration and exploitation in territory that has been set aside for preservation.

Large sums of money are invested annually in locating and evaluating mineral deposits. The successes are neither frequent nor spectacular, and mining does not rank high in capital return rate. Various exploration approaches can be used in the search for economically viable deposits. Some techniques examine the crust for geophysical or geochemical signatures that suggest the presence of mineralization by identifying fundamental distribution controls. Others define targets within a broad geological context, including direct observations of unmineralized rocks that can serve as outer rings around an unseen bull's-eye of ore. The effective-

ness of today's technology depends on the application of tools and the nature of the target. In all cases the greater the understanding about the nature of the crust, the better the chance of discovering worthwhile mineral deposits.

At the present time, an oversupply of many nonfuel commodities exists despite the fact that minerals are being consumed globally at an unprecedented rate. This disparity is partially due to demands that fluctuate more rapidly than supplies produced by the mining industry. Inevitably, the rate of discovery will decline, and shortfalls will emerge once more. While each new technique or concept revives portions of the Earth as prospecting targets, the remaining virgin territories are disappearing.

Understanding Mineral Deposits

Ore deposits can be subdivided into two major classes, epigenetic and syngenetic. The former involve addition of minerals to existing rock; the latter are mineral concentrations that form when the host rock forms. Epigenetic deposits precipitate from migrating aqueous solutions when mineral solubility changes through variations in pressure, temperature, or chemistry. These ores tend to form at contacts between rock types or near the margins of intrusive rocks, generally within a few kilometers of the surface, where metal-rich solutions can mix with cooler surface waters. Epigenetic deposits are associated with near-surface environments. Their host rocks are extremely variable and may lack genetic affiliation with the epigenetic minerals, and they are less likely to provide clues about the mineral deposit.

In contrast, syngenetic deposits, which are generated along with their host rocks, are spread throughout the crust. Any attempt to inventory the crust's mineral wealth must recognize that although surface samples may be valid representations of the abundance of *syngenetic* mineralization in the crust, such samples are likely to underrepresent *epigenetic* mineral accumulations.

Recent major advances in understanding the genesis and nature of mineral deposits fit into four general groupings. The first is acceptance of the fact that ore deposits are fundamental parts of the crust. Theoretically, this concept is neither novel nor complex, but its application to ore exploration has been acknowledged only recently. The second advance is the exploration of less hospitable environments—sea-bottom thermal vents, the subsurface of the crust below active hot springs, and lava lakes and flows—that have revealed active ore-forming processes. Third, a less adventurous but equally significant frontier involves the rapid manipulation of masses of data that can identify intricate correlations and analyze complex processes—leading to improved ore deposit models. Finally, the host of modern tools and techniques, such as stable isotope measurements, ion microprobes, fluid inclusion studies, and age dating systems, allow geoscientists to analyze and select interpretations from the myriad of heretofore untestable hypotheses regarding the formation and distribution of mineral deposits. The application of these four kinds of advances to problems in mineral exploration and exploitation has provided a number of deposit models, which represent application of new ideas to a wealth of new data.

Crustal Processes and Ore Deposits

Plate tectonic concepts have been applied to mineral exploration as associations between tectonic environment and certain types of mineral deposits have become clear. For example, general links between porphyry copper deposits and volcanic arc systems are well established. However, recognition of this relationship does not indicate why one 1,000-km² tract is more favorable to mineral concentrations than an adjacent area. Often the only effective method for local exploration is a detailed study in the immediate vicinity of a mine or prospect. Combining plate tectonic concepts on a large scale with detailed geological, geophysical, and geochemical data on the local scale represents a constructive advance in understanding mineral deposits.

As the Earth evolved, the abundances of various types of mineral deposits also changed (Figure 4.8). Until about 2.8 billion years ago, magnesium-rich lavas known as komatiites erupted on the seafloor and, as they crystallized, precipitated rich deposits of nickel that are now preserved in Canada, Australia, and South Africa. During the intervening eons, komatiites have been essentially absent from the numerous varieties of rocks forming on our planet.

Between 1.8 billion and 2.8 billion years ago, huge sedimentary banded-iron deposits formed in what are now areas of Australia, Brazil, the former Soviet Union, and the Lake Superior region of the United States and Canada. At the same time, gold and uranium were accumulating as placer deposits. These iron deposits contain the bulk of the world's

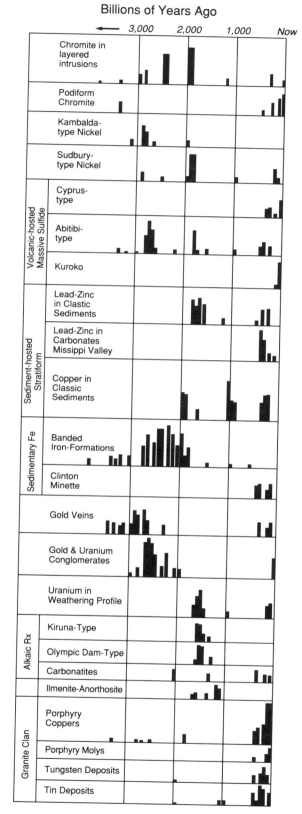

FIGURE 4.8 Distribution through geologic time of several types of ore deposits. The width of each vertical bar represents approximately 50 million years; the length is a measure of the relative quantity of that type of ore formed during that time period.

economic iron resources, and the paleoplacers include more than one-half of the total known gold resources. No accumulations rivaling in size those banded-iron formations and gold-bearing paleoplacers have been discovered in rocks that have formed since that ancient period.

These cases of mineral formation—distinguished by time of genesis—are relatively straightforward, and there is little likelihood of statistical aberration. An explanation of their formation could be complex, however. For example, it is agreed that the mantle of the younger Earth was hotter than it is today. The very high temperatures, approaching 1600°C, required to keep komatiites molten could have occurred close enough to the surface to permit eruption of the rock magmas. Formation of the large iron deposits and the gold-uranium-rich placers has been attributed to a different composition of the early atmosphere.

Examples of other time-dependent processes are little more than as yet unexplained correlations. Some well-documented associations between deposit type and geological time may result from a higher probability of exposure and weathering of deposits formed near the surface in continental environments. Erosional surfaces can be recognized, but there is no way of knowing what has been eroded. And because some deposits can be found only in rocks that have formed recently does not prove that the conditions necessary for their formation have occurred only recently. Deposits such as the Mississippi Valley type appear only in relatively young formations, but similar deposits formed in the past may have been subsequently removed from the geological record.

The ores of the Mississippi Valley type, as exemplified in southeast Missouri and the Joplin region of Oklahoma, Kansas, and western Missouri, are prime sources of zinc and lead within the conterminous United States (Figure 4.9). The ores result from the filling of solution cavities in carbonate rocks with metal-rich material at relatively low temperatures (90° to 125°C). Recent developments that have contributed to understanding this process include the discovery of brines rich with zinc and lead in deep petroleum exploration wells drilled throughout the region, recognition that metal-bearing brines have permeated carbonate rocks extensively through the midcontinent regions, and advanced modeling of fluid flow in compacted sedimentary basins. These scientific developments have produced a refined model of mineral deposition based on the expulsion of metal-bearing brines from a sedimentary basin that concentrates miner-

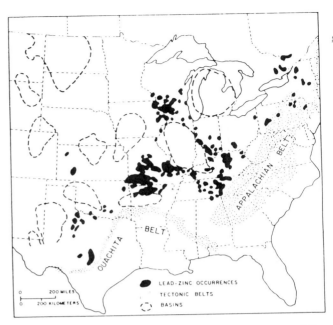

FIGURE 4.9   Map showing the relationship between Mississippi Valley-type lead-zinc deposits and tectonic belts and basins. After J. E. Oliver, NRC, 1990, *The Role of Fluids in Crustal Processes*.

alization around the margins of the host basin. The more eastern Mississippi Valley-type mineralization was probably emplaced as a discrete pulse during the collision of an island arc with North America about 450 million years ago. The larger western episode of mineralization was related to the collision of Gondwana with Laurentia, during the assemblage of Pangea about 300 million years ago. In both dramatic events, mountains depressed the North American continental margin and drove subsurface fluids toward the interior of the continent.

Researchers suspect that the cessation of komatiite eruptions resulted from the long-term cooling of the Earth; other changes, such as the closing of ocean basins during continental assemblage, are cyclic. Another example of cyclic change that affects mineral deposition is the global warming that dissipates polar ice caps. Such a warm period occurred 100 million years ago. Sea level was high; circulation of cold, well-oxygenated polar water toward the equator was diminished; and the deep oceans were warmer than the near-freezing temperatures of today. This caused stagnation in the deep basins; with a continual supply of organic matter from dying organisms near the surface, the deep waters became chemically reduced. This permitted the reduction of manganese oxides on the seafloor to form the relatively soluble divalent manganese. Solution progressed and the manganese content of the deep seawater increased. As upwelling manganese-rich water crossed shoals near the continents, it mixed with oxygenated surface water and reprecipitated the manganese, yielding the giant manganese deposits of Chiatura and Nikopol in the former Soviet Union; Groot Island, Australia; and Malango, Mexico.

At convergent plate boundaries, a plate passes beneath another overlying plate and descends into the mantle. The lower plate heats enough to drive off water—water that lowers the melting temperature of the rocks above the site of dehydration, which produces a series of volcanoes parallel to the plate boundary. The volcanoes ringing the Pacific Ocean form an excellent example of this type of pattern. In a terrestrial setting the roots of the volcanoes can produce a complex of porphyry copper, skarn, and replacement deposits of both precious and base metals. These deposits are part of the crust that is elevated, so they tend to become eroded and destroyed in a geologically short time. Thus, even though such deposits may have continually formed over much of earth history, the preserved record is almost entirely young.

Genetic Studies of Ore Deposits

The record of the geological past is cryptically retained in the rocks of today. The geologist's challenge in interpreting mineral deposits is to determine what happened, when it happened, how it happened, why it happened, and where else it might have happened. Investigations aimed at answering these questions are termed genetic studies.

Genetic studies attempt to decipher distinguishing characteristics that will provide details about the processes involved in ore deposit formation. Over the past 30 years one of the great accomplishments in the area of genetic studies has been in the field of geochemistry. The original temperatures, pressures, and chemical compositions of the fluids that deposited ores can be determined by detailed interpretation of existing mineral assemblages, particularly from fluid inclusions, by the use of isotopic and other chemical tracers, and by the analysis of organic interactions.

Fluid inclusions are tiny amounts of fluids or gas trapped within a mineral. They are very common; jewelers distinguish between natural and synthetic gems on the basis of the presence or absence of fluid inclusions. Primary inclusions are those trapped as the host crystals grow. Secondary inclusions form as scars after a crystal cracks and is healed by subsequent crystallization. The hot liquid from which the crystals formed cools, shrinks in volume, and releases gas—generally steam—to form a bubble. When crystals are heated on a microscope stage, the liquid within the fluid inclusion expands. When the fluid fills the cavity and the bubble disappears, it indicates the temperature of mineral formation. Crystals form within the briny fluids. Sodium chloride and other salts precipitate as the crystal cools and are present in the inclusions as daughter minerals. If the crystal is chilled and the liquid frozen, the freezing temperature reveals the total salt content of the liquid. Microchemical and electronic instruments have been devised to analyze the fluids chemically and isotopically. The results of such analyses may indicate the physical and chemical nature of the host rock's genetic environment and thus possible sources of the fluids.

The isotopic and trace chemical signatures of mineral deposits are complex, and that chemical complexity is used by geoscientists to identify possible sources of deposits. Stable and radiogenic isotopes are the most useful tracers, although other indicators such as rare earths and the platinum-group metals are also used to constrain models of mineral genesis.

Stable isotopes exhibit patterns that are readily distinguished in terms of various processes and sources, thereby providing fingerprints that enable geoscientists to decipher processes. As an example, water escaping from magmas is depleted in deuterium relative to the amount in seawater, and meteoric water—rain—of the American cordillera is depleted by comparison to the deuterium amounts from magmatic water. Thus, by measuring the deuterium fraction of hydrous mineral fluid inclusions of cordilleran mineral deposits, it can be determined whether the original fluid was seawater, magmatic water, or rain.

Radiogenic isotopes, principally lead and strontium, are important in geochronological methods, but they also possess distinctive signatures that record their heritage. The rare earths share similar chemical properties and tend to remain closely associated through most geological processes. Thus, the ratios of rare earths to each other can be measures of processes related to mineralization. Ratios between platinum-group elements constitute a similar sort of signature.

The role of organic compounds in the formation of ore deposits serves as an outstanding example of how studies of mineral deposits are enhanced by other sciences. Organic matter can mobilize or fix metals directly; it can also control the oxidation or hydrolysis state of the system. Organic matter may

FIGURE 4.10 Processes active at the scale of individual vent fields. Hydrothermal venting (plumes) and low temperature, diffuse venting (at tube worm/clam beds) release significant heat to the deep ocean. Massive sulfide-sulfate precipitation can result in the growth of complex vent structures or simple chimneys when high-temperature fluids mix rapidly with seawater.

supply sulfur that is subsequently incorporated into sulfides. Thus, organic matter has a powerful control over the mobility of metals in the crust, especially under near-surface conditions. One of the most exciting research frontiers in organic geochemistry is the development of engineered bacterial species that leach metals from ores—dubbed biometallurgists—or that precipitate and isolate metals from waste—biotrashmen, of course. The challenges, opportunities, and rewards in this field are great and are being vigorously pursued.

Organic matter also has a passive role in the search for and recovery of minerals, as a guide to conditions instead of as an active agent. For example, the thermal maturation of organic matter can provide a guide to related ore-forming thermal events. In addition, living or fossil species provide clues to the environment prevailing at sites of ore deposition. One of the best guides to active deep-sea smoker sites is the distinctive biota of giant white clams, tube worms, and deep-sea crabs. Fossils of that distinctive biota provide telltale evidence that mineral deposits formed at ancient ocean spreading centers may lie nearby (Figure 4.10).

Finally, geochemists have refined laboratory ex-

periments that attempt to duplicate natural processes. Experiments in genetic studies use two broad but contrasting approaches. The first is to design and reconstruct a natural process and then use the reconstruction as a model. The second is to assemble the assumed necessary thermodynamic and chemical data and then calculate just how nature might perform under those conditions. With the second method, developing a model is a process of trial and error. The first method is most effective when the natural system is well established and reactions are rapid. Unfortunately, nature presents major obstacles to the study of certain natural processes: unrestricted time spans and complex environments. Nature has taken millions of years to weather a rock or to crystallize a glass, while experiments requiring more than a few months are difficult to maintain, and studies lasting years are rarely funded. Natural environments simultaneously involve high temperatures, high pressures, and corrosive chemistries; only some of these combinations that alter and dissolve rocks can be duplicated in the laboratory.

The solution to this problem has been to combine several experimental approaches in the laboratory. Artificial systems can be simplified to eliminate extraneous factors for short periods. For instance, experiments can be performed at high temperatures, where the reaction rates provide definitive results in relatively short time frames. Thermodynamic data can also be acquired that permit extrapolation of high-temperature results from experiments performed at lower temperatures or to a range of pressures and chemical reactivities.

Models in the Study of Mineral Deposits

Humankind has searched for minerals throughout recorded history and undoubtedly developed prospecting theories very early. Perhaps the approach was to search for another gold nugget in a stream in which one nugget had already been found. Then came the idea that other streams might have similar nuggets—the first germs of an analog model. Intensive prospecting in and around a stream might have led to an outcrop of gold-bearing quartz vein or perhaps to finding a nugget still attached to a piece of vein quartz, and eventually to the idea that there might be more than one place to look for gold. In this way the idea of alternative descriptive models would develop, one for stream placers and another for veins. Such models continue to develop today, and just as the first gold-bearing vein was a new kind of deposit to ancient miners, modern researchers continue to discover and develop new deposits. An ore body is defined as a new class of deposit if, in the absence of knowledge of its mineral content, an experienced exploration geologist would not recognize its economic potential. Even with current supposedly sophisticated knowledge of the Earth, new classes of deposits continue to be discovered. Classes recognized for the first time during the past 30 years include unconformity-related uranium, Olympic Dam type of ore, Carlin-type gold, salt-dome sulfides, and sediment-hosted tungsten.

Unconformity-related uranium deposits contain the largest known concentrations of uranium. The world's richest and largest known deposit is in Canada at Cigar Lake, Saskatchewan. It contains 150,000 metric tons of uranium metal in the form of high-grade ore—8 percent U_3O_8. This type of deposit also occurs in Africa and Australia but was not recognized until 1968.

The Olympic Dam deposit, an iron-rich breccia in South Australia, contains more than 1 billion tons of ore rich in copper, gold, uranium, and rare earths. No portion of the deposit was exposed, and it was discovered by drilling located on subsurface regional lineaments and magnetic signatures.

The Carlin-type gold deposits of Nevada are low-grade, large-volume deposits containing very fine-grained gold. Now the source of most of the current U.S. gold production, they were recognized only recently, although a few high-grade zones had been mined in the region during the past century. A major contributor to the successful production from such low-grade resources has been the advent of heap or pad leaching, which extracts the gold from the ore at minimal cost.

The cap rock of salt domes in the Texas Gulf Coast has been found to contain lead, zinc, and silver in minerals whose quantities and grades approach the profitable. The existence of similar sulfide concentrations in other salt-dome cap rocks is anticipated.

The largest known tungsten deposit in Europe, at Felbertal, Austria, was discovered relatively recently in metamorphosed sedimentary and volcanic rocks. This deposit of sediment-hosted tungsten is remote from the igneous rocks with which other tungsten mineralization is closely associated.

Thus, nonconventional ore deposits continue to be found, and prospects continue for discovering additional unique mineral concentrations. Such ore bodies provide a stimulus for continually revising the conceptual models of mineral deposits. Revisions either validate hypotheses embodied in existing models or generate completely new models and hypotheses to be tested.

Studies that build models for mineral exploration have several similarities with petroleum studies—questions about source, migration, and traps. However, mineral deposits are generally more complex because there is not a single source material, as there is in petroleum formation. Minerals may originate in the mantle or in the crust and may represent concentrations in a particular environment, such as dissolved salts in the ocean, which are concentrated by evaporation until they form salt deposits called evaporites. Transport processes are equally varied, as are controls on the deposition of mineral accumulations.

The variety of possible formation environments for copper deposits can illustrate the complexity of mineralization processes in contrast to the relative simplicity of petroleum development. Major copper deposits occur in primary or altered sedimentary environments, in veins within rocks of all types, in products of seafloor hot-spring vents, in segregates from mafic lavas as native copper, and in other distinct geological settings. This multiplicity of host environments suggests a variety of deposition and transport processes, which complicates the task of inventorying copper resources. Deposit types for other metals are just as diverse, requiring the design of various models and consideration of many alternatives. The source, transport, and accumulation processes that concentrate minerals in valuable deposits are seldom fully understood. New ideas that seem to answer outstanding questions form the bases for new models.

Transport, or migration, involves two basic steps. The first is mobilization of materials by means of a carrier, such as the solution of metals from source rocks into aqueous fluids or igneous melts. The second step involves movement of the carrier to the site of deposition. Research into transport by aqueous fluid is based on the recognition that the common ore minerals are extremely insoluble in pure water and that complex ions are required to increase solubility. Most of the ore minerals in vein deposits are sulfides, so sulfur chemistry is particularly important. Aqueous sulfur chemistry is strongly influenced by an environment's oxidation-reduction status and by its acidity. Reactions with wallrock minerals, water from sources including groundwaters, and loss of boiling gases are important factors in metal transport and eventual deposition. At temperatures above 400°C, ionic links tend to be replaced by molecular links, and solution chemistry becomes simpler in concept but experimentally very challenging.

Transport by means of igneous melt is even more

complicated. Magmas consist of partly molten rock, and they transport a variety of elements, including tin, molybdenum, iron, nickel, copper, the rare earths, and the platinum group. Such metals may then either precipitate directly from the magmas into magmatic deposits or be incorporated into aqueous fluids as the magmas solidify.

Whether aqueous fluid or igneous melt acts as the carrier in transport, the mineral must reach a trap and accumulate in significant concentrations before a deposit can be considered economical. These traps are the targets of mineral exploration. A great deal of progress has been made in quantitatively modeling individual accumulation processes, such as boiling, cooling, mixing with dilute waters, reactions with wallrock, adiabatic expansion, and sulfurization of metal-bearing magmas. These individual process models require considerable additional data before they can be integrated into a comprehensive form that can handle all the known variables together.

Computers, using data bases on both fluid and mineral characteristics, have enabled modeling the complex geochemical and hydrologic processes related to ore formation. For example, the flow paths, flow rates, and residence times for fluids can be predicted, given a geological starting point such as a sedimentation interval within a basin or the intrusion of a granite magma into a sequence of rocks. The anticipated temperatures, pressures, and minerals can be calculated for every point within a flow system. This exercise is analogous to watching a mineral deposit form. It is also an experiment ideally suited to predicting the behavior of chemical wastes injected into rocks and of fluids used for in situ mining.

Mineral Exploration and Exploitation

Mineral deposits display great diversity in material; grade; size; and style of localization, accessibility, and minability. These largely independent variables complicate the search for profitable mineral concentrations. The ideas developed by early geoscientists about the origin of mineral deposits were based on surmise because only a few cases of ore formation could be observed directly. These few included placers forming in streams, volcanic fumaroles yielding sulfur and metallic sublimates, and dried playa lakes and marine lagoons depositing salt. Additional examples of direct observation of ore formation have been recognized in recent years as exploration of our planet has reached into even the most hostile environments (see Figure 4.10).

Since 1979 spectacular discoveries of huge thermal springs have been made at spreading centers along the mid-oceanic ridges in the Pacific and Atlantic oceans. Heat from subseafloor basalt creates a massive circulation of hot seawater, and reactions between the water and the basalt form acidic solutions. Leached metals and sulfur from the rock are deposited from vents on the seafloor above the spreading center to form cones containing iron, copper, zinc, gold, and other metals. Active vents, which range in temperature up to 350°C, are characterized by rising plumes of rapidly precipitating sulfides, graphically called black smokers. These precipitates resemble the deposits hosting the copper ores that gave Cyprus its name as well as other minerals valued since ancient times. These long-familiar mineral deposits are now recognized as ocean-floor rocks formed at extinct spreading centers and then incorporated into the continents at arc and continental collision sites. Variants occur in Mexico's Gulf of California and off the continental shelf west of Washington and Oregon. Close to continents, active sedimentation buries the spreading center, resulting in mineralization that is disseminated throughout the sediments as well as around the black smokers on the seafloor.

Of even greater economic interest are the vents associated with rifts along the crests of island arcs, such as that in the China Sea west of Okinawa; they are associated with magmas that are richer in silica than ocean basalts. These magmatic heat engines are apparently larger and longer lived than those immediately above oceanic crust and could produce deposits similar to the Kuroko-type massive sulfides, rich in copper, zinc, lead, silver, and gold. The Kuroko deposits in northern Honshu, Japan, formed about 13 million years ago when a rifting episode split the axis of the island arc. Oceanic waters flooded the rift to a depth of more than a kilometer, and the massive sulfides were deposited where seawater and magma interacted.

Another mineral deposition environment occurs in deep depressions along the axis of the Red Sea. The water descending to the heat source along that axis dissolves large amounts of rock salt from deposits buried along the margins of the sea. The resulting brine forms dense pools at the bottom of the Red Sea. Large amounts of zinc and iron are precipitating in the brine pools, demonstrating that a deep stratified brine environment may have been a progenitor of the ancient strata-bound ore deposits exposed on the continents.

Active thermal springs and solutions encountered in the course of drilling for geothermal power sources locally contain erratic quantities of ore-grade gold and silver mineralization similar to the deposits of the Comstock silver and gold lode in Nevada and the McLaughlin gold-bearing region of California. Chemical and physical observations from these geothermal operations have shown how water is heated by igneous bodies, alters host rocks, mixes with surface water, and precipitates minerals. We are provided with a peek at an ore-forming environment originating tens to hundreds of meters below geysers and bubbling springs. Gold has been discovered, in concentrations equivalent to relatively high-grade ores, in muds at the bottom of hot springs in Japan, New Zealand, and Yellowstone Park.

In particular types of ore bodies the concentrations of metal grades—the relative quantities of mineral in the ore deposit—are likely to extend over a limited range. The total tonnage of an ore body is also characteristic of particular classes of deposits. Both grade and tonnage can vary considerably, but if statistically sufficient examples of a particular type of ore body are available, the likely grade and tonnage limits for a yet-to-be discovered mineral deposit can be estimated. This is an extremely useful capability for the mining industry because it can supply guidance prior to the application of standard procedures for determining grade and tonnage of an ore body such as drilling, trenching, and extensive sampling. For instance, more than 400 Kuroko-type ore bodies have been recognized in rocks ranging widely in age in many parts of the world. Data plotted from the known Kuroko ore bodies show that the median size of a Kuroko ore body is 1.5 million tons, at an average grade of 1.3 percent copper (Figure 4.11). The tabulation also predicts that few Kuroko ore bodies can be expected to exceed 18 million tons or to contain grades of more than 3.5 percent copper.

Grade and tonnage models can also be applied to ore bodies whose origin is not as well understood as the Kuroko-type deposits. This is illustrated (Figure 4.12) for the Carlin-type ore bodies, which consist of mineralization emplaced in carbonate rocks. The population considered in this model is virtually all from Nevada and is considerably smaller than that used for the Kuroko estimates.

Exploration is a risky business, and those who minimize the risks are generally the most successful. Traditionally, exploration methods have emphasized empirical field-documented analogies with known deposits, supplemented by diagnostic geophysical and geochemical signatures. These methods have offered the lowest feasible exploration

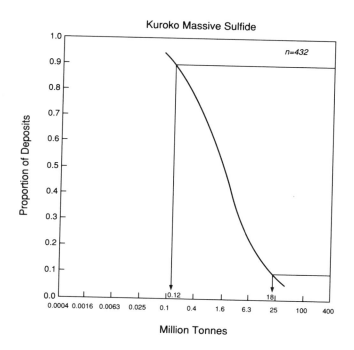

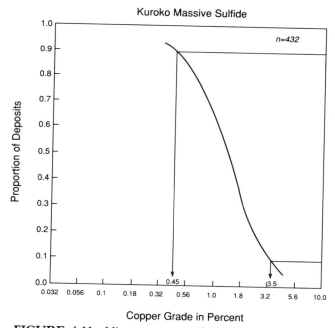

FIGURE 4.11 Ninety percent of Kuroko-type deposits have between 0.12 million and 18 million tonnes of ore (upper) and have ore grades (lower) between 0.45% and 3.5% of copper.

risks and have proven very successful, as evidenced by continuing discoveries in recent years. However, exploration has been literally skimming off the top of the crust, discovering and developing the deposits that are most easily detected and exploited. Few prospecting programs have succeeded in extrapolat-

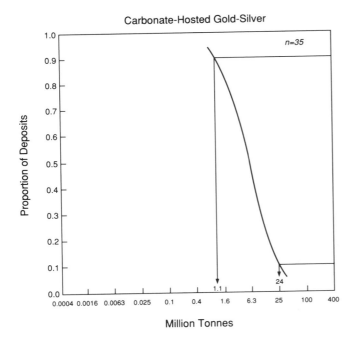

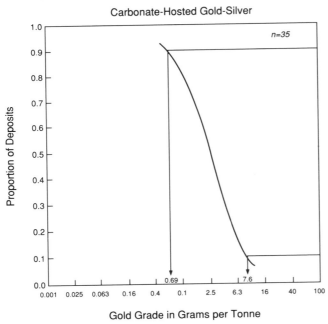

FIGURE 4.12 Eighty percent of Carlin-type gold deposits in Nevada are between 1.1 million and 24 million tonnes of ore (upper) and have ore grades (lower) between 0.69 and 7.6 grams of gold/tonne.

events. Petroleum is formed from organic matter in sediments by gentle thermal maturation, so the parts of basins that have been simmered offer the best prospects for finding petroleum. In a similar way, metal deposits are often characterized by the superimposition of thermal events on rocks that, both before and after the mineralization, were relatively cool. An examination of rocks for evidence of a former thermal event may prove useful in locating buried mineralizing systems.

In mineral exploration, genetic concepts have been augmented by empirical methods that observe what has been overlooked at the surface, seek clues about deposits that are only thinly veiled, and attempt to detect the presence of deeply buried deposits having no surface expression. Geophysical methods are foremost among the subsurface sensing techniques, particularly where the geology is well known and the nature of the target can be identified by surface examination.

Geological observations of the surface serve as the foundation for the science of mineral deposits. Geophysical techniques can sense subsurface characteristics of the crust through various methods. Ground and airborne geophysics have their most dramatic successes in the direct discovery of ore deposits through detection of particular physical properties. Magnetic surveys have discovered numerous deposits bearing magnetite (Fe_3O_4) and pyrrhotite ($Fe_{1-x}S$), and negative anomalies have identified magnetite-destructive alteration in igneous intrusions associated with other types of deposits. In the search for massive sulfide and gold-bearing deposits, increased use has been made of deeply penetrating pulsed electromagnetic surveys and of airborne surveys using a combination of airborne geophysical methods. Electrically conductive bodies have been found using induced polarization, audiomagnetotelluric, and controlled-source audiomagnetotelluric methods. There is an amazing variety of instrumentation and signal and detector array configurations.

In areas that are concealed beneath thick sedimentary deposits—such as the basement of the North American midcontinent—geophysical characterization of the region is an important adjunct to direct detection surveys. Both techniques can identify features related to mineralization. Even where the geological picture is well known, the geophysical characterization often provides provocative new information to upgrade the geological interpretations. High-speed computers permit the processing and graphic display of geophysical data, making geological interpretation and ore search easier. The

ing concepts into uncharted ground, applying more sophisticated models to predict unseen targets, and following up with persistent exploration.

One approach utilized in exploration for concealed deposits is shared by both the fuels and metals industries: the search for paleothermal increased resolution and penetration of geophysical methods will enhance the search for, and evaluation of, mineral resources.

Remote sensing from either space platforms or aircraft offers another tool to consider for both reconnaissance and detailed examination of terrains. Techniques include side-looking radar and synthetic aperture radar, which reveal information about the geological structure in poorly exposed terrains, and a range of optical and infrared detection systems. A variety of multispectral instruments, including Landsat, can detect differences in vegetation that reveal nearby mineral concentrations and are capable of distinguishing the mineralogy of some altered rocks.

The examination of natural materials such as soils, stream sediments, and stream waters for traces of economically valuable metals or, more commonly, for pathfinder elements associated with the sought-for metals, is a widely used exploration technique. The method identifies primary anomalous patterns generated by the mineralization, as well as secondary patterns resulting from dispersion due to weathering. Soil-gas geochemistry applicable to mercury, sulfur gases, and radon is a developing area of mineral exploration.

The actual physical mining of ore bodies is certainly one of civilization's oldest primary industries. Surprisingly, the techniques used from earliest times barely changed until after the Industrial Revolution, well into the nineteenth century. Since the advent of the Iron Age, miners have descended into shafts that invade the crust. Working in closed spaces with poor ventilation and high temperatures, miners have borne conditions that threaten lesser souls on a primal level. The traditions of miners and smelters are still with us in names such as nickel, taken from Old Nick, and cobalt from Kobold, a goblin, and in the only recently violated superstition that forbade women to enter mines. Many cultures relate tales of mining activities—that of the seven dwarves in one famous story or that of searching for ores with special power to work into swords, rings, or other talismans.

At first miners dug the ore from the walls with picks and shovels. They piled the rock in baskets and carried it out on their backs, sometimes up ladders of dizzying heights. Then they dug the holes a bit larger and trained draft animals to carry the piles of rock to central shafts where pulleys and ropes were used to haul the ore up to the surface. Picks and shovels were used to dig all the rock until the late nineteenth century, when hydraulic methods, engine-powered haulers and lifters, railroad cars, and elevators were developed. But even today there are many tight spots underground that can be reached only by a miner with a tool in hand. It can be a dangerous job. Dangers can include the collapse of roofs, bucking of floors, and seepage of gas, which could explode at the slightest spark, into the caverns. Because of such dangers, subsurface hand mining has become limited today to situations where no other method can be applied.

The discoveries and developments of the past quarter century have transformed modern mines into a different type of operation. Now surface mining produces a large proportion of society's mineral material. Heap leaching is a typical approach. Whole chunks of the surface are torn loose by a machine-driven shovel that fills a dump truck or railroad car in one scoop. The rock is transported to a crusher where it may be sieved and mixed with chemical catalysts. The crushed material is dumped onto a pad and treated with a spray of another chemical that will bond with the desired mineral, leach through the entire heap, and drain along the impermeable pad, leaving the unwanted material behind. The altered but separated mineral then collects in another receptacle where further treatment, either chemical or more commonly electrolytic, isolates the valuable metal.

Surface mining and separation methods such as heap leaching produce large amounts of minerals with negligible risk to the on-the-job welfare of miners. But this method could leave tracts of scarred landscapes, vast piles of rock debris, reservoirs of toxic liquid, and drastically altered environments as its by-products. As these are unacceptable, the mining industry has developed approaches that can ameliorate the effects. The industry's efforts to repair damage resulting from its activities have led to research forming the core of environmental engineering. One of the goals of such environmental engineering and related subjects is not limited just to amelioration. With enough foresight in planning complete mining operations and with dedication to prevention of hazardous and wasteful by-products, surface mining—when necessary—can be a benign activity within a reasonable time scale.

At the close of the twentieth century subsurface and surface mining are providing the bulk of needed mineral resources, but researchers are looking for breakthroughs with in situ mining techniques in the future. In situ mining is analogous to the hydrothermal mineralization process but works in reverse. It uses fluids to extract minerals without actually destroying the host rock. Strictly interpreted, this method is not new, because steam has been used to melt and extract native sulfur, and chemically tailored extractive solutions have been used to dissolve rock salt, potash, copper, and uranium. In situ mining of other deposit types has been characterized by poor recovery, but the minimal environmental impact as well as the ability to mine under extreme temperature and pressure conditions make in situ techniques attractive for some deposits. The two major problems are reaching the sought-after mineral grains, which may be widely dispersed or chemically inactive within a rock, and finding the appropriate solvent to release ore minerals while not adversely affecting the environment. Successful applications of in situ mining have been limited so far to naturally permeable rocks, because attempts to create fracture permeability have generally failed, and to disseminated rather than massive mineralization. Cultivating bacteria to facilitate metal extraction represents a current and challenging opportunity for applications to in situ mining, heap leaching, and by-product recycling.

ENERGY RESOURCES

The human species depends on energy. Advanced societies consume fuel on a scale that represents one of the greatest single changes on Earth over the past century. Modern consumption habits have developed because geoscientists have been able to discover and exploit energy resources with remarkable success. However, this success entails some adverse consequences. Few resource developments have affected the environment more than the burning of coal, gas, and oil.

The two main classes of energy sources are related to the external and internal engines of the Earth. A combination of external and internal engine components produces hydroelectric power. The topographical relief caused by cycling of the Earth's internal engine collects and channels river systems, which are elements of the externally driven hydrologic cycle. Dams are built to interrupt the downward river flow and harness the ensuing energy potential. The Sun, the Moon, and the Earth's spin—components of the external engine—generate solar energy, tidal energy, and wind energy. Wind power and solar power have recently come of age. Large wind farms can produce electricity at a profit. They are especially suited to flat open spaces such as western North America, northern Europe, North Africa, India, and the former Soviet Union. Similar environments, with the added characteristic of negligible cloud cover, also can support large thermal solar collection installations. Small-scale solar collectors are becoming more common, especially in other parts of the world—India reports the use of solar energy systems in some 6,000 villages. Tidal energy is still in the planning stage, with trial installations testing the suitability of various designs.

These energy sources are renewable and are generally harmless in a global environmental sense, although the concentration of collection facilities does alter the local landscape. However, in most situations, they are not yet economically competitive with fossil fuels. Fossil fuels are our main energy source. They are produced primarily by the external engine of the Sun, but they are also affected by the slow consistent heat of the internal engine. Oil, gas, and coal are all derived from organisms that relied on solar energy millions of years ago. Accumulations of such organisms provide the most economical source of energy today, but the burning of fossil fuels releases abundant amounts of carbon dioxide into the atmosphere. The quantity of carbon dioxide produced by industrialized society may cause significant global warming, which in turn could produce undesired environmental effects.

The interior of the Earth is hotter than the surface, although the rate at which heat reaches the surface from the Sun is at least 10,000 times greater than the rate at which heat emerges from inside our planet. The source of the internal heat is largely the slow decay of naturally radioactive nuclides. Gaining access to this natural decay of radioactive nuclides, with the goal of harnessing the resultant excess energy, remains a great challenge of the future.

Uranium is the most abundant natural source of available radioactive nuclides. The controlled fission of ^{235}U in nuclear power plants has been used as a source of energy for more than 30 years. The most vexing problem involving nuclear power is long-term disposal of the radioactive waste. Although there are several proposed methods for the treatment and isolation of such waste, none have been licensed in the United States, both because of technical uncertainties and social and political resistance. Without an appropriate solution to the problem, public utilities are not investing in developing new nuclear power plants nor are operational licenses being granted.

The heat emanating from the Earth's interior is an attractive energy source. This geothermal energy can be used for electrical generation or direct space heating. It is a relatively clean source of energy, and large amounts are potentially available. Present use of this source is limited to areas where partly molten, or recently crystallized (in a geological sense), rock heats groundwater that carries heat toward the surface where it can be harnessed to generate electricity. Few locations can claim the coincidence of both hot rocks and groundwater; consequently, geothermal resources have been developed only locally. Research in geothermal energy includes the use of magma as a direct energy source and the extraction of energy from hot dry rock by injecting water and recovering it to extract the heat.

Earth scientists will continue to address the nation's energy needs through the discovery of energy resources, their assessment, and their economic development in an environmentally sound manner. Fulfilling these responsibilities calls for an adequate scientific understanding of the environment in which the resource occurs. An appreciation of the past—how that environment and the resource have evolved over geological time—complements the anticipation of the future—how resource development will affect the environment.

Growth of the worlds population and rising per capita energy use will lead to increased energy demands in the next century. Despite concerns about the environmental effects of increasing carbon dioxide levels, the world simply cannot withdraw from its dependence on fossil fuels in less than several decades. Substitution, the development of new technologies, and conservation can reduce this reliance.

The development of solar, tidal, and wind energy requires continued study, as does increased use of geothermal sources. The continued use of nuclear energy is dependent on the efficient and responsible handling of radioactive wastes. Future energy resources may include forms that seem exotic today: hydrogen fuels and nuclear fusion are subjects of ongoing research. Regardless of the mix of eventual sources, society has become more conscious of the problems of integrated development, consumption, and waste disposal—the activities constituting the use of any energy or mineral resource.

Petroleum Resources

Significant accumulations of oil and gas are geological anomalies, and they constitute only a minute fraction of the total material within the sedimentary basins where they are found. These two factors create the scientific and technical complexity of the geoscientist's role in petroleum recovery and of the industry itself. Even in relatively well-understood sedimentary basins, such as the Gulf of Mexico, the

success rate for exploration drilling is only about one in six. In the true frontier areas that risk might be increased by a factor of 10 or more. The cost of a single exploration well can be in the tens of millions of dollars. Appropriately, great effort and expense are invested in geoscientific research that increases the probability of successful strikes and effective extraction.

The historical development of petroleum resources has reached a critical point. In the conterminous United States, additional discoveries are sought within the framework of developed resources. Recent increases in reserves have resulted predominantly from enhanced recovery methods in known fields and from in-fill drilling between successful strikes. Some frontier contributions can be expected from the deep-water Gulf of Mexico and from Alaska. Worldwide, however, new and undeveloped resources are sought. The two searches are complementary, and they present a remarkable opportunity for geoscientists. Researchers can apply a full spectrum of approaches and technologies to the search for, and the exploitation of, petroleum deposits. These range from the use of traditional interpretations of oil genesis to the application of innovative quantitative and genetic models of producing reservoirs. Vast quantities of new data from both satellite imagery and seismic surveys offer unmatched opportunities for gaining insights that were unimaginable only a few decades ago. These new methods are being applied all over the world but especially in the Middle East.

According to the most recent estimates, the countries of the Middle East overlie more than one-half of the world's undeveloped petroleum resources. This accumulation came about as part of the cyclical process involving the opening and closing of the ocean basins, which is the consequence of plate tectonics.

About 100 million years ago the area that is now Arabia was located along the north coast of the supercontinent of Gondwana, forming an Atlantic-type continental margin at a tropical latitude (Figure 4.13). Anoxic episodes, characterized by the absence of free oxygen, encouraged accumulation of organic debris in sediments deposited at water depths of a few hundred meters along the margin of the continent. In shallower water close to sea level, abundant limestone reefs grew; with oscillating sea levels, evaporite salt deposits precipitated among the limestones. Uplift to the west, in what is now Africa, resulted in erosion that episodically carried sand into the offshore area. These related events provided the carbon-rich source rocks in which petroleum would

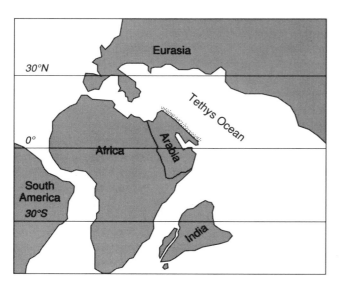

FIGURE 4.13 Organic-rich source rocks formed along the north coast of Gondwana. As the Tethys Ocean closed and with the collision of Arabia and Eurasia to form the Zagros Mountains, the oil migrated and was trapped in structures that now form the huge oil fields of the Middle East.

later form, the porous coarse limestones and sands that would act as future reservoirs, and the evaporites and fine-grained limestones that would trap the petroleum beneath their seal.

Approximately 15 million years ago, this well-prepared Arabian shore collided with Asia and thrust up the Zagros Mountains. Today, the Zagros range continues to build over the Arabian plate. The now deeply buried oil source rocks are being heated to generate oil that migrates upward into huge retentive structures sealed by the fine-grained limestone and evaporites. Close to the Zagros, these structures are long anticlinal arches parallel to the mountain ranges in Iran, Iraq, Syria, and Turkey. Farther south, these structures are the huge domes underlying the Persian Gulf, Kuwait, Saudi Arabia, and the United Arab Emirates.

The exceptional petroleum resources of the Middle East result from a set of conditions excellent for forming oil and trapping it within structures ideal for its accumulation. Because this process has occurred within the geologically recent past—during the past 15 million years—little of the oil has had time to escape from the traps, and it has not been degraded to gas by thermal cracking—the breakup of chemical compounds into simpler forms.

The petroleum resources of the Persian Gulf area have been discovered and exploited within the past 80 years. This same time span has witnessed the expansion of society's dependence on fossil fuel, the

understanding of plate tectonics as an explanation for genesis and location of resources, and the development of vast data-processing abilities within the geological research community. Progressive understanding of the Persian Gulf oil fields has engendered an invaluable model that provides a basis for future exploration efforts and more efficient extraction techniques.

Origin of Petroleum Within Sedimentary Basins

Significant petroleum resources result from series of events that fall into four categories: genesis of oil and gas from preserved organic debris within a source rock, primary and secondary migration of the generated hydrocarbons over distance, entry of the hydrocarbons into a properly prepared reservoir, and finally entrapment of the hydrocarbons (Figure 4.14). All of these events occur within the confines of a specific basin. During the past century, geoscientists have studied the evolution of such basins closely, producing models such as that of the

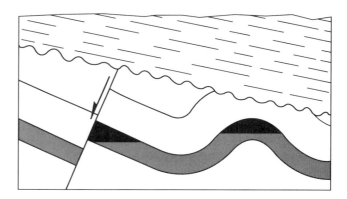

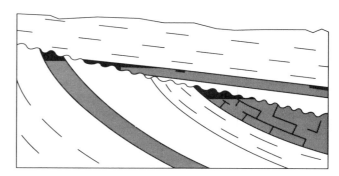

FIGURE 4.14 Idealized section (upper) shows characteristic pre-unconformity traps that are obscured by post-unconformity formations. A diagrammatic section (lower) shows traps commonly associated with surfaces of unconformities.

Persian Gulf. Each of the four developmental categories, while overlapping in time and location within a basin, has been studied separately, and in the final assessment all are combined into an integrated approach for exploration.

Organic material is oxidized at the surface by processes such as bacterial decay. Under some conditions the bacterial decay is incomplete, and burial of the remains segregates carbon-rich material within sedimentary rocks. It is this sequestered organic matter that, upon further burial, is capable of generating fossil fuels. Coal consists of solid plant matter preserved where it was originally buried, while oil and gas are fluids formed from buried organic matter that subsequently migrates from its source area.

Researchers try to establish exactly how distinct fluids are generated and what kinds of oil and gas relate to what source materials. This knowledge can help direct the search for new oil deposits. For example, although 50 years ago it became obvious that lake beds, such as the Green River Formation of Utah, may contain significant quantities of oil and gas source materials, it is only in the past decade that lake beds worldwide have been emphasized in oil and gas exploration.

Sedimentary basins are depressions on the surface where loose particles accumulate and eventually turn into solid rock. These depressions result from the effects of thermal convection on the lithosphere and from the application of loads generated by a variety of processes, including mountain building and rifting. The process of building a mountain belt puts a heavy load at the surface and forms depressions on either side; rifting thins the lithosphere and forms a depression where dense material, usually consigned to a deeper position, is brought closer to the surface. Lighter or buoyant loads are locally associated with rifts in volcanic areas where hot or light rock is abundant near the surface. These buoyant loads, which correspond to elevations on the crust, are distributed sporadically along the lengths of active rift systems like that in East Africa. An environment in which basins are particularly well developed is the rifted Atlantic-type margins of continents where juxtaposition of dense oceanic lithosphere against less dense continent generates an elevation difference of several kilometers. This is the environment in which most active sedimentation on Earth is concentrated, much of it building the world's great deltas.

The thickness of material accumulating in sedimentary basins may become greatly amplified in comparison with the depth of the initial depression.

For every unit of sedimentary load deposited at the surface, a corresponding mass of material, with a somewhat higher density, flows away in the underlying asthenosphere. In this way further subsidence is induced; the fill in sedimentary basins may become as thick as 15 km.

Petroleum source rocks are accumulations that are rich in organic matter. Most are shales, but siltstones, limestones, evaporites, algal mats, and even coals can be important. The essential characteristic likely to create petroleum resources is rapid accumulation of organic material—fast enough to exceed the oxidation of carbon due to bacterial decay. Typical environments are primarily aquatic—such high biological activity takes place in lakes, swamps, lagoons, bays, and estuaries. These anoxic, or nonoxidizing, conditions have developed episodically in deep ocean basins—at times and places of unusually high nutrient flux and limited bottom-water circulation. Restricted ocean basins—those just beginning to open (such as the Red Sea), or nearing closure (like the modern Mediterranean), or left as remnants of former oceans (such as the Black Sea)—are strong candidates for anoxia. Through analogy, this recognition guides oil exploration to sites of ancient anoxic conditions. Climate exerts a strong control over the processes of nutrient flux and anoxia, so studying the distribution of organic-rich rocks contributes to an understanding of ancient climatic processes—and vice versa.

Buried organic material changes within sedimentary basins. Temperature and the duration of burial control complex modifications, called maturation, producing a diverse group of organic compounds. The most important compounds, for the purpose of forming fossil fuel resources, are the kerogens. Different kinds of plant matter yield different kerogens that, in turn, are capable of producing different proportions of oil and gas. The kinetics of the maturation process vary with the complex organic compounds of the specific kerogen involved, but generally oil is produced from less mature kerogens than gas. Some kerogens are specifically oil prone and some specifically gas prone from the beginning. According to their maturity, source rocks may be incapable of producing any fluid hydrocarbons at shallow depths, capable of producing oil at intermediate depths, and capable of producing gas at greater depths. At very great depths kerogens become overmature, having released all their mobile hydrocarbons, and eventually become hydrogen-free graphite.

In oil and gas exploration the optical properties of vitrinite, a dark glassy material occurring in most coals and in some kerogens, provide a valuable index to the level of maturity. Vitrinite reflects more light as it becomes more mature, so the comparison of measurements with standard samples can indicate whether a particular rock is immature, mature, or overmature. There is no point in looking for oil where the rocks are still immature, where no hydrocarbon has been generated, or where all hydrocarbons have been driven off. Both surface outcrop samples and drill-hole cuttings are used in this way to help guide exploration. A variety of other thermal indicators of sediment maturity are also used, ranging from the color of contained microfossils such as conodonts and the annealing of fission tracks to the proportion of radiogenic argon driven from fragmented microcline crystals after deposition.

The formation of oil and gas from buried organic matter is thermally controlled, so in oil exploration it is important to be able to characterize the thermal state of sedimentary basins through time. A useful distinction can be made between basins formed by extension, a process that thins the lithosphere and generally brings hot deep-lying rock nearer to the surface, and those formed by shortening, a process that thickens the lithosphere and initially produces a cooler crust. In the former circumstance, the first sediments deposited in the basin are likely to mature rapidly through the oil-generation phases, but in the latter circumstance an initially cold pile of sediments gradually warms, with oil maturation taking place later in the basin's evolution. Sedimentary basins formed at continental ruptures—rifts where oceans have failed to open and Atlantic-type margins where oceans have successfully formed—are likely to contain petroleum source rocks that attained maturation within a few tens of million of years of deposition. The main producing basins in China are of this type, as are the Gabon and Angola basins of West Africa. For sedimentary basins formed in association with the shortening of continental collision, the maturation process is much slower. In the case of the Anadarko basin in Oklahoma, for example, a 380-million-year-old source rock began to generate abundant oil and gas only when deeply buried 100 million years later.

The relative integrity of geological records preserved in sedimentary basins makes them ideal natural laboratories for multidisciplinary studies, including depositional processes, earth history as recorded in sediments, biological evolution including extinction, climatic variation in space and time, thermal structure and evolution, and lithospheric structures that vary with depth. The opportunities for addressing these issues in sedimentary basins are exceptional because it is within those basins that petroleum, coal, uranium, and water resources are largely concentrated. Seismic reflection data are available in explored sedimentary basins, and oil wells and their geophysical logs complement and permit calibration of seismic results.

Basins can be classified in many different ways. One simple way is to recognize that basin development is controlled by the plate structure of the lithosphere. For example, the Arabian oil-bearing basin lies at a convergent plate boundary where Asia and Arabia are colliding, and the Suez basin lies in a rift zone where Africa and Arabia are beginning to separate. Extending this idea to cover plate evolution through time, and recognizing that the dominant result of plate tectonic processes at the surface is the cyclic opening and closing of ocean basins, lead to a simple classification of basins related to that cycle. Thus, sedimentary basins develop by various means: in rifts as continents begin to rupture, at Atlantic-type margins when continents have ruptured, and at convergent margins in association with the island-arc and continental collisions that reduce the areas of ocean basins and lead to their eventual closing.

Several hundred sedimentary basins are recognized worldwide, but most oil and gas are produced from fewer than 50 basins. The double challenge now is to establish which of the remaining basins will become major producers in the next decades and how to produce more from those already in an advanced state of development.

The question of the origin of petroleum leads to the study of organic matter likely to be involved in petroleum formation and to the examination of source rocks in various stages of maturity. Particular oils and gases formed from the maturation of particular source materials are available to the investigator for complementary study near well-documented sites of exploitation. A challenge lies in establishing what happens to the fluids between the place of generation and the place of eventual entrapment—during migration.

Migration of Petroleum Within Sedimentary Basins

Oil and gas migrate from their places of generation through permeable rocks and through fractures. Globules of petroleum move when their buoyancy exceeds the minimum capillary pressure necessary to displace the water already filling the pores and fractures of the rock. The globules are less dense than water and are driven by buoyancy and

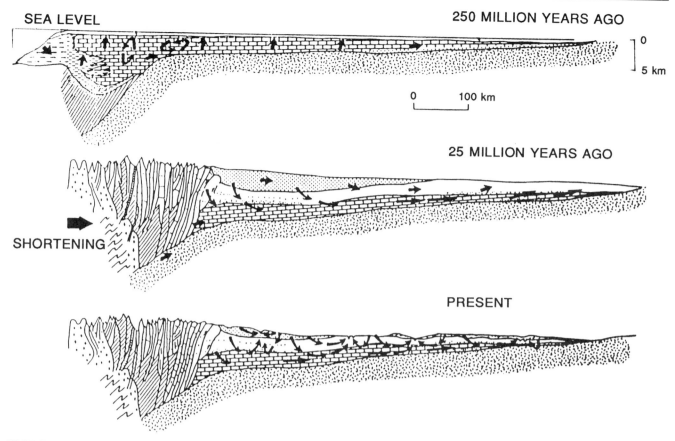

SEA LEVEL   **250 MILLION YEARS AGO**

25 MILLION YEARS AGO

SHORTENING

PRESENT

FIGURE 4.15 Three cross sections showing the development of the Alberta Basin; arrows show the sense of fluid flow in these sections.

hydrodynamic forces. Dominant fluid-driving forces are the gravitational head created by topographic relief above sea level and the compactional head generated by subsidence and compression below sea level. Oil and gas stop moving when they reach a capillary pressure equilibrium or when they are trapped by capillary seals that reduce or prevent further migration.

The Alberta Basin, including its southward extension into Montana, provides an example of how fluid flow has evolved as a sedimentary basin developed (Figure 4.15). Some 250 million years ago the west coast of North America was the site of an Atlantic-type margin. Sediments were deposited and various fluids moved through the deposits, both processes occurring largely in response to sea-level changes or down-warping of the coast. From about 200 million years ago, a complex succession of island arcs colliding from the west built a range of mountains along that margin, and a foreland basin formed in front of the rising mountain belt in response to the loading of the crust by the mountains. Sediments eroded from the mountains, filling

the basin, and fluids moved through the basin accordingly. Compressional compaction of deeply buried sediments beneath the shortening mountains further enhanced fluid migration.

During the first phase, thermally driven free convection of waters in contact with seawater dissolved some minerals and deposited others, leading to a variety of changes that together are called diagenesis. Within the Alberta basin these changes included replacing some of the calcium in calcite with magnesium to form dolomite, a process that often reduces rock volume and yields more pore space to form potential reservoirs, and cementing sands to form sandstones, a process that reduces porosity. Sulfide mineral deposits were formed locally in the sediment wedge by fluid flow under this regime. In areas where rapid deposition of impermeable sediments confined water-filled porous materials, overpressures—which eventually are dissipated through faulting or seepage—may have developed. Oil formation in the Atlantic-type margin took place as cold, organic-rich sediments were deeply buried and warmed in the Earth's conductive gradient. Oil and gas migration would have taken place over short distances.

The construction of a mountain load on the edge of the continent buried fluid-bearing rocks to great depth, and buoyancy carried the fluids long distances—hundreds but not thousands of kilometers—through permeable beds and along thrust faults in the foreland basin to join locally generated oil and gas. Finally, at least in part in response to flexural relaxation after the shortening in the mountain belt had ceased, the mountains were elevated to about their present heights, and the active, regionally extensive, gravitationally driven flow system was established.

During burial in a subsiding sedimentary basin, rocks are measurably altered, largely by inorganic chemical processes. Flow patterns, temperature fields, and patterns of chemical reaction between rock and fluid are all interdependent. The flow of fluids through sedimentary rocks modifies the rock matrix, changing porosity and permeability as some minerals dissolve and others are deposited. Elevated temperatures foster the maturation of organic material, producing organic acids that can dissolve matrix, thus enhancing porosity. And coating of a mineral grain surface with iron oxide minerals may prevent further matrix deposition and thus maintain porosity. The processes and interactions that can happen are very diverse, and the sequence of events may prove critical to the formation and preservation of exploitable oil and gas resources.

Geologists treat all these changes as part of a single process, diagenesis. Diagenesis is important in the exploration for energy resources because there is a need to predictively model the distribution of mature source rocks in a basin and to understand the patterns of change through time that may alter porosity and permeability in reservoir rocks. Just as maturation of source rocks depends on time and temperature, so apparently do the diagenetic changes that reservoir rocks undergo during subsidence and burial. Aided by seismic and sequence stratigraphy, the understanding of diagenesis provides a powerful exploration tool.

Mathematical modeling of fluid flow in sedimentary basins represents an important complement to the study of diagenesis. The tradition of modeling began decades ago in groundwater research, when simple electrical conductivity analogs of water flow were successfully developed. Modeling studies have been extended to the migration of oil and gas only within the past 35 years. Modern computational facilities are being used to simulate transient fluid flow and heat transport with a continuum approach. Current challenges include coping with fractured rock, scaling heterogeneities, and assessing rock physics data such as the quality of available permeability. Scale invariance, recognized as plausible in many properties of sedimentary basins, is leading to successful application of fractal methods.

Two-dimensional models of coupled fluid and heat transport have been used to develop pictures of ore-forming processes, geothermal processes, and regional oil migration. Two-dimensional models of paleohydrology have been published for important basins such as the Alberta, Williston, Illinois, Gulf Coast, North Sea, Suez, and Rhine graben and, on a smaller scale, for the Denver and Palo Duro basins. Geochemical change has been addressed in a few studies, including quartz cementation of sandstone, hydrothermal mineral vein deposition, and uranium ore deposition within sediments.

Fully three-dimensional study of flow in sedimentary basins lies in the future, as access to supercomputers and the establishment of appropriate codes permit the incorporation of compositional and physical variations.

Exciting though the possibility of a full description of fluid flow in a sedimentary basin may be, there is also a need to model the full development of the basin itself as a response to loading of the lithosphere. Basin subsidence, both under mechanical load and in response to cooling—as well as compaction and deformation of the basin fill—and variation in rate of fill in response to sea-level change and uplift of neighboring mountains are all being accommodated in partial and two-dimensional sedimentary basin models. The properties of the fully integrated three-dimensional dynamic basin model are beginning to be defined, and realization of such models should help in revolutionizing many aspects of oil exploration.

Traps: Where Oil and Gas Stop

The critical link in the accumulation of hydrocarbons is entrapment. After maturation, expulsion from the source rock, and migration through buoyant fluid flow, hydrocarbons will resort to any available path to the surface and escape—unless they are trapped. Porous material in which hydrocarbons can accumulate, surrounded by an impermeable seal that prevents further updip flow, constitutes a trap.

The simplest of traps is an anticlinal fold, an arch-like flexure of rock layers. The hydrocarbons migrate to the crest of the fold, because they are less dense than formation waters and are trapped by an impermeable seal—often a layer of shale or salt. Anticlines were recognized to have the capability of trapping fluid hydrocarbon early in the development of the petroleum industry. At first, surface anticlines were sought and drilled, but two developments rapidly expanded the possibilities. Anticlines beneath the surface began to be located by geophysical methods, and other entrapping structures were recognized, particularly structures in which faults contribute to enclosure of the hydrocarbons.

The large majority of discovered hydrocarbon reserves have been found in structural traps. Exploration in frontier basins today continues to be dominated by the drilling of structural prospects, once the potentially favorable source and reservoir stratigraphy have been identified. Despite a long history of experience and success, many unsolved problems of extreme structural complexity remain. For example, significant potential accumulations are associated with thrust belts (such as those found in Wyoming), which may yield giant oil and gas fields. Thrust belts are relatively unexplored, or unsuccessfully explored, because of the great difficulty in acquiring and interpreting the seismic data to form a recognizable subsurface picture. Major industrial and academic research programs in structural geology and seismic technology are aimed at this unsolved problem.

Improvements in seismic data are also important in aiding the discovery of hydrocarbons in formations more subtle than the familiar structural traps, such as the stratigraphic trap. One of the world's largest oilfields lies in a stratigraphic trap, East Texas, discovered in 1930. Stratigraphic traps depend on a wedge-shaped termination to a reservoir unit at a seal. The revolution in seismic stratigraphy over the past decade has promoted the location of stratigraphic trap geometry. Stratigraphic relationships can now be interpreted directly on a seismic reflection record, and inferences relative to the position of reservoir-prone and seal-prone lithologies can be made. The power of this approach in defining the subtle stratigraphic trap remains very promising, and some substantial new discoveries have been made.

Combined structural and stratigraphic traps provide very attractive exploration targets. Representative of such a trap would be a submarine fan reservoir sand along the flank of an anticline, downdip from the crest of the structure. In years past the submarine fan would not have been recognized, and prevailing exploration practice would not consider drilling a well anywhere but on the crest of the anticline. We now have the technical ability to search for potential stratigraphic traps defined by position relative to structure. This should greatly broaden production possibilities from structures that were formerly explored only in the conventional way, many of which are now productive despite the absence of reservoirs at the crest of the structure.

The timing of the formation of the trapping configuration relative to the timing of generation and migration of the hydrocarbons is critical. Traps are of no value unless they exist prior to the migration of hydrocarbons. The integration of maturation modeling, which concerns predicting the timing of hydrocarbon generation, with the delineation of drillable traps is vital to a sound and successful exploration program. It is also a further instance of the importance of the integrated basin analysis approach to petroleum exploration.

Oil Shales and Tar Sands

The hydrocarbons used to fuel the industry and transportation networks of twentieth-century societies have migrated from source rocks and been trapped beneath the surface before they could be altered into unproductive compounds. Scientists have identified, and are attempting to utilize, those hydrocarbons still sequestered within the source rocks as well as those that have changed too much to be exploited by known methods. Examples of these are oil shales that are exposed source rocks and tar sands that are fossil compounds altered by exposure to near-surface agents.

Some shales contain exceptionally large amounts of kerogenic organic matter, and for more than a century small quantities of oil have been distilled from oil shales in various parts of the world. In the United States, interest has focused on the 50-million-year-old oil shales in the Green River lacustrine deposits of Colorado, Utah, and Wyoming that contain an estimated 2 trillion barrels of oil in the richest beds. The oil is in the form of solid kerogen, however; without the slow gentle thermal maturation necessary to fluidize the kerogen, the desirable components remain locked in a mixture of sandstones and clays and in layers of precipitated carbonates and evaporites. Production of oil from oil shale is not economical at present, but interest for the future persists because the quantities of oil locked in these rocks represent many times the world's reserves of conventional petroleum. Any recovery method would require the application of high temperatures and large quantities of water, however, and tremendous amounts of spent shales would accumulate on the landscape. One calculation suggests that recovering all the oil from just U.S. sources would leave enough spent shale to cover the entire globe to a depth of about 3 m.

If fluid hydrocarbons escape entrapment, they migrate toward the surface. Once the compound approaches the level at which it interacts with the atmosphere, groundwater, or soil biota, it may accumulate in sedimentary deposits and form tar sands—thick concentrations of viscous degraded crude oil that has saturated detritus until it became an asphalt. Traditionally researchers assumed that the oil's fluid nature was altered through oxidation, with lighter fractions escaping on contact with the atmosphere. Recent studies, however, suggest a bacterial influence acting at depth that removes particular components from the migrating fluid. While research into biological agents that transform hydrocarbons may not solve the problems inherent in exploiting tar sand deposits, it may lead to solutions for cleaning up oil spills and other environmental dangers associated with the use of petroleum products.

Exploration for Oil and Gas: The Integrated Approach

The earliest oil and gas discoveries were accidental. Natural gas wells several hundred meters deep were drilled in Sichuan, China, nearly 1,000 years ago, following gas discoveries during salt production. The oldest wells in North America were drilled near oil seeps. Technical understanding has subsequently played a progressively larger role in exploration, ranging from recognition of anticlinal traps at the turn of the century through increasing application of geophysical methods to the sophisticated array of instruments now used in the sondes lowered into drill holes to log wells. The use of micropaleontological, geochemical, structural, stratigraphic, sedimentary, and diagenetic understanding over the past 80 years illustrates the growing role of diverse branches of geology in petroleum exploration. Geophysics has grown even more rapidly. Gravity and magnetic measurements continue to be useful, but the field is dominated by seismic techniques closely linked to advanced computational capabilities. These high-technology applications have become universal and critical in modern oil and gas exploration.

Much current research is directed at the analysis of unconformity-bounded depositional sequences (see Figure 4.14). Work of this kind, termed se-

quence stratigraphy, establishes a hierarchy of such unconformity-defined units to be identified in many basins. Understanding the way in which sedimentary rocks are layered has proven extremely valuable in exploration for petroleum and coal resources. Mounded shapes are diagnostic of submarine fan deposits, and nested sigmoidal packages often indicate advancing delta complexes at the mouths of ancient river systems. Both of these environments may contain sandstones that have high potential as reservoir rocks for the entrapment of oil and gas.

Unconformity-bounded depositional sequences at all scales are associated with cycles in which the position of the ancient shoreline is observed—within the sedimentary record—to move alternately in a landward and then a seaward direction. Some scientists have related these oscillatory variations to global sea-level change, but others suggest that vertical motions of the crust, variations in sediment supply, and climatic fluctuations are just as important. The importance of sequence stratigraphy to this controversy is that the formation of sequence boundaries requires changes of relative sea level, not merely of sediment input. By comparing the ages of sequence boundaries in numerous basins in different parts of the world, it should be possible to distinguish sequence boundaries of eustatic, or global sea-level, origin from boundaries of only local significance. One of the major research opportunities afforded by the emerging field of sequence stratigraphy is, therefore, the attempt to define the history of sea-level change.

The importance of this new technique to all petroleum exploration efforts cannot be overemphasized. It provides geoscientists not only with a tool for interpreting the depositional systems within a basin, including potential reservoir and source intervals, but also with a means of making temporal correlations where no other control exists. And where independent calibration does exist, precise and internally consistent correlations can be more firmly established.

The need to characterize the basin environment as fully as possible in oil exploration together with the wealth of information in seismic and well records, has propelled oil company geologists and geophysicists to the forefront in developing a new approach to understanding sedimentary basin fill. That approach is termed seismic stratigraphy.

Seismic stratigraphy is a way of interpreting the sedimentary fill of a basin. The fundamental concept is to project into the Earth a known controllable source of seismic energy at a precisely known location and then to record the scattered seismic waves returning to the surface. As the seismic waves penetrate the crust, a small fraction of their energy is scattered back to the surface as they encounter differences in acoustic properties related to changes in rock formations. In principle, these returned signals, which contain detailed information about the subsurface geology, can be processed by computer to produce an acoustic image—a seismic reflection profile—of the subsurface down to depths of several kilometers. The profile is not a true geological cross section, but the gross geometry of stratal layers can be determined from it because these surfaces bound rocks with sharply contrasting acoustic properties.

Sedimentary rocks are composed of wedge-shaped, almost tabular, bodies bounded by discontinuities or unconformities that may extend regionally. In many cases an unconformity is associated with the pinching out of overlying and underlying layers. Stratal or layer discontinuities are subtle and difficult to observe, even in outcrop exposures, but often can be identified from good-quality seismic data. By piecing together information from a grid of intersecting seismic profiles and incorporating data from wells and boreholes, stratigraphers can often interpret the three-dimensional layered geometry in the sedimentary basins. Recognition of regional unconformities is especially significant because rocks above an unconformity are everywhere younger than rocks below the unconformity. The geometric shapes recognized in seismic reflection data can, therefore, be interpreted as depositional units, commonly referred to as depositional sequences (see Figure 4.14).

Reflection seismic data represent the single largest class of scientific data collected in digital form. More than 1 million data tapes are recorded annually worldwide, and the processing and analysis of this immense volume are a major commercial driving force behind the development of the advanced scientific computer industry in the United States. Even with the recent downturn in the petroleum industry, reflection seismology is still the single largest scientific and technical activity of seismologists worldwide.

In modern seismic data acquisition, the signal emanating from the energy source, or shot point, is digitally recorded at hundreds of locations on the surface. Because the scattered signals are very weak, the shot point is moved, and the shot is repeated hundreds, even thousands, of times, resulting in millions of data points being recorded per linear kilometer. Conventional seismic data are thus col-

lected in long linear segments called seismic lines, which give a two-dimensional cross-sectional picture of the sedimentary section. To approximate the three-dimensional Earth, a network or grid of lines is recorded. A single regional marine survey, for example, might include 5,000 km of such data—an enormous volume of data by any standard.

Probably the most significant technical advance in petroleum earth sciences in the past decade has been the extension of reflection seismology from two dimensions to complete three-dimensional volumes. The processing of these data produces subsurface images of unprecedented geological detail. In every case where this technique has been applied, substantial revisions have been made to the geological interpretation—with subsequent economic effects. The volume of data recorded for three-dimensional surveys, however, is much larger than that for conventional two-dimensional grids covering the same area. State-of-the-art surveys now collect more than 1 billion data points per square kilometer. Processing these data stresses the largest computer systems available and often takes up to 1 year for large surveys.

Despite the cost and volume associated with three-dimensional seismology, even larger surveys are being recorded, with some exceeding 1,000 km^2 each. It is predicted that by the mid-1990s the Gulf of Mexico will have been covered completely with three-dimensional surveys. This data set, once assembled, will provide an unparalleled view of an entire major producing basin and could revolutionize detailed understanding of the complete petroleum system.

Well logging is the term for acquiring complex measurements in a drilled borehole. Vast arrays of sophisticated devices are lowered into the well, in effect creating a remote-controlled downhole laboratory. Data are continuously collected on the electrical, acoustical, and nuclear properties of the rock formations that have been penetrated by the bit.

These data are processed and analyzed to yield a very detailed description of the type of rock and its porosity, and the presence or absence of oil or gas. This complex analysis, termed formation evaluation, is the primary means by which geologists determine whether a well will be a dry hole or a producer. All exploration wells are extensively examined using these critical data as well as selected samples or cores.

Well-log measurements and reflection seismic data comprise the essential subsurface measurements collected for modern hydrocarbon exploration and development. Integrated into the modern geological framework of basin analysis, they form the observational backbone for predicting detailed sedimentary structure and stratigraphy.

More Oil from Producing Fields

Oil and gas exploration, development, and production proceeded rapidly in the United States by comparison with the rest of the world. Consequently, the oil fields of the conterminous states are now more depleted than their counterparts elsewhere. Most U.S. oil reserves are in discovered fields and are considered to be nonrecoverable resources. This presents a unique opportunity and challenge for improved technology, because recovery from most hydrocarbon fields in the world today ranges from only 15 to 55 percent. An immediate need to develop a better understanding of reservoir heterogeneity, in situ rock and fluid properties, and enhanced oil recovery techniques is apparent.

The most significant trend in the U.S. oil industry has been the shift in effort toward recovery of a maximum amount from existing fields and established areas. In the major oil companies a majority of earth scientists are now engaged in production- or exploitation-related activities instead of the more traditional exploration-related areas. New giant fields—those with recoverable reserves greater than 100 million barrels—in the United States are likely to be found only in remote areas and have associated greater costs. However, there are numerous old giant fields that have yielded only a fraction of their original oil. Hence, billions of barrels of domestic reserves could be added by the use of enhanced recovery techniques. Horizontal drilling has increased reserves in some areas. Complemented by discoveries from more detailed exploration within existing producing regions, the motivation for the shift of effort to established areas is clear.

New types of well-logging tools and interpretational techniques are developing constantly. These developments now allow the location of previously unrecognized oil and gas zones in both actively producing and abandoned wells. The potential for increasing reserves is enormous. The current focus of the domestic oil industry is commonly referred to as reservoir management, and most major companies have actually restructured their organizations in this direction—establishing interdisciplinary teams that include geologists, geophysicists, and reservoir engineers. The teams are charged with characterizing the established fields and reservoirs and then using the information to manage further development drilling and the application of enhanced recovery techniques. In addition, such teams explore for untapped reserves that become economically viable because of their proximity to existing producing facilities.

The most significant technological development in the past decade supporting reservoir management has been the advance of three-dimensional seismology. The large investment for three-dimensional seismic acquisition and processing is often easily justified on economic grounds because of its leverage against the much larger costs of development and field production. The technical base for three-dimensional seismology should continue to grow, along with the capability to record ever larger surveys and produce more detailed images with ever larger computers.

Entirely new seismic technologies are also now in the research and development phase. The most promising is the area of cross-well seismology, in which the active seismic experiment is carried on between wells instead of on the surface. Producing an image from these data is conceptually similar to medical CAT scan technology. The key advantages of the technique are that much higher seismic frequencies can be propagated between wells than from the surface, and measurements can be made within an existing and operating oil field. Moreover, time-lapse repeat measurements can be made with great precision and have been tested for monitoring the advance of hot steam fronts in enhanced production projects that attempt to flush extra hydrocarbons from an otherwise depleted field. Although still in the research phase, an entirely new industrial application of seismology is developing.

Characterization of complex reservoirs for the development of geological and fluid behavior models could lead to optimal hydrocarbon recovery from existing reservoirs. The problems are not different from those generally addressed in modern oil field research and embrace such general considerations as the flow of fluids through porous media. General practice has been to expect such research to be conducted by the 10 or so major oil companies. However, major companies are focusing increasingly on larger targets, and the needed research is lagging.

Coal: An Abundant Fuel Resource

The United States is rich in coal and its reserves are sufficient to support consumption at the present rate for centuries. Most U.S. coal is used in the generation of electricity; it accounts for more than one-half of all electricity produced. The future of the industry depends not only on traditional market variables but also on coping with the environmental effects of mining and burning coal. Public perception about coal and technological advances in efficient utilization are likely to have major influences on the future of the industry.

Geoscientists contribute to the progress of the coal industry by addressing the origins of coal beds and the evolution of their genetic environments. For example, researchers have observed that although sulfur content varies greatly within regions of the country, there is a tendency for the younger coals in the western United States to be lower in sulfur than are the older coals in the Midwest and the Appalachians. The higher-sulfur coals are often associated with marine strata, suggesting that the sulfur originated as sulfate in seawater. The sulfate was reduced to sulfide by bacteria and then incorporated into the peat that eventually developed into coal. Coals that formed in river systems far removed from marine conditions tend to be lower in sulfur content.

Although researchers understand the origin of some sulfur-rich coal, they have not yet worked out how best to cope with the problems it presents. Much of the sulfur is contained as organic sulfur in very fine-grained iron sulfides, and to separate the coal from the sulfur-bearing minerals would require grinding it into very small particles. Such a procedure is not economical at present. Coal research is dominated by the quest to identify the coal-forming environment and trace its subsequent history. Characterizing the coal physically and chemically on scales ranging down to the submicroscopic is essential for interpretation of its origin and history as well as determination of its technological properties. Such characterization is vital to determine whether coal can be economically mined, cleaned, and used either in conventional combustion or in some other way that may prove economical.

Origin and Development of Coal

Coal is composed largely of nonmarine plant material deposited in swamps. Understanding the depositional environments leading to peat formation and the subsequent diagenetic processes that determine coal type is important in predicting coal characteristics in advance of exploitation. The properties of coal are determined, in part, by the kinds of plants or parts of plants that dominated the original peat accumulation. As with petroleum source rocks, the study of coal distribution has been an important part of paleoclimatological research.

The salinity, pH, sulfate, and bacterial content of water in which peat accumulates largely determine the chemical characteristics of the resulting coal, including its sulfur and nitrogen content. Some of the mineral matter, the ash—which includes part of the sulfur—is derived from plants, but most consists of stream- or wind-transported material, and a large amount of it may enter the deposit long after the swamp has disappeared. Peat deposits in broad coastal freshwater swamps may result in thick coal beds of considerable areal extent. Swamps or marshes in an area of active delta deposition are episodically inundated by flood waters that deposit clay, silt, or sand on previously accumulated peat. The resulting coals contain splits—thin coal seams separated from the main seam by a layer of different sediments—that may reduce their minability. Abandoned stream channels sometimes accumulate thick peat deposits of elongate geometry.

Plant material is further altered chemically and physically by heat and pressure. Coals are ranked progressively upward, from lignite to anthracite, according to the degree of such metamorphism. Rank increases as the fixed carbon content increases and the percentage of volatile matter and moisture decreases.

Considerable research on the origin of coal involves study of the depositional environments of modern peat formation and is aimed at identifying physical and chemical changes during the coal-forming process. Discoveries of peatlands with incipient coal formation in Indonesia were detected through extensive mapping and coring. Studies are developing more sophisticated models for future investigations. Detailed outcrop and subsurface studies of individual coal basins are similar to the basin analysis methods used in the petroleum industry; they aim at defining depositional environments and subsequent coalification processes for model building.

Exploration for Coal

Coal exploration does not have the search component inherent in mineral and petroleum exploration because the limits of most coal-bearing beds in U.S. fields have long been known. Nonetheless, enhanced time-stratigraphic depositional models aid in predicting the location, geometry, and quality of coals. Despite these advantages, the inherent heterogeneity of the material makes sampling of coal deposits a difficult task; thus, development of accurate sampling techniques and analysis of statistics are important in the mining of coal. Where coal beds are nearly horizontal and laterally continuous over large areas, conventional methods can be used to ascertain the thickness and character of beds. Samples, cores, and geophysical logs are analyzed to determine basic characteristics. Where coal beds have complex geometries, other evaluation techniques may be necessary, such as seismic mapping, closely spaced drilling, and exploratory pits. During exploration, the environmental data required to obtain permits for mining are established, including the trace element content of the coal and surrounding formations, groundwater parameters, mineralogy, flora and fauna content, and soil properties.

More than one-half of the coal mined in the United States is produced from surface mines. Contour, area, or open-pit mining methods are used where coal seams of sufficient thickness and quality are near the surface, which is generally within 50 to 70 m. Surface mining of steeply inclined seams requires special planning that is specific to the conditions at the site. Where coal seams are deeper, shafts must be constructed to reach the coal and allow room-and-pillar or longwall mining methods. The latter method usually results in eventual subsidence of the land surface. An important research problem involves the correlation of geological parameters with the extraction method to allow optimal coal recovery while still mitigating the effects of subsidence on the hydrology and productivity of overlying soils.

Coal can be converted to liquids and gases by a number of different chemical, thermal, combination, or other treatment processes. Liquefaction produces feedstocks for chemicals such as benzene, ammonia, methanol, and acetic acid, as well as sulfur-free petroleum. Some of these processes are available commercially, while others are in various stages of laboratory or pilot plant testing.

Some coal beds contain large quantities of methane gas, which can be produced directly. There is current research into using bioengineered microbes and other methods to produce methane gas from the coal. This form of energy exploitation may be applicable to coal beds that have exceptionally large quantities of contained gas or that are too deep or too thin to be mined economically by conventional methods.

Limitations of Coal

Direct combustion of coal to produce electricity or steam heat requires further research toward the development of higher efficiency standards and cleanup techniques for hot and cold gas combustion and conversion facilities. Research in clean coal technology involves scientists and engineers specializing in combustion chemistry and mechanical engineering, but geologists have a critical role in characterizing what coals will burn most efficiently. Advanced combustion technologies to minimize sulfur dioxide and nitrogen oxides, as well as particulate emissions, have been under active development. These technologies include fluidized-bed combustion, staged slagging combustors, and limestone injection multistage burners. Combustion technology research must be pursued to gain a better understanding of capture mechanisms and to obtain reaction rates. Geoscientists continue studies to select sorbents with optimum reactivity, such as lime, limestone, dolomite, and other natural and man-made carbonates. In addition, desulfurized fuels produced by coal conversion processes require combustion testing to develop acceptable burning properties.

In the past 10 years considerable advances have been achieved in hot gas desulfuration using mixed metal oxides, which absorb sulfur and can be regenerated in a multicycle operation. Efforts should continue to develop reliable and cost-effective processes for cleanup of gas-borne constituents such as chlorine, sulfur, nitrogen, alkali compounds, and fine particulates, especially from fluidized-bed combustors, high-pressure gasifiers, and other technologies that promise higher efficiencies in electricity generation.

Complementary studies are needed for development of an integrated approach to solving problems of coal use such as the control of disposal and the use of waste materials from coal cleaning, fluidized-bed, and other advanced combustion processes. Problems in user equipment that are attributed to the corrosive effects of noxious elements such as chlorine need to be minimized; methods for fine grinding and dewatering of coal need to be improved; and methods for particle-size enlargement or pelletization of fine coal need to be refined.

Preparation of coal for the consumer market includes removal of contaminants such as mineral matter, sulfur, or other undesirable elements without substantially changing the general organic structure of the coal. Physical cleaning, such as washing, gravity separation, or flotation, remains the most economical and widely used technique. Geoscientists study the form, size, and distribution of the mineral matter contained in coals to guide the efforts of mineral preparation engineers in evaluating the efficiency of different processes, including oil agglomeration and column flotation, and of different reagents. Research continues on innovative wet and dry methods, such as magnetic or electrostatic removal, to maximize removal of iron sulfides. Considerably more research is needed to improve the removal of organic sulfur. Chemical, thermal, and microbial cleaning that breaks carbon-to-sulfur bonds and that can be combined in a two-step process with physical cleaning is being sought. Other approaches being investigated include genetic engineering to design efficient, organic, sulfur-consuming microbes and the use of microbes to alter the surface of iron sulfides to allow for its removal by physical cleaning processes, such as flotation.

Termination of the government-supported Synthetic Fuels Corporation in the early 1980s and recognition of the high cost associated with upgrading of single fuels from coal has led to the research concept of the coal refinery. In a coal refinery a combination of solid and multiphase products and fuels would be produced from coal by chemical, thermal, or other treatments. To date, research into production of a clean solid fuel, with simultaneous coproduction of gases and/or liquids, has not produced an economically viable and technically acceptable product for combustion in conventional utility equipment. Issues such as agglomeration of some coals during heating also must be addressed. Future coal refinery efforts may well depend on a better understanding of organic composition and the role of mineral matter; these will help identify the most selective and mildest forms of treatment necessary to produce an appropriate mix of marketable solid and multiphase products.

Every stage of coal development, from exploration to consumption, affects the environment. For example, acid mine waters can be neutralized, but predicting the movement of near-surface groundwater in surface or underground mining is more difficult. The effects of such elements from fluidized-bed combustion wastes on groundwater quality is not known. Other research should focus on the technical viability of disposing of scrubber sludge wastes in abandoned or existing underground mines, considering the potential effects on groundwater and surface subsidence.

Sources of Energy from the Internal Engine

All of the previously discussed energy sources are driven primarily by the external engine influencing earth systems. But the Earth's internal engine also produces vast amounts of energy that invite exploitation. Early humans were well aware of the heat coming from below the surface—volcanoes were the chimneys ventilating the forge of the god Vulcan, and thermal springs and geysers were considered sure signs of a sacred presence. Many societies attribute healing powers to naturally heated springs; popular spas have grown around such locations as the retreat established by the ancient Romans at Bath, England, and the early-twentieth-century hot spot, Saratoga Springs, New York. Attempts to use the warm waters for heating homes and workplaces have traditionally been small in scale. Since the discovery of radioactive energy resulting from the decay of radioactive elements within the Earth, scientists have hoped to use internal heat generated by this process and offer a new, cheap, unlimited energy source to society. Research continues to perfect, and expand, both methods of harnessing energy from within the Earth.

Nuclear Energy

A theoretically promising method of harnessing the internal heat generated by the natural radioactivity of the Earth is to mine uranium minerals, extract uranium from such concentrations, and assemble a critical mass in a reactor. Heat generated by the fission process (the spontaneous or induced splitting of atomic nuclei with consequent release of energy) can, when controlled within a reactor, be used to generate electricity. This method of generation is widely used in the world today. In France nearly three-quarters of all electricity is nuclear generated. In the United States the proportion is much smaller and is currently declining. Uranium-powered reactors remain important to the United States, and a few other countries, for powering nuclear submarines and other vessels.

One reason for its limited use in the United States is that nuclear power may not be economically competitive. This lack of competitiveness stems from a variety of causes, some of which are relevant to earth science research. Uranium ores were mined on a substantial scale in the United States for about 30 years but are not marketable today due to the availability of higher-grade, lower-cost ores from abroad. Worldwide uranium exploration over the past 50 years has profited from government interest in access to fissionable material and from the fact that the natural radioactivity of the element makes concentrations unusually easy to detect. Uranium occurs in both oxidized and reduced states and is highly mobile in aqueous solutions in its oxidized state. Most uranium mined in the United States has come from ancient stream channels on the Colorado Plateau, in Wyoming, and on the Texas coastal

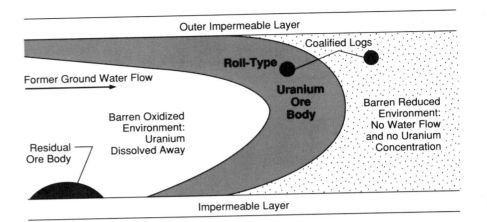

FIGURE 4.16 Schematic cross section of uranium-ore roll and surrounding rocks in Wyoming. Because uranium-ore rolls are of variable width, no scale is given.

plain. In these areas deposition of uranium minerals was localized where oxidized groundwater, carrying uranium in solution, came in contact with organic reducing materials, such as plant matter in the form of logs or leaves. Anaerobic sulfate-reducing bacteria may have been important in precipitating uranium minerals at the reduced surface (Figure 4.16). The transport and deposition of uranium present yet another example of the importance of fluid flow through porous media in geologically significant environments. Research related to uranium deposits is, in many ways, similar to the basin analysis that dominates research in the petroleum industry.

The nuclear power industry faces a continual challenge in seeking options to dispose of waste materials. This costly need is another reason for the decline in the use of uranium as a means of electric power generation in this country. Geoscientists are active in research related to this problem, since the currently preferred disposal option is to deposit radioactive waste materials beneath the surface. The possible interaction of fluids with buried radioactive waste and the subsequent migration of those fluids through porous rocks to the biosphere must be projected over the long intervals during which hazardous material can remain radioactive. Related research opportunities occur where radioactive wastes have already been emplaced. Both of these areas of research involve fluid flow and chemical interaction in porous media. While some policy makers strive for complete isolation of nuclear wastes over very long durations, some researchers feel that such materials must be regarded as by-products that will be used as resources in the future. This opinion demands accessibility to radioactive wastes; many environmentalists agree that accessibility is desirable but for the purpose of monitoring—not with an eye toward exploitation.

Nuclear power plants add no carbon dioxide to the atmosphere, in contrast to all fossil fuel plants. As our understanding of the possible effects of carbon dioxide emissions emerges over the next decade, it is possible that nuclear power plants may become more acceptable. There is a need to be ready for this eventuality, and additional research on uranium mineralization and waste isolation should be undertaken now.

Geothermal Energy

Another method of harnessing the energy generated within the Earth by radioactive decay is the direct extraction of heat from hot rocks that remain buried (Figure 4.17). Energy generated from this source is called geothermal energy, and research on its extraction is an active field in the geosciences. Most of the geothermal energy used in the world today comes from volcanic areas and areas of active extensional faulting where hot, or partly molten, rock approaches the surface closely enough to be accessible by drilling. Natural steam or hot water, confined by an impermeable seal and heated by the magma or hot rock, is extracted from holes drilled into the geothermal system and is converted at the surface to electricity.

The combination of crustal features required for a geothermal energy supply occurs rarely: hot rock near the surface with a reservoir of confined water that is large, permeable, shallow, adequately recharged for a life of decades, and low in dissolved solids. Proximity to an energy market is also an advantage. Geothermal energy has been used successfully for generating electricity in California, Mexico, Japan, Italy, New Zealand, Iceland, and Tibet, but large-scale economic geothermal production is not widespread. Problems arise because of depletion of reservoirs and undesirable environ-

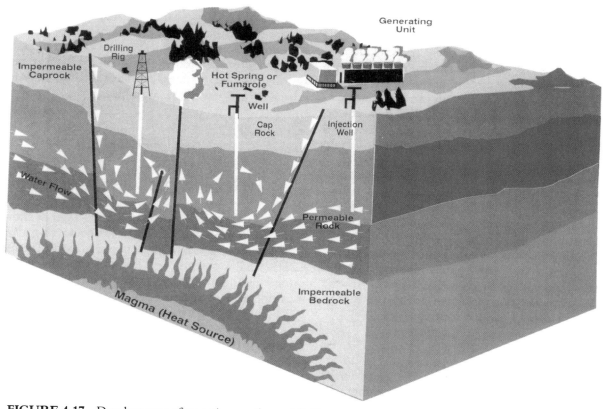

FIGURE 4.17 Development of an active geothermal field.

mental effects. An exciting challenge for the geoscientist is to locate exploitable geothermal resources lacking obvious surface manifestations. Remote sensing (especially by infrared techniques that reveal heat), geochemical, and geophysical methods are being developed and tested for this purpose.

Other uses of geothermal energy have been developed. Hot waters are used for space heating, agriculture, and mariculture in Iceland. Apartments in the suburbs of Paris and elsewhere in Europe are warmed by groundwaters. Boron from geothermal waters has been marketed in Italy. Possible developments in the future extend to extracting heat from hot dry rocks by injecting cold water and recovering heated water or steam.

RESOURCE DEPENDENCY

Geothermal energy use is an obvious example of how developing better comprehension of rock-fluid interactions can lead to efficient resource exploitation. Throughout this chapter more subtle interactions have been discussed. The concentration of traditional minerals, the accumulation of fossil fuels, and the deposition of uranium minerals all result from rocks and fluids interacting chemically,

thermally, and mechanically. As scientists study this crucial and fundamental field, the problems associated with each of these resources will be solved one by one, and society will become better equipped to live on the Earth's surface without permanently harming it.

The need for resources has always been a major stimulus for scientific advance, providing both economic reward and intellectual challenge. The discoveries made and the problems presented by the resource industry have kept scientists busy explaining the hows and whys. Once an explanation becomes established, further discoveries and data accumulation by industry either ratify or reject prevailing opinions about how crust has evolved from the matter and energy of the core and mantle. An explanation that holds provides a model from which the resource industry can speculate confidently on the locations and forms of future discoveries.

Historically, advances in knowledge gained through collaborations between scientific theory and social necessity have led to greater understanding of natural systems. More recently, scientists have considered the reciprocal relationships between natural systems and human activity within those systems. Consideration of these relationships may result in predictions of natural disaster potential and in responsible planning that can avoid suffering and loss of life. But scientific consideration of the relationship between natural systems and human endeavors can also provide assessments of those activities that threaten the environment—disasters caused by humankind.

RESEARCH OPPORTUNITIES

The Research Framework (Table 4.1) summarizes the research opportunities identified in this chapter and in other disciplinary reports and references. These topics, representing significant selection and thus prioritization from a large array of research projects, are described briefly in the following section.

Resource recovery is restricted to the top 10 km of the crust, and most of it is from much shallower depths. Understanding of the structure of the shallow crust and of the processes operating to form and modify that crust, especially those involving the flow of water through rock, provides a certain unity to resource research. It is convenient to define three arbitrary depth zones (with gradational boundaries) from the surface downward. The most widespread and generally used resources are land and water; the most dominantly used water is at or near the surface (0 to 10 m). At intermediate depths (10 to 100 m) groundwater development and surface mining are major activities, and remote sensing of properties becomes increasingly important. The shallow crust, at depths of 100 m to 10 km is the domain of deep mining, deep groundwater, and the petroleum industry. It is here that remote (geophysical) sensing from the surface by a wide range of techniques and advanced logging of boreholes become critical in understanding the crust.

Making the best use of mineral and energy resources calls for understanding the earth systems of which they form a part and for appreciating the need for waste isolation and the environmental consequences of mineral development. For these reasons, the research opportunities identified for earth systems outlined at the ends of Chapters 2 and 3, and those related to environmental considerations at the end of Chapter 5, are relevant to much research on depletable resources. The focus for this chapter is deliberately restricted to topics identifiable as important under Objective B: "Sustain Sufficient Resources." Many research topics relate to more than one Research Theme. The comprehensive topic of "Sedimentary Basin Analysis," for example, involves all the research themes. Similarly, the subject of "Water-Rock Interaction" relates to at least themes I-IV and has particular importance in basin analysis and in modeling of ore formation.

Sedimentary Basin Analysis

The material carried out of drainage basins is deposited in the sedimentary basins of the world. It is from these sedimentary basins—both active basins currently being filled and ancient basins that were filled long ago—that the world's oil, gas, coal, and groundwater, together with many of its mineral deposits, including its uranium deposits, are produced. Research in resources should continue to stress a multidisciplinary basin analysis approach, emphasizing detailed studies of specific depositional environments in outcrops and cores in conjunction with physical measurements of rock properties.

Organic Geochemistry and the Origin of Petroleum

The processes of maturation of organic material and diagenesis merit continued study by observation, experiment, and modeling. Such phenomena as natural cracking and migration require emphasis. The whole forms an important activity in relation to the broad field of sedimentary basin analysis.

Kinetics of Water-Rock Interaction

Understanding the kinetics of water-rock interaction is a research frontier critical to sustaining both water and mineral resources. Mineral surface catalysis, poisoning effects, and complex mineralogy make this understanding hard to achieve in systems at temperatures from 100° to 500°C; at less than 100°C, organisms, organic material, and materials added to groundwater by humans play corresponding roles.

Water Resources

Analysis of Drainage Basins

Surface waters are best understood in terms of drainage basins that allow such processes as erosion, deposition, and water and sediment transport to be addressed in an appropriately integrated way. An immediate resource challenge is presented by the

TABLE 4.1 Research Opportunities

Objectives

Research Areas	A	B. Sustain Sufficient Resources Water, Minerals, Fuels	C	D
I. Global Paleoenvironments and Biological Evolution		■ Evolution of mineral deposits through time ■ Coal petrology and quality		
II. Global Geochemical and Biogeochemical Cycles		■ Most of the opportunities listed under III below relate to integral parts of biogeochemical cycles ■ Organic chemistry and the origin of petroleum		
III. Fluids in and on the Earth		■ Kinetics of water-rock interaction ■ Analysis of drainage basins ■ Water quality and groundwater contamination ■ Modeling water flow and hydrocarbon migration ■ Source-transport-accumulation models ■ Numerical modeling of depositional environments ■ In situ mineral resource extraction ■ Crustal fluids		
IV. Crustal Dynamics: Ocean and Continent		■ Sedimentary basin analysis ■ Surface and soil isotopic ages ■ Prediction of mineral resource occurrences ■ Concelaed ore bodies ■ Intermediate scale search for ore bodies ■ Exploration for new petroleum reserves ■ Advanced production and enhanced recovery methods ■ Coal availability and accessibility ■ Concealed geothermal fields		
V. Core and Mantle Dynamics				

Facilities - Equipment - Data Bases
- Deep seismic studies, three-dimensional data, complementary advanced well logging
- Advanced supercomputing facilities
- Continental scientific drilling
- Oceal drilling
- Space-borne measurements, LANDSAT, SPOT, HIRIS, magnetometer, gravity survey
- Geophysical techniques, side-looking radar from aircraft
- Laboratory equipment with high spatial resolution and high-precision analytical capability
- Geographic information systems
- Data bases on topography, geology, geochemistry, and geophysics

drainage basins of the western United States. Policy decisions on the allocation of surface waters are driven by jurisdictional considerations. Recognition of the drainage basin as the fundamental scientific unit, provides a basis for effective communication with decision makers in providing information useful for the allocation of scarce surface waters.

Surface and Soil Isotopic Ages

The development and application of new techniques will allow ages to be assigned to surfaces and shallow rock materials, which should improve understanding of both the processes and the rates of landscape modification. Improvement in the ability to quantify these ages in years will revolutionize understanding, which is becoming more urgent as appreciation of the possibility of anthropogenically modified climatic change spreads.

Water Quality and Groundwater Contamination

To reach a realistic baseline on water quality nationally, chemical and physical measurements must be made at a vast network of stations. To assess change, frequent repeat measurements will be needed. Part of the national water quality problem relates to the relatively small volumes of rock where toxic and radioactive waste materials have entered the groundwater. Challenges in this frontier area range from asking whether there may be some places where the wisest course may be to do nothing to assessing possible roles for novel techniques, such as remediating organic wastes using microorganisms, including special genetically engineered microbes.

Modeling Water Flow

Thermal, chemical, and fluid transport models for water need to be integrated into four-dimensional models—that is, models that involve variation with time.

Mineral Resources

Source-Transport-Accumulation Models for Mineral Resources

Mineral resource studies are poised for dramatic advances through broad application of the models that guide modern petroleum research and exploration. Integrated studies, involving all the essential superposed geological processes, offer great opportunities, particularly as they can now draw on a wealth of new global tectonic and geochemical concepts and predict new types of mineral deposits. Quantitative geological mapping offers one of the greatest opportunities for significantly advancing our understanding of mineral resources and their relations to the geological features and processes. New quantitative mine-scale and regional-scale maps are both needed.

Prediction of Mineral Resource Occurrences

Mineral deposit models must be built that contain and identify the most important information for resource prediction, and practical methods must be pioneered for making resource predictions for which the confidence and accuracy can be estimated. Incorporation of organic geochemistry is required. These models are the fundamental building blocks for the advancement of mineral resource sciences. The challenge of preparing sufficiently reliable models is immense.

Numerical Modeling of the Depositional Environment of Ore Bodies

Modern computers have recently made possible numerical modeling of the complex hydrologic, geological, and geochemical systems related to ore formation. For the first time, drawing on reasonably comprehensive thermodynamic/kinetic data bases for water-rock interactions and fluid flow models, it is possible to develop integrated models that reasonably reflect natural systems for flow through media with irregular fractures, accommodating conditions under which liquids boil and allowing fluid compositions to change with time. Progression toward comprehensive computational systems could provide screening of the maze of geological variables to help indicate which are most important to ore formation.

In Situ Mineral Resource Extraction

The extraction of metals from mineral deposits by the circulation of fluids offers the potential for economic mining with potentially lower costs and less environmental disruption. The great potential and challenge of in situ mining are the integration and application of decades of advances in the diverse fields of mineral deposit geology, hydrogeology, geophysics, rock mechanics, geochemistry, structural geology, geoengineering, and chemical engineering. Determination of the effects of organisms and organic geochemistry is important.

Crustal Fluids

A major research objective is to identify and characterize each type of fluid, its distribution, and how it came to be. It would be best to start with fluids in modern environments, including modern ore-forming environments (such as the black smokers and the Red Sea) and shallow crustal settings (i.e., Smackover and Salton Sea brines). Opportunities abound for experimental research into alternate reaction paths, which reactions occur, why those paths are followed, how fast they occur, and what governs their rates. One critically important objective is the description of metal complexing and mineral solubilities under a variety of geologically realistic conditions. Development of thermodynamic models for appropriate trace elements and volatile components will permit numerical modeling of magma differentiation and possible identification of critical variables affecting ore-bearing fluids.

Evolution of Mineral Deposits Through Time

The evolution of deposits presumably signals changes in the types and/or intensities of processes as well as in the atmosphere, oceans, thermal regimes, climate, and tectonics. Study of the evolution of resource types through time and their relations to other features should therefore guide our understanding of earth processes. In turn, this understanding should sharpen our knowledge of ore genesis and our ability to predict known and unknown deposit types that are associated with particular periods of earth history and particular geological features.

Concealed Ore Bodies

Even the largest of ore bodies occupies a tiny area of the land surface, so production of mineral wealth has been concentrated in those places where surface indications of ore bodies are strongest. A challenging research frontier is to find ways of locating buried ore bodies in the many areas where surface signs are lacking. This is rather like looking for a needle in a haystack, and part of the challenge is to develop coherent exploration strategies that are not prohibitively costly.

Intermediate-Scale Search for Ore Bodies

Mineral explorationists identify this level of search as a research frontier. The characterization of environments likely to be hosts to particular classes of ore body on the basis of recognition of ancient and modern plate tectonic environments has proved to be powerful in distinguishing blocks of land that are promising for exploration on the scale of thousands of square kilometers. On a very local scale new deposits can often be found in close proximity to existing mines. The challenge is to establish criteria permitting recognition of potential ore-bearing areas at scales intermediate between these extremes.

Energy Resources

Exploration for New Petroleum Reserves

In much of the world where the petroleum industry is less mature than in the United States, the challenge lies in exploration for new reserves in new oil fields, calling for an integrated approach to basin analysis that ranges from establishment of the tectonic environments in which basins have developed; through interpretation of depositional, structural, and thermal evolution; to an understanding of oil and gas generation within the basin, of fluid migration, and of related diagenetic change; and finally to definition of the types of traps in which oil and gas are held.

Advanced Production and Enhanced Recovery Methods

Because of the maturity of the petroleum industry in the United States, the challenge is to produce more oil from existing fields in well-known basins. Drilling, well testing, sampling, logging, advanced seismic methods, sedimentology and organic geochemistry, as well as modeling, need to be developed toward an integrated understanding of reservoir structure and dynamics. Advanced production methods such as those involving in situ bacterial modification will call for close interaction between solid-earth scientists and reservoir engineers. If bacterial techniques are to prove useful, they will depend greatly on current frontier research into the organic chemistry of petroleum.

Coal Availability and Accessibility

Total coal resources in the United States are large, but knowing only their size is not sufficient information. Some coals are not minable because their combustion produces unacceptable amounts of toxic material; others are less accessible to mining than appears from simply mapping the distribution of coal beds. The amount of coal that is available for mining and the accessibility of those "available coal resources" are much less than the overall deposit size. This is because of land-use restrictions, mining methods, and the thickness of individual beds. Such studies are being conducted in the Appalachian and Illinois coal basins. For a more realistic assessment of potentially available coal resources, they should be expanded to include all major coal-producing regions of this country.

Coal Petrology and Quality

The increased use of coal in power generation and in other more sophisticated uses requires knowledge of the quality of coal to be used. Microscopic study of the different materials that make up coal is proving important. Different materials characterize different environments of origin and different combustion responses. Information about chemical elements, mineral macerals (organic components), and their interrelationships will allow for more environmentally acceptable and economically efficient use of coal.

Concealed Geothermal Fields

Hot rock, abundant water in a sealed reservoir, and access to a market for the electricity produced must all coexist for a geothermal field to be exploitable. We may have failed to discover geothermal reservoirs with little surface manifestation. The development of remote sensing methods of detection, especially infrared survey from the air, geochemical methods, and geophysics, together make up a challenging field.

Facilities, Equipment, and Data Bases

Facilities

Activities that are too large to be operated by a single agency, university, or national laboratory can be considered facilities, and there are several examples of these in resource research. Access to advanced supercomputing capabilities is needed for testing models of processes in the shallow crust. Deep seismic studies using the consortia approaches that have been so successful in the past decade have a continuing role, and, where continental scientific drilling is perceived as an essential part of research, there is an established mechanism for bringing it into play under the existing interagency agreement. Ships of the University-National Oceanographic Laboratory Systems's (UNOLS) fleet and the Ocean Drilling Program are involved in experiments relevant to resource research, for example, under the RIDGE program. Satellite data from SPOT and Landsat are important in mineral research, and it is important that these be available to researchers on a continuing basis and at reasonable cost. Instruments with the capabilities of the High-Resolution Imaging Spectrometer (HIRIS) and the Advanced Spaceborne Thermal Emission and Reflection Radiometer (ASTER), proposed for as instruments for National Aeronatics Space Administration's (NASA) Earth Observing System, would be marvelous for resource research, as would a low-orbit magnetometer mission, such as ARISTOTELES. Gravity and magnetometry data obtained from aircraft for poorly known continental areas could also be important, but plans for development are at a very early stage.

Equipment

Most of the laboratory equipment that is needed for research in resources is similar to that used in the kinds of laboratory investigation identified in Chapters 2, 3, and 5. Advanced chemical analytical equipment and facilities for isotopic analysis, including mass spectroscopy of organic materials, has become a general need. New methods with high spatial resolution and high-precision analytical capability such as laser-ablated inductively coupled plasma-source mass spectrometry seem likely to be capable of yielding large amounts of useful information. Bacterial research may find a focus in resource and environmental research at this time. The largest field for advanced equipment development is in exploration for petroleum and minerals. Expenditures on mineral and petroleum exploration in the United States annually run at more than $1 billion, much of which is spent on the acquisition and processing of reflection seismic data. Three-dimensional seismic surveys are expected to become much more widely used in the coming decade, and processing and archiving the large quantities of data generated by this practice will present a challenge. The availability of three-dimensional seismic data will contribute greatly to our understanding of the shallow crust and will call for complementary advanced well logging and well-to-well measurements.

Mineral explorationists use advanced geophysical research techniques, including audiomagnetotelluric methods, and remote sensing from space and aircraft by methods such as side-looking radar and high-wavelength-resolution infrared spectroscopy are beginning to be used. Geochemical prospecting is emphasizing the use of pathfinder elements as complementary to direct search for metals, and soil gas exploration is developing. All these instrumental developments have a part to play in the integrated effort to understand the properties and processes of the shallow crust, which is the concept unifying resource research at this time.

Data Bases

In many ways the preparation and skilled use of data bases provide the greatest frontier challenge in resource research. This comes about because most resource issues involve handling a diverse variety of spatial data together. For example, hydrologists commonly need good digital topographic data. They also need geological maps of both surficial and bedrock distribution and may need data on water quality and quantity for both surface water and groundwater over an area of interest. The ability to access this kind of information at will and to use it as separate "layers" is the essence of the geographic information systems approach. Once the data are accessible, the challenge is to interpret them by constructing and testing models of the systems under study. In the United States access to important data bases is improving all the time, but there are opportunities for improvements that could revolutionize both the way resource research is done and the likelihood of attaining useful results. For example, mineral explorationists are compiling data bases that enable them to use pattern recognition with advanced computational techniques, including expert systems. Some high-priority data bases for resource research follow.

■ Topographic Data Base. The digital topographic data for the United States available from the U.S. Geological Survey (USGS) is as good as any in the world, but there is a continuing need to update the data base for change and to review its quality, which, simply because of its vast size, is not uniform. On a worldwide scale, digital topographic data are only very locally as good as those of the United States; we draw attention to the recommendations of other Academy committees, which, for rather different reasons, have recommended obtaining better data in various ways, culminating in a dedicated space mission to acquire a coherent high-resolution topographic data set for the land surface of the Earth.

■ Geological Map Data Base. Geological maps of the United States are of varied resolution and uneven quality, and relatively few of them are accessible in digital form. The Association of American Geologists, working with the USGS, has defined an approach to obtaining the kind of geological map coverage that is needed. The need to move more rapidly to improve geological maps of bedrock and surficial deposits in the United States and overseas is regarded by resource geologists as well as environmental and research geologists as of the highest priority.

■ Geophysical Data. Data bases for geophysical data are varied and needs are diverse. A major distinction can be made between data acquired for commercial purposes such as petroleum exploration, which is not normally within the public domain, and data acquired by state and federal agencies, which is commonly freely available. Commercially acquired data are to a considerable extent traded within industry, and some publicly held information is useful for commercial activity. In the long run some commercially acquired data may enter the public domain. A serious question is whether nationally useful information may be lost through these current practices, especially for data sets where the cost of archiving in usable form is substantial. To give a single example of a publicly acquired data set that is likely to be important in exploration, airborne magnetic data at high spatial resolution will be important in the search for hidden ore bodies.

BIBLIOGRAPHY

National Research Council Reports

NRC (1981). *Studies in Geophysics: Mineral Resources: Genetic Understanding for Practical Applications*, Geophysics Study Committee, Geophysics Research Board, National Research Council, National Academy Press, Washington, D.C., 119 pp.

NRC (1982). *Studies in Geophysics: Climate in Earth History*, Geophysics Study Committee, Geophysics Research Board, National Research Council, National Academy Press, Washington, D.C., 198 pp.

NRC (1983). *Opportunities for Research in the Geological Sciences*, Committee on Opportunities for Research in the Geological Sciences, Board on Earth Sciences, National Research Council, National Academy Press, Washington, D.C., 95 pp.

NRC (1983). *Fundamental Research on Estuaries: The Importance of an Interdisciplinary Approach*, Geophysics Study Committee, Geophysics Research Board, National Research Council, National Academy Press, Washington, D.C., 79 pp.

NRC (1984). *Studies in Geophysics: Groundwater Contamination*, Geophysics Study Committee, Geophysics Research Board, National Research Council, National Academy Press, Washington, D.C., 179 pp.

NRC (1987). *Geologic Mapping in the U.S. Geological Survey*, Committee Advisory to the U.S. Geological Survey, Board on Earth Sciences, National Research Council, National Academy Press, Washington, D.C., 22 pp.

NRC (1988). *Scientific Drilling and Hydrocarbon Resources*, Committee on Hydrocarbon Research Drilling, Board on Mineral and Energy Resources, National Research Council, National Academy Press, Washington, D.C., 89 pp.

NRC (1989). *Technology and Environment*, Advisory Committee on Technology and Society, National Academy of Engineering, National Academy Press, Washington, D.C., 221 pp.

NRC (1989). *Margins: A Research Initiative for Interdisciplinary Studies of Processes Attending Lithospheric Extension and Convergence*, Ocean Studies Board, National Research Council, National Academy Press, Washington, D.C., 285 pp.

NRC (1989). *Prospects and Concerns for Satellite Remote Sensing of Snow and Ice*, Committee on Glaciology, Polar Research Board, National Research Council, National Academy Press, Washington, D.C., 44 pp.

NRC (1990). *Rethinking High-Level Redioactive Waste Disposal: A Position Statement of the Board on Radioactive Waste Management*, Board on Radioactive Waste Management, National Research Council, National Academy Press, Washington, D.C., 38 pp.

NRC (1990). *Studies in Geophysics: Sea-Level Change*, Geophysics Study Committee, Board on Earth Sciences and Resources, National Research Council, National Academy Press, Washington, D.C., 234 pp.

NRC (1990). *Studies in Geophysics: The Role of Fluids in Crustal Processes*, Geophysics Study Committee, Board on Earth Sciences and Resources, National Research Council, National Academy Press, Washington, D.C., 170 pp.

NRC (1990). *Competitiveness of the U.S. Minerals and Metals Industry*, Committee on Competitiveness of the Minerals and Metals Industry, National Materials Advisory Board, National Research Council, National Academy Press, Washington, D.C., 140 pp.

NRC (1990). *Ground Water Models: Scientific and Regulatory Applications*, Water Science and Technology Board, National Research Council, National Academy Press, Washington, D.C., 302 pp.

NRC (1990). *A Review of the USGS National Water Quality Assessment Pilot Program*, Water Science and Technology Board, National Research Council, National Academy Press, Washington, D.C., 153 pp.

NRC (1990). *Ground Water and Soil Contamination Remediation: Toward Compatible Science, Policy, and Public Perception: Report on a Colloquium*, Water Science and Technology Board, National Research Council, National Academy Press, Washington, D.C., 261 pp.

NRC (1990). *Surface Coal Mining Effects on Ground Water Recharge*, Water Science and Technology Board, National Research Council, National Academy Press, Washington, D.C., 159 pp.

NRC (1991). *Toward Sustainability: Soil and Water Research Priorities for Developing Countries*, Committee on International Soil and Water Research and Development, Water Science and Technology Board, National Research Council, National Academy Press, Washington, D.C., 65 pp.

NRC (1991). *Opportunities in the Hydrologic Sciences*, Committee on Opportunities in the Hydrologic Sciences, Water Science and Technology Board, National Research Council, National Academy Press, Washington, D.C., 348 pp.

NRC (1991). *Undiscovered Oil and Gas Resources: An Evaluation of the Department of the Interior's 1989 Assessment Procedures*, Board on Earth Sciences and Resources, National Research Council, National Academy Press, Washington, D.C., 108 pp. plus appendices.

NRC (1991). *Managing Water Resources in the West Under Conditions of Climate Uncertainty*, Water Science and Technology Board, National Research Council, National Academy Press, Washington, D.C., 344 pp.

NRC (1992). *Water Transfers in the West: Efficiency, Equity, and the Environment*, Water Science and Technology Board, National Research Council, National Academy Press, Washington, D.C., 359 pp.

Other Reports

Department of Energy (1991). *National Energy Strategy*, U.S. Government Printing Office, Washington, D.C., 217 pp. plus appendices.

U.S. Congress, Office of Technology Assessment (1991). *U.S. Oil Import Vulnerability: The Technical Replacement Capability*, OTA-E-503, U.S. Government Printing Office, Washington, D.C.

USGS (1987). *Geologic Applications of Modern Aeromagnetic Surveys*, U.S. Geological Survey Bulletin 1924, 106 pp.

5

Hazards, Land Use, and Environmental Change

ESSAY: A FRACTION OF THE EARTH'S SURFACE

The surface is the interface between the Sun-powered processes dominated by erosion and deposition and the tectonic processes driven by the Earth's internal energy. Most of the surface lies at two general levels: nearly 60 percent is between 1 and 5 km below sea level, and about 25 percent lies within a kilometer above or below sea level (Figure 5.1). The remaining 15 percent of the surface is concentrated in tectonically active mountain belts, continental slopes, and oceanic trenches.

The part of the surface most heavily populated is usually less than 1 km above sea level; most activities are concentrated in areas not far above sea level. Mountainous areas usually are sparsely populated because they are less favorable to agriculture, construction, and transportation and are more susceptible to hazards. Such areas tend to be used for specialized agricultural activities, mining, and recreation. In places where large populations concentrate in or close to structurally active mountainous regions—for example, those in Armenia, Chile, Nepal, and California—the residents are faced with special problems in developing the land resource because of the risk of seismic, volcanic, and landslide hazards.

Just at sea level—within the zone affected by tidal and storm-generated oscillations—spreads the biologically diverse region of environmentally sensitive terrains: marshes, swamps, tidal flats, and fens. Historically, such expanses were considered wastelands or were drained, filled, or surrounded by sea walls to increase their agricultural worth. Research conducted within the past 50 years has disclosed the inadvisability of such projects because they can disturb the natural breeding habitats of wildlife and the cleansing functions of the world's wetlands. Close study shows that wetlands constitute whole ecosystems supporting vast populations, not only of waterfowl, reptiles, and amphibians but also microscopic creatures. Coastal wetlands abound in unfamiliar fungal and bacterial species that perform the invaluable tasks of isolating and neutralizing the toxic compounds flushed through the system in the hydrologic cycle.

The shallow-water area of the continental shelf is recognized as essential to fisheries and as a source of significant oil and gas production. The most recent ocean-waters flooding of the continental shelves took place within the past 10,000 years, and the nature of this geologically transient environment is as yet

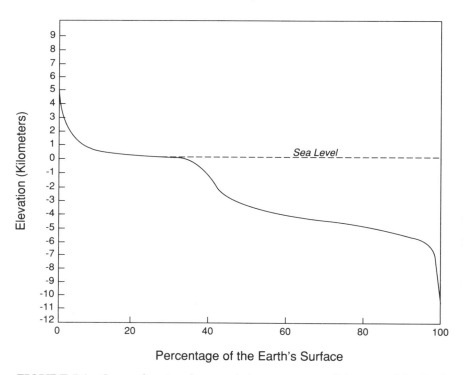

FIGURE 5.1 Curve showing the cumulative percentage of the area of the Earth that is higher than any particular elevation. Most of the surface lies either 3 and 5 km below sea level or within 1 km above sea level. Human activities are concentrated in the areas that are less than 1 km above sea level.

poorly understood. Research techniques such as advanced side-scanning sonar systems are aiding in the study of the continental shelves. If sea levels rise more rapidly over the next century, interest in the continental shelf environment will intensify, and the use of land surfaces below sea level may extend beyond the Netherlands, where it is now focused.

Great attention is given to sudden landform changes such as abrupt changes of elevation during earthquakes, yet more money is spent yearly in attempts to retard the slow changes of landform development than on mitigation of the effects of sudden change. The incremental changes produced by erosion and deposition lead to soil loss; silting of reservoirs; and destructive transformations of hillslopes, rivers, and coastlines. Such evolution is natural and often inevitable. The activities of humankind have commonly accelerated the transformations, catapulting natural systems over thresholds and producing immediate environmental threats. The geomorphological processes affected by humans cover a staggering range of scale. From local denudation caused by livestock overgrazing, a significant component of the process of regional desertification, humankind has evolved into a major geomorphological agent. Proper understanding of progressive geomorphological changes can forestall precipitous transformations and prevent the loss of landform stability.

Landforms are not random features; they are the consequences of the interplay of constructive and destructive geological and hydrologic forces requiring careful study before they can safely be artificially modified. Our baselines for understanding processes at the surface are disturbingly short, although existing landscapes provide important information about the magnitudes and return frequencies for many natural processes. Only in the past century have detailed observations been made concerning such features as flood intensities and frequencies, debris-flow distribution, subsidence, and landslides. Thus, predictions are based on recent to current conditions that are known to have been altered by human actions. Predictions of longer-term events, such as 1,000-year floods, are very unreliable because they must be based on models with uncertain numerical characteristics.

Landforms are sensitive to climatic change because the operating rates of the processes that mold landforms vary dramatically with climate. Fluvial processes dominate in sculpting landforms in semiarid regions, while wind processes are more significant in arid regions. Under very wet climatic conditions, landslides and downhill movements of surface rocks are dominant. Each of these geomorphic processes leaves distinctive evidence in the geological record. When climate changes, the dominant land-forming processes change in response. Threshold values of rainfall and temperature can be defined at which a change from the dominance of one land-forming process to another is likely to occur. It is therefore possible to forecast how agricultural regions might shift size and location in response to global warming and related changes in rainfall.

On an even longer time scale, low-lying coastal areas are subject to episodic flooding as sea level oscillates. Most remarkable are the paleo-landscapes locally exposed by erosion beneath extensive blankets of sedimentary rocks that have been deposited on the continents during flooding episodes. Glacial valleys 450 million years old are evident at central Saharan sites now occupied by desert wadis, a 170-million-year-old sea stack lies fallen on a modern beach in Scotland, and a tropical beach 450 million years old can be visited on the outskirts of Quebec City. The long-term durability of low-lying continental surfaces—less than 1 km above sea level and less than 0.2 km below sea level—that is demonstrated by landscape exhumation can be seen on much shorter time scales by the slow rates of erosion that characterize such flat areas.

The occurrence of these extensive areas of low relief, coupled with a suitable climate, makes regions such as the American Midwest prime land resources. The lush agricultural production of such regions can be maintained only if the landscape is treated with the same conservation ethic that inspires reverence for parks and wilderness areas. Prevention of soil erosion and respect for natural ecological balances in low-lying, low-relief regions can ensure their productivity far into the future. Geological characterization forms an essential basis for planned preservation and maintenance.

Human society exists in the biosphere, which thrives at the boundary layer between the solid earth and its fluid envelopes, perched between the two engines of mantle convection and solar energy that drive geological processes. The biosphere, composed of chemical elements that are cycled among the reservoirs of the atmosphere, hydrosphere, crust, and mantle, also contributes to cycles of rapid chemical turnover and thus functions as a part of the geological processes. The more vigorous manifestations of those processes, however, regularly destroy the parts of the biosphere—and its human constructions—located in their paths. As society has increased its utilization of earth resources and expanded its population and area of colonization, the frequency of its encounters with vigorous geological processes has increased. Society has adopted a term, *geological hazards*, for these perfectly normal processes that began occurring long before humans arrived on the scene.

Many of the most tragic episodes in the natural history of humans have been related to geological hazards such as disastrous floods, earthquakes, sea waves, landslides, and volcanic eruptions. The fact is that most geological hazards can be avoided or mitigated through proper land-use planning, engineered design and construction practices, building of containment facilities such as dams, use of preventive measures such as stabilization of landslides, and development of effective prediction and public warning systems. In many parts of the world such measures have already significantly reduced human suffering from geological hazards, although major challenges remain. To further mitigate these hardships, it is essential that a better fundamental understanding of each hazard-causing geological phenomenon be gained and widely disseminated.

Before the development of agriculture, the effects of human beings on the Earth were comparable to the effects of other species. But the onset of crop cultivation and animal domestication, with the subsequent growth of urban civilizations, introduced a new set of forces. Today, humans are changing basic earth processes in unprecedented ways and to unfamiliar degrees. At present, every person in the United States is responsible on average for the consumption of 16 metric tonnes—about 35,000 lb—of minerals and fossil fuels each year. This use does not include the tremendous volume of material moved during the construction of homes, parking lots, office buildings, factories, dams, highways, and other structures. On a worldwide basis, the human population uses nearly 50 billion metric tonnes of earth materials each year. This amount is more than three times the quantity of sediment transported to the sea by all the rivers of the world. Clearly, human beings have become a geological agent that must be taken into account in considering the workings of the earth system.

The various materials moved by human society are perturbing not only the physical cycles of the Earth, by increasing mass transfer, but also the chemical aspects. The biogeochemical and geochemical cycles that convert elements into living creatures, and into ore deposits and other geological concentrates, now have new aspects. The chemicals generated by manufacturing and the disposal of materials, including toxic compounds, occur in concentrations and combinations never before involved in natural systems. The consequences of such contaminations are poorly understood.

Some earth systems operating at the boundaries of the geosphere, hydrosphere, atmosphere, and biosphere are very fragile, and every human effort toward survival or improvement of the human condition necessarily results in repercussions on those systems. We dispose of our wastes in the same sedimentary basins that supply us with the bulk of our groundwater, energy, and mineral resources. Through our social, industrial, and agricultural activities, we are changing the composition of the atmosphere, with potentially serious effects on climate and terrestrial and marine ecosystems. The human population is expanding into less habitable parts of the world—steeper mountainsides, more ephemeral deltaic and barrier islands—which increases vulnerability to natural hazards and strains the biological and geological systems that sustain life. In this sense, geological conditions control the quality of human life.

People all over the Earth are constantly moving from less economically viable rural areas to the crowded cities in hopes of achieving a better livelihood. Wastes produced in and around the cities further compromise the quality of surrounding lands, and consequently there is a strong need for urban renewal or recycling of crowded urban spaces. With the crowding and densification of urban living comes an increased need for more uses of recycled land space. Buildings and other structures for human habitation, transport, or manufacturing are made taller and heavier. They may be founded on sites that are less than optimal for resisting the physical stress induced by their presence. Geology is the main interconnecting element between the craft of civil engineering and the intricacies of nature. It is being used in efforts addressing water quality, resource supply, waste isolation, and disaster mitigation to accommodate and reduce the adverse effects of societal growth.

If present trends continue, the integrity of the more fragile systems on which human society is built cannot be assured. The time scale on which these systems might break down may be decades or it may be centuries. Human beings are unique among the influences on the Earth—we have the ability to foresee possible consequences of our activities, to devise alternative courses, to weigh the pros and cons of these alternatives, to make decisions, and then to behave accordingly.

Understanding the natural systems acting at the land surface presents a major scientific challenge because of the enormous social implications of those systems. Land resources, as basic elements of the global ecosystem, profoundly affect the lives of every human being on the face of the Earth. The loss of topsoil and forests, the loss of life and property caused by human-induced geomorphic change, and the pollution of air, soil, and water all result in growing consequences for both national and international welfare and security.

GEOMORPHIC HAZARDS

Landforms are continually changing, but except for a few spectacular instances their change is so gradual that it is scarcely noted. As population increases, more people are exposed to the effects of processes that have been going on for hundreds of millions of years. In many cases the increase in human population contributes to instability of the physical landscape. When human populations are threatened by geomorphic processes, those processes become geomorphic hazards. Great attention is given to hazards representing abrupt changes, such as earthquakes and volcanic eruptions, but more damage is caused and more money has to be spent annually in attempts to retard ongoing hazards such as landslides, debris flows, and the normal slow progression of erosion and redeposition that leads to soil loss; reservoir infilling; and river, coastline, and hillslope changes.

The surface of the land is shaped by internal forces—folding and faulting with consequent elevation or subsidence—and by erosion—wind and water weathering driven by solar energy and gravity. Wind and water accomplish erosion by forcibly loosening, removing, and transporting solid material. That eroded material becomes the sediment deposited elsewhere. Erosion and sediment production result from the exposure of earth materials and from variations of climate, vegetation, and topographic relief. For materials of similar strength, natural sediment production reaches a maximum at about 33 cm of rainfall per year. Below that amount, less runoff causes less removal of material; above that amount, increased vegetation protects the soil so the amount of erosion and sediment production decreases. Modern erosion rates can be

very high, because both urban and agricultural development require removal of the original vegetation.

Geomorphic hazards may involve a slow progressive change in a landform (Figure 5.2) that, although in no sense catastrophic, can become a significant hazard involving costly preventive and corrective measures. There are three types of geomorphic hazards that combine different spans of time, different degrees of damage, and different energy expenditures. The most obvious type is a sudden event that produces an abrupt change—a landslide caused by monsoonal rains, an earthquake, or human activity such as removal of toe support from a stable slope (Figure 5.3). A second type is progressive change that leads to an abrupt result, typified by weathering breakdown of soil or rock that initiates slope failure, gullying of a steepening alluvial fan, meander growth and cutoff, or stream channel shifting. The final type of progressive change gradually produces a slowly developing geomorphic hazard—for example, gradual hillslope erosion, gradual meander shift, channel incision, or channel enlargement. Misidentifying or wrongly estimating the pace of progressive changes may result in incurring pointless protective or remedial costs. Accurate identification of impending problems attributable to slow progressive change can aid in the choice of remedial action and in prudent allocation of money toward engineered hazard prevention. Geomorphologists have quantified ways of identifying potential hazard sites that suggest future change. When historical information is available, it may be possible to make a crude estimate of the timing of landform failure. But, usually, too many influences operating on too fine a scale make it difficult to predict when a failure will occur.

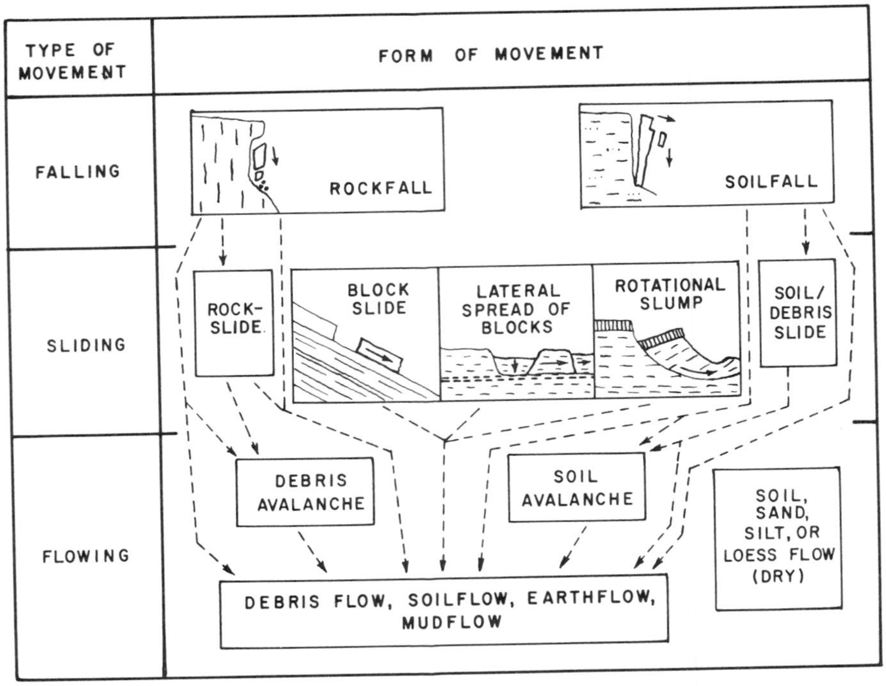

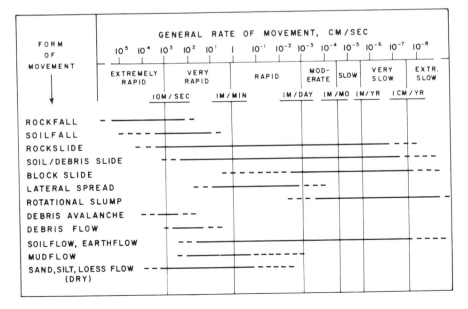

FIGURE 5.2 Principal forms of mass movement, as correlated with dominant mechanisms of falling, sliding, and flowing. Each block represents a form that can characterize single or multiple events of ground failure; in many occurrences one form can give way to another, as indicated by dashed lines and arrows. Blocks in center and at left represent failure in bedrock, those in center and at right failure in surficial deposits. After R. H. Jahns, NRC, 1978, in *Geophysical Predictions*.

Increased erosion of productive agricultural soils has grown into a serious problem, both for the farmer and for those downstream who must cope with siltation. Soil erosion is accelerated by a variety of agricultural practices, including cultivating slopes at too steep an angle and irrigating with too much water or water under pressure. Under certain circumstances, erosion can proceed so rapidly and over so wide an area that remote sensing techniques may be the best way to monitor it. In situations that require estimates of slow erosion rates from significant topographic features, radioactive isotopic anal-

ysis can help in establishing chronologies of erosion surfaces and stratigraphic horizons.

Historically, the greatest amount of erosion in the eastern United States and resulting sediment transport by streams probably occurred in the eighteenth century as agriculture first became widespread. Estuaries became severely silted at that time. In the nineteenth century the western United States experienced a similar development; the process was documented in the Colorado River Basin during the 1880s as extensive channel incision developed in tributary valleys and created the characteristic ar-

FIGURE 5.3 Ranges in general rates of movement for landslides and related features. Dashed parts of horizontal bars represent relatively uncommon or poorly documented occurrences. After R. H. Jahns, NRC, 1978, in *Geophysical Predictions*.

FIGURE 5.4 April 1983 Thistle, Utah, landslide. The landslide flowed down the valley damming Spanish Fork Canyon. Photograph courtesy of Gerald F. Wieczorek, U.S. Geological Survey.

royos. Sediment samples taken since 1930 show a significant decrease in sediment loads, suggesting that the incised channels have reached a new state of relative stability. This new equilibrium may result from a combination of conditions: less material is being eroded, and more of the sediment that is produced is being held up in the valleys, where it is deposited in newly developing floodplains. This trend has been enhanced by dam construction. The reservoirs extending behind the dams are clogging up with sediment, drawing attention to the long-term obsolescence of such facilities. In time the renewable resource of water for hydropower and agriculture may become severely compromised. On the lower Mississippi River, a 50 percent decrease in sediment transport has been associated with dam construction upstream on the Missouri River and bank stabilization elsewhere (see Figure 1.11). The Mississippi delta is being modified as the rate of sediment delivery decreases, and the natural slow subsidence due to basin deformation and sediment compaction outruns sediment accumulation, thereby permitting the sea to encroach onto the land.

Landslides and Debris Flows

In the 1970s, landslides—all categories of gravity-related slope failures in earth materials—caused nearly 600 deaths per year worldwide. About 90 percent of the deaths occurred in the circum-Pacific countries. Annual landslide losses in the United States, Japan, Italy, and India have been estimated at $1 billion or more for each country.

Landslide costs include direct and indirect losses affecting both public and private property (Figure 5.4). Direct costs can be defined as the costs of

replacement, repair, or maintenance of damaged property or installations. An example of direct costs resulting from a single major event is the $200-million loss attributed to the 21-million-m^3 landslide and debris flow at Thistle, Utah, in 1983. The slide severed three major transportation arteries—U.S. highways 6 and 89 and the main line of the Denver and Rio Grande Western Railroad—and the lake it impounded by damming the Spanish Fork River inundated the town of Thistle, resulting in the destruction of businesses, homes, and railway switching yards. The indirect costs involved the cutoff of eastbound coal shipments along the railroad line. In 1983 oil was expensive and coal was crucial for generating electricity. With supplies from the west severed, eastern coal normally exported to Europe had to be rerouted. European industry, in turn, had to adjust to lowered supply. Ultimately, the landslide affected the international balance of payments.

Destructive landslides have been noted in European and Asian records for over three millennia. The oldest recorded landslides occurred in Hunan Province in central China 3,700 years ago, when earthquake-induced landslides dammed the Yi and Lo rivers. Since then, slope failures have caused untold numbers of casualties and huge economic losses. In many countries, expenses related to landslides are immense and apparently growing. In addition to killing people, slope failures destroy or damage residential and industrial developments as well as agricultural and forest lands, and they eventually degrade the quality of water in rivers and streams. Landslides are often associated with other events: freeze-thaw episodes, torrential rains, floods, earthquakes, or volcanic activity. The bulging of the surface of Mount St. Helens over a rising magma body

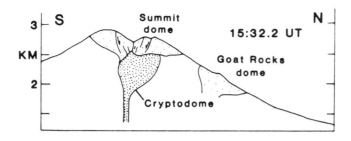

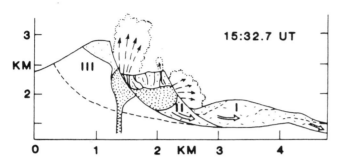

FIGURE 5.5 Sections through Mount St. Helens showing early development of rockslide-avalanche into three blocks (I, II, III) and explosions emitted primarily from block II; fine stipple shows preexisting domes and coarse stipple shows the 1980 cryptodome. After J. G. Moore and C. J. Rice, NRC, 1984, in *Explosive Volcanism: Inception, Evolution, and Hazards*.

led to a massive air-blast landslide—2.8 km^3 of rock, the largest slide in recorded history. Loosened material slipped off the side of the growing dome, unroofing the magma and permitting it to degas in a spectacular and locally disastrous eruption (Figure 5.5). The volcanic ash, dust, and pumice, mixed with rain and snowmelt, caused widespread debris flows in local valleys. A minor volcanic event high on the slopes of Nevado del Ruiz volcano in Colombia in 1985 melted enough glacial snow and ice to produce a debris flow that killed 25,000 people in the valley below. An earthquake off the coast of Peru in 1970 initiated a rockfall on Mt. Huascaran in the high Andes. The rockfall turned into a debris avalanche that moved at speeds approaching 300 km per hour and killed more than 20,000 people in the towns of Yungay and Ranrahirca.

Very large slides also are found on slopes below sea level. In the area around the Hawaiian Islands, recently discovered slide debris covers about 15,000 km^2 and contains single blocks more than a kilometer thick that slide as much as 235 km away from the shallower water. A major problem in the Gulf of Mexico is slumping, which can disrupt seafloor pipelines and the foundations for drilling platforms

worth hundreds of millions of dollars. In Hawaii the debris is well-consolidated basaltic lava, while in the Gulf of Mexico it is unconsolidated, or at best semiconsolidated, clastic sediments. In the geological record, boundaries of rock masses representing such major slides could very well be confused with the effects of tectonic faults; indeed, distinctions between the largest landslides and gravity-driven faults may be in the eye of the beholder.

Despite improvements in recognition, prediction, mitigative measures, and warning systems, worldwide landslide losses—of lives and property—are increasing, and the trend is expected to continue into the twenty-first century. Some of the causes for this increase are continued deforestation, possible increased regional precipitation due to short-term changing climate patterns, and, most important, increased human population.

Demographic projections estimate that by 2025 the world's population will number more than 8 billion people. The urban population will increase to 5.1 billion—more than the total number of humans alive today (Figure 5.6). In the United

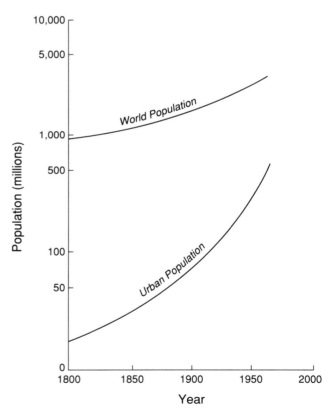

FIGURE 5.6 Urbanization compared with world population growth since 1800. Modified after Davis, 1965, *Scientific American*.

States the land areas of the 142 cities with populations greater than 100,000 increased by 19 percent in the 15-year period from 1970 to 1985. By the year 2000, 363,000 km² in the conterminous United States will have been paved or built on. This is an area about the size of the state of Montana. Accommodation of this population pressure will call for large volumes of geological materials in the construction of buildings, transportation routes, mines and quarries, dams and reservoirs, canals, and communication systems. All of these activities can contribute to the increase of damaging slope failures. In other countries, particularly developing nations, the urbanization pattern is being repeated but often without adequate land planning, zoning, or engineering. Not only do development projects draw people, but the projects themselves as well as the people who settle the surrounding area often occupy just those hillside slopes that are susceptible to sliding. At present, there is no organized program to provide the geological studies that could prevent the worst scenarios posed by this threat.

To reduce landslide losses, research efforts should encompass more than investigations of physical processes in hazardous areas aimed at understanding the nature of slope movement. Earth scientists also need to perfect methods for identifying areas at risk and for mitigating contributory factors. These goals are attainable. Scientists can predict areas at risk and advise means to avoid or moderate danger, but much of the research needed has yet to be done.

For the past half century, geologists have relied primarily on aerial photography and field studies—ideally in combination—for identification of vulnerable slopes and recognition of landslides. In recent years, since multispectral satellite coverage has become available for much of the world, an additional tool is available that can provide images in black and white or color as well as spectral bands through red, green, and near-infrared wavelengths. The coverage, scale, and quality of multispectral imagery is expected to improve considerably within the next decades and provide valuable information that can lead to improved identification of landslide-prone locations.

The information gathered from satellite reconnaissance can contribute to the growing store made use of in geographic information systems, which are digital systems of mapping spatial distribution that can be applied to the preparation of landslide susceptibility maps (Figure 5.7). These modern data-handling systems facilitate both pattern recognition and model building. Patterns and models that suggest landslide susceptibility can be tested, revised and improved, and then tested again against large numbers of observations.

As a result of these gains in knowledge, progress has been made in determining appropriate types of landslide mitigation. The most traditional mitigation technique is avoidance: keep away from areas at risk. When occupation of a site warrants risk, engineered control structures may be required, including surface water diversion and subsurface water drains, the construction of restraining structures such as walls and buttresses, and devices such as rock bolts. The establishment and enforcement of site grading codes calling for appropriate slope stabilization instituted in 1952 by Los Angeles County have worked well. Cut-and-fill grading techniques involving the removal of material from the slope head, regrading of uneven slopes, and hillslope benching are all of proven value. Consideration of such factors has had a major effect on reducing landslide losses in the United States, Canada, the European nations, the former Soviet Union, Japan, China, and other countries. Landslide research today is focused not so much on locating where landslides are and what hazards they represent as on figuring out how to cope with the potential hazard that they represent. This is an area of close cooperation between solid-earth scientists and geotechnical engineers.

Mitigation has also benefitted from substantial progress in the development of physical warning systems for impending landslides. Significant improvement will result as better instrumentation and communication systems are developed. Of particular importance will be continuing advances in computer technology and satellite communications. Hazard-interaction problems require a shift in perspective from the incrementalism of individual hazards to a broader systems approach. Earth scientists, engineers, land-use planners, and public officials are becoming aware of interactive natural hazards that occur simultaneously or in sequence and that produce cumulative effects that differ from those of their component hazards acting separately. In the case of landslides, research is particularly needed on cause-and-effect relationships with other geological hazards. For example, in the 1991 Mount Pinatubo (Philippines) eruption, the thick accumulating ash-fall and ash-flow deposits proved particularly liable to generate landslides and debris flows during typhoons. Research on the social aspects of such relationships in terms of warning systems and emergency services is necessary. At what point should people evacuate and abandon their little piece of the Earth?

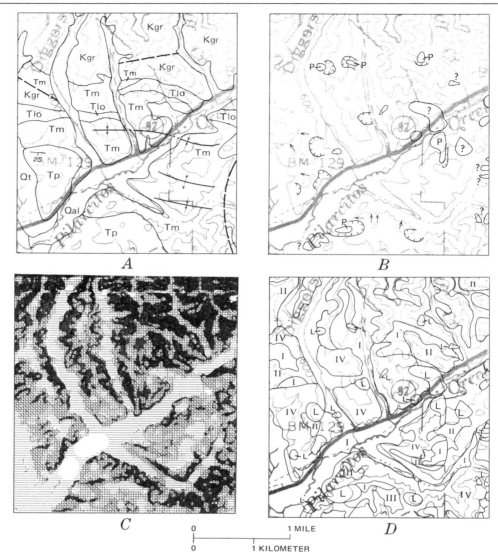

FIGURE 5.7 Landslide susceptibility map (D) produced using geographic information systems methods of a part of San Mateo County, California, near the town of La Honda. A landslide inventory map (B) was compared with a geological map (A) and a slope map (C) to determine the percentage of each geological unit that has failed by landsliding in the past, and the slope important for the failure. These data formed the matrix for the susceptibility analysis (D). The higher the roman numeral, the more susceptible the slope is to landsliding in the future. Landslide deposits (L) are shown as a separate category (highest). This map was used by San Mateo County to reduce potential development in landslide-prone areas and to require detailed geological studies to determine the safety of building sites. Figure from Earl Brabb, U.S. Geological Survey.

Human intervention can reduce landslide risk by influencing some contributory causes. Projects that undermine slopes in marginal equilibrium or destabilize susceptible areas by quick drawdown of reservoirs can be avoided. Among projects that can lay the groundwork for disastrous landslides are road building, mining, fluid injection, and building construction that entails clearing vegetation. Planning and designing such projects with the local landslide potential in mind is absolutely essential. While these activities may not individually cause a landslide, they can increase the likelihood of slope failure as preconditions to which cloudbursts or earthquakes are added. Wherever hillsides receive precipitation over days and weeks, the pore-water pressure can build in rock fractures and decrease bulk shear strength, which can then induce displacements under less force than would be needed to shear a drier material. A proven mitigation technique in such cases is for geologists to locate the water surface in fractured rocks and drain off destabilizing water by drilling horizontal wells.

Then there are the regional-scale contributory causes of increased landslide susceptibility such as deforestation. According to the World Resources Institute, approximately 109,000 km^2 of tropical forest is being destroyed annually—an area the size of Ohio. Removal of the forest cover increases flooding, erosion, and landslide activity. This deforestation is causing serious landslide problems in many developing countries.

Over a period of about 3 years in the 1980s, during the course of an El Niño episode, regional weather changes in the western United States resulted in much heavier than average precipitation in mountainous areas. That increased precipitation caused a tremendous increase in landslide activity in California, Nevada, Utah, Colorado, Washington, and Oregon. Scientists are coming to understand such cycles through integration of collected data with information found in the historical and geological records. Cycles such as El Niño form the background variation of the climate pattern. But earth scientists do not know what to expect with additional perturbation from a changing greenhouse effect. Will the predicted temperature increase cause a decrease in precipitation, as occurred in mid-America during the summer of 1988? Will it increase storm activity throughout the mid-latitudes? Will it disrupt global climatic patterns, resulting in droughts in some areas and increased precipitation in others? Documented cause-and-effect sequences such as those related to El Niño episodes suggest that if areas prone to landslides are subjected to heavier than normal precipitation, susceptible slopes are likely to fail.

Land Subsidence

Land subsidence can be currently observed in at least 45 states; it is estimated to cost the nation more than $125 million annually. Subsidence can have human-induced or natural causes; both are costly. In the United States at least 44,000 km^2 of land has been affected by subsidence attributed to human activity, and the figure is probably higher. As for natural subsidence, one event—the 1964 Alaskan earthquake—caused an area of more than 150,000 km^2 to subside as much as 2.3 m. This event was extreme but not atypical of past or probable future disturbances.

The causes of subsidence are various but well known, and the hazard presented is well recognized; however, the indications of specific imminent danger and the possible cures or preventives are not clear. Subsidence can be induced by withdrawing subsurface support—by removing water, hydrocarbons, or rock without a compensating replacement. In many instances, oil or water is removed from porous host sediments that compact as the interstitial fluid is removed. In such cases, collapse is slow and gentle. More dangerous are situations that leave voids—withdrawal of water from cavernous limestone or mining of coal, salt, or metals. These voids can collapse gradually or suddenly. Not all subsidence is unexpected—ground over longwall coal mines is supposed to subside gradually to a new elevation that is both safe and stable. The ground surface above subsiding land is not a good place for a shopping center or school building, but it may be quite suitable for crops or recreation.

Natural subsidence occurs for several reasons. A basin surface may warp downward in response to recent sediment loading or by dewatering and compaction of sediments; both processes presently affect the Mississippi River delta. Tracts may subside because of folding or faulting, as in the Alaskan example above. Regions such as the Texas Gulf coast are triply vulnerable because they overlie a downward-flexing part of the crust; are above a thick pile—up to 10 km deep—of compacting sediments; and are being mined for groundwater and petroleum, which accelerates deflation of the sedimentary pile.

The primary cause of the most common, and most dangerous, subsidence in the United States is groundwater extraction through water supply wells. In California's San Joaquin Valley, 13,500 km^2 of land surface has sunk as much as 9 m in the past 50 years because of removal of groundwater for irrigation. The danger, of course, is in more populated areas, especially those close to sea level. Some cities that are already struggling because of groundwater extraction include Houston-Galveston, Texas; Sacramento and Santa Clara, California; and Baton Rouge and New Orleans, Louisiana. The problem of induced subsidence is international and also threatens London, Bangkok, Mexico City, and Venice. If sea level continues to rise, the cities that are literally on the edge now will be fighting to stay above water.

Sinkholes, another common source of land collapse, can occur unexpectedly on a more local scale than wholesale subsidence. Sinkhole collapse generally results from the slumping of poorly consolidated surficial material into underground caverns.

Collapse over underground caverns can usually be attributed to a recent lowering of the groundwater level and consequent loss of pore-water pressure. As long as a cavern is filled with water, the material covering it receives enough support to stay in place. Actual collapse of bedrock caverns is less common. Sinkhole collapses are rapid but very local and can be either natural or human induced. Similar events result from the dewatering of abandoned mines that penetrate close to the surface. Ground-penetrating radar is used to identify shallow caverns, and seismic tomography has promise for evaluating the stability of mine pillars—volumes of unmined rock left to support the roof over adjacent mined areas.

Subsidence in one region, as a consequence of loading by former ice sheets, may be accompanied by compensating elevation in another region. The most recent glaciation of North America isostatically depressed the parts of the continent covered by ice—4.0 km thick in places—and caused upward bulging of land along the margins of the depression. Following deglaciation, the process reversed, and today the area once covered by ice is slowly rising at rates as large as about 3 mm/year as the glacial forebulge subsides. This process is tilting the Great Lakes region, and in a few thousand years it will divert Great Lakes drainage from the Niagara River through Chicago to the Illinois River, drying up Niagara Falls. In the meantime, some of this uplift may play a part in inducing small-magnitude earthquakes in the north central and New England states.

Floods

Any relatively low-lying area is subject to flood. Hurricanes and typhoons, tidal surges, and tsunamis can deliver too much water from the direction of the ocean. Snowmelt and ice dams, cloudbursts, and prolonged rainstorms can deliver too much water from inland areas. Some of the most frightening floods descend on mountain settlements when their watersheds receive cloudburst rain, and flash floods completely scour out valleys. In canyons of the mountainous western United States, warning signs read, "In case of flash flood, climb straight up." In the United States, rainstorms and their accompanying flooding and debris flows accounted for 337 of the 531 federally declared disaster areas from 1965 to 1985. Human activities also cause or contribute to flooding—one of the most tragic floods struck Johnstown, Pennsylvania, in 1889 when a dam failed and 2,200 people were killed. Urbanization augments flooding. Studies revealed that in Houston, Texas, the creation of impervious

surfaces increased the magnitude of the 2-year flood by nine times. Paving and stream channelization remove water quickly from one area, but they deliver more water more quickly to other areas. Factions argue about responsibility for the flood that struck Rapid City, South Dakota, in June 1972. That flood, which killed 245 people and caused $200 million in damage, followed an exceptional rainfall that was preceded by cloud-seeding efforts.

Floods can be terrible, but they are also accepted as few other hazards are. Humans have come to realize that access to transportation, fresh water, and rich alluvial soils is the reward for surviving floods. Long experience with floods has led to flood management practices. Because different management theories evolved in different environments, policy makers cannot, and perhaps should not, agree on standardized practices. Methods of management include land-use regulations; structural measures such as dams, levees, and floodwalls; land-treatment measures such as reforestation or terracing of stream banks; zoning ordinances and building codes; and warning systems.

Floods are not discussed at length here because they are considered in the 1991 National Research Council (NRC) report *Opportunities in the Hydrologic Sciences*, in which specific research activities such as short-term forecasting are identified. In that report emphasis is placed on the responsibility for reduction of flood loss that lies with public policy makers who can regulate development in flood plains. Estimating the risk in flood plains depends on knowing the probability of a repetition of events of a particular magnitude. In areas such as the western states where the historical record is short, this is not easy. Quaternary geologists are able to use the geological record to estimate both the timing and the scale of events in an area prior to written history.

Coastal Fluctuation

The coastline is a major battleground in the competition between the forces of deposition and erosion. It is also a region of concentrated human activities. Thus, coastal processes constantly affect populations and have long been the subject of conjecture. The causes of coastline fluctuations are numerous; they result from both internal and external earth processes.

One major factor is the uplift or subsidence of the coastal regions caused by tectonic forces. Most of the west coast of the United States is currently rising and has been for the past half-million years.

Uplift stages are typically marked by successively elevated terraces. Each terrace was once a wave-cut bench below a sea cliff, indicating a brief pause for a steadily rising coastline. The coastal terrain is riddled with landslides, an outcome accentuated by occasional earthquake shaking and by the infrequent cloudbursts. The problem is apparent, but identification of the most vulnerable slopes requires the careful scrutiny of engineering geologists.

The present site of the southeastern seashore of the United States stood well above sea level during the most recent continental glaciations. Rivers entering the sea did so farther to the east, through valleys incised into the coastal plain. As the ice sheets melted, the sea level rose and flooded the valleys, forming estuaries such as the Chesapeake Bay. The Patapsco, Potomac, and James rivers were all once tributaries of a Susquehanna River that flowed out past the present bay and formed a delta on what is now the continental shelf. As drastic as such changes might seem, they are relatively recent, having occurred over the past 10,000 years, and have cycled back and forth as ice sheets waxed and waned repeatedly over the past few hundreds of thousands of years.

Within the past 10,000 years of high sea-level stand, currents and wave action along the shores have piled up barrier beaches such as those near Cape Hatteras, protecting shallow lagoons on the landward side. These are ephemeral and fragile creations, even without human dredging, building, and destruction of the vegetative cover. The lagoons and barrier islands record a complex history of deposition, erosion, and redeposition. A 1-m variation in sea level will change the sea-land interface substantially. In areas such as the Texas Gulf coast, where a thick section of sediments is gradually consolidating, scientists anticipate that shoreline features will be severely affected, probably with major losses of valuable property (Figure 5.8). Proposals to recover energy from geopressured fluids from southeast Texas aquifers would almost surely result in further subsidence of the surface and encroachment of the gulf on present-day shores.

As rivers erode the land, deltaic and estuarine environments become the first sites of major sediment deposition—except for constructed reservoirs and rare natural lakes. These brackish tidal waters are nursery grounds for many marine organisms as well as invitations for the establishment of human populations because of access to those nursery grounds, to transportation routes, and to freshwater sources just up river. Many, if not most, of the major estuaries of the country have become con-

taminated by natural or human-introduced pollutants. Boston Harbor is, unfortunately, an example of a contaminated estuary. The fish populations have declined to the point of disappearance, and the beaches are nearly deserted throughout the summer because of pollution. Chesapeake Bay pollution, first detected by the Public Health Service before World War I, has contributed to a decline of striped bass. Pollution of estuaries has an extremely deleterious effect on fish spawning, and encroachment often leads the way in destroying the estuarine habitat, the eventual result being diminution of potential food supplies.

Deltas, barrier islands, lagoons, tidal flats, and estuaries support a number of dynamic physical, chemical, and biological processes that operate in complex association. Although scientific understanding encourages informed management, only a few of the world's populated coastal environments have been studied in the required detail. The United States has engaged in a major effort in coastal-zone planning, but there has been no systematic effort at data collection to determine either baseline information or trends in essentials such as estuarine quality. Study of estuaries and other coastal environments should prove most productive when approached by interdisciplinary teams that can appreciate the complexities of the physical, chemical, and biological processes. The sensitivity and significance of coastal processes warrant a systematic integration of biological, sedimentological, hydrological, and geochemical data collection and analysis. Such an effort should result in a product that will inform decision makers in the near future.

TECTONIC HAZARDS

Earthquakes

Hazards such as landslides, subsidence, and floods are often exacerbated by human activities. But they are also often triggered by violent tectonic upheavals—earthquakes or volcanoes, or by the former's menacing effects, tsunamis. There is nothing that society can do at present to prevent tectonic upheavals, but there are methods to monitor, predict, mitigate, and avoid potential tectonic disasters (Figure 5.9). In the United States, progress in these strategies has involved many geologists over the past 20 years because of the need to protect population centers from threats such as the 1980 Mount St. Helen's eruption and the 1989 Loma Prieta earthquake. Scientists and policy makers know that

FIGURE 5.8 Impact of coastal erosion from Hurricane Andrew on Raccoon Island, Isles Dernieres, Louisiana. The top photograph was taken on July 9, 1992; the bottom photograph on August 30, 1992. Photographs courtesy of Jeff Williams, U.S. Geological Survey.

major natural disasters can be forestalled only by continuing research into the nature of such upheavals; responsible planning of engineered works and land-use facilities in areas at risk; and methods for warning, evacuating, and providing emergency relief. Only cautious, realistic planning can prevent overwhelming tragedies such as when hundreds of thousands of people were killed by the 1976 earthquake in Tangshan, China, or 3,000 were killed in the Philippines by a tsunami only 3 weeks later. The more knowledge that scientists gain about the causes and effects of earthquakes, the more able they will be to predict and plan for all these potential disasters. For example, the seismological techniques used to monitor earthquake zones can be used to monitor potential volcanic activity and potential tsunami threat. And assessment of earthquake, volcano, and tsunami hazard potential can help plan-

ners predict dangers from the associated landslides and floods.

Earthquake hazard evaluation involves determinations of the specific location, frequency of occurrence, and intensity of energy release—which, in turn, require characterization of the space, time, and size distribution of the earthquakes that give rise to the hazard (Figure 5.10). Decision making in the face of serious earthquake threat requires that the hazards and risks be neither overestimated nor underestimated because of the great consequences for life, safety, and economic security. Administrators cannot expect citizens to abandon their daily lives for anything but imminent danger. Therefore, earth scientists have an obligation to acquire relevant data and pursue research aimed at reliable depiction of an earthquake threat.

Quantitative experience from many earthquakes

FIGURE 5.9 Earthquake damage to a precasting plant in Axtoran (near Leninakan), Ukraine. This plant made building components for the precast, reinforced-concrete buildings in Leiniakan. Note the collapsed precast-prestressed concrete trusses. Photograph by Fred Krimgold; reproduced from *Earthquakes and Volcanoes 21*(2), p. 73.

provides an accurate and useful understanding of an earthquake's seismic hazard parameters, the nature of the earthquake source, the maximum size of the resulting seismic waves, the qualitative characteristics of those seismic waves, and the potential effect along the surface in response to the wave's energy. The study of these parameters involves diverse disciplines, ranging from sedimentology and seismology to geotechnical and civil engineering. To arrive at a realistic picture of the damage that may result from an earthquake, decision makers must apply scientific knowledge of the seismic hazard to the specific characteristics of engineered works. Only then can they estimate the seismic risk—the threat that earthquakes present to human lives and property.

Great concentrations of population have settled close to major plate boundaries. Of those cities whose populations are projected to exceed 2 million by the year 2000, 40 percent are within the 200-km earthquake shock radius of a plate boundary zone. While most deadly earthquakes are related to plate boundaries, those boundaries are not always sharp,

and the geometry of subduction-zone dip amplifies the breadth of potential damage. Studies of seismicity and active deformation have shown that the regions affected by plate interactions can be large. For example, the entire belt of mountains formed by the Indian-Asian collision stretches over an area of more than 6 km^2. And the effects of the diffuse plate boundary in the western United States extend over 1,000 km into the continent to the active Wasatch Fault (Figure 5.11), which passes through Salt Lake City. The distribution of earthquakes in the United States reveals that broad zones can be involved in plate boundary deformation. In the western states occupying these zones, many faults underlie urban areas, where earthquakes of magnitude 6 or larger pose serious threats. Some of these dormant faults are concealed by basin sediments or overlying rock that responds to tectonic forces by folding rather than faulting; such landscapes may provide few visible geomorphic clues about potential hazard. And while these faults seem to approach the surface at high angles, geophysical surveys find correlations with regionally extensive reflection sur-

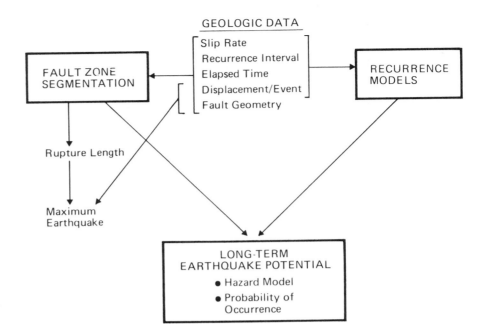

FIGURE 5.10 Relationship between geological data and aspects of seismic hazard evaluation. After D. P. Schwartz and K. J. Coppersmith, NRC, 1986, in *Active Tectonics*.

faces that are nearly horizontal and relatively deep within the crust. Analysis of geophysical data suggests that the faults are curvilinear in cross section, with one arm angling toward the surface and the other laterally underlying extensive areas. Possibly these *listric* faults may act as earthquake sources, although no significant earthquakes have so far been demonstrated to have occurred on the flat-lying segments of such faults.

With the continuing development and empirical refinement of plate tectonic theory, a first approximation of potential earthquake source location can be estimated reliably from the regional tectonic setting. Recent studies have shown that the total seismic energy released in plate boundary regions is more than 99 percent of the total worldwide seismic energy release. The problem is that most plate boundary regions have earthquakes quite frequently, which relieves stress incrementally, while the 1 percent of seismic energy released in intraplate events occurs only occasionally. Those singular events may therefore be extremely violent spasms. Nevertheless, the basic plate tectonic setting provides invaluable clues. One example is the Cascadia subduction zone in the Pacific Northwest, which marks the boundary between the Juan de Fuca and North American plates. Historically, this part of the plate interface zone has been seismically quiescent. But the possibility of powerful earthquakes occurring along this zone, which includes the cities of Vancouver, Seattle, and Portland, is suggested on the basis of its plate tectonic setting as well as neotectonic geological studies that indicate the oc-

currence of major subduction-zone earthquakes here within the recent prehistorical past. Earthquakes comparable in size to the 1964 Alaskan earthquake—which devastated Anchorage—may well occur along this coastline of the Pacific Northwest.

Intraplate earthquakes can pose a threat comparable to plate boundary events, as shown by the three very large earthquakes that rocked the New Madrid, Missouri, region in late 1811 and early 1812. Those earthquakes drained swamps, altered reaches of the Mississippi River, and created new lakes in parts of Tennessee. Chimneys toppled and stone walls cracked in St. Louis, Louisville, and even Cincinnati; church bells rang in Washington, D.C. Another major destructive intraplate event struck Charleston, South Carolina, in 1886. The considerable increase in population density throughout the eastern United States since these great intraplate earthquakes guarantees that comparable events would devastate communities in the eastern two-thirds of the country.

A valuable tool for identification of earthquake sources is the study of earthquakes described in the historical record. Modern seismometers have been useful only since the end of the nineteenth century, but thousands of earthquakes that predate such instrumentation are well documented in the historical literature. Systematic assessment of historical descriptions provides estimates of the sizes and locations of earthquakes in China, the Mediterranean, and the Middle East. Earthquakes have been cataloged in these regions of the world for a few

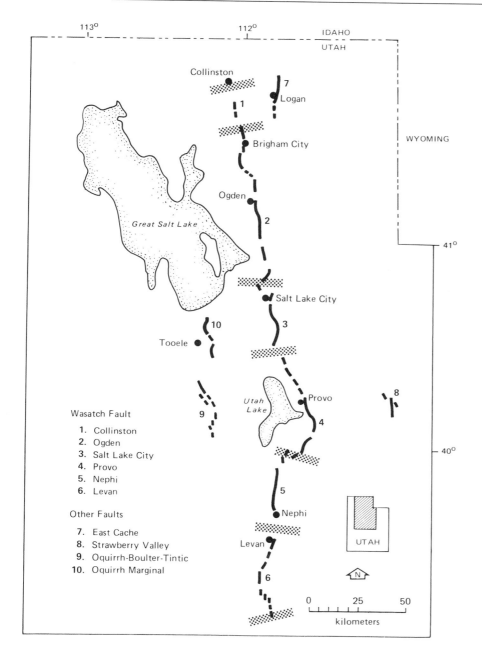

FIGURE 5.11 Segmentation model for the Wasatch Fault zone, Utah; stippled bands define segment boundaries. After D. P. Schwartz and K. J. Coppersmith, NRC, 1986, in *Active Tectonics*.

thousand years—intervals long enough to embrace rare larger events that recent instrumental data often do not include. Clearly, these historical events can indicate earthquake sources that might not otherwise be evident. The extensive historical catalogs may also provide indications of short-term fluctuations in the locations, sizes, and rates of earthquake activity that have not left evidence in the much longer geological record. The Chinese historical record bears witness to apparent spatial and temporal cycles having periods of a few hundred years. Such seismic cycles may also occur in Japan with regular intervals that range from 100 to 200 years.

But there are no cases of instrumentally recorded data covering periods long enough to provide any real understanding of seismic cycles—or even to offer any certainty in predicting earthquake recurrence.

Major advances have been made in identifying prehistorical earthquake sources on the basis of geological field studies. Displacements along active faults can be established by combining the historical record with stratigraphy, geomorphic analyses, and age-determination techniques in the new subdiscipline of paleoseismology. A good example of how the new techniques have given insight into the

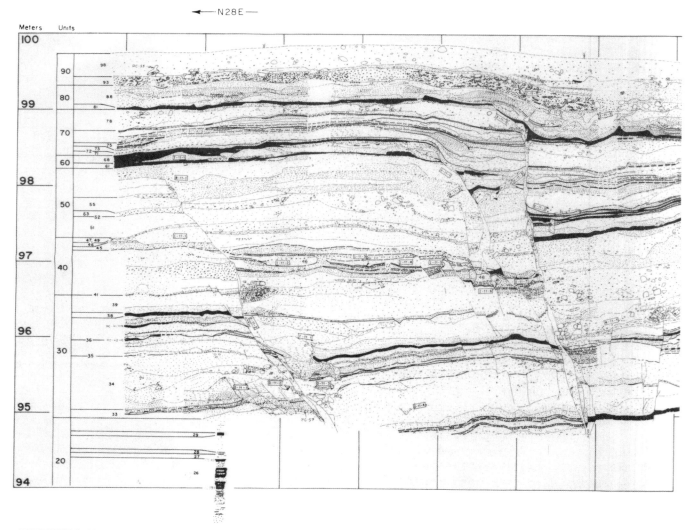

FIGURE 5.12   San Andreas Fault exposed in a trench at Pallett Creek, California. Black horizins are datable peat layers, which are progressively offset by greater amounts at greater depths (greater ages), the cumulative effect of earthquake faulting. From K. E. Sieh, 1978, *Journal of Geophysical Research 83*, pp. 3907–3939.

long-term seismic record is provided by studies of that segment of the San Andreas Fault in California that broke in 1857 (Figure 5.12). The segment currently is seismologically very quiet, but paleo-seismological evidence indicates repeated large pre-historical displacements, which presumably generated large earthquakes. The average intervals between these large paleoearthquakes taken together with the time of the latest event in 1857 are used by geologists to make statistical extrapolations as to how likely similar events may be in the near future.

Relatively large significant earthquakes are sometimes caused by ruptures on faults that do not break through the surface, as illustrated by the Coalinga and Whittier Narrows earthquakes in California.

Investigation at such localities, however, shows that these events do in fact leave their marks in the geological record. Measurement of changes in the shape of the surface provide geodetic evidence of continuing folding, such as slope steepening, and exposure of cross sections can reveal evidence of strong ground shaking, such as liquefaction—the transformation of a saturated soil into a fluid mass—and of continuing, long-term, centimeter-scale deformation within deposits overlying the faults. Detailed investigations using field geology techniques are essential for locating buried seismogenic faults.

The largest earthquake that a particular seismic source is capable of generating occurs very infrequently, so that the historical record of seismicity at any locality probably does not include the largest earthquake that ever occurred at that locality during the past few thousand or tens of thousands of years. Consequently, estimates of maximum magnitude, or energy release, for seismic sources are based on the geological record. During individual earthquakes, faults typically do not rupture over the entire fault length; instead they slip over only a few segments of the entire length. Total rupture lengths increase with earthquake magnitude. Similarly, fault rupture area—the size of the plane that has ripped loose—increases with magnitude. The lengths and areas of old ruptured segments (see Figure 5.10) place constraints on the maximum sizes of earthquakes that might reasonably be expected at that location. Current work is aimed at identifying physical, geometric, and fault behavioral characteristics that are diagnostic of individual fault segments. Detailed geological studies along the length of whole fault systems help to identify segments at risk of future ruptures. The problem comes in estimating these parameters for a given fault *before* the occurrence of the maximum event. A wide variety of geological studies—geophysical, seismological, and geomorphic—aid in the estimation of parameters related to magnitude. Methods used to gather information include interpretation of geometric constraints determined from geophysical analysis, investigation of fault-scarp dimensions as paleoseismic indicators, and integration of stratigraphic relationships in exploratory trenches excavated across the fault (Figures 5.12 and 5.13). All of these approaches may, in combination, provide a clearer picture of the sizes and dates of paleoearthquakes exhibited in the geological record and help to predict whether a large earthquake is imminent.

The first known instrument for earthquake detection was invented nearly nineteen hundred years ago by a Chinese mathematician and astronomer. It was designed with eight dragons poised around the circumference of a hollow globe with unattached balls perched in their mouths and eight open-mouthed toads waiting below them. An earthquake of even the slightest magnitude would allow a ball to fall from the mouth of a dragon into the mouth of a toad, setting off an alarm. It was thought that the earthquake would have originated in the direction of the empty-mouthed dragon. For two millennia after this invention, however, earthquake detection remained an elusive goal for researchers in Asia and Europe.

In the late nineteenth century, advances in both theory and engineering led to the development of the first seismometers—instruments that could reliably detect and record earth tremors. Modern seismometers continually monitor the Earth's vibrations and, being rigged with magnification equipment, produce a sensitive record of those vibrations. When an earthquake hits, anywhere on the globe, seismometers distributed over the surface fluctuate according to their reception of body waves and surface waves. There are two types of body waves: compressional—similar to the push and pull of a punch—and shear—similar to the writhing of a whiplash; both of these travel through the Earth. Surface waves travel in a complex array and cause most of the damage. To locate the source of the earthquake, both its focus deep within the lithosphere and its epicenter on the surface above the focus, seismologists need three seismological readings of body wave onset. They calculate the difference in arrival times between the compressional waves and the shear waves, which indicates the distance through the Earth from the seismometer to the focus; they can then pinpoint the fault rupture through triangulation. The techniques developed for recording natural earth shaking have produced a completely independent field of inquiry; seismic technology is a geophysical technique that induces mechanical shock waves in the Earth, records their reflections, and produces images from the reflections that enhance understanding of the interior structure. The technology has come full circle as an essential tool for studying earthquake characteristics.

Recent advances in geophysical techniques are providing better descriptions of the three-dimensional geometry of earthquake sources. Deep crustal images—such as those constructed from data produced by the Consortium on Continental Reflection Profiling, or COCORP—can reveal the regional scale of low-angle faults as well as previously unsuspected local fault geometries. Seismic tomography—comparable to the remarkable CAT-scan images used for medical investigations—contributes dense numbers of consecutive slices through the Earth and promises the ability to detect ancient lesions and scars in crustal structure with a level of detail previously thought to be unattainable.

Earth scientists have also improved their ability to use mathematical inversion techniques. Inversion analyzes the effects to gain knowledge about the nature of the cause. Seismic inversion, using the data from seismometers, studies patterns produced by an earthquake's body and surface waves and determines the type of earthquake that could generate them. Geodetic inversion measures the physical changes on the surface produced by an earthquake and characterizes the sort of earthquake that

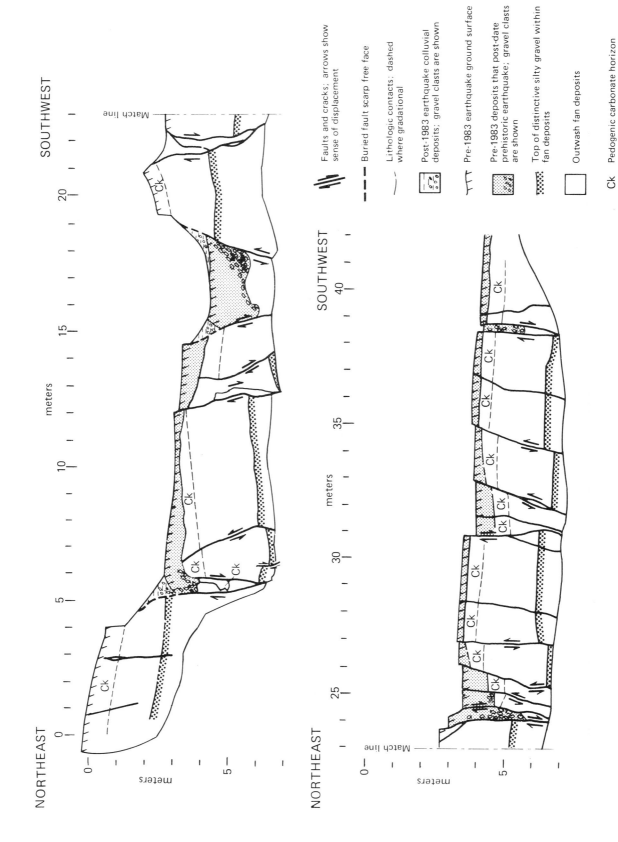

Faults and cracks; arrows show
sense of displacement

Buried fault scarp free face

Lithologic contacts: dashed
where gradational

Post-1983 earthquake colluvial
deposits; gravel clasts are shown

Pre-1983 earthquake ground surface

Pre-1983 deposits that post-date
prehistoric earthquake; gravel clasts
are shown

Top of distinctive silty gravel within
fan deposits

Outwash fan deposits

Ck   Pedogenic carbonate horizon

FIGURE 5.13 Generalized log of trench across the Lost River Fault and surface
rupture from the 1983, Borah Peak, Idaho, earthquake. After D. P. Schwartz and
K. J. Coppersmith, NRC, 1986, in *Active Tectonics*.

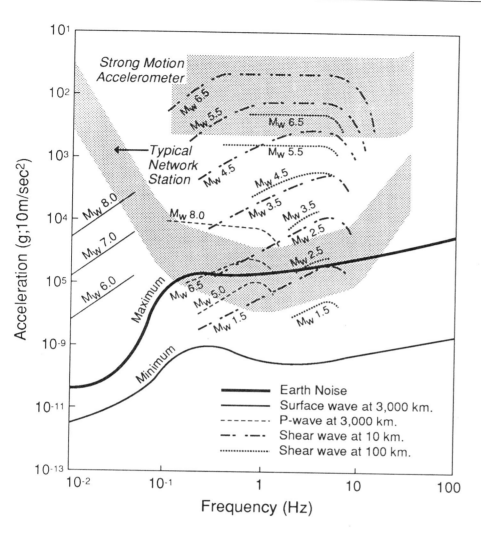

FIGURE 5.14 Expected accelerations produced by a range of magnitude earthquake as a function of frequency. The ranges of operation of the typical existing California network station and typical strong motion station are indicated by the stippled areas. The curves marked −120 and −160 dB indicate constant power spectral levels of acceleration of 10^{-12} and 10^{-16} $(ms^{-2})^2/Hz$, respectively. Note that many earthquake motions are not within the range of existing instruments in the networks. From NRC, 1991, *Real-Time Earthquake Monitoring*.

would be capable of making those changes. Successful seismic and geodetic inversions may construct remarkable pictures of the fault rupture surface at depth and the associated complexities in both fault geometry and the dynamic rupture process. Powerful examples of the application of inversion techniques can be found in the studies of recent earthquakes at Imperial Valley and Coalinga, California, and the earthquake at Borah Peak, Idaho.

During the past 20 years efforts to obtain records of strong ground motion have increased, and the data bank has grown with each successive earthquake. The data have been recorded, for the most part, on networks of relatively simple accelerometers with minimal magnification capabilities. These records have been invaluable to the engineering community in designing structures to withstand earthquakes. Within the past few years a new generation of instruments has been developed, and they are now being deployed to supplement the older

instruments (Figure 5.14). This new generation of instruments can digitally record a wide spectrum of frequencies over a wide dynamic range. The new data so gathered permit the analysis not only of strong ground motion, as did the older generation of instruments, but also details of the faulting process. Such instruments, and the data recorded by them, are revolutionizing our understanding of rupture mechanics.

Theoretical models have been developed for predicting the shapes of curves that represent earthquake ground-motion intensity, called seismic wave spectra. Although there is still some controversy over the physical interpretation of the models, different investigators have made predictions that are in general agreement with observed records, using data recorded in regions such as western North America where substantial numbers of strong-motion records are available. With that agreement as encouragement, theoretical models have been proposed that make ground-motion estimates for intraplate regions where strong-motion records are sparse to nonexistent. Research using accelerograph records is supplemented by the study of data from seismometers located at great distances from an earthquake epicenter, called teleseismic data.

Techniques have been developed for simulation of near-source ground motion. These techniques are still the subject of some controversy, but they are being improved and have been used for the design evaluation of such critical facilities as nuclear power plants and nuclear waste repositories. Such simulations are routinely used to estimate the damage potential of postulated subduction-related earthquakes in the Pacific Northwest.

Several recent earthquakes have emphasized the critical importance of soil conditions to earthquake ground motions at a particular site. On September 19, 1985, a strong earthquake ruptured a fault along the Pacific coast of Mexico, west of Mexico City. Resonance of seismic wave energy within the lake sediments beneath Mexico City caused amplified ground motion and resulted in tremendous damage to high-rise buildings in the city even though it is located more than 350 km from the earthquake's epicenter. On October 17, 1989, the damage to buildings in San Francisco and Oakland, located more than 75 km from the epicenter of the Loma Prieta earthquake, occurred almost exclusively at locations underlain by unconsolidated man-made fill or soft sedimentary deposits. Saturated soils that are supporting heavy loads may transform into a fluid slurry when given a jolt—the process of liquefaction (Figure 5.15). Soil structure disruption from this process is identifiable in the geological record as a sign of past earthquake activity. Ironically, in both Mexico City and San Francisco, the susceptibility was known: the problems had been clearly identified in advance and were accurately shown on seismic hazard maps. Photographs taken in the Marina District following the 1906 San Francisco earthquake show evidence of liquefaction-induced soil failures identical to those revealed in photographs taken in 1989. These regrettable facts underscore the futility of hazard research and accurate risk assessment if communities are incompletely informed or choose to ignore known geological threats.

A fundamental goal of seismic hazard research and risk assessment is accurate prediction of potentially damaging ground motion. Seismic hazard is assessed using deterministic and probabilistic statistical analyses, and a dynamical systems approach is developing rapidly. The deterministic approach concentrates on the maximum earthquake that a given seismic source is believed capable of producing during a specified time period. Assuming the maximum earthquake will occur, this approach determines ground motions that can be empirically associated with a known fault of specific seismogenic characteristics. Deterministic seismic hazard analyses are useful for engineering applications at critical facilities, such as dams and nuclear power plants, where there is a need to develop a conservative seismic performance design and to compare the results with those of other approaches.

Probabilistic seismic hazard analyses that attempt to incorporate a more complete picture of the seismic environment affecting the site of interest are now in routine use for virtually all types of structures. This approach delineates the individual uncertainties associated with all aspects of the site and incorporates these uncertainties into the analysis. Key parameters, such as maximum magnitude, are expressed as probabilistic distributions in time and space; the final results are expressed as the probability of exceeding various levels of ground motion at the site. At this point an informed decision can be made about appropriate design requirements for the particular facility. An acknowledged disadvantage of the probabilistic approach is the need to estimate the frequency of occurrence of various levels of earthquake magnitude rather than to merely assume that the maximum earthquake will occur during the lifetime of the engineered works.

Since the 1970s, attempts at short-term deterministic prediction have been augmented by probabilistic forecasts as a pragmatic means of quantifying seismic risk in a socially useful manner. Along the San Andreas Fault in California, for example, 30-year forecasts of earthquake activity form the basis for earthquake hazard zonation and mitigation activities on both the state and local levels. The 1980 forecast for a 2 to 5 percent per year probability of a great earthquake along the fault in southern California led directly to a decade-long program of structural retrofitting or removal of the entire class of buildings with greatest life safety risk in Los Angeles. Cooperation between providers and consumers of earthquake predictions proves to be highly desirable when information flows in both directions. The earthquake prediction experiment at Parkfield changed from a purely scientific investigation to an operational short-term prediction program through the direct participation of the emergency response community in planning and funding the experiment.

FIGURE 5.15 Sand boils resulting from liquefaction during the 1989 Loma Prieta earthquake near Santa Cruz, California. Photograph courtesy of Gerald F. Wieczorek, U.S. Geological Survey.

As a consequence of emphasizing the inherently statistical nature of prediction, our understanding of the prospects of damaging earthquakes has been revolutionized. Through the development of a framework encompassing all that is known about a particular region, from the historical and instrumental record to its plate tectonic setting and the short-term influence of specific geophysical events, seismologists can work toward making useful predictions. For example, a decade ago the prospect for the generation of great earthquakes in the Cascadia subduction zone off the Pacific Northwest coast was a hotly debated subject. Evidence for prehistorical large-magnitude events found by paleoseismologists has redefined the subject of debate from the likelihood of a future event to its timing and the severity.

Society is increasingly dependent on location-specific scientific information, expressed quantitatively and qualitatively, for decision-making purposes. Earthquake prediction information must be communicated accurately and articulately in order to avoid panic while preparing for highly probable potential crises. Whether long-term or short-term predictions, such messages should provide maps of earthquake damage potential, including such near-instantaneous changes as landsliding and liquefaction. Within this framework, short-term predictions can provide crucial information, provided the users are properly prepared to receive it; preparation must be made through public education programs. California has begun to issue short-term earthquake advisories based on probabilistic models of fore-shock activity. These low-probability advisories—a 2 to 5 percent chance of a larger event in a 3-day window—have proved their worth by spurring individuals and institutions toward earthquake mitigation actions they had previously ignored. More precise short-term predictions may, some day, again revolutionize the technical ability to give earthquake warnings. Even if and when this goal is achieved, the translation of scientific knowledge into planning, preparation, and action will remain the most important task for scientists and public officials alike.

Modern computer-networking capabilities promise a new form of hazard mitigation involving real-time seismology that can immediately identify the most severely damaged areas and assign emergency relief priorities. It is now technically possible to determine earthquake source parameters such as size, depth, and direction of rupture propagation immediately after a large earthquake. Large-scale deployment of the new generation of broadband seismic instruments with satellite or other telemetry capabilities is making this a reality. The intensity distribution of earthquake ground shaking often exhibits a very irregular spatial pattern because of variations in the crustal structure near the epicenter, in site response, and in the force mechanisms—such as strike-slip, along the San Andreas Fault, or dip-slip, in western Mexico. When seismic networks are supplemented by portable instruments, generic path effects and site responses in an earthquake-prone region can be defined. It then becomes possible to quickly determine the spatial distribution of ground shaking and the damage exposure in an entire epicentral region.

Instantaneous communication capabilities offer further opportunities to mitigate earthquake damage. In cases where the earthquake is centered some distance away, it is possible to warn a region of imminent strong shaking as much as several tens of seconds before the onset of damaging shaking because of the relatively slow speed of seismic waves. Endangered regions could respond to the warnings in time to shut down delicate computer systems, isolate electric power grids and avoid widespread blackouts, protect hazardous chemical systems and offshore oil facilities, and safeguard nuclear power plants and national defense facilities.

A simple warning system using this concept has already been incorporated into the Japanese railroad system. A similar strategy is being implemented in the form of a tsunami warning system that will alert areas around the Pacific when any large submarine earthquake occurs in the Pacific basin. Real-time seismic and geodetic systems are also critical components of volcano monitoring systems that can provide fairly short-term warning of impending explosive eruptions.

Tsunami Hazards

A tsunami along several hundred kilometers of the coast of Nicaragua on September 2, 1992, resulted in over 100 deaths as a 25-foot wave inundated coastal areas (Figure 5.16) and in places had run-ups of up to 1,000 m. Tsunamis are large ocean waves most commonly generated by the uplift or depression of sizable areas of the ocean floor during large subduction-zone earthquakes; significant tsunamis have also resulted from volcanic eruptions and large landslides or submarine slides. Like earthquakes, great sea waves are of little consequence in remote regions. In the open ocean they are hardly noticed, but as they approach the shore the waves increase in amplitude as they move into shallower water, depending on the nature of the local submarine topography. The resulting tsunami hazard in many coastal areas is far greater than is often appreciated. For example, during the great Alaskan earthquake of 1964, the loss of life from the tsunami generated in the offshore area was more than 15 times as great as the loss of life directly attributable to earthquake shaking; much of it occurred far from Alaska. Indeed, during the past 50 years, significantly more people have been killed in the United States by tsunamis than by other effects of earthquakes—although these statistics could change radically overnight with a major earthquake in a metropolitan area.

The areas of the United States most affected by tsunamis are Alaska, Hawaii, and the Pacific Northwest. Hilo, Hawaii, has been hit repeatedly by tsunamis originating as far away as southern Chile, and the same Chilean earthquake that produced tsunami devastation in Hilo in 1960 caused 200 deaths in far more distant Japan. Very recent geological field studies suggest that large prehistorical tsunamis occurred along the Oregon-Washington-British Columbia coast, probably generated by

FIGURE 5.16 Damage resulting from the September 2, 1992, tsunami that struck Nicaragua. Photograph courtesy of Mehmet Celebi, U.S. Geological Survey.

large earthquakes in the Cascadia subduction zone. Because no large earthquakes—or locally generated tsunamis—have occurred there within the short historical record, there has heretofore been essentially no planning by public agencies for such a contingency. Because the effects of a large tsunami could indeed be devastating in many low-lying, highly populated areas throughout this region, the issue has now become one of great societal as well as scientific interest. Perhaps the most revealing lines of evidence have come, and will continue to come, from geological field studies of the direct effects of recent abrupt changes in elevation and the resulting tsunamis in the very young sediments of coastal areas as well as changes in the coastal morphology.

Closely related to the tsunami hazard is seiching, or the oscillation of closed and partially open bodies of water caused by long-period surface waves, which are often produced by the same large earthquakes that generate significant tsunamis. For example, the 1964 Alaskan earthquake caused waves as high as 2 m in bays and channels along the Gulf of Mexico, with sizes and shapes suitable for resonating with the wavelength of the arriving long-period seismic energy. In addition to a tsunami that might be caused by a large subduction-zone earthquake off the Pacific Northwest coast, damaging seiches might be generated by the same seismic event on water bodies such as Lake Washington in Seattle, in the Straits of Juan de Fuca and Georgia, on individual arms of Puget Sound, and in large bays such as Grays Harbor and San Francisco Bay.

Although the basic physics of tsunami generation, wave propagation, modification, and run-up

on the shallowing shore are generally understood, the oceanic waves and their effects have had little systemic and comprehensive examination. A tsunami research planning group commissioned by the National Science Foundation in 1985 recommended, with highest priority, that a major effort be made to gather meaningful field data related to the generation and propagation of tsunamis. Specifically, the group suggested that arrays of seafloor pressure sensors be placed in two areas where there is a high likelihood of large earthquakes that could result in tsunamis—the Shumagin and Yakataga seismic gaps in Alaska. Their purpose would be to capture the onset of a tsunami and provide critical data on the time histories of shallow-water surface elevation and relative bottom displacement. As part of the same experiment, one or more deep-water arrays of bottom-pressure sensors should be established at greater distances, perhaps north of the Hawaiian Islands and off the California shore, for measuring water-surface elevation, wave directions, and seismic spectra characteristics. Such data are essential for evaluating theoretical models of tsunami generation, propagation, and run-up.

Tsunamis are generated by long-period seismic processes, and it is therefore important to determine the long seismic period nature—those in excess of 100—of earthquakes in areas that might generate tsunamis. Whereas abundant long-period records of such earthquakes are available from instruments located at great distances from the source, almost no close-in, long-period records are currently available because no appropriate instrumentation is in place. The research planning group recommended that two low-gain, long-period seismometers be established in each of the two Alaskan seismic gaps.

Additional important research areas were identified by the research planning group. They include the fundamental dynamics of long, weakly nonlinear, three-dimensional waves; refraction-diffraction and other nonlinear transformations of long waves in the near-shore region; run-up, run-down, and overland flow; interaction of waves and engineered structures; and basic research pertaining to tsunami warning systems.

Tsunami research involves a number of different government agencies and encourages, even demands, international cooperation. The warning systems that alert population centers around the Pacific Ocean basin of potential and approaching tsunamis are a model for international disaster-mitigation efforts. The tsunami warning system involves three levels of alert. First, it notes the occurrence of an earthquake strong enough to trigger a dedicated alarm located in Honolulu, in the center of the Pacific. Second, if the epicenter is found to be in an area of the Pacific basin capable of generating a tsunami—under or near the ocean—a tsunami watch is established, and communication with areas around the epicenter is attempted to verify the existence of a wave. Finally, if a tsunami has been witnessed, the Honolulu station issues a formal warning along with estimated arrival times for specific locations. Once the formal warnings have been made, individual government agencies proceed with appropriate action.

This system worked quite well for the Alaskan earthquake on March 28, 1964, except for two problems. First, the earthquake damage was so severe in the immediate vicinity of the epicenter that all communication systems were knocked out. Honolulu knew an earthquake had occurred that was large enough to set off the alarm, but the location was not established until an hour later. After another hour the first tsunami report arrived from Kodiak Island, 650 km from the epicenter. Kodiak Island reported two more crests, and a full warning with estimated times of arrival was issued by the Honolulu station. The second problem came from the public response. In Crescent City, California, citizens left their homes when alerted, but they grew impatient and returned before the third wave moved through. Seven people died. In San Francisco and San Diego only the weakness of the waves when they arrived averted disaster. Thousands of people headed down to the beach to watch the big waves, proving, once again, that all the wisdom in the world cannot save those who choose to ignore it.

Volcanic Hazards

The energy of a major volcanic eruption is well beyond what can reasonably be expected to be controlled by engineering. Consequently, volcanic phenomena can best be adapted to by accurately predicting the occurrence and the likely results of an eruption. Fortunately, volcanic eruptions have many precursory phenomena (Figure 5.17) that can readily be detected with modern instrumentation and techniques.

The reduction of volcanic risk over the next decade or two will involve three distinctly different efforts: basic volcanological research, monitoring of high-risk sites coupled with public education, and study of volcanic effects on climate. Basic research is required on how volcanoes function; particularly needed are investigations into the mechanisms that trigger eruptions and better determinations of the average time spans between explosive eruptions of large magnitude. The research on recurrence intervals should consider both global and regional frequencies of eruption. This basic research could lead to accurate assessment of specific risks. Known techniques must be expanded, and new ones developed, for assessing volcanic hazards and monitoring active and potentially active volcanoes. A high priority must be given to public education; as with earthquake and tsunami warnings, a fine line must be followed between offering correct information about probability and risk and diluting the importance of a warning by false—or misunderstood—alarms.

A better understanding of the interaction of volcanic emissions with the atmosphere and hydrosphere is necessary. Historical records show that some volcanic events dramatically modified climate for several years following eruption by introducing large volumes of dust and gas into the atmosphere. Recent research has presented evidence that large submarine eruptions along ocean ridges may alter ocean temperatures. This mechanism has even been suggested to play a part in the short-term El Niño sea-warming condition with its subsequent implications for climatic variations that fluctuate over periods of a few years. Understanding these correlations with climate will depend on global monitoring of volcanism, most likely through satellite-based remote sensing. These new data will yield a better theoretical understanding of climatic response to heat and mass transfer from the interior into the hydrosphere and atmosphere.

Deciphering the workings of volcanoes is an eclectic challenge, involving contributions from

FIGURE 5.17 More than a dozen eruptions have been successfully predicted at Mount St. Helens, Washington. One of the important data sets for prediction is the rate of thrusting in the floor of the crater adjacent to the lava dome. The crater formed during the explosive eruption of 1980 and the lava dome that subsequently developed are shown in this photograph of April 17, 1981. The graph shows contraction across a thrust fault on the floor prior to the September 1981 eruption. The black rectangle is the period within which the eruption was predicted to occur. The prediction was made at the arrow, and the eruption occurred at the vertical dashed line. Figure from NRC, 1986, *Active Tectonics*.

many areas. Data are obtained both directly and through association. Direct methods include continuously monitoring the geological, geophysical, and geochemical changes that occur on active volcanoes and geological mapping of ancient volcanoes whose internal structure has been exposed by eruption and/or erosion. Associations can be determined by analyzing the character and sequence of historical and prehistorical eruptive products from various types of volcanoes and then matching these and other data to conceptual models of how volcanoes work (Figure 5.18).

Present understanding of the dynamics of active volcanic systems is still in its infancy—particularly those volcanoes related to convergent plate boundaries, like the Cascade Range, the Andes, and the Aleutian Islands. In the past 40 years research on active volcanoes has expanded from field observation and description to include experimental and theoretical studies. The remaining challenge is the monumental task of assembling all the parts to create a better understanding of how volcanoes work. Two aspects of basic research that would lead to specific reductions of volcanic risk are discovery of the mechanisms that trigger volcanic eruptions and determination of the frequency of large explosive eruptions on a regional and global basis.

Knowing what triggers an eruption could lead to

more successful searches for detectable precursors. Several precursors have already been recognized. These include a dramatic increase in earthquake activity beneath a potentially active volcano; a swelling of the ground surface near the volcano's summit (see Figure 5.5); and a continuous ground vibration, called volcanic tremor, which is detectable on sensitive seismometers. These signs indicate the injection of molten rock into the shallow roots of the volcano and definitely point to the potential for eruption.

Sometimes, when researchers monitoring a volcano have alerted the public to an imminent eruption, the shallow intrusion of molten rock will stabilize and cool without reaching the surface. When this occurs, the scientists may be accused of issuing a false alarm. A better term for this scenario would be an *aborted eruption*. More precise methods are needed to distinguish between shallow intrusions of molten rock that will erupt to the surface and those that will not. In addition, scientists need to learn how to communicate the uncertainties involved in forecasting eruptions to the governing officials and the public at risk near potentially active volcanoes.

Volcanoes in Hawaii, especially Kilauea, have been thoroughly monitored over the past several decades. Research indicates that molten rock rises

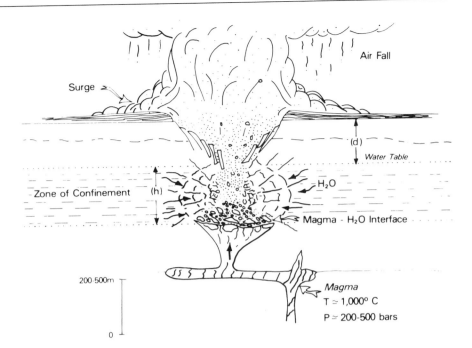

FIGURE 5.18 Diagrammatic cross section of a hydrovolcanic eruption in which magma interacts with near-surface groundwater. The interaction fragments the magma and country rocks, vaporizes the water, and produces steam explosions that excavate a crater and eject tephra. From K. H. Wohletz and R. G. McQueen, NRC, 1984, in *Explosive Volcanism.*

slowly but nearly continuously from a depth of about 65 km and accumulates in a shallow chamber about 3 km beneath Kilauea's summit. When that chamber becomes filled but the delivery of molten rock continues, the rock walls of the reservoir split, and molten rock either erupts to the surface or fills underground fractures in the volcano's flank.

Many questions about the Hawaiian volcanoes remain unanswered, but their general structure and dynamics are understood relatively well by comparison with those of the Cascade volcanoes like Mount St. Helens and Mount Rainier, which are less continually active. This understanding comes from familiarity; the spattering eruptions of Hawaiian volcanoes are part of everyday life. As convergent plate volcanoes, the Cascade volcanoes rise frigid and silent above the jumbled ridges clustered around them—Mount Shasta and Mount Rainier are over 4,000 m high; Mount Hood and Mount Baker reach heights over 3,000 m; Mount St. Helens had grown to only 2,550 m before its eruption in 1980. At these volcanoes, typical of those around the Pacific rim, earthquake swarms and ground deformation episodes are much more intermittent than at Kilauea. Long periods of quiet are suddenly interrupted by intense episodes of shallow intrusion and eruption. When these volcanoes erupt, they can be violently explosive.

Although violent eruptions are easily perceived as a major hazard, so can associated mudflows or lahars. About 4,000 years ago a massive mudflow having a volume of over 3 km³ originated on the edifice of Mount Rainier and partially filled many surrounding valleys and inlets of Puget Sound to hundreds of feet. Such a mudflow can be a major hazard. They might not necessarily be directly associated with a volcanic eruption but can be triggered by a nearby earthquake that might destabilize the glaciers on the volcanic edifice. Mount Rainier currently has over 4 km³ of perennial ice on its peaks; such mudflows are poised to occur again. Because of the potential mudflow and volcanic hazards at Mount Rainier, it has been designated as a "Decade Volcano" within the context of the International Decade of Natural Disaster Reduction. The intent is to focus hazard-related research activities; these activities are currently being developed.

So far, earthquake and deformation data on the convergent-boundary volcanoes reveal little about their dynamic behavior. A possible reason for this may be that much significant activity at these volcanoes is beyond the penetration capabilities of deformation-monitoring instruments, perhaps 25 to 35 km beneath the surface. When the sources of potential ground-surface deformation are so deep, the changes at the surface are either too small or too widely distributed around the volcano's summit to be measured by conventional surveying techniques.

The advent of satellite-mounted surveying systems, such as the Global Positioning System (GPS), may help to resolve this problem. With an accuracy of ± 2.5 cm over 50 km, satellite surveys may be able to detect such small local or widely distributed changes in the elevations and horizontal positions of benchmarks around potentially active convergent plate margin volcanoes.

Continuous monitoring complements geological mapping of potentially active volcanoes, which is generally the best way of determining their past eruptive habits. History often repeats itself in the natural world, so this type of geological assessment can provide very useful data for long-range forecasting of the activity of an individual volcano. A good example is the Nevado del Ruiz volcano in Colombia. An eruption and mudflow in 1845 killed 1,000 field hands at a tobacco plantation near its base. The summit of Nevado del Ruiz is covered by a large snow and ice cap, and even a small eruption can generate massive floods of meltwater and debris that course down the canyons on the volcano's flanks. Sweeping up soil and debris, the floods become destructive mudflows. When earthquake swarms and small eruptions began at Nevado del Ruiz in 1985, geologists warned that mudflows similar to those in 1845 were likely to occur again. They did—killing 25,000 people. The reasons for this tragedy are complex, but one was that people in Armero—the principal town that was destroyed—did not comprehend what a massive mudflow was. Nothing like this had happened for 140 years—long enough for everyone but a few alarmed geologists to completely forget what such a threat could mean.

Only about 10 percent of the world's 1,300 potentially active volcanoes have been geologically mapped to assess their past eruptive habits. This leaves 90 percent yet to be studied. Efforts to monitor and map these threats using field geologists can be accomplished with a high degree of return for a low hazard-assessment budget investment.

Understanding how a bomb works does not eliminate its danger, but if something is known about its size and the way in which it is detonated, the danger can assuredly be reduced or avoided. The same holds true for geological hazards—the better a phenomenon is understood, the more likely it is that its threat can be mitigated. At the same time, urging the establishment of simple but effective educational programs to inform governing officials and the public at risk remains the responsibility of the scientist studying the hazard. How might such an education program work? As in advertising, keep the message simple, and present it over and over again.

HAZARDS OF EXTRATERRESTRIAL ORIGIN

Within the past dozen years the idea of catastrophic terrestrial impact has been revived. From being a topic that was hardly considered respectable, it has become accepted as something that certainly happens occasionally and that may have had global consequences at various times in the past. Recent theories attribute both the origin of the Moon and the extinction of the dinosaurs to impacts of extraterrestrial objects.

The historical record, which goes back less than 3,000 years, contains no reference to anyone killed by a meteorite fall—although injury from a meteorite that penetrated a house is reported, and early in the century the Naklha meteorite that fell in Egypt may have hit a dog. On the time scale of current societal interest, the danger from impacts is insignificant. However, if a large extraterrestrial object did collide with Earth, the consequences could be devastating.

A distinction should be made between the near-field consequences of impact, dominated by the crater and its surrounding debris field, and the far-field consequences, which for larger impacts would be catastrophic. Research related to the former includes the need for better understanding of the flux of impactors on the Earth. Astronomical observations are essential, while the search for and study of impact craters on Earth yield complementary information about the flux in the past. Astronomical monitoring has produced estimates of impact flux that indicate there must be many unrecognized impact structures waiting to be discovered. At present, attention is focused on the search for evidence of an impact that occurred 66 million years ago and was large enough to have caused the extinction of the dinosaurs by means of the far-field effects. The Chicxulub structure in the north of the Yucatan Peninsula is considered a strong candidate, as discussed in Chapter 3.

Far-field effects become important if the impactor is large enough to make a crater more than a few tens of kilometers in diameter. Dust and aerosols propelled to great heights would induce fluctuations in weather patterns as well as produce acid rain. Wildfires might ignite over huge areas, and part of the atmosphere could be blasted away. In such catastrophic circumstances, many organisms would perish and many species would face extinction. The geological record should contain evidence of such drastic events. The challenge of impact theory is twofold: to seek evidence of past impacts in the geological record and to model patterns of atmospheric and oceanic systems that could produce distributions matching the evidence.

Giant terrestrial impacts are capable of perturbing the earth system rapidly and to an extent that few other phenomena can rival. Within the past half century human beings have acquired the ability, through nuclear explosions, to produce sudden perturbations of the earth system on a scale approaching that of the greatest volcanic and impact events. In the distant future it even seems possible that nuclear explosives might be used to divert incoming asteroids. Perhaps equally significant is the ability, also developed in this century, of the human race to modify its environment on an equally grand scale but more slowly: global change, largely a consequence of intensive fossil fuel consumption.

PROBLEMS RELATED TO POPULATION CONCENTRATION

Human interactions with the Earth can disrupt the background equilibria, but that disruption is usually confined to a very limited extent of area and time when compared to the Earth's size and history. The effects of small-scale farms and industries, such as gristmills or kilns, are absorbed into the earth system with little long-term influence. But human activity has reached a critical level that is beyond small-scale local equilibration. The twentieth-century metropolis creates intense, perhaps irrevocable, disruption of earth systems.

Modern engineering has created the most technologically advanced and safely built environment in history. For many the quality of life is enhanced through improved transportation methods, resource development, and engineered structures. The successes of this built environment have led to expectations that can be met only with constant maintenance and strict building codes.

In a resource-critical environment, exemplified by the modern urban center, humans will have to adapt to the local landscape and learn to use it to their advantage without destroying it. In the United States, general land-use decisions and land-use regulations are established at the city and county planning commission level and approved by local governments. Local government creates land-use policies, which become the legal authority that guides growth and development. To create an environmentally responsible policy, the socioeconomic objectives decided on by a community should balance against the constraints and opportunities offered by the physical setting. Urban geology therefore needs to be carried out to address local planning needs. For economic viability and public safety, local governments must be environmentally informed. Earth scientists have the opportunity, perhaps the duty, to provide that information.

Significant issues for the future will involve assessments of possible land use, concentrating on reuse, as affected by geological constraints. Society must act as part of the environment—foregoing attempts to master it—because human actions have far-reaching, often unforeseen, effects on the natural system. Land-use decisions should remain within the limits of our economy while appreciating both existing and future public needs. Many parts of the country are encountering development limits, those points at which the economic and environmental costs of dealing with geological constraints exceed the benefit of a desired development.

Every type of land-use plan could benefit from a preliminary geological assessment. The assessment should generally address the entire geological spectrum: surface topography, composition of subsurface material, arrangement of underlying rock, and character of the groundwater system. This general geological assessment would define foundation cost and would allow estimation of recurrence intervals for floods, landslides, subsidence, and, in some areas, earthquake and volcanic activity. In urban and metropolitan areas the physical assessment should focus specifically on the characteristics of the earth materials. This will lead to the selection of safe and appropriate construction sites, to the conservation of natural resources essential to the city's development, and to the preservation of natural aesthetic qualities. Any proposed land use is likely to produce its own effects on the environment; negative consequences of development should be integrated into the assessment and suitable adaptations to geological conditions suggested.

Urban geology studies and interprets the geological conditions underlying the most intense concentrations of human population. Its immediate goals are to determine the physical properties of individual sites and to locate the deposits of materials necessary for construction projects. In population centers scientific understanding is just one more factor among social, economic, and political components. To the scientist these other factors seem too prone to subjective interpretations and appear to be based more on emotion than logic, but the habit of thinking in geological terms is certainly unfamiliar to the nonscientist. Whatever the constraints of theoretical recurrence intervals, lives to be led in the here and now are more likely to focus on the supply of necessary raw materials and the efficient removal of waste.

While waste management concerns even the smallest rural communities in which waste production outstrips disposal capability, it is a major problem of today's urban centers. The urban expanse of pavement, with limited capacity to absorb rainwater, collects runoff and delivers it episodically in quantity to storm sewers. Traditionally, urban storm runoff has been allowed to mix with municipal sewage, contaminating the whole volume. That entire volume of contaminated water then requires treatment prior to discharge to the environment— hence the multibillion-dollar outlay for storage tunnels and water treatment plants in Chicago and Milwaukee—two cities that have conscientiously responded to federal requirements to control the quality of urban storm-water discharge.

Throughout the design, construction, operation, and maintenance of waste management facilities, consideration of local geology helps to ensure structural integrity and containment of contamination. Toxic liquid wastes require extremely complex treatment before the cleaned effluent can be discharged to the surrounding environment; during all stages such wastes must be prevented from leaking into groundwater. Solid waste disposal is another major headache for all city management. There is no way to absolutely prevent small amounts of hazardous waste from becoming mixed with municipal garbage waste. For example, used motor oil, photographic developing chemicals, household cleaners, pesticides, and paint products all go to the garbage. Given the huge volumes of municipal wastes and the shortage of suitable disposal spaces, an understanding of the options available within the geological constraints is essential for reducing the negative environmental impact. Modern sanitary landfill design depends on geological characterization of proposed sites. Implementation of the 1991 Resource Conservation and Recovery Act's solid waste regulations in the immediate future is meant to ensure groundwater protection through leachate collection and detection of any bottom liner failures. Finally, radioactive waste management has placed society in an unusual quandary: these wastes, most of which are produced in the course of providing electricity, conducting medical research, and maintaining national defense, are both a hazard and a valuable resource for future reprocessing. Underground storage/disposal appears to be the solution of choice—though it will be critical to locate the proper geological environment to prevent any long-term leakage of radionuclides, in liquid or gaseous form. Retrievability from storage for reprocessing may be an important consideration in the future.

Population centers have developed through accumulation, with little or no attention to geological consequences. Our cities are in a constant state of redevelopment. As more and more people have moved to urban centers, land use has become a very complex problem: there is little unused space available. The intricacies of redevelopment are compounded for several reasons. Inadequate existing facilities must be maintained, available surface space is minimal, construction debris accumulates, waste-disposal areas shrink, new transportation routes are not easily created, and often short-sighted decisions are made over acceptance of construction-related disruption and change. Too frequently, political expediency is the deciding factor in development design. But the geological constraints themselves are becoming more evident, as is the need to analyze them more thoroughly—not only for their effects on redevelopment projects but also for their effects on the physical stability of existing buildings and potentially large cost overruns associated with unexpected geological conditions, especially weak earth materials and large inflows of groundwater.

Engineered Structures

Geological site interpretations are needed for all forms of engineered construction, such as dams, nuclear power plants, hydropower plants, pumping plants, factories, sewers, canals, levees, high-rise buildings, harbors, locks, highways, bridges, tunnels, airports, and port facilities. Increasingly more detailed geological data provide the basis for each step of design and construction. The process begins with a feasibility study, based on knowledge of regional geology and enough specific site information to ensure that major geological features are identified and considered in cost analyses. A key element of the feasibility study is identification of geologically based flaws that could defeat individual projects. The national need for good geological and topographic maps is nowhere more clear than at construction sites. The relatively small scale of generally available maps can never serve all the geotechnical data requirements. Site-specific geological mapping and subsurface exploration are required, and preservation of site geological maps showing conditions at the time of construction is a necessary safeguard in case problems requiring repair or rebuilding arise later.

The selection of sites suitable for the location of major facilities, especially those for which public safety considerations are an overriding issue, is generally accompanied by geological site investigation. But even site selection of homes and parking lots must involve concern for the stability of foundations and for off-site disruption of drainage, water supplies, stream sediment, and air quality. The fact that the foundations of structures do fail means that attempts to characterize the sites prior to construction have in some cases been inadequate and that recognized risk was accepted. Although it is impractical to evaluate all sites exhaustively, an experienced engineering geologist can usually detect potential fatal flaws such as major faults, landslides, subsurface cavities, and poor soil conditions quite quickly.

The strengths of natural rocks and soils, and of the things we build from them, are generally predictable, but there is a need to anticipate, understand, and correct the failure of these materials. Understanding the nature of materials may lead not only to enhancing strength but also to decreasing strengths temporarily so that urban excavations can be made more efficient.

Use of Earth Materials

Earth materials are linked to land use and environmental management in many ways. Engineered works utilize earth materials both as the foundations, soils or underlying rock, and as raw material for construction.

The physical characteristics of foundation soil, weak rock, and rock are carefully incorporated into engineered design. The engineering characteristics of earth materials greatly affect many design features and the overall cost of all projects. Soil, in the engineering sense, consists of any unconsolidated overburden material. Most engineered structures are founded in soil, and design characteristics include soil moisture content, strength, and compressibility as well as the depth of the groundwater surface.

Foundations of large, heavily loaded structures are usually made to bear on rock, if it is accessible. Rock is normally much stronger than soil, thus making it more competent to support the loads imposed by large structures. Engineering features, such as piles and excavation of basements, are used to transmit structural loads to the underlying rock. Cost-effective alternatives are always sought, such as development of less expensive ways of controlling groundwater, stabilizing excavation slopes, and increasing the performance of soils through admixture and injection. Increasingly, manufactured materials such as synthetic membranes and geotextiles, grouts, and soil nails are being used to strengthen relatively weak bodies of earth material.

Earth materials are the base component of a significant amount of most construction. Building materials are made from common rock types that are easily obtained and relatively inexpensive. Since the 1970s, however, the source areas—stone quarries and sand and gravel pits—have been under increasing pressure from urban siting, regulatory permits, and operational constraints and reclamation requirements. All of these are profoundly affected by site geological conditions.

Quarried rock is a major source of construction material. Rock is used to make aggregate for both concrete and asphaltic concrete, roofing granules, filter stone for drainage, crushed stone for roadway base course and railway ballast, riprap for erosion protection, rock fill, building stone, ornamental and dimension stone, and monuments. Each type of use calls for a different set of properties, selected from durability, density, spacing of rock fractures, hardness, color, physical appearance, abrasiveness, and shape of particles when crushed. These properties are controlled largely by the mineralogy and fabric of the rock, and quarry sources are selected to meet the requirements for the particular use intended. Engineering petrography is routinely used to assess such properties and characteristics.

Construction demands for highways, roads, streets, parking lots, sewers, dams, levees, dikes, and airfield runways specify dimension stone aggregate, granular fillers, and pervious and impervious soils. Gravel and sand pits supply large quantities of material for concrete aggregate, while clay pits supply the material needed for manufacturing brick. Engineering geologists assess the quality and quantity of these products and provide the basic information for permitting and design.

Raw materials for construction—sand, gravel, crushed stone, earth for fill, clay, building stone, and cement—are first in tonnage and second in dollar value only to fuels as the major mining and quarrying products of the United States; they also involve a minimum level of geotechnical oversight and present few scientific problems. Sources for these bulk materials are selected on the basis of convenience to users and environmental insensitivity of the sites. They leave large holes in the ground that because of their size cannot be restored to their original condition but that usually offer attractive opportunities for imaginative subsequent use. Scientific research opportunities include establishing optimum formulations for cheap durable concrete and working out ways to excavate earth materials and process the spoil in a manner least damaging to the environment and most easily accommodated close to users.

Tunnels and Underground Openings

As cities become ever more crowded, and as mineral resources are sought at greater depths, the ability to create and maintain tunnels and underground openings becomes more and more important. Underground openings are essential. They serve as sewers and accommodate storm drainage. They carry water from distant sources to cities and provide efficient access for many transportation systems, including highways and subways. And large underground caverns, often left by mining enterprises, are increasingly being used to store petroleum and natural gas as well as radioactive waste. Greater Kansas City is now a world leader in conversion of mined underground space to low-energy, secure commodity storage. Underground excavations also house electricity-generating stations and defense command posts. Underground car parking is vital to many communities. The use of underground space will grow further as older population centers are redeveloped. Already the city of Austin, Texas, has a subterranean utility and freight supply trafficway, and Toronto, Canada, has an underground mall network and pedestrian trafficway.

Use of underground space requires geological understanding; the nature of the surrounding earth material—its composition, fractures, and engineering properties—is the primary consideration for space utilization and design. The stability of an excavation, as well as that of adjacent structures, depends on the earth material characteristics and groundwater control. Transportation and infrastructure routes follow excavations, embankments, and tunnels; all require geological analysis for structural integrity and to alleviate effects on adjacent facilities.

Engineering geophysics was born in the late 1950s, the child of technologies developed for petroleum and mining exploration. It was the advent of nuclear power plant siting, which required detailed subsurface structural and positional accuracy, that led to the use of improved oil field geophysical techniques to detect near-surface stratigraphic anomalies, the presence of groundwater, and faults. An amazing array of engineering geophysical technology is now available—based on the refraction and reflection of sound and electromagnetic waves, electromagnetic properties of earth materials, electromagnetic fields, natural or induced radioactivity, and gravitational attraction of earth materials—as are borehole seismic and television imaging devices. These innovations contribute to the abilities of the engineering geophysicist, but, as with so much in the earth sciences, they need to be interpreted in light of visually observed geology. Boreholes and outcrop observations are still necessary to confirm remotely sensed data.

Health Risks from Geological Material

In the continuing search for new sources of industrial minerals, the possible health hazards from occupational exposure to mineral dusts have become an important factor in deciding what minerals can safely be exploited. More recently, concern has grown that exposure to various mineral dusts even in nonoccupational environments may cause disease or death. This concern has focused on asbestos dusts (or dusts perceived to be asbestos-like), radon, and zinc mining. Each of these is given below as examples, and each illustrates how inadequate comprehension about the nature and use of geological materials can lead to profound and quite unnecessary regulatory problems.

Asbestos

Asbestos is a nonscientific commercial term for certain silicate minerals that separate into flexible, heat-resistant, chemically inert fibers. Their particular mechanical, chemical, and especially thermal properties have been valuable in yarns, cloth, paper, paint, brake linings, tiles, insulation, fillers, filters, putty, and cement—applications requiring incombustible, nonconducting, or chemically resistant material. The world consumption of asbestos is about 4 million metric tonnes per year; that of the United States is disproportionately small, just over 80,000 metric tonnes. Of the six types of asbestos that have been mined commercially, only three are quantitatively important—a serpentine mineral called chrysotile and two amphibole minerals called amosite and crocidolite. The three other minerals, which are rarely used as asbestos, are the amphiboles—anthophyllite, actinolite, and tremolite.

Since 1972 the Environmental Protection Agency (EPA) has taken steps to limit human exposure to asbestos on the grounds that the fibers are carcinogens. This conclusion is based on epidemiological observations that former asbestos miners and fabricators experience an increased incidence of lung cancer and mesothelioma, a tumor of the chest cavity lining. Several mineralogists have questioned the EPA's classification of all these different minerals together and consideration of them all as equally hazardous. Recent medical research suggests that chrysotile asbestos is not hazardous outside the workplace—and perhaps not even there.

The serpentine mineral chrysotile is a layer silicate that curls into strong, flexible, needle-like fibers and is chemically and thermally stable because of a misfit between the layers. The overwhelming bulk of asbestos used for commercial purposes is chrysotile, mined principally in Canada. Close study has shown that the incidence of mesothelioma and lung cancer among townspeople in the major chrysotile-producing town, Thetford, Quebec, is about the same as among townspeople in Zurich, Switzerland. Thetford has pervasive chrysotile dust pollution, while Zurich is essentially free of it. From these and other data the environmental hazard of chrysotile asbestos is difficult to prove.

Crocidolite is an amphibole mineral having double chains of silicate tetrahedra linked by sodium, iron, and magnesium. It was mined principally in South Africa and Australia, but its high association with lung cancer incidence in former miners, millers, and fabricators has essentially terminated its use. Amosite is another amphibole mineral, but it lacks the significant sodium or iron found in crocidolite. It too was mined mainly in South Africa, but as with crocidolite the incidence of lung cancer among workers has ended its production and use.

Epidemiological studies have shown that the amphibole asbestos is much more dangerous than chrysotile asbestos; low-to-moderate exposure to chrysotile shows minimal effects on human health (e.g., *Report of the Royal Commission on Matters of Health and Safety Arising from the Use of Asbestos in Ontario*). However, EPA and other regulatory agencies have assumed that all forms of asbestos are equally dangerous.

Unfortunately, the arbitrary combination of needle-like shapes and a specified fine grain size has been chosen by regulators to characterize asbestos. Not only harmless chrysotile but also a wide variety of other fine-grained, needle-like minerals that are useful and probably of negligible hazard have been nearly banished from use. Minerals also are being judged guilty by association: there is no human health study that shows that the ubiquitous nonfibrous amphiboles are harmful. Nonetheless, the Occupational Safety and Health Administration has proposed, in pending federal regulations, that dust particles of the nonfibrous tremolite, actinolite, and anthophyllite amphiboles be classified as asbestos. A ban on the use of any serpentinite rock—of which noncarcinogenic chrysotile is one type—for commercial purposes is now pending in California; at the same time, serpentine is California's state rock. It should perhaps be noted that in 1969 the common mineral quartz was placed on California's list of carcinogens because tumors were generated in

rats breathing quartz dust, and the International Agency for Research on Cancer has since designated quartz as a carcinogen. Quartz, the crystalline form of silica, or SiO_2, is the second most common mineral found on the surface of the Earth; it is the major component of most sands.

In view of the force of the federal initiative with regard to mineral dusts, the prognosis for the continued vitality of the U.S. mining industry is ominous. Asbestos regulations and the fear of asbestos are having a crippling effect on the talc and vermiculite industry and on important segments of the mineral aggregate industry. In the future, operations that mine rocks in which amphibole, serpentine, and possibly even quartz are found could be terminated.

Radon

Radon is an invisible, odorless, radioactive, chemically inert gas resulting from the decay of radium, which is itself produced by the radioactive decay of uranium and thorium. There are three isotopes of radon. Two of the isotopes, one with a half-life of a minute and the other with a half-life of less than 4 seconds, decay rapidly to less mobile elements before they can escape from their sites of origin. The third isotope, with a half-life of about 4 days, is the environmental hazard.

Uranium and thorium are radioactive elements contained in minerals that are unevenly distributed within rocks and soils, and some of the minerals break down through weathering. These elements and their radioactive daughter products may be leached from rocks by groundwater and transported into soils. In ordinary soils radium is usually present at a concentration of a few parts per million. If the soil is impermeable, the radon may be trapped in the ground, where it creates no hazard. If the soil is permeable and is not water saturated, the radon daughter product, in the form of a chemically inert gas, may migrate away from its original location and into houses through their foundations. Although radon is the mobile step, it is the subsequent chemically active airborne decay products of radon—radioactive polonium, bismuth, and lead—that can lodge in lung tissues, causing damage through alpha particle emissions. Radon itself is inhaled and exhaled without significant buildup.

Estimates of risk from environmental exposure to radon, like those for asbestos, are extrapolated from excess mortality detected by epidemiological studies of uranium miners, but the extrapolations are highly uncertain. The EPA estimates a 1 to 5 percent chance of excess lung cancer from a lifetime exposure to radon at the danger level of 4 picocuries of radon per liter of air. The International Commission on Radiological Protection is less stringent, recommending a danger level of 15 picocuries per liter of air for existing structures. The EPA also claims that 20 to 30 percent of the houses monitored over the past 2 years had radon above the EPA danger level. If these numbers are correct and representative of the whole country, the environmental risk is enormous—but epidemiological studies done on the general population have not demonstrated excess risk. Nevertheless, there must a be a significant number of houses in the country containing radon at or above the present control level in uranium mines. These houses should be identified for remedial action.

High household radon levels can be generated from very-low-radium-bearing but very permeable soils. Conversely, high radium levels are not always dangerous if soil permeability is poor. A more direct approach to evaluating hazard is the simple and cheap measurement of household radon levels, as urged by the EPA. Such determinations should be made in those areas of homes that are occupied much of the time, instead of little-used basements and storage areas. There is already significant mandated compliance as more and more jurisdictions and mortgage lending institutions require that the radon level be determined—just as a termite inspection is required—before a house can be sold or mortgaged.

In summary, ongoing direct determinations of household radon levels make any further geological effort to identify high-risk areas redundant. The role of the earth science community is probably limited to urging more people to voluntarily monitor their household radon levels and to pointing out that, in the past, too much emphasis was placed on soil radium concentrations and too little on soil permeability. High-radium soils may not lead to radon hazards in the home, and low-radium soils do not guarantee safety.

Zinc

A final example of poorly thought-out control measures for potentially hazardous geological materials is the standard recently set by the EPA for zinc mill and mine effluent. The agency set the allowable zinc content of mill effluent at 0.2 parts per million and the allowable zinc content in mine effluent at 0.5 parts per million. Throughout the U.S. central zinc-mining districts, the ambient groundwater contains an average of about 1.5 mg of zinc per liter, or 1.5 parts of zinc per one million parts of water.

Not surprisingly, removal of zinc from the groundwater used as influent was not feasible. As a consequence, the affected mines and mills were shut down; cleanup to the standards set by the EPA is costing huge amounts of money; and zinc supplies must be sought elsewhere, often outside the United States. At the same time, zinc supplements are fed to livestock in those same districts to promote growth, development, and healing. Zinc deficiencies lead to skin lesions, retarded growth, hair loss, emaciation, and loss of appetite in livestock and humans. To meet the minimum requirement for zinc, a person would have to drink 100 liters of EPA-standard mine effluent per day.

Agriculture

Concentrations of population require vast networks to provide even minimal needs, including food and fiber. The production of these commodities in rural environments stresses natural equilibria. Agriculture annually moves more mass and influences a greater area on the Earth than all other human endeavors combined. The chemical input to the environment through agriculture is tremendous; as populations rise, agricultural activity increases, as do its chemical byproducts. During 1988 the use of fertilizer in the United States alone exceeded 10 million tons of nitrogen, 31 million tons of phosphate rock, and 4.6 million tons of potash. The rest of the world used several times those quantities. It is paradoxical that the success of solid-earth scientists in finding and developing potash and phosphate deposits for fertilizer manufacture has helped generate the problems of agricultural waste now being addressed by some of their colleagues. While domestic and industrial disposal of phosphate detergents is controlled to improve the quality of streams, agriculture puts 10 times the controlled amount of phosphate into the environment. Long-lived persistent organic-chemical pesticides and herbicides, and their trace contaminant dioxin enter the groundwater system daily. Tillage and overgrazing foster dramatically increased erosion by water and wind and are major contributors to desertification. The ancient technique of slash-and-burn clearing of tropical forests is front-page news today. The effects of wholesale forest burning are widely dispersed, and so are a challenge to evaluate. The practice is difficult to influence directly because so many individuals in developing countries are economically bound to agricultural activity. This simple method for clearing land enables a farmer to establish a foothold as other farming methods could not. The projected population increases over the next five or six decades probably will demand tremendous expansions in productivity, from more use of marginal croplands to genetic engineering of crop plants. Agriculture, the major human activity, is the major source of anthropogenic influence on the environment. The solid-earth scientist is involved with agriculture today in land-surface and soil research, in finding and developing raw materials for fertilizer, in surface and groundwater hydrology, and in waste isolation. As in so many other aspects of our science, there is a growing need for an integrated approach to the system and for handling different kinds of data together in geographic information systems. The variety of temporal and spatial scales involved in soil development affords a good example of the need for an integrated approach.

Soil Development and Soil Degradation

Human society is dependent on the abundance, fertility, and integrity of soils. Soils lie at the interface between the geosphere, biosphere, hydrosphere, and atmosphere. They have unique properties that derive from the intimate mixing of partly disintegrated rocks; dead organic matter; live roots and microorganisms; and an atmosphere high in carbon dioxide, nitrogenous gases, and moisture. Chemical reactions in soils link the geology and biology of the biogeochemical cycles that support life. Ions of potassium, sodium, calcium, magnesium, sulfur, and phosphorus are released from minerals by hydrolytic weathering, while carbon and nitrogen are released from dead organic matter by microbial digestion. The composition of the parent rock influences soil biota through ionic deficiencies or the release of elements that are toxic to microorganisms. Soil biota affect weathering rates through respiration, mechanical disruption, and production of organic acids.

Soils have changed throughout earth history. The oldest preserved soils have been interpreted as reflecting the existence of an atmosphere poorer in oxygen than that of today, and less ancient soils indicate responses to the evolution of land plants and soil faunas. However, these ancient changes hardly rival in speed or extent those that have occurred since agriculture began and the human population started to rise. Soil erosion has accelerated enormously, and a new process, soil degradation, has developed.

Soil erosion has become a major threat to humanity's ability to feed itself. In North America sediment production may be lessening as new equilibria are established by dams constructed in the upper reaches of watersheds, but this trend involving the loose particles that are weathered from rock is only part of the story. The deep, fertile, ancient soils that support abundant fields of waving grain are flushing away. The humus, litter, minerals, and particles that make up soil are separating and being redistributed into different reservoirs. Yes, new soils will form on deposits that do not flow into the ocean, but new soil develops by complex chemical and biological processes only over geological time scales. In terms of human lifetimes, soil is hardly a renewable resource.

Although soil erosion has long been identified as a serious global problem, soil degradation is now becoming recognized as comparable. Soils become degraded not by bodily removal but by selective depletion in nutrients, including water, or by the buildup of toxic materials, such as insoluble residues from fertilizer and pesticide or from salts after irrigation.

The vulnerability of the soils of a particular place to erosion or to one or more of the varieties of degradation depends on many variables. Surface slope, rainfall, vegetational cover, soil history, and agricultural practice are only a few of the factors that act together. Earth scientists need to work with agricultural scientists to understand the complex processes of soil degradation and soil erosion.

Waste Management

Advanced societies generate huge masses of diverse wastes that need to be minimized and then isolated in environmentally acceptable ways with a view toward possible reuse. At present, efforts in this direction are poor, as are witnessed by the dissension concerning all levels of nuclear wastes; by vagabond ocean-going garbage vessels vainly seeking a disposal site; by billions of discarded vehicle tires; and by polluted streams, lakes, and even seas.

An integrated approach is the key. Without it attempts to achieve a specific end frequently create unforeseen or unconsidered problems elsewhere. Planners and policy makers have come to appreciate that the efficient use of land and the protection of existing resources require respectful treatment of the natural systems enveloping the site. After concentrated exploitation, either reclamation must be undertaken or the site must be isolated in a way that will protect the unsuspecting from whatever toxic byproducts remain. In the United States, Congress has appropriated significant funds and stimulated innovative methods for cleaning up large-scale environmental problems resulting from prior activities. But the demand for such cleanup far exceeds the substantial resources already allocated. This is a political, legal, health, economic, emotional, educational, technological, and scientific arena. Science is essential for understanding the system and for estimating what might be the consequences of various courses of action, but society will act or not act on the basis of a broader range of considerations.

Earth scientists are called on to determine the potential suitability and durability of proposed waste repositories. Seismic disturbance, fluctuations of groundwater levels, and leakage should all be analyzed, and, where appropriate, provisions to counter them should be incorporated into facility designs. Understanding earth processes involves determining the interaction of earth materials with existing or evolving contaminants and the consequences of their integration. Earth scientists are thus well qualified to establish and verify the criteria for selection of waste repository sites and for the operation and strategic protection of such sites.

Five general regulatory categories of waste are recognized: solid, special, hazardous, low-level radioactive, and high-level radioactive. The solid category includes nontoxic debris of mixed nature, generally municipal wastes such as garbage and construction debris, and aqueous wastes, such as municipal sewage and sewage-treatment sludge. The special category embraces high-volume, low-toxicity wastes of predictable chemistry, including those of utility generation, dredge spoil, cement kiln dust, and other nontoxic industrial byproducts. The hazardous category refers to toxic wastes produced by industrial or mineral recovery activities and includes asbestos and asbestiform minerals. Low-level radioactive waste is composed of the residues of manufacturing, medical and industrial research, and components of nuclear power generation. High-level radioactive waste includes the residues of fuels for nuclear power generation and defense weapons production.

Every category of waste has a spectrum of physical and chemical properties and characteristics that affect the manner in which it should be managed. If not effectively managed, each waste type tends to produce an aqueous or organic liquid effluent, known generally as leachate. Leachate is released from bodies of wastes that are improperly disposed of into the environment; it is this contaminated

liquid that characteristically moves into the atmosphere and into groundwater and then migrates from disposal sites, all too often reaching human or other environmental receptors. At that point the waste can inflict serious public health or environmental damage. Remediation of the waste has been attempted. One common approach is removal of the contaminated material from one site to another whose characteristics might retard the movement of the contaminants. Other approaches include treatment, often of contaminated waters by a variety of "pump-and-treat" methods, while still others attempt to treat the contaminants in situ using microbial methods to neutralize certain toxic compounds.

Some parts—fortunately small ones, such as a number of older industrial plants and uncontrolled dump sites—of the planet are already so severely contaminated that no reasonable amount of time or effort can return them to a pure predisposal state. Other parts are eminently suitable for total reclamation. Probably the vast majority deserve a treatment somewhere between total abandonment and total reclamation. The earth sciences input is particularly critical for determining the most suitable course of action for each problem area. Blind demand for total reclamation in some instances is unreasonably expensive, unlikely to succeed, and probably more punitive than constructive for society as a whole.

A 1990 NRC report, *Rethinking High-Level Radioactive Waste Disposal*, suggested that the present regulatory structure in the United States may seriously inhibit attainment of the desired aims in the disposal of radioactive waste because the present structure requires procedures to be decided on before excavation begins. Experience has taught geological engineers that it is only during the preparation of a site for mining or waste disposal that the site can be fully characterized. The existing rules are inappropriate and cannot be expected to produce a successful outcome. Conversion of radioactive wastes to ceramics, which are then encased in additional outer containers of ceramic material, has been shown to be a satisfactory way of isolating waste and in many cases may be sufficient. Storage in accessible facilities in dry areas that are well above the groundwater surface and are surrounded by rock having ion-exchange fixation capability offers a second level of assurance.

The outstanding example of intensive site characterization is the current situation at the potential first high-level reactor waste isolation facility at Yucca Mountain, Nevada. On May 28, 1986—39 years after the first National Research Council study of the problem—this site was recommended by the secretary of the Department of Energy (DoE) and approved for detailed study. DoE subsequently prepared a site characterization plan, in accordance with the requirements of the Nuclear Waste Policy Act, to summarize existing information about geological conditions at the site, to describe the conceptual design, and to present plans for acquiring the necessary geological information. To date, detailed geological, seismic, and regional geophysical mapping has been accomplished, and 182 boreholes and 23 exploration trenches have been excavated within a radius of 9.5 km around the prospective site. A 1992 NRC study, *Ground Water at Yucca Mountain, How High Can It Rise?*, assessed the long-term outlook for this site and is a good example of the need to engage scientists with a broad range of geological and geophysical expertise. The report found no likelihood of a postulated environmental risk from future tectonic and hydrologic changes.

Contaminated Water, Air Pollution, and Acid Rain

Complex changes in water chemistry take place throughout the hydrologic cycle, from the time rain falls on the Earth to the time when it flows into the ocean. Some water moves quickly to streams, providing flood flows and moving large loads of sediment. Some moves through the soil and the shallow saturated groundwater system, sustaining the base flow of streams. Some water moves deeper into the crust, circulating for long periods. The deeper water becomes a major transportation system for mass and heat in the Earth and influences the complex chemical changes taking place within the crust.

Since the first irrigation projects, humans have affected the flow and chemistry of water as it circulated along the surface. During this century, such influence has greatly increased. Dams have been built on many of the streams of the world. Water quantity and quality are now seriously at risk, and societally generated contamination may be the most serious of water problems. The consequences of agricultural and resource-extraction practices threaten humans by contaminating the water supply and posing a threat to the natural environment by changing ecosystems.

Wastes also are released into the atmosphere, increasing particulate matter and adding greenhouse gases that prevent heat from escaping the atmosphere. Modifying the greenhouse effect has the potential to seriously disturb climate, which means a change in rainfall. Much of hydrology, as it has

been practiced, and many statistics are based on the assumption that the hydrologic cycle does not change within spans of decades to centuries, in contrast with geological time. A precipitously changing climate alters the basic assumption, making a large proportion of analyses suspect. A vacilating hydrologic cycle makes deeper scientific understanding of the basic phenomena much more important. Hydrologic instability, with more intense extremes such as floods and droughts, appears to be more likely during a period of change. Predicting future climate and the associated rainfall and runoff is a great challenge to science.

The atmosphere is composed principally of nitrogen and oxygen, with minor amounts of carbon dioxide and water vapor and traces of many other gaseous materials. The airborne waste products of civilization include gases such as sulfur dioxide and sulfur trioxide, nitrogen oxides, carbon dioxide, and ozone. Rainfall dissolves these gases, from the atmosphere, transporting them to the ocean. The sulfur and nitrogen species and the carbon dioxide are hydrolyzed to yield weakly acidic solutions that may still be substantially more acidic than ordinary surface waters. Eventually, these solutions will increase the acidity of the surface waters, especially in those lakes and rivers where the water acidity is poorly buffered by rocks and soils. The enhanced solubility of many major and trace constituents of ordinary rocks in acid waters releases unfamiliar elements and compounds into the water. These increasing amounts of rock-derived chemical constituents can threaten biological systems as well as the increasing acidity of the water.

Much sulfurous gas comes from electricity-generating plants that burn sulfur-bearing coal and from smelters processing sulfide ores. Coal contains sulfur both as sulfur-bearing organic compounds and as iron sulfides. Although it is possible to remove some of the iron sulfide prior to burning, the organic sulfur is an integral part of the coal. Corrective approaches for coal have relied on using the lowest-sulfur coals and on scrubbing the effluent gases. Corrections for smelter emissions have principally taken the form of prohibitory regulations, which have put the smelters out of business and led to transferral of the mining-smelting industry to other countries. Limited studies of alternatives to thermal smelting—such as dissolution mining or solvent extraction of finely crushed ores—have not yet yielded widely acceptable technologies, but they offer significant scientific and technical challenges as well as long-term economic, environmental, and national security incentives.

Although civilization's contribution to atmospheric acidity is a serious problem, our role can seem puny when compared with that of nature. Even modest volcanic eruptions can inject vast amounts of sulfur into the atmosphere very quickly, and they may remain airborne for several years. The sulfur disperses globally as sulfur dioxide in stratospheric aerosols and returns to the surface gradually. Scientific experience in observing such events has lately taken a leap forward with observations of the plume from the 1991 volcanic eruption of Mount Pinatubo in the Philippines. Understanding the global consequences of a major eruption is likely to be important for volcanology, for understanding waste emission, and for global change studies.

Atmospheric emissions and their consequences have been reviewed repeatedly by the NRC over the past decade, and with the passage of the Clean Air Act in 1990 it became clear that the national perspective on these matters was in the process of change. Hydrologic research was the topic of a recent full-scale assessment by the NRC. Solid-earth scientists are closely involved in hydrology, and their research dominates both surface drainage and underground water. Underground hydrology plays an important role because of the need to ensure the isolation of radioactive waste and untreatable toxic fluids, both of which are considered for deep disposal only. In the past, shallow groundwater was often studied as a separate resource, assumed to be isolated from deeper waters. Modern theory recognizes the importance of shallow and deeply circulating groundwater in the hydrologic cycle and thus in environmental pollution scenarios. For example, fluid contaminants introduced into a deep formation by a waste-injection well might someday reach the biosphere through shallow aquifers, unless precautions are taken in the design of such facilities and enough research is invested to understand the hydrogeological setting.

Much current research in groundwater hydrology and environmental engineering is centered on the study of contaminant transport in shallow aquifers, with respect to both fluid flow and chemical reactions. Significant progress has been made with computational and laboratory simulation of transport processes, although the problems of accurately predicting contaminant transport over human or geological time scales remain largely unsolved. In this light, studies of hydrothermal processes such as sediment diagenesis and ore mineralization may provide insight and natural analogs for verifying models of chemical transport.

There are problems of cleanup and prevention of contamination in both surface water and groundwater. These involve the transport of chemical constituents by water, often with simultaneous chemical reactions. It is essential that more robust and accurate analytical models be developed for predicting the chemical fate and transport of contaminants in groundwater. Certain contaminants can be chemically remedied by means of absorption or caging, and many of the chemical reactions are controlled by microorganisms. Understanding the kinetics of complex chemical reactions in a heterogeneous medium populated by microorganisms poses an exciting, though daunting, scientific challenge. That challenge must be met.

GLOBAL CHANGE

Recorded history affords but a brief glimpse of environmental change, one that fails to shed light on events and conditions that may confront us in the decades ahead. Global climates may soon be warmer than at any time during the past 1 million years, and sea level may stand higher than at any time during the past 100,000 years. Within decades, we may find ourselves in the midst of a biotic crisis that rivals the most severe mass extinctions of the past 600 million years. Nature is a vast laboratory that we can never manipulate or duplicate artificially on the scale necessary to test theories of global change. In developing plans for the future, we must therefore rely heavily on observations of nature itself.

Change is universal throughout the earth system, and has been since the origin of the Earth, but the term "global change" has come to be used much more narrowly to refer to human-induced changes affecting the atmosphere, hydrosphere, and biosphere. These changes are beginning to be recognized as likely to modify our environment in unprecedented ways within the next century. While many of the processes causing the changes are not unusual, it is an *accelerated rate* of change that human activities have introduced into the earth system. That accelerated rate of change may prevent normal adaptation mechanisms in the atmosphere, hydrosphere, lithosphere, and biosphere from working without major consequences—at least to humans. Public interest has focused mainly on global climatic change, but the scientific community recognizes that expected changes are much broader, extending to changes in such phenomena as sea level, groundwater quality, pollution, and biodiversity. Exactly what phenomena are included as elements of global change varies, depending on the breadth of the interests of the individual scientist or policy maker concerned.

Understanding Global Change

The worldwide scientific community has risen to the global change challenge. Over the past decade it has begun to address such questions as how the world is changing, what might happen in the next decades, and what might be done to avert possible unwanted changes. The problems are great because the world is always changing, and the differences we would like to be able to measure are very small, especially when seen against the background of general fluctuation.

The NRC has identified a set of issues and suggested possible lines of global change research (NRC, 1983, 1986). In a complimentary effort, NASA established the Earth System Science Committee (ESSC), which attempted a comprehensive assessment of how the earth system works and how satellite missions might address how it operates and how it might be changing (ESSC, 1988). Space provides the right environment from which to obtain an overall view of the planet's behavior, so it is not surprising that the NASA-sponsored study identified an important role for current and future measurements from space. Recognizing the scale and expense of the necessary programs, in the mid-1980s an interagency Committee on Earth and Environmental Sciences was established under the Federal Coordinating Council on Science, Education, and Technology with a working group on global change. In recent years, budget plans worked out through that structure have accompanied the president's budget when it has been presented to the Congress (e.g., *Our Changing Planet*, Office of Science and Technology Policy, 1992). These plans, which represent close cooperation among science planners in numerous agencies, reflect an integrated view of the earth system similar to that stressed in this report.

The federal government's planners have assigned the highest priority to the fluid parts of the earth system—clouds, ocean circulation, and biogeochemical fluxes—but solid-earth science research has not been neglected. Two of the sets of research priorities identified are of special importance to solid-earth scientists because they relate to earth system history and solid-earth processes. The latter category includes volcanic activity, sea-level change, coastal erosion, and frozen ground; earth system history focuses on paleoclimatology, paleoecology, and ancient atmospheric composition.

When the scale of the national effort that would be needed for addressing the problems of global change became clear in the mid-1980s, the NRC established the Committee on Global Change (now the Board on Global Change), which has issued a number of reports, of which *Research Strategies for the U.S. Global Change Research Program* (NRC, 1990) is of special interest because it has a section devoted to strategies for the study of earth history. Study of the most recent past is emphasized; the past 1,000 to 2,000 years, the earlier Holocene, the most recent glacial cycle, and the past few glacial cycles are all discussed as offering specific opportunities for important research into global change. Environments of extremely warm periods and climate-biosphere connections during abrupt changes are also considered as particularly likely to yield significant research findings.

Science planners in the United States have been closely involved with developments in the International Council of Scientific Unions (ICSU). An International Geosphere-Biosphere Program (IGBP) was set up under the ICSU in the mid-1980s to study global change, and the NRC's Board on Global Change acts as the U.S. National Committee for the IGBP. The IGBP defined initial core projects in 1990 (IGBP Report #12) that both supplement and complement U.S. initiatives. The international body has set up a project on past global changes; its "two-stream approach" emphasizes earth history over the past 2,000 years and the glacial-interglacial cycles in the Late Quaternary.

The scientific community has clearly defined the kinds of research needed to understand global change, and the Bush administration recognized the scale of the effort by submitting programs to Congress that amounted to roughly $1 billion a year. Because measurements from space are critical and because observation over a long time is essential, costs of about $2 billion a year over 20 years could be involved.

Mitigation and Remediation

The NRC has lately looked at the question of what the policy implications are of global change (*Policy Implications of Global Change*, 1991). Among its recommendations, some specific roles for solid-earth science research may be discerned—for example, "Make greenhouse warming a key factor in planning for our future energy supply mix. . . ." Action items include:

■ Encouraging broader use of natural gas. On both a national and a global scale, there is a clear opportunity not only for seeking out more natural gas supplies but also for applying geology and geophysics to efficient development and production.

■ Developing and testing a new generation of nuclear reactor technology. Understanding the solid earth will be important for nuclear fuel, waste isolation, and hazard assessment.

■ Accelerating efforts to assess the economic and technical feasibility of carbon dioxide sequestration from fossil-fuel-based generating plants. Efforts in "clean coal" technology have concentrated strongly on high-temperature combustion and related aspects. Solid-earth scientists can identify what coals from what areas can be efficiently used most in the new technology.

Other recommendations relate to forestry, agricultural research, and making the water supply more robust by coping with present variability. Solid-earth scientists through their understanding of the land surface, its drainage, and its groundwater clearly have much to contribute in all basic aspects of mitigation and remediation.

Three Roles for the Solid-Earth Scientist

There are three aspects of global change in which the solid-earth scientist is particularly involved. The first is understanding global change. The record of the past reveals both the extent and the pace of change. This information is useful not only in revealing the extent of past changes but also in showing how fast they have happened and in throwing light on how remote parts of the earth system have accommodated themselves to perturbations. Understanding the approaching changes will be easier if we understand what has already happened. Models of the future can be tested for validity by seeing how well they can reproduce past conditions. The past record is also important in helping to distinguish between anthropogenic change and what is sometimes called natural variability, although the involvement of the human race is hardly unnatural. The role of the solid-earth scientist in increasing understanding of global change is to document the background rates, ranges, and intensities of the environmental kaleidoscope without the factor of anthropogenic changes. That documentation can be obtained in the ways outlined in Chapter 3, The Global Environment and Its Evolution.

The second aspect of global change in which solid-earth scientists will play a part is assistance in reducing the extent of future change. Increasing the

world's reserves of natural gas, for example, could make more readily available a fuel that produces less carbon dioxide than most other forms of fossil hydrocarbon. Better understanding of aquifers worldwide could lead to informed management practices less likely to lead to pollution from waste disposal or to depletion of the water supply.

Because the world's population is so large and is continuing to increase, some kinds of global change are inevitable. More coal burning in populous India and China over the next 20 years can hardly be avoided and cannot fail to increase the carbon dioxide content of the atmosphere faster than plausible reductions in carbon dioxide output in the most advanced communities. Some consequent climatic change in the next 50 years appears unavoidable. Because of this prospective change, there is a third useful role for solid-earth scientists in global change research: study of how to mitigate and ameliorate the effects of possible changes. The following examples illustrate how such studies could be effective. Research in areas where sea level is rising today, such as the Gulf Coast where water extraction, reduced sediment supply, and other factors have induced rapid local subsidence, is likely to pay dividends if sea-level rise becomes more general in the next century. Research on the hydrology of arid lands—for example, on how transitions from desert to more moist conditions take place—can be carried out now in areas such as the Sahara-Sahel boundary. The results of such studies could prove useful if, as seems not unlikely, there are changes in the distribution of the present climatic zones of the continents in the next century.

At present, models of global change are not capable of suggesting the possible extent of changes in the future, partly because of limited spatial resolution and partly because they cannot accommodate all the controlling variables. Solid-earth scientists have the opportunity to work with other earth scientists, including social scientists, in testing and refining models of the earth system that will be needed for understanding global change sufficiently well to permit informed decision making. The roles of the solid-earth scientist in understanding change, managing change, and mitigating the effects of unavoidable change are all likely to increase in importance in the next decades.

RESEARCH OPPORTUNITIES

The problems and the research discussed in this chapter are concerned with the major interactions between the earth sciences and society and involve many aspects of the earth system that impinge directly on the quality of life. The problems affect the daily lives of the average American more than those discussed in any other chapter of this report. The Research Framework (Table 5.1) summarizes the research opportunities identified in this chapter with reference also to other disciplinary reports and recommendations. These topics, representing significant selection and thus prioritization from a large array of research projects, are described briefly in the following section.

The main topics are geological hazards and changes in the environment on global and local scales. Some changes occur naturally, and others are caused by the activities of society. Human beings are now the most significant geological agents. The topics are identified in Table 5.1 under Objective C, mitigating geological hazards, and Objective D, minimizing and adjusting to global and environmental change, but they also involve some aspects of Objective B, finding, extracting, and disposing of natural resources, and certainly depend on Objective A, understanding the processes in research themes I-IV. These themes extend from the surface into the interior and are all interrelated. Research projects commonly involve more than one theme, as was discussed in connection with the research opportunities for Chapters 3 and 4. In particular, the fluxes of material involved in the biogeochemical cycles (II), largely transported in fluids (III), have influenced changing paleoatmospheres and paleoceans (I).

Research conducted during recent decades on geological processes and the causes and mechanisms of geological hazards has provided unprecedented opportunities to reduce the potential for disaster, and the International Decade for Natural Disaster Reduction is concerned with converting the opportunities into science-wrought realities. High national priority has been attached to the comprehensive U.S. Global Change Research Program, which encompasses the full range of the earth system, and many research opportunities have been identified.

Objective C: To Mitigate Geological Hazards

Although understanding of hazards and the engineering capacity to control them are both growing, hazard losses continue to increase because this knowledge is not reflected in engineering design and in public regulation, private policies, and investment decisions. Achievement of the mitigation objective therefore requires not only continued basic research but also attention to matters such as social science and governmental issues. These issues are the concern of the U.S. program for the International Decade for Natural Disaster Reduction. They include such items as the following:

■ Scientific and engineering efforts aimed at closing critical gaps in knowledge to reduce loss of life and property.

■ Guidelines and strategies for applying existing knowledge.

■ Physical adjustments for avoiding the impacts of hazards, such as land-use planning, building site evaluation, building to withstand hazards, predicting occurrences, and preventing hazards.

■ Social adjustments for avoiding the social effects of hazards, including land-use controls and standards, public awareness campaigns, emergency preparedness programs, and financial arrangements to spread economic loss among a larger population.

■ Education of the public and of public officials to raise the level of awareness about how to plan for and respond to natural hazardous events.

■ Governmental issues, including the strengthening of communication links among federal officials and among federal, state, and local levels of government and the development of efficient lines of authority for decision making, especially in multiple-hazard events.

■ A concerted effort to identify research on any hazard that has applications to other hazards, which is very cost effective.

There is clearly commonality in issues related to geological hazards, and there is therefore a strong overlap between the research needs and opportunities listed under the following subheadings. There are extraordinary opportunities in geoscience for scientists and engineers to make valuable and lasting contributions to public welfare.

Earthquakes, Volcanoes, and Landslides

■ *Earthquake Prediction.* Both the prediction of events and the assessment of associated hazards emphasize cross-disciplinary research that draws on seismology, geodesy, geochemistry, hydrology, tectonics, and geomorphology. The most recent large events in the best-instrumented seismogenic regions are particularly important as sources of potential understanding and of new ways of thinking, such as that represented by a dynamical systems approach. There is a continuing need to improve dialogue about seismic risk between solid-earth scientists and engineers and decision makers, as well

TABLE 5.1 Research Opportunities

Objectives

Research Areas	A	B	C. Mitigate Geological Hazards—Earthquakes, Volcanoes, Landslides	D. Minimize Perturbations from Global and Environmental Change—Assess, Mitigate, Remediate
I. Global Paleoenvironments and Biological Evolution				■ Environmental impact of mining coal ■ Past global change ■ Catastrophic changes in the past ■ Solid-earth processes in global change ■ Global data base of present-day measurements ■ Climatic effects of volcanic emissions
II. Global Geochemical and Biogeochemical Cycles			■ Soil processes and microbiology	■ Earth science/materials/medical research ■ Biological control of organic chemical reactions ■ Geochemistry of waste management
III. Fluids in and on the Earth			■ Seismic safety of reservoirs ■ Precursory phenomena and volcanic eruptions ■ Volume-changing soils	■ Isolation of radioactive waste ■ Groundwater protection ■ Waste disposal: landfills ■ Cleanup of hazardous waste ■ New mining technologies ■ Waste disposal from mining operations ■ Disposal of spent reactor material
IV. Crustal Dynamics: Ocean and Continent			■ Earthquake prediction ■ Paleoseismology ■ Geological mapping ■ Remote sensing of volcanoes ■ Quaternary tectonics ■ Soil cohesion ■ Landslide susceptibility maps ■ Landslide prevention ■ Age-dating techniques ■ Real-time geology ■ Systems approach to geomorphology ■ Extreme events modifying the landscape ■ Geographic information systems ■ Land use and reuse ■ Hazard-interaction problems ■ Detection of neotectonic features ■ Bearing capacity of weathered rocks ■ Urban planning and underground space ■ Geophysical subsurface exploration ■ Detection of underground voids	
V. Core and Mantle Dynamics				

Facilities, Equipment, Data Bases

■ Global data base of present-day measurements for detection of future changes
■ Geographic information systems
■ Remote sensing from satellites
■ Dating techniques
■ Geophysical techniques for subsurface exploration in engineering
■ New mining technologies
■ New methods for fracture sealing
■ Methods to densify soils
■ Geophysical techniques for subsurface exploration
■ Global data base of present-day geochemical and geophysical measurements

Education: Schools, Universities, Public

■ Essential to devise ways to inform the public and policy makers about the scientific basis for understanding geological hazards and environmental changes as to become fully aware of how risk is perceived in the community at large. Many societies cannot afford earthquake-resistant construction, and even those that can retain large inventories of structures that are far below the state of the art in design and construction.

■ *Seismic Safety of Reservoirs*. Reservoir safety needs continued research on topics such as proximity to active faults, reservoir-induced seismicity, and the known seismicity of particular areas.

■ *Paleoseismology*. Development of data in the new subdiscipline paleoseismology has provided a sufficiently long historical sample to permit the identification of patterns and rates of occurrence of large earthquakes. For example, great earthquakes in intracontinental regions are now known to recur at intervals as long as several thousand years, and seismically quiescent periods between clusters of events on a given fault system may be hundreds of thousands of years long.

■ *Precursory Phenomena and Volcanic Eruptions*. Basic research in geology, petrology, geochemistry, and geophysics (see Chapter 2) is required in order to understand what triggers volcanic eruptions and how to predict them. Eruptions are generally preceded by seismic activity as volcanoes are reactivated, and the activity may include destructive earthquakes.

■ *Geological Mapping*. Mapping of potentially active volcanoes to determine their past eruptive history and their eruptive frequency, with the help of new dating methods, is an effective, low-cost way of reducing risk from volcanic eruptions themselves and from the associated devastating mudflows. Our present ability to predict along which of the radial directions from a volcano's center the brunt of explosive or mudflow destruction is likely to be directed is especially useful.

■ *Satellite-Based Surveying Systems and Remote Sensing*. These observation techniques offer great potential for monitoring the hundreds of potentially active volcanoes that have not been mapped and for measuring the increased temperatures associated with rising magma below the volcanoes.

■ *Quaternary Tectonics*. Fundamental research focusing on Late Quaternary history (roughly the past 500,000 years), including tectonic geology and geomorphology, paleoseismology, neotectonics, and geodesy, presents an unusual opportunity to catalog and understand ongoing active tectonic processes. This understanding could become the basis for assessing the potential to predict tectonic activities and other changes up to several thousand years in the future. Predicting or forecasting the future is

essential for designing ways to minimize the disastrous effects of earthquakes and volcanic eruptions and for selecting the most stable sites for long-term disposal of hazardous and radioactive wastes.

■ *Soil Cohesion*. Improved methods of densifying or increasing cohesion of soil materials can increase the shear strength of susceptible soils and thereby eliminate potential liquefaction during earthquake and other cyclic vibrational loadings.

■ *Volume-Changing Soils*. Much of the western United States is dotted with patches of land that are susceptible to significant changes in volume, activated by slight increases in moisture content. The changes, whether collapse or swelling, are sufficient to bring about structural damages to engineering works. Such damage is diffuse and rarely life threatening but amounts to as much as $2 billion per year. State coverage maps of areas with inherently weak soil fabrics or certain clay minerals could raise public awareness to the point that such damage could be nearly eliminated.

■ *Landslide Susceptibility Maps*. Maps for landslide detection and evaluation have been formulated by application of new concepts such as geographic information systems, a digital system of mapping. We can expect major strides in decision making as applied to human activities around landslides.

■ *Landslide Prevention*. Alternative methods of preventing landslides and of stabilizing active landslides are a prime area for research, for the purpose of reclaiming otherwise unusable land for high-value human use.

Geomorphic Hazards

■ *Age-Dating Techniques*. The greatest research need is for data management and physical models to quantify rates of active tectonic processes. Public policy decisions, for example, need such data arrays to determine what levels of earth hazard risks are acceptable. Decisions rest largely on evaluations of the rates (and the variability of the rates) of processes, the frequency of events, and the prediction of when hazards might become critical.

■ *Real-Time Geology*. Computers and electromechanical instrumentation now make possible the monitoring and instantaneous analysis of geological processes that occur very rapidly, and this capability leads to exciting new possibilities for prediction and forecasting. In addition, the documentation and understanding of rapid geological processes are an entirely new dimension for earth scientists to explore.

■ *Systems Approach.* A systems approach to geomorphic and engineering problems must be developed if serious and unforeseen consequences of human actions are to be avoided. The geomorphologist has the long-term perspective of landform change that is needed in meeting and dealing with many engineering and land management programs, and familiarity with the effects of time, landform evolution, and thresholds of instability are critical for prediction of future events and landform responses.

■ *Extreme Events.* The effect of extreme events in modifying or conditioning the landscape is a developing field. For example, rainfall-runoff floods and floods from natural-dam failures may be accompanied by massive erosion and sedimentation. A related field is the interdependency of events, as illustrated by the 1980 eruption of Mount St. Helens. The eruption and huge landslide formed several large lakes, and groundwater disruption became the cause of new instability of the natural dam. Sediment eroded from the debris avalanche in the Toutle River valley continues to plague downstream channel capacity, fish habitat, and water quality. A huge and expensive dam will be required to mitigate this problem.

Land-Use and Urban Planning

■ *Geographic Information Systems (GISs).* The way in which solid-earth science information is used with other kinds of societally important information requires the information-layering capability that is the essence of GISs. The solid-earth scientist has an important role in establishing appropriate standards for the way in which specialized data are used in GISs, as well as to ensure that such information is up to date and of high quality.

■ *Land Use and Reuse.* The effects of geological constraints on land use and reuse will continue to be a significant issue for the future. Society must become a deferential part of the environment, not its master; human actions have profound and far-reaching impacts on the rest of the earth system.

■ *Hazard-Interaction Problems.* A shift in perspective is needed from the incrementalism of individual site-specific hazards to the broader systems approach, dealing with single or multiple hazards common to broader physiographic regions. Increasingly, earth scientists, civil engineers, land-use planners, and public officials are noting the existence of interactive natural hazards that occur simultaneously or in sequence and that produce synergistic cumulative impacts that differ from those of their separately acting component hazards.

■ *Detection of Neotectonic Features.* Identifying these features, which are key data links to a better understanding of a long-term seismicity, requires geological mapping of each seismically active region of the country. Such assessments are critical to the siting and design of safe facilities that are of critical-performance nature or serve housing-dependent populations (hospitals and schools) and to the establishment of realistic building codes.

■ *Soil Processes and Microbiology.* New theories of the relations between microbiology and soil processes, which are important because food production is tied to soil science, can be established with new chemical and instrumental techniques. Large-scale remote sensing can monitor growth patterns on different soil types and enhance land management to preserve our precious soil bank.

■ *Rock-Bearing Capacity.* Research on the bearing capacity of the weathered rock zone together with better definition, identification, and classification of strong soil/weak rock could save considerable money, both in design and construction of buildings.

■ *Urban Planning and Underground Space.* If we are to unclog our cities with the limited financial resources at hand, planning must start to take into account utilization of the subsurface. The use of underground space is an undeniable necessity of the future, to be undertaken as population centers are redeveloped. There is only so much room on the land, and the prime space must be used for human habitation. With the predicted increase in usage of underground space, new technologies for characterization and sealing of rock fractures against groundwater inflow will become a necessity.

■ *Geophysical Subsurface Exploration.* Improved geophysical techniques for subsurface exploration and computer-based methods for data processing offer new opportunities to predict and control human-induced land subsidence. Techniques such as ground-penetrating radar are now commonly used to detect underground cavities, such as caverns and abandoned mines at shallow depths, and seismic tomography is being experimented with to evaluate the conditions of pillars in abandoned mines of such urban areas as Pittsburgh, Birmingham, and Kansas City.

■ *Underground Void Detection.* The detection of underground voids by indirect methods needs to be greatly improved. Many subsidence problems are associated with either carbonate-solution caverns or abandoned underground mines, and it is usually impractical to drill out suspect areas. Prediction of limestone sinkhole collapse will remain elusive until the triggering mechanisms are better understood.

Objective D: To Minimize Perturbations from and Adjust to Global and Environmental Change

Mineral Products and Health

■ *Interdisciplinary Earth Science/Materials/Medical Research*. Interdisciplinary research in these apparently diverse fields can provide a substantial opportunity to address major societal issues regarding health. This involves research into and public education about acceptable levels of risk in relationship to costs. Three examples are (1) the "asbestos hazard" where the relation between crystal size, shape, structure, and composition and chemical/biological interaction is particularly important; (2) the radon concern; and (3) the presence and hazard to humans of trace elements present in the environment, especially if they may interact.

Contamination of Aquifers

■ *Radioactive Waste Isolation*. To ensure the isolation of radioactive waste, which has been designated by the Congress to deep disposal only, hydrologic research on sedimentary and volcanic deposits plays an important role. Sedimentary deposits contain aquifers, which are the single most important reservoir for fresh water.

■ *Groundwater Protection*. Protecting groundwater quality requires monitoring and sampling, such as vadose zone sampling, sampling of volatile organic components in groundwater, and sampling volatile organic components that might affect employee safety. Improved methods for testing field permeability and in situ methods of measuring the physical and chemical properties of soil and rock are needed.

■ *Organic Chemistry Control*. Biological control of organic chemical reactions, especially in groundwater, is a rich area for research—research with a high potential monetary payoff in terms of current and ongoing expenditures for waste site remediation. Chemical reactions involving the fate of in situ organic compounds are controlled by soil microorganisms, and understanding the kinetics of such reactions poses a scientific challenge. Introducing an organism that produces a benign byproduct may be the cheapest way to remediate some forms of groundwater contamination. In many instances

groundwater moves so slowly that colonies of microorganisms can actually move with the plume of contamination.

Waste Management and Environmental Change

■ *Waste Management: Landfills*. The earth science community is equipped to develop management expertise for those materials that must be consigned to land burial.

■ *Waste Management Geochemistry*. A new field of geochemistry of waste management can be produced by recycling metals from waste.

■ *Hazardous Waste Cleanup*. In situ cleanup of uncontrolled hazardous waste sites vitally needs attention. The importance of fracture sealing of radioactive waste repositories in the subsurface cannot be overemphasized. Future generations should not be left with isolated pockets of hazardous material whose limits will undoubtedly become unknown with time.

■ *Waste Disposal from Mining Operations*. The problems with disposal of the waste from mining, milling, and in situ leaching need much greater emphasis, both for better understanding and, where possible, for mitigation of the potential adverse effects of exploiting and utilizing energy resources. Research and technological development are needed. Also, methods for site selection, characterization, and design of proposed repository sites for radioactive wastes have still not been developed.

■ *Coal Mining and the Environment*. The environmental impact of mining coals and their environmental acceptability as fuel are the most important aspects of coal research. Every stage of coal development, from exploration to consumption, has potential environmental impacts. Mine safety and mine site rehabilitation will also require research.

■ *Nuclear-Energy-Related Wastes*. Disposal of spoil waste at mine sites and of spent reactor material is the most important research area for nuclear energy. Recent concern about atmospheric carbon dioxide buildup, as well as about the increase in acid rain resulting partially from coal usage, may lead to a reevaluation of policies relating to nuclear power.

■ *Global Change*. The solid-earth scientist is the custodian of the past record of global change, and the research areas recommended in Chapter 3 are related to that record. Here the emphasis is on the most recent past.

■ *Past Global Change*. The record in ice cores, tree rings, deep-sea cores, lake-bed deposits, and surface features like sand dunes and glacial deposits is generally the most complete record of the recent past, but other environments preserve some information. Integrated pictures of what the world was like at specific times—for example 5,000, 10,000, or 20,000 years ago—such as those compiled by the CLIMAP and COHMAP consortia can help in understanding how the world has changed. They can also be used to test the usefulness of the models constructed by geophysical fluid dynamicists.

■ *Abrupt and Catastrophic Changes in the Past.* Episodes of sudden environmental change have been identified from times as recent as 10,000 years ago to 66 million years ago and more. If we can work out what happened under the peculiar circumstances of those times, it may help in understanding the likely effects of the current changes.

■ *Solid-Earth Processes in Global Change.* We need a better understanding of how sea level might be changing at present and the effects of volcanism, both above and below sea level, on the global environment. The volume of the ice sheets should be monitored on the decadal scale, and changes in permafrost areas should be observed. Monitoring global change at the surface from space, using the ongoing high-spatial-resolution imagery of Landsat and SPOT-type instruments, is essential for seeing what is happening in such environments as the Sahel, central Asia, and the Altiplano.

■ *Present-Day Geochemical/Geophysical Data Base.* A global data base of present-day geochemical and geophysical measurements needs to be established for direct detection of future changes. Such data are lacking for remote and underdeveloped regions of the world.

■ *Climatic Effects of Volcanic Emissions.* The effect on climate modification of volcanic emissions, and their interactions with the atmosphere and hydrosphere, can be better understood from global monitoring of volcanism.

BIBLIOGRAPHY

NRC Reports

NRC (1978). *Studies in Geophysics: Geophysical Predictions*, Geophysics Study Committee, Geophysics Research Board, National Research Council, National Academy Press, Washington, D.C., 215 pp.

NRC (1983). *Fundamental Research on Estuaries: The Importance of an Interdisciplinary Approach*, Geophysics Study Committee, Geophysics Research Board, National Research Council, National Academy Press, Washington, D.C., 79 pp.

NRC (1983). *Toward am International Geosphere-Biosphere Program*, National Academy Press, Washington, D.C.

NRC (1983). *Safety of Existing Dams: Evaluation and Improvement*, Water Science and Technology Board, National Research Council, National Academy Press, Washington, D.C., 384 pp.

NRC (1984). *Studies in Geophysics: Groundwater Contamination*, Geophysics Study Committee, Geophysics Research Board, National Research Council, National Academy Press, Washington, D.C., 179 pp.

NRC (1984). *Studies in Geophysics: Explosive Volcanism: Inception, Evolution, and Hazards*, Geophysics Study Committee, National Research Council, National Academy Press, Washington, D.C., 176 pp.

NRC (1985). *Safety of Dams: Flood and Earthquake Criteria*, Water Science and Technology Board, National Research Council, National Academy Press, Washington, D.C., 224 pp.

NRC (1986). *Global Change in the Geosphere-Biosphere: Initial Priorities for an IGBP*, U.S. Committee for an International Geosphere-Biosphere Program, Commission on Physical Sciences, Mathematics, and Resources, National Research Council, National Academy Press, Washington, D.C., 91 pp.

NRC (1986). *Studies in Geophysics: Active Tectonics*, Geophysics Study Committee, Board on Earth Sciences and Resources, National Research Council, National Academy Press, Washington, D.C., 266 pp.

NRC (1987). *Confronting Natural Disasters: An International Decade for Natural Hazard Reduction*, Advisory Committee on the International Decade for Natural Hazard Reduction, National Research Council, National Academy Press, Washington, D.C., 60 pp.

NRC (1987). *Recommendations for the Strong-Motion Program in the United States*, Committee on Earthquake Engineering, National Research Council, National Academy Press, Washington, D.C., 59 pp.

NRC (1987). *Responding to Changes in Sea Level: Engineering Implications*, Marine Board, National Research Council, National Academy Press, Washington, D.C., 148 pp.

NRC (1988). *Space Science in the Twenty-First Century: Imperatives for the Decades 1995 to 2015: Mission to Planet Earth*, Task Group on Earth Sciences, Space Science Board, National Research Council, National Academy Press, Washington, D.C., 121 pp.

NRC (1988). *Estimating Probabilities of Extreme Floods: Methods and Recommended Research*, Water Science and Technology Board, National Research Council, National Academy Press, Washington, D.C., 141 pp.

NRC (1989). *Reducing Disasters' Toll: The United States Decade for Natural Disaster Reduction*, Advisory Committee on the International Decade for Natural Hazard Reduction, National Research Council, National Academy Press, Washington, D.C., 40 pp.

NRC (1989). *Margins: A Research Initiative for Interdisciplinary Studies of Processes Attending Lithospheric Extension and Convergence*, Ocean Studies Board, National Research Council, National Academy Press, Washington, D.C., 285 pp.

NRC (1989). *Irrigation-Induced Water Quality Problems*, Water Science and Technology Board, National Research Council, National Academy Press, Washington, D.C., 157 pp.

NRC (1989). *Great Lakes Water Levels: Shoreline Dilemmas*, Water Science and Technology Board, National Research Council, National Academy Press, Washington, D.C., 167 pp.

NRC (1990). *Assessing the Nation's Earthquakes: The Health and Future of Regional Seismograph Networks*, Committee on Seismology, Board on Earth Sciences and Resources, National Academy Press, Washington, D.C., 67 pp.

NRC (1990). *Rethinking High-Level Radioactive Waste Disposal: A Position Statement of the Board on Radioactive Waste Management*, Board on Radioactive Waste Management, National Research Council, National Academy Press, Washington, D.C., 38 pp.

NRC (1990). *Research Strategies for the U.S. Global Change Research Program*, Committee on Global Change, U.S. National Committee for the IGBP, National Research Council, National Academy Press, Washington, D.C., 291 pp.

NRC (1990). *Studies in Geophysics: Sea-Level Change*, Geophysics Study Committee, Board on Earth Sciences and Resources, National Research Council, National Academy Press, Washington, D.C., 234 pp.

NRC (1990). *A Safer Future: Reducing the Impacts of Natural Disasters*, U.S. National Committee for the Decade for Natural Disaster Reduction, National Research Council, National Academy Press, Washington, D.C., 67 pp.

NRC (1990). *Ground Water Models: Scientific and Regulatory Applications*, Water Science and Technology Board, National Research Council, National Academy Press, Washington, D.C., 302 pp.

NRC (1990). *Managing Coastal Erosion*, Committee on Coastal Erosion Zone Management, Water Science and Technology Board, Marine Board, National Research Council, National Academy Press, Washington, D.C., 182 pp.

NRC (1990). *Ground Water and Soil Contamination Remediation: Toward Compatible Science, Policy, and Public Perception: Report on a Colloquium*, Water Science and Technology Board, National Research Council, National Academy Press, Washington, D.C., 261 pp.

NRC (1990). *Surface Coal Mining Effects on Ground Water Recharge*, Water Science and Technology Board, National Research Council, National Academy Press, Washington, D.C., 159 pp.

NRC (1991). *Policy Implications of Greenhouse Warming*, Committee on Science, Engineering, and Public Policy, National Research Council, National Academy Press, Washington, D.C., 127 pp.

NRC (1991). *Real-Time Earthquake Monitoring: Early Warning and Rapid Response*, Committee on Seismology, Board on Earth Sciences and Resources, National Research Council, Washington, D.C., 52 pp.

NRC (1991). *Toward Sustainability: Soil and Water Research Priorities for Developing Countries*, Committee on International Soil and Water Research and Development, Water Science and Technology Board, National Research Council, National Academy Press, Washington, D.C., 65 pp.

NRC (1991). *Opportunities in the Hydrologic Sciences*, Committee on Opportunities in the Hydrologic Sciences, Water Science and Technology Board, National Research Council, National Academy Press, Washington, D.C., 16 pp.

NRC (1992). *Ground Water at Yucca Mountain: How High Can It Rise?*, Board on Radioactive Waste Management, National Research Council, National Academy Press, Washington, D.C., 231 pp.

NRC (1992). *Restoration of Aquatic Ecosystems: Science, Technology, and Public Policy*, Water Science and Technology Board, National Research Council, National Academy Press, Washington, D.C., 485 pp.

Other Reports

Hanks, Thomas C. (1985). *The National Earthquake Hazards Reduction Program—Scientific Status*, U.S. Geological Survey Bulletin 1659, U.S. Government Printing Office, Washington, D.C., 40 pp.

Ziony, J. I., ed. (1985). *Evaluating Earthquake Hazards in the Los Angeles Region—An Earth Science Perspective*, U.S. Geological Survey Professional Paper 1360, U.S. Government Printing Office, Washington, D.C., 505 pp.

Earthquake Engineering Research Institute (1986). *Reducing Earthquake Hazards: Lessons Learned from Earthquakes*, Earthquake Engineering Research Institute, El Cerrito, California, 208 pp.

Decker, R. W., T. L. Wright, and P. H. Stauffer, eds. (1987). *Volcanism in Hawaii*, U.S. Geological Survey Professional Paper 1350, U.S. Government Printing Office, Washington, D.C., 1,667 pp.

ESSC (1988). *Earth System Science: A Program for Global Change*, Earth Systems Sciences Committee, NASA Advisory Council, National Aeronautics and Space Administration, Washington, D.C., 208 pp.

Wallace, R. E., ed. (1990). *The San Andreas Fault System, California*, U.S. Geological Survey Professional Paper 1515, U.S. Government Printing Office, Washington, D.C., 283 pp.

The International Geosphere-Biosphere Programme: A Study of Global Change (IGBP) of the International Council of Scientific Unions (1990). *The Initial Core Projects*, IGBP Report No. 12, International Council of Scientific Unions, Stockholm, 232 pp. plus appendices.

The International Geosphere-Biosphere Programme: A Study of Global Change (IGBP) of the International Council of Scientific Unions (1992). *The Pages Project: Proposed Implementation Plans for Research Activities*, Pages Scientific Steering Committee, 110 pp.

Office of Science and Technology Policy (1992). *Our Changing Planet: The FY 1992 U.S. Global Change Research Program*, Committee on Earth and Environmental Sciences, 90 pp.

Wright, T. L., and T. C. Pierson (1992). *Living with Volcanoes*, U.S. Geological Survey Circular 1073, U.S. Government Printing Office, Washington, D.C., 57 pp.

6

Ensuring Excellence and the National Well-Being

ESSAY

The previous chapters have documented the tremendous opportunities that exist in the solid-earth sciences, in terms of both greater scientific understanding and potential benefits to society. This chapter examines the educational, material, informational, and institutional resources that will be needed to ensure excellence in the earth sciences and thus maximize the scientific and societal returns. Funding for the solid-earth sciences and a prioritization of basic and applied research activities are discussed in Chapter 7.

Given the many pressing problems that involve the earth sciences, excellence among earth scientists is more important now than ever before. But scientific excellence does not happen by chance. It requires an adequate number of well-educated scientists, a strong scientific infrastructure, careful planning, and sustained effort—or, stated more directly, better training, better instruments and facilities, better access to information, and strong management. The leadership necessary to achieve these ends must come from geoscientists themselves, from government, and from industry. Investments in the long-term health of the profession must be made today, and planning must begin now to address problems that can be foreseen for the years ahead.

Scientific excellence also requires a social environment that nurtures the spirit of inquiry, is enthusiastic about innovation, and respects curiosity and intellectual talents. Instincts for inquiry, innovation, and curiosity are inherent in human nature, it is true, but experience shows that although science thrives in some situations it languishes in others. Natural instincts can be nurtured or neglected. Today's society cannot afford to be neglectful.

Geologists have played a pivotal role in societal growth and health for the past century. Through the efforts of geologists and other earth scientists, great deposits of underground water and mineral and energy resources have been found and made available. The composition and dynamics of the solid earth have been explored, leading to insights of scientific, aesthetic, and economic value. Links with the defense community have been forged, organized around a common interest in the nature of the Earth and a concern over ensuring adequate supplies of strategic materials.

If advances in the solid-earth sciences are to continue to occur at the rate demanded by societal needs, the profession must have access to adequate resources. The first and most important such resource is a sufficient number of well-qualified professionals engaged in studying the Earth and applying their knowledge about it. According to a recent survey, there are approximately 80,000 individuals who can be classified as solid-earth scientists in the United States. Over half of these people are employed by the petroleum and mining and minerals industries. Government employs about 12,000 geoscientists, and about 10,000 work in academia.

The supply of and demand for earth scientists have historically been out of phase, and that remains true today. Because of the dramatic decline in petroleum prices in the mid-1980s, following aggressive hiring by the industry during the 1970s, petroleum-related hiring decreased by about a third. The mining industry was also in a long period of depressed commodity prices, which led to the loss of jobs by thousands of mineral resource earth scientists. Decline in employment in the extractive industries was concurrent with increased demand in other areas. Environmental legislation in the early 1980s dealing with waste disposal sites was enacted and enforced. Employment projections indicate that opportunities in the earth sciences are growing again, with emphasis on issues of groundwater, the siting of waste repositories, and environmental cleanups. Because of this shift, the retirement of current earth scientists, and the eventual recovery of the oil and gas industry, the demand for solid-earth scientists can be expected to continue to grow.

Undergraduate enrollments in the earth sciences have tracked the ups and downs of employment in the extractive sector. The prospects for a recovery of enrollments are cloudy despite the changing emphasis of the field toward environmental issues. In general, the college-age population has been declining in the United States, and declining numbers of freshmen express an interest in science in general and the earth sciences in particular. Also, women and minorities are greatly underrepresented in the earth sciences, and these are the groups that will make up nearly 70 percent of new entrants into the work force in the 1990s. Growth in the enrollment of women, at both the graduate and undergraduate levels, has been increasing in the past decade.

Much of the reason for lack of interest in the earth sciences among college freshmen is that relatively few of them are taught earth sciences in elementary and secondary schools. The resulting paucity of teaching opportunities has contributed to a corresponding shortage of qualified earth science teachers. If more precollege students are to be exposed to the opportunities and rewards of earth sciences, these subjects must be taught more widely, by a larger number of better-prepared teachers, and in a more exciting manner.

At the undergraduate level, attention must focus on introductory geoscience courses, on courses for future earth science teachers, and on the preparation of earth science majors. Introductory courses must be among the best that an earth science department offers, because a significant fraction of majors will emerge from this group and because, for most of the other students, this will be their only formal exposure to earth sciences. Earth science departments must also collaborate with education departments in designing programs for future precollege earth science teachers, because the best-prepared teachers are those who have completed most, if not all, of the courses required of earth science majors.

For both undergraduate and graduate earth science students, flexibility, versatility, and a firm foundation in such allied sciences as mathematics and computer science are crucial. All of these students should be involved in research under the guidance, but not strict direction, of teachers. Fundamental principles must be emphasized, because a narrow focus on job training will eventually require substantial levels of retraining as national needs change.

A number of steps can be taken to strengthen precollege, undergraduate, and graduate education in the earth sciences, including greater involvement of professional societies in education, government programs directed at science and mathematics education, and the efforts of individual earth scientists who resolve to pass on their knowledge and enthusiasm to others. Not only will these initiatives strengthen the earth science field, but they also will promote greater public awareness of earth sciences, with corresponding benefits for public decision making and public policy.

However, even if earth science education in the United States is thoroughly reformed, personnel shortages are likely to occur. Given that likelihood, geoscientists must be ready to modify their activities to remain as productive as they have been.

One possible way to do this is to make the fullest possible use of modern instrumentation. In recent years, technological progress has transformed the analytical tools available to earth scientists. Satellite measurements, high-pressure experimental instruments, microanalytical techniques, digital seismometers, and high-performance computers have greatly expanded the range and sensitivity of modern instrumentation.

The development of ever more sophisticated (and usually more expensive) instruments has raised a number of difficult questions in the earth sciences. How best can small groups of researchers gain access to instruments that are too expensive for a single group, or even an earth science department, to afford? What is the optimum division of support between individual researchers and the infrastructure needed to answer vanguard research questions? How should instruments be operated and maintained once they have been acquired?

The course of science is largely unpredictable, but the needs of a discipline for instruments and facilities usually must be planned in advance. Chapter 7 looks at a number of highly promising research areas and touches on the instruments and facilities that will be needed to advance in those areas.

Over the years the use of these various instruments and facilities has resulted in enormous amounts of information becoming available to solid-earth scientists. Maps, text, physical samples, aerial and space-based imagery, well logs, and seismic data are all just small parts of the array of data from which earth scientists can draw.

Over the past several decades, the acquisition, retention, dissemination, and use of data have been undergoing a fundamental change because of the advent and rapid development of the computer. Today, an ever-increasing fraction of the data used in science is in digital form. Digital data can be readily stored, accessed, distributed, edited, and presented in various forms. Even more important, digital data can be processed, analyzed, modeled, and evaluated quickly, automatically, and quantitatively.

However, the onslaught of digital data in the earth sciences threatens to overwhelm more traditional methods of data management. Coordination within the profession in the areas of retention and distribution of data is now limited. Incompatible data formats, lack of knowledge about the existence of data, proprietary and national security concerns, and the lack of centralized archives all potentially limit the use of data in solving important problems.

Greatly improving the availability and utility of earth science data requires a national earth science data policy or set of guidelines dealing with a wide range of issues. One important element of this policy should be the establishment of a distributed national data system built on existing data centers as appropriate. A policy should also deal with issues of digitization of original data, incentives for data retention and dissemination, data-base standardization, exchange formats, the provision of data directories, research into data systems, and the training of students and professionals in data management.

Questions of data management also extend internationally. Study of the Earth is intrinsically global. The earth sciences, by nature, have always had a global orientation, but an opportunity is now at hand for global activities that will advance the earth sciences in unprecedented ways.

International collaboration in the earth sciences takes place both through formalized programs and individual scientists working with colleagues from other countries. In addition, many international contacts take place through the overseas operations of multinational corporations. These interactions should not only be continued but also increased, given the ever more global orientation of much work in the earth sciences.

An emerging trend that will inevitably influence the earth sciences in the years ahead is the growing awareness that problems long considered to be local are really global problems. Humans are altering the environment at an ever-increasing rate. Anthropogenic effects include erosion of the land, deforestation, pollution and exhaustion of water resources, destruction of atmospheric ozone, and changed composition of the atmosphere. Understanding and altering these trends demand international cooperation and planning. The earth sciences, through emphasis on international research and communication, can help show the way toward this new era of global cooperation.

ROLES, NUMBERS, AND BACKGROUNDS OF SOLID-EARTH SCIENTISTS

Civilization has reached a critical stage on this planet. Earth scientists, because of their training, must play a prominent role in solving some of society's most pressing problems, even more than they have in the past. They must help locate adequate energy and mineral resources; safely dispose of toxic chemical and radioactive wastes; contribute to responsible land use for an expanding and increasingly mobile world population; maintain safe and adequate water supplies; and reduce the danger from volcanoes, earthquakes, landslides, and other natural hazards.

Future roles for earth scientists will be even more important. Not only will geologists be called on to help find new resources, they also will play an essential role in assessing the consequences of using those resources. In this way, understanding of the Earth will provide an essential underpinning for sound public policy. The earth science community must be able to pursue research in response to demands from policy makers, and it must do so on a schedule that anticipates societal imperatives and facilitates transfer of scientific information to decision makers and the general public.

As in other areas of science, the specific roles of earth scientists have become increasingly specialized as the field has evolved. Table 6.1 gives a breakdown of the major groupings within the solid-earth sciences. Other breakdowns are possible, and individual specialties can be further divided. In addition, many solid-earth scientists work in areas that span two or more specialties, and many problems must be addressed through the combined contributions of more than one specialty.

The earth science community still lacks a full understanding of the many factors that contribute to the supply and demand for personnel in specific specialties. **Analyses of the factors affecting the work force should continue, both through surveys and broader studies of the need for scientific and technical personnel in the United States.** Once the dynamics of supply and demand are better understood, the profession can begin to address its education, training, and hiring mechanisms with a view toward reducing these imbalances.

In addition, **more detailed and accurate data will be needed to assess personnel needs and trends in the solid-earth sciences**. The National Science Foundation (NSF) could become a much

TABLE 6.1 One of Many Classifications of Disciplinary Specialities Within the Solid-Earth Sciences

Economic and mining geology
Engineering geology and engineering geophysics
Environmental geology
General geology and earth sciences
Geochemistry
Geomorphology
Geophysics
Hydrology and hydrogeology
Marine geology/marine geophysics
Mathematical geology and geostatistics
Mineralogy
Paleobiology
Petroleum geology
Petroleum geophysics
Petrology
Planetary geology
Sedimentology
Soil geology
Stratigraphy
Structural geology and tectonics
Volcanology

SOURCE: American Geological Institute (1987).

more valuable source of data if it recognized the geosciences as an independent discipline, with appropriate distinctions made between specialties in the earth sciences. The professional societies, which are conducting an increasing number of surveys, are another valuable source of information. Their activities should be coordinated, perhaps through an independent body such as the Commission on Professionals in Science and Technology.

Solid-Earth Sciences and National Security

In addition to their civilian activities, a subset of earth scientists have forged close links with the defense community over the past half century because of a common interest in understanding the operation of the earth system, or parts of the system, on varied spatial and temporal scales. For example, mapping of the magnetic stripes on the ocean floor, which was important for submarine operations, provided information that enabled geophysicists to establish how the ocean floor had evolved over the past 170 million years. Also, installation of the World-Wide Seismic Network in the early 1960s was related to a need to discriminate between underground nuclear tests and natural earthquakes. This network permitted identification of the precise location of earthquake epicenters, so the sharp boundaries of tectonic plates were soon recognized. These data were used by the founders of plate tectonics to show how the lithosphere forms and evolves, thus radically changing the character of the earth sciences.

In return, earth scientists have provided the defense community with sophisticated methods to detect underground nuclear testing, to map the Earth's magnetic field to assist submarine navigation, and to gauge the Earth's gravity field to assist in satellite tracking and navigation—to name but three contributions. Today, the commonality of interest extends from the Department of Defense's involvement in new and more sophisticated seismic networks (which will contribute to growing understanding of the structure of the deep interior) to detailed mapping of the topography of the ocean floor and determination of the gravity field from satellite altimetry.

The interests of the earth science community and the defense community often overlap, but they by no means always coincide. For example, studies of altimetry from the Sea Satellite (SEASAT) and the exact repeat mission orbits of the U.S. Navy's Geodetic Satellite (GEOSAT) that repeated the SEASAT orbits are throwing light on the structure of the ocean floor. But more information on scientifically interesting topics, such as the structure beneath active hot spots and hot-spot tracks, is not currently available to the earth science community, although it exists in data from the GEOSAT orbits that lie between the SEASAT orbits. These data are, however, being declassified for much of the Southern Hemisphere, which is a welcome development. The Navy has released the full data set for regions south of 30°S and is considering release of data for the entire Southern Hemisphere.

Differences of this kind are unavoidable. But as the recent declassification of topographic data from the Exclusive Economic Zone illustrates, efforts to bring scientifically important information into the scientific domain can be successful. Maintenance of dialogue between the two communities is vital.

Another way that solid-earth scientists interface with national security concerns is with raw materials. In particular, many questions surround the strategic minerals, those that are largely or entirely imported into the United States. Are supplies assured? Can alternative sources be found, either within the United States or elsewhere? What can be done if the cost of imported raw materials rises rapidly and greatly? How feasible is their conservation or recycling? Is there a need for stockpiling reserves?

The operation of the free market does much to answer these questions, but in some cases the government has taken nonmarket actions, such as establishing the strategic petroleum reserve. However, the United States still lacks a comprehensive mineral policy, and responsibility for minerals is spread among a number of government agencies. The 10 regional resource offices in the State Department are an important link in the complex chain of understanding resource availability, but they have weakened in recent years.

The issue of strategic minerals is like many in the earth sciences in its complexity and in the close interconnections of economic, technical, and political matters. Geoscientists have the technical knowledge to respond to national needs in whatever ways are appropriate. Limitations on this ability are set by the availability of professionals and by current levels of activity. It is not feasible to greatly increase levels of research, exploration, or production suddenly when skilled personnel are unavailable, as is discussed below.

Demographic Characteristics of Solid-Earth Scientists

Solid-earth scientists in the United States have a wide range of educational backgrounds. Many earth scientists received degrees from traditional geology or earth science departments in colleges and universities. Others received degrees in chemistry, physics, biology, mathematics, or some other discipline and later came to specialize in an earth science field.

The multidisciplinary character of geoscience is one of its great strengths, but it makes it difficult to precisely delineate the outlines of the field. For example, the NSF designs, conducts, and supports a number of surveys that collect information on scientific and engineering personnel, including the Survey of Experienced Scientists and Engineers, the Survey of Doctorate Recipients, and the Survey of Recent Science and Engineering Graduates. However, these surveys do not break out the solid-earth sciences and are therefore not very useful for assessing the numbers of scientists working in various areas of the field.

To meet the need for better information about the field, the American Geological Institute (AGI), an umbrella group of 20 earth science professional societies, conducted a survey in 1989 on the composition and characteristics of the solid-earth science community in the United States and Canada and plans to do so again in the future. The survey results are relatively rough because only two previous such surveys were completed, but they give approximations of the size and characteristics of the geoscience community.

TABLE 6.2 Employment Levels (as of 12/31/89) by Employer Category and Degree Level

Employer Category	Degree Level			Total	% of Total
	B.A./B.S.	M.A/M.S.	Ph.D.		
Domestic oil and gas industry	10,976	19,959	5,013	35,948	51
Domestic mining and minerals industry	3,092	2,465	441	5,998	8
Federal/state government agencies	3,661	2,876	1,943	8,480	12
Research institutions/DOE-funded national laboratories[a]	582	827	1,409	2,818	4
Geoscientific consulting firms	3,391	3,798	814	8,003	11
Academia	45	491	9,238	9,774	14
TOTAL	21,747	30,416	18,858	71,021	100

[a]DOE, Department of Energy.
SOURCE: American Geological Institute (1989).

The survey was sent to a sample of 310 companies, agencies, and universities. On the basis of these and previous returns, AGI estimated that in 1989 there were 120,000 individuals, including petroleum engineers and mining engineers, who could be classified as solid-earth scientists in the United States. If those two categories are excluded, the total number of geoscientists at the end of 1989 was about 71,000. About half of the 71,000 geoscientists in the United States were employed by the petroleum industry, with an additional 6,000 employed by the mining and minerals industries (Table 6.2). Government was the second largest employer, with some 12,000 geoscientists, followed by academia, with 10,000.

Figure 6.1 shows the occupational objectives of respondents to an earlier survey. As might be expected from the large number of geoscientists employed by the petroleum and mining and miner-als industries, finding and developing oil and gas deposits were the primary occupation of more than half of the survey respondents. Geotechnical applications were the next largest category, followed by finding and developing other resources and then by geoscientific education, basic research, and communications.

Table 6.3 gives a cross-tabulation between employer category and the specialty practiced for 1987. Thus, about 42 percent of the geoscientists employed by the petroleum industry specialized in petroleum engineering, while almost 50 percent of those in engineering, construction, and consulting specialized in engineering geology. Geology, geochemistry, geophysics, and other earth sciences are heavily represented among those in research and related fields.

The levels of training of solid-earth scientists in the United States show a similar diversity. Among

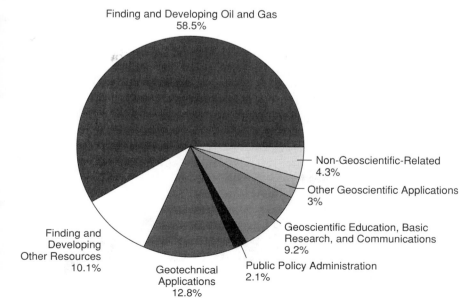

Finding and Developing Oil and Gas
58.5%

Non-Geoscientific-Related
4.3%

Other Geoscientific Applications
3%

Geoscientific Education, Basic Research, and Communications
9.2%

Public Policy Administration
2.1%

Geotechnical Applications
12.8%

Finding and Developing Other Resources
10.1%

FIGURE 6.1 The occupational objectives of employed solid-earth scientists. From American Geological Institute, 1989.

TABLE 6.3 Estimated U.S. Population of Geoscientists by Employer Category and Occupational Category (rounded to nearest thousand)

	Geologists, Geochemists, Earth Scientists	Geo-physicists	Petroleum Engineers	Mining Engineers	Engineering Geologists	Hydrologists/ Hydro-geologists	Total	% of Total
Academia	5,000	2,400	600	600	[a]	200	9,000[b]	7.5
Government	5,500	3,000	800	700	2,000	2,000	14,000	11.7
Petroleum industry	16,500	15,000	28,000	[a]	[a]	200	60,000	50.0
Mining/minerals industries	3,500	1,000	[a]	6,500	[a]	[a]	11,000	9.2
Engineering, construction, and consulting	1,500	700	[a]	600	3,000	2,200	8,000	6.7
Research and related	1,500	1,500	[a]	[a]	200	[a]	3,000	2.5
Other employers with geoscientific requirements	2,000	1,000	[a]	[a]	[a]	[a]	3,000	2.5
Retired and unemployed	4,000	3,000	2,000	1,500	500	500	12,000	10.0
TOTAL	40,000[b]	28,000	31,000	10,000	6,000	5,000	120,000	100[c]

[a]Too few to estimate.
[b]Totals within categories are not necessarily the sum of each cell due to rounding and open categories.
[c]Percentage totals do not add to 100% due to rounding.
SOURCE: American Geological Institute (1989).

the respondents to the surveys, slightly more than 70 percent reported that their most advanced degree was in a geoscientific specialty. The remainder had degrees in other specialties, including chemistry, mathematics, physics, and mechanical engineering.

Of the 1989 survey respondents, 27 percent reported having a doctoral degree and 43 percent a master's degree. Therefore, about 30 percent of the individuals employed as geoscientists have only a bachelor's degree.

Figure 6.2 shows the age distribution for survey respondents in 1987, and Table 6.4 shows the age distribution by occupational category and employer category. The largest numbers are in the 25 to 39 age group, with a secondary maximum in the 50 to 59 age group. This bulge is particularly notable in academia, where about 50 percent of the respondents were 50 and over and only 12.5 percent were under 35. As this older group begins to retire, new geoscientists must be found to replace them, as discussed under the section Education in the Solid-Earth Sciences later in this chapter.

Figure 6.3 gives the distribution of incomes for survey respondents in 1987, and Table 6.5 gives the approximate median annual incomes by employer categories. The highest median incomes were earned by those working for other employers with geoscience requirements, but this group makes up only about 2 percent of the total population. The next highest earnings went to those in the petroleum industry.

Future Demand for Solid-Earth Scientists

The supply of and demand for solid-earth scientists tend to be out of phase. On the demand side, the problem has been a chronic oscillation in earth science employment—the so-called boom-and-bust cycle—related to employment needs in the petroleum and mining and minerals industries. Although such oscillations are universal under a free market economy, they have been of exceptional amplitude in the solid-earth sciences. The high amplitude results from the important role of exploration in

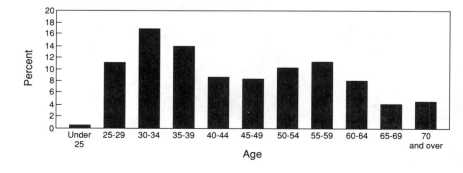

FIGURE 6.2 Age distribution for solid-earth scientists. From American Geological Institute, 1989.

TABLE 6.4 Age Distribution and Employer Category in 1987 (as percentage of employer categories)

Employer Category	Age		
	% Under 35	% 35–49	% 50 and Over
Academia	12.5	38.1	49.4
Government	15.2	38.4	46.4
Petroleum industry	38.7	31.0	30.3
Mining/minerals industries	22.1	39.6	38.3
Engineering, construction, consulting	19.7	37.0	43.3
Research and related	24.0	46.0	30.0
Other employers with geoscientific requirements	27.3	36.4	36.3
Employers with no geoscientific requirements	35.7	38.1	26.2
Not employed	16.5	13.6	69.9
Total	28.8	31.6	39.6

SOURCE: American Geological Institute (1987).

TABLE 6.5 Approximate Median Annual Income by Employer Category

Employer Category	Approximate Median Income
Academia	$49,000
Government	44,000
Petroleum industry	56,000
Mining/minerals industries	45,000
Engineering, construction, consulting	51,000
Research and related	51,000
Other employers with geoscientific requirements	63,000
Employers with no geoscientific requirements	<20,000
Not employed	<20,000
Total	51,000

SOURCE: American Geological Institute (1987).

professional employment. Exploration for oil and gas as well as minerals is expensive and does not yield immediate income. Companies have therefore commonly reduced or suspended exploration in hard times. Many geoscientists involved in exploration are fired at those times. The education system responds to employment fluctuations, but, because there is a lag between education and employment, serious mismatches between the supply and demand for earth scientists can result.

As shown in Figure 6.4, the supply of solid-earth scientists generated by American colleges and universities is closely related to production in the oil and gas industry, since the petroleum industry employs about half of all geoscientists in the United States. Production is in turn partly related to price,

so that future changes in oil and gas prices will be one determinant of future supplies of students.

The number of earth scientists employed in the extractive industries has declined in recent years and may never again match its earlier peaks. But the continued use of oil, gas, and mineral reserves and the aging population of the earth scientists employed in these industries guarantee that more geoscientists will be needed in those areas.

Fortunately, the decline of petroleum-related jobs has been concurrent with increased demand for other types of solid-earth scientists. In the early 1980s, environmental legislation dealing with the proper siting and characterization of new waste disposal sites and cleanup of existing sites was enacted and strictly enforced. This has created demands for hydrologists, geophysicists, and low-temperature geochemists. During 1988 to 1989, according to hiring surveys conducted by AGI, while petroleum-related hiring in the earth sciences decreased by 33 percent, hydrogeological and engi-

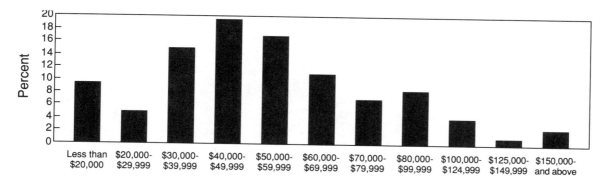

FIGURE 6.3 Incomes for solid-earth scientists. From American Geological Institute, 1989.

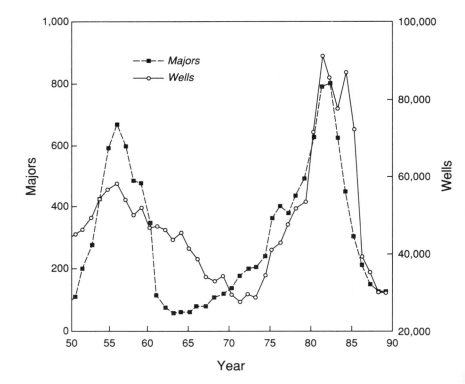

FIGURE 6.4 The number of oil and gas wells drilled in the United States corresponds closely with the number of geology majors at the University of Texas, emphasizing the close relationship between the petroleum industry and geological education. From W. L. Fisher, University of Texas at Austin.

neering geology hiring increased by 43 percent. For example, virtually any student who holds a graduate degree in an "environmental geology" field receives numerous job offers, and some of the best graduate students are being hired by private industry before finishing their degrees.

Several surveys indicate that there will be a greatly increased demand for geoscientists trained in specific applied fields in the future. Employment projections indicate a sixfold increase in many areas of the solid-earth sciences during the 1990s for Superfund sites alone. Costs associated solely with the cleanup of Department of Energy facilities are estimated by some to be over $150 billion, a large part of which must be directed at geoscience issues. When other waste cleanup demands and future legislation dealing with local, regional, and global environmental issues are considered, there is no question that employment opportunities in the geosciences will grow significantly. If the solid-earth science community cannot provide the critical expertise, the work will be performed by others who are not necessarily qualified.

These trends, plus the aging of the earth science community and brighter prospects for the oil and gas industry, indicate that there will be a substantially increased demand for geoscientists in the 1990s and into the twenty-first century. Yet the U.S. education system now appears incapable of

meeting that demand. This looming personnel crisis in the solid-earth sciences is discussed in more detail in the section below.

EDUCATION IN THE SOLID-EARTH SCIENCES

As pointed out above, the continued vitality of the solid-earth sciences will be critically dependent on a continuous supply of well-prepared geoscientists to satisfy national needs in both basic and applied research. Yet the prospects for finding personnel in the solid-earth sciences, as in other areas of science, are extremely troubling. Since 1983 the number of undergraduate students enrolled in the approximately 800 solid-earth science programs in the nation's colleges and universities has dropped by more than half (Figure 6.5). Graduate enrollments and degrees granted in the approximately 125 schools offering doctoral programs in the solid-earth sciences have remained more stable, but declines can probably be expected as the number of undergraduate majors diminishes (Figure 6.6).

Furthermore, many of the students in graduate programs are not U.S. citizens, and a substantial fraction of them will return to their native countries after graduation. Those who remain, if they do not become citizens, may not be able to work on at least some environmental projects. Curricula in graduate

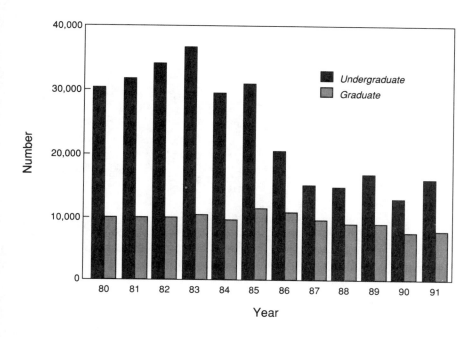

FIGURE 6.5 Enrollments in the solid-earth sciences between 1980 and 1991. From American Geological Institute.

and undergraduate programs are also a focus of concern. Although some geoscience programs are attempting to increase offerings in high-demand fields (such as hydrology), the changes are occurring slowly, and there are few qualified scientists to teach in these programs because those who are qualified can make much more money in the consulting field.

An economic recovery of the oil and gas industry will likely increase undergraduate enrollments. But long-term demographic trends may still limit the number of individuals trained in the solid-earth sciences. There are about 20 percent fewer college-aged students now than there were during the mid-1970s, when the peak of the baby boom was moving through its college years. Also, levels of interest in science and technology have been declining, as has the level of precollege exposure to the earth sciences. Of the 18 year olds entering college in the fall of 1989, only 1 in a 1,000 cited earth science as his or her probable major field of study, according to the Cooperative Institutional Research Program at the University of California at Los Angeles. (Only 2.2 percent listed any area in the physical sciences, with an additional 3.7 percent listing areas in biology.) Even among the best-qualified high school students (those with mathematics SAT scores above the 90th percentile), interest in the natural sciences has fallen to about 15 percent, and those interested in the earth sciences constitute less than 1 percent.

The composition of younger age groups poses

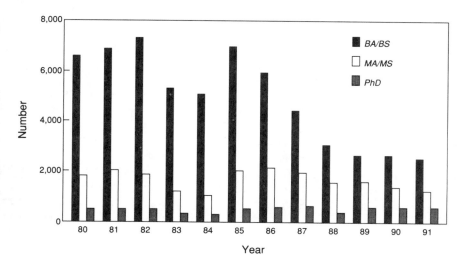

FIGURE 6.6 Numbers and types of degrees granted in the solid-earth sciences between 1980 and 1991. From American Geological Institute.

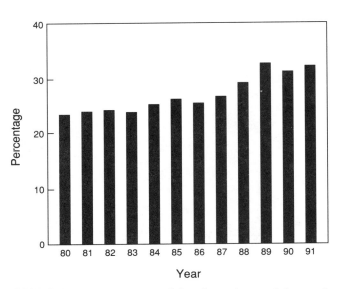

FIGURE 6.7 Percentage of female students of the total solid-earth science enrollment between 1980 and 1991. From American Geological Institute.

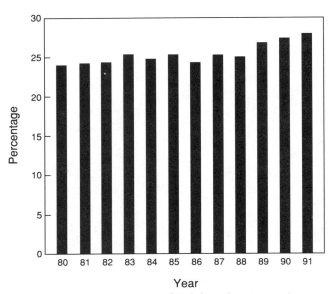

FIGURE 6.8 Percentage of solid-earth science degrees awarded to females between 1980 and 1991. From American Geological Institute.

further challenges to the solid-earth sciences. Between now and the year 2000, nearly 70 percent of the new entrants to the labor force will be women and minorities, yet these groups are poorly under represented in the solid-earth sciences. In the AGI survey discussed earlier, only about 5 percent of survey respondents were women and only about 5 percent minorities. Sampling biases could distort these figures, but it remains a fact that women and minorities are not entering the earth sciences in numbers anywhere near their proportion in the general population. If the geosciences are to have sufficient numbers of new practitioners in the years ahead, efforts must be made to attract much greater numbers of women and minorities. The past decade has shown percentage rises in both female enrollment and degrees awarded, which can be interpreted as representing some progress (Figures 6.7 and 6.8).

Formal Education in the Solid-Earth Sciences

Precollege Education in the Earth Sciences

The entire country is awakening to the need for improved education in science and mathematics, and a number of initiatives have been undertaken by the private sector, government, and universities. For example, the American Association for the Advancement of Science (AAAS, 1989), through its Project 2061 (named for the year in which Halley's comet will return), is seeking to foster a high level

of scientific awareness in the general public. The project involves a complete rethinking of teaching methods, curricula, and content in the sciences, with pilot programs established in a number of schools and outreach programs to spread those examples more widely.

Other professional societies are also redoubling their education efforts. The National Science Teachers Association (NSTA, 1992) has proposed changes in secondary school science education through its Scope, Sequence, and Coordination Project. In place of the traditional years devoted to biology, chemistry, and physics, NSTA has proposed 6 years of science for all students in grades 7 through 12 in four "distinct and well-coordinated" subject areas: chemistry, physics, biology, and earth-space science.

Many science museums have recently mounted aggressive precollege and public education programs. The geosciences are a significant component of many of these museum activities, since the earth sciences are a natural avenue for a more general introduction to science.

The geoscience community is contributing to these broader efforts and coordinating more specific actions to enhance general appreciation for the solid-earth sciences and to improve the educational position of the discipline. For example, the earth science community is working with the AAAS to help students achieve literacy, the development of logic and creative thought, and the ability to test hypotheses.

The AGI has traditionally had a large role in these areas. As part of Project 2061, it has coordinated activities directed toward a complete restructuring of the K-12 earth science curricula. In addition, the Geological Society of America has organized the SAGE program (Scientific Awareness through Geoscience Education), which will at first be focused on the K-12 level and at public awareness.

Despite these efforts, the earth sciences are not widely taught at the precollege level. Earth science courses are currently available in fewer than 5 percent of the nation's high schools, whereas 99 percent teach biology, 91 percent chemistry, and 81 percent physics. Some states have recently dropped earth sciences as a subject that will satisfy college entrance requirements, and getting the subject into the curriculum is often difficult. For example, a school superintendent or school board may add a required second course in biology rather than an introductory course in earth sciences.

Finding good teachers for earth sciences is also difficult. Because of the paucity of teaching opportunities, there are few qualified precollege earth science teachers. Only 15 percent of precollege earth science teachers say they feel qualified to teach the subject, and 22 percent of earth science teachers in grades 7 through 9 never had an earth science course in college (over 50 percent had two or fewer courses). As a result, earth science cannot be adequately taught and promoted in most elementary and secondary schools.

Getting more precollege students interested in science, particularly the earth sciences, will require better trained and more dedicated science teachers plus more exciting science curricula. Many in the earth sciences have recognized the problem of inadequate numbers and preparation of teachers in earth sciences, and the various professional societies have begun to take steps to deal with it. In these efforts it must be recognized that precollege science teachers need to have a fundamental grasp of how best to teach science. Teachers of earth sciences at the secondary level preferably should have an earth science degree. A science degree, supplemented by course work in education, is appropriate at the elementary teaching level. Thought processes and observational skills need to be emphasized rather than memorization of facts. Real-world societal problems have much more relevance to students than strictly subject-related approaches. **Earth science departments in colleges and universities and their respective schools of education need to develop collaborative programs in training earth science educators.**

College Education in the Earth Sciences

Earth science education at the undergraduate level has three overlapping components: the introduction of undergraduates to problems and issues that involve understanding the Earth, the training of earth science teachers, and the preparation of earth science majors.

Introductory courses for undergraduates must be among the best that an earth science department offers and should involve its best teachers. They must contain accurate science, provide an introduction to the way scientists work and think, and address the needs of society. Most students who take an introductory course will not go on to become earth science majors, but at least they will become familiar with earth science concepts and gain an appreciation of the extent and limitations of the knowledge. In addition, because of a lack of exposure to the earth sciences in high school, a significant fraction of future earth science majors will emerge from this group to join those who begin their college education with more advanced earth science courses.

Environmental geology should especially appeal to an undergraduate audience that will form the next generation of educated adults. They will find relevant the geological fundamentals that pertain to land-use planning, the limits of resources (fuels, metals, water), and the management of wastes. Other subjects such as planetary geology, plate tectonics, and biological evolution are scientifically exciting and can be integrated into introductory courses.

Although introductory courses can be made superficially more appealing by various packaging and marketing techniques (e.g., eye-catching course titles), their success depends on the excellence of the teaching and the involvement of students through laboratories and field trips in a hands-on investigation of their world.

Undergraduate programs for future teachers of precollege earth sciences should involve both earth science and education departments. The best-prepared teachers at the high school level are those who have completed most, if not all, of the courses required of earth science majors. Universities and colleges can also run summer institutes for precollege teachers, involving teachers in field or laboratory research.

Undergraduate earth science departments are currently vastly undersubscribed. One way to maintain the continuity of the earth sciences is to help prepare future geoscience teachers at all levels. All departments should assess how they can contribute to this essential aspect of geoscience education.

For many decades the undergraduate training of future geologists has been strongly influenced by the needs of the petroleum and mining industries. Statistics on future employment opportunities indicate that these industries will not be so dominant in future hiring, and undergraduate earth science programs are, in fact, increasingly emphasizing environmental aspects.

Given the shifting trends in geoscience careers and the need for adaptability as well as creativity, undergraduate departments should produce majors who are as versatile as possible, whether they are headed for graduate studies or directly to employment. To achieve this versatility, curricula will have to become more flexible. New courses in geology reflecting emerging fields and greater cooperation with related fields (e.g., civil engineering) will often be needed. A firm foundation in the allied sciences, especially mathematics and computer science, is more important than ever as the earth sciences become more quantitative and interdisciplinary. If earth science graduates do not have the necessary quantitative skills, jobs will go to people outside the geosciences who do. Faculty must therefore be capable of teaching in these areas, and students must enter college with the necessary base of quantitative training.

Participation in research should be a key element of undergraduate education. Only by becoming involved in research can students really understand how scientific thinking proceeds and how science is carried on. For smaller departments, consortia of students and researchers can carry out intercollegiate research projects, thereby broadening their exposure to problems and techniques. Field experience should also be a critical element for all earth science degrees.

Importance of Research in Science Education

Basic research is the primary process for generating new concepts and technological development. The training of the most intuitive and perceptive students for basic research requires changes in the present geoscience education system. All students should be given the opportunity to formulate problems based on their own observations and to deduce relationships amenable to testing.

This training for basic research can start before college. Students can be prepared to propose a senior high school independent studies project, a process continued in a bachelor's thesis and ulti-

mately a thesis for an advanced degree. It is essential that all students interested in basic research learn to choose their own problem with the guidance, but not strict direction, of teachers.

Because of the complexity of geoscience problems, exposure of students to a wide range of observations is essential, as is the opportunity to ascertain their relative importance. In other words, not only must the scientific process be acquired, but the tools for observations also must be mastered. In addition, to find solutions of lasting value for many of society's larger problems, the fundamental scientific questions that underlie the problems must be identified. For example, extraction of the organic residuum remaining after an oil field has been depleted of the easily pumped petroleum requires a knowledge of organic geochemistry, the dynamics of fluid flow in porous media, heat and mass transfer, rock mechanics, and a host of other topics not now available to most students of geology.

A narrow focus on job training will eventually require a massive effort to retrain graduates as the needs of the nation change. In contrast, a strong background in the basic sciences will give a student the breadth and flexibility to acquire skills in related developing areas.

Continuing Education

As the average age of the scientific work force increases, some consideration must be given to revitalization of previously trained research workers. Because of the dynamism and rapid evolution of the solid-earth sciences, shifts of emphasis in professional fields that require some degree of retraining will almost inevitably occur. This situation is not confined to industry; it also applies to academia, where major shifts in curricular emphasis can occur.

This role has traditionally been fulfilled by professional societies. As demands for specialized knowledge in particular fields shift, professional societies can take the lead in retraining for continued employability.

The Geological Division of the U.S. Geological Survey (USGS) offers an excellent example of a continuing education program. Programs range from 1-hour lectures on new developments in the geosciences to full years of formal academic training. These opportunities are intended to promote the scientific currency of employees as well as to foster the exchange of information and ideas among scientists.

As the earth sciences become more complex and

fragment into specialty groups, professional societies have an important responsibility for integrating the field as well as focusing on disciplinary matters. This can be done through a number of mechanisms, including short courses and workshops, conferences, field trips, seminars, lecture series, symposia, and publications.

Public Awareness of the Earth Sciences

In addition to the challenges of professional education and training, it remains the case that society as a whole does not have much awareness of the importance of the earth sciences in societal issues. This situation aggravates the educational problems, because an educated public is the pool from which future geoscientists will come.

When nonscientists are asked to name the scientific disciplines, "earth sciences" or "geology" are rarely mentioned. People mention chemistry, biology, physics, and perhaps astronomy but often fail to realize that the study of the Earth is as much a science as are the more familiar disciplines. Even other scientists sometimes make this mistake.

There are many reasons for this common misperception. One is the lack of earth science courses in elementary and secondary schools. Also, geoscientists have perhaps been less vocal in promoting their discipline than have physicists, biologists, and chemists.

A lack of recognition of the earth sciences adversely affects many issues of widespread social importance. A good example is the asbestos problem. Inadequate mineralogical input to the legislative and regulatory process has resulted in the lumping together of many noncarcinogenic fibrous materials with the dangerous asbestos varieties. As a result, billions of dollars is being spent to clean up problems that probably do not exist.

Other areas where the solid-earth sciences intersect with the legislative process include earthquakes, waste disposal, resource usage, groundwater protection, global change, and land use. More generally, citizens need to know about natural phenomena (earthquakes, floods), resources (water, minerals, fuels), and environmental-economic concerns (acid rain, groundwater contamination, global change) to make informed decisions.

Recent events in the United States, such as the Landers and Big Bear earthquakes of 1992, Hurricane Andrew in 1992, the Loma Prieta earthquake in 1989, Hurricane Hugo in 1989, and the eruption of Mount St. Helens in 1980, help to enhance awareness of earth hazards, but the opportunity is much broader. The many links between the earth sciences and everyday life create a unique chance to promote greater awareness of the sciences. The essential nature of the earth sciences needs particular emphasis. Significantly and rapidly elevating public awareness of the earth sciences is therefore a major opportunity as well as a substantial challenge.

Coping with the Supply of Solid-Earth Scientists

Even if all of the committee's recommendations addressing educational shortcomings in the earth sciences are implemented, there are likely to be problems of maintaining an adequate work force in the future. Geoscientists therefore need to consider how the field can continue to advance despite these shortcomings.

One requirement is that practicing earth scientists must be ready to modify their interests and activities. For example, petroleum geologists are moving in increasing numbers into hydrogeology as job opportunities become available. However, the diversion of earth scientists from one field to another can introduce difficulties. For example, the teaching of mining geology is now rare in the United States, and, apart from some gold operations, metal mining is greatly reduced. Within 10 years many industry leaders will have retired, leaving very few individuals to teach the young mining geologists that will be needed when the field recovers.

Earth scientists also need to make the fullest possible use of modern instrumentation and technology so that a smaller or level population of earth scientists can achieve more. In particular, because of the extensive and diverse nature of the data used in the earth sciences, there is an exceptional opportunity to develop advanced data-handling and data-fusion systems, such as the geographic information systems. However, instrument development has declined in the United States compared with some other countries, suggesting that perhaps an important opportunity is being missed. These issues are discussed in more detail in the Instrumentation and Facilities section below.

Finally, there are many opportunities for improving the participation of the United States in international earth science activities. U.S. multinational corporations have established overseas facilities specifically designed to keep abreast of advances in the geoscientific disciplines. Individual programs range from major investments in overseas laboratories to closer monitoring of scientific journal articles. In addition, other countries have much larger cooperative and technical aid programs in the earth sciences than does the United States. Not being involved to a comparable extent could prove costly if the United States is left out of important new developments. These and related issues are explored further in the Global Collaboration section later in this chapter.

INSTRUMENTATION AND FACILITIES

Science and scientific equipment advance hand in hand. As with science in general, progress in the earth sciences is critically dependent on improvements in the ability to observe and measure. For research and teaching to thrive, appropriate instrumentation and facilities must be available.

Earth scientists observe and measure in situ, or in the field. They also carry out measurements in the laboratory, where selected samples of earth materials or their synthetic analogs can be studied under controlled conditions.

Earth scientists have used both field observations and laboratory experiments for many years, but technological progress has greatly expanded the range and accuracy of measurements available. Today, the inventory of instruments and analytical tools used bears little resemblance to what supported state-of-the-art research as recently as 10 years ago. Combined satellite, airborne, and ground-level geodetic measurements are capable of detecting both lateral and vertical ground movements as small as a few millimeters a year on global as well as local scales. Laboratory facilities can reproduce pressures and temperatures equal to or greater than the extremes encountered within the Earth. Rare elements can be measured at extremely low abundance levels, and stable isotopic ratios can be measured in individual zoned crystals and microscopic fluid inclusions. Computers with ultrafast interactive graphic capabilities make it possible to grapple with complex nonlinear phenomena such as convection in the mantle, the behavior of the magnetic field, fluid flow in sedimentary basins, and the equations of state for minerals.

The discussions in this section do not try to inventory all of the instruments and facilities that are important in the solid-earth sciences. Such an inventory would be virtually impossible, since instrumentation is constantly developing and the needs of geoscientists for instruments and facilities are as varied as the earth sciences. Rather, the following material examines several particular examples of instruments or facilities and seeks to draw conclusions that apply more broadly in the field.

Issues involving instruments and facilities in the solid-earth sciences are difficult but are also representative of similar situations in many scientific disciplines. Only by continued close examination of those issues can a proper balance be maintained between the support of individual researchers and the support of the many instruments and facilities they need.

Some examples of the earth-based critical instrumentation and facilities follow.

Global Positioning System

From the first attempts—nearly 3,000 years ago—to measure the circumference of the Earth to the present day, geodetic measurements have always been essential tools for earth scientists. Their applications range from static mapping to the direct observation of active deformation.

For the immediate future, many geodetic questions will be answered by data from the global positioning system (GPS), which uses satellites and ground stations for position determinations precise to a few millimeters over distances up to 1,000 km. Using GPS, it is possible to establish the rate at which Houston and Perth are approaching each other (about 7 cm a year) and to prove that plate tectonics is a fact and no longer a theory. The GPS also generates high-resolution data used to determine the rigidity of plate interiors and to evaluate velocity changes at zones of plate convergence. Use of this resource to measure sea-level changes is anticipated as interpretive methods are refined.

In tectonically active areas, such as those around volcanoes and in earthquake zones, there is a need for rapid, or even continuous, monitoring of spatially dense networks. However, such needs are not well satisfied by GPS receivers. To make rapid, dense, extensive, and remote monitoring more affordable and practical, a different technology is needed. Microwave techniques, using inexpensive transmitters on the ground in conjunction with orbiting transponders, seem to be particularly promising. Optical ranging techniques using ground-based retroreflectors and orbiting lasers have also been proposed. Microwave transponder satellites could also be used to collect many kinds of telemetered data from geophysical instruments.

Modern, highly precise, distance-measuring techniques can be expected to play a progressively larger role in the practice of the solid-earth sciences and in scientific research. This role will extend beyond the traditional geodetic field into surveying and mapping of all kinds. For example, field geologists will be able to determine their precise position whenever they need it. As recommended in a 1991 National Research Council (NRC) report, an initial step would be to establish a global network of precisely monitored geodetic sites; specifically, the report addresses appropriate strategies for implementing and operating such a network. Such a network would be vital in sea-level and active tectonic research and would complement the roles of satellite laser ranging and very-long-baseline interferometry.

Digital Seismology

Both earthquakes and controlled ground motions generated by explosions or vibrations provide means of imaging the Earth's interior. Earthquakes, which are unrivaled in energy except by the largest nuclear explosions, provide a snapshot in time of the major dynamic processes that shape the surface of the planet, from the sliding of tectonic plates to the movement of magma beneath soon-to-erupt volcanoes. Controlled sources now provide unprecedented resolution, accuracy, and depth of penetration in viewing the structure of the crust and the lithosphere, thanks largely to advances in computing capability and numerical techniques.

Gathering the data generated by both earthquakes and controlled ground motions requires seismic networks and their associated data management systems. These data can be gathered on both regional and global scales. On the regional scale, many more modern portable seismographs are needed. The small number of such instruments now available greatly limits the effectiveness of our "telescopes" for looking inside the Earth.

In the future, large arrays of portable seismographs will be used in programs that integrate all seismological research techniques. They will exploit the high resolution of reflection profiling, the velocity information of wide-angle reflection and refraction data, and the greater depth of penetration available from surface waves and teleseismic studies. The instruments will have large storage capacities and be triggered by seismic events to exploit earthquake sources that are rich in shear waves. Key technical attributes of the new generation of instruments now becoming available include digital recording; highly accurate phase-lock timing; variable recording bandwidths; multichannel recording; high-density recording media; and event-triggered recording for local, regional, and teleseismic waves.

On a global scale, advances in electronics and telecommunications technology, as well as developments in seismic sensors, now make it possible to obtain reliable and continuous global seismic coverage in real time. The collection and rapid distribution of high-quality seismic data over the complete frequency and amplitude of ground motion induced by earthquakes are now possible. The key technical requirements of a new global network include digital data acquisition; a bandwidth from hours to approximately 10 Hz; broad dynamic range; low instrumental and environmental noise; standardization and modularity of design; and, most important, seafloor as well as land stations.

Since the 1960s, seismology has been one of the most important forces driving the development of larger and faster computers. Even today the three-dimensional modeling requirements exceed current computer capability by an order of magnitude and are driving the development of new approaches in parallel computing techniques. These driving forces originate from both the resource exploration industries and academic research needs.

Furthermore, once the data have been gathered, a corollary requirement is efficient data management and dissemination. For example, a fully operational global network will produce as much data per year as are presently housed in a major university library, requiring a well-designed data management system.

High-speed computer communication links like INTERNET, which so far are virtually free to users, have become the backbone of research and information exchange. If this resource were not kept widely available for users, or if users should have to spend scarce research dollars to maintain this link to centralized data centers, the research capability of parts of the solid-earth science community might be jeopardized. The information needs of earth scientists are discussed more fully in the Data Gathering and Handling section of this chapter.

Instrumentation in Earth Science Laboratories

Instrumentation for earth science laboratories has been addressed in two recent NRC reports *Earth Materials Research* (1988) and *Facilities for Earth Materials Research* (1990). The latter report distinguishes a level of resources adequate to keep laboratories in the forefront of science from a level that would only allow research programs to survive. A consideration that was not addressed at length in these studies is the way in which resources can be stretched by making use of state and local funds and amenities, as well as by involving industry and private foundations and by sharing facilities among institutions. There is and always will be a challenge in developing new instruments, applying newly developed techniques, and training and maintaining a pool of scientific and technical staff. A current factor, whose implications are not yet fully clear, is the recognition by the Department of Energy (DOE) and its national laboratories of its changing role. DOE's commitment to waste isolation and kindred research will surely demand a stronger emphasis on instrumentation for earth science laboratories.

Chemical and Isotopic Analysis

There is a critical need to know the chemical and isotopic composition of rocks, minerals, and solutions to pursue research in such areas as mineral exploration and groundwater movement and, more generally, to understand the processes that created and continue to modify the Earth.

Not long ago most chemical analyses of earth materials were performed with classical wet-chemical techniques that were tedious, offered relatively poor detection limits (in the parts per thousand range), and consumed large quantities of samples. In the past few decades, geochemistry has been transformed by the development of a variety of instrumental techniques that can probe for a very large number of elements at vastly improved detection limits. In many cases the limits now approach the parts per trillion level, with a spatial resolution of a few micrometers.

The more common geochemical studies include elemental analysis, isotopic analysis, and spatial resolution analysis.

■ *Elemental Analysis*. Detecting and measuring the elements in earth materials require techniques that are sensitive to a wide range of elements at concentration levels that range over many orders of magnitude. The most commonly chosen techniques for determining concentrations of major elements (present at levels greater than parts per thousand) and minor elements (parts per thousand to parts per million) are x-ray fluorescence spectrometry (XRF) and inductively coupled or direct-current plasma emission spectrometry. Atomic absorption spectrometry (AA) is used in more restricted cases.

For trace elements (parts per million to parts per billion), the conventional analytical technique used in earth science research has been neutron activation analysis. A versatile alternative method to activation

analysis, in its early stages of development, is the inductively coupled plasma mass spectrometer (ICP-MS). Although it is initially more costly and requires some destruction of the sample, ICP-MS offers greater sensitivity for a wider range of elements, avoids the complication of handling radioactive samples, and offers higher precision when combined with isotope-dilution techniques.

High-precision analysis, necessary for many radiometric dating techniques, is currently performed by isotope-dilution mass spectrometry. This well-established, but slow, procedure uses the thermal ionization mass spectrometer and requires a chemical separation of the element that is to be analyzed. The method cannot be used for all elements, but where applicable it offers higher precision and better sensitivity than other methods.

Modern methods for determining molecular structure and the amounts of organic compounds in the geological environment rely on coupling various chromatographic separation techniques with spectroscopic detection. Gas chromatography and mass spectrometry are routinely used to detect parts per billion to parts per trillion of an individual component. Gas chromatography coupled with matrix-isolation Fourier-transform infrared spectrometry has similar sensitivity.

■ *Isotopic Analysis*. Isotopic analysis has many applications in the earth sciences, including age determination through measurements of isotopic variations caused by the decay of naturally occurring radioactive isotopes, determination of the temperatures of formation of minerals and fossils, and tracing of isotopically distinct constituents through geochemical cycles (e.g., rock-water interactions, ore formation, groundwater circulation, or the subduction of sediments followed by upward migration of partial melt products in subduction-zone tectonic cycles).

The primary tool for high-precision isotopic analysis in the earth sciences is the magnetic sector mass spectrometer. Recent improvements in the design of detector systems and detection electronics have led to increased sensitivity, improved precision, and greater speed of analysis through more efficient simultaneous collection of the mass-separated ion beams. Current techniques allow determinations of isotopic ratios to precisions of better than 0.002 percent on sample sizes as small as a billionth of a gram, or even 10 percent on samples of literally only a few thousand atoms.

Increasing the detection limits for measuring very-low-abundance isotopes that are masked by neighboring mass isotopes present in high abundance is the goal of a new generation of mass spectrometers now being developed. These high-dynamic-range isotope-ratio mass spectrometers offer the promise of being able to make direct measurements of rare short-lived radioisotopes such as carbon-14, helium-3, beryllium-10, and thorium-230. As currently measured by decay-counting techniques, the study of these isotopes at natural abundances requires long counting times (days to months) and large samples. The improvement, by several orders of magnitude, in sensitivity offered by the new generation of mass spectrometers will allow for greatly improved precision in dating geological processes that occur on the time scale of a few hundred to several million years. The advent of accelerator mass spectrometry (AMS), based on high-voltage accelerators, offers ultra-high dynamic range and the capability of measuring isotopic ratios as small as 10^{-16}. AMS has had an important and exciting impact on the earth sciences. It has enhanced the utility of a great variety of cosmogonic radioisotopes in geological studies and has opened up entire fields to quantitative study, such as the age and circulation patterns of groundwater, the rate of physical and chemical erosion in surface and near-surface crustal environments, and the role of subducted sediments in island-arc volcanism. As one example, AMS is providing an understanding of the evolution of island arcs and the role of marine sediments in that evolution through the measurement of beryllium-10 in island-arc volcanic rocks. The unique capabilities of AMS have also allowed extension of the carbon-isotope dating technique to smaller and older samples. Many exciting applications involving other cosmogonic isotopes have been identified and need to be developed systematically.

■ *High-Spatial-Resolution Analysis.* The primary tool for measuring major and minor elements in individual mineral grains is the electron microprobe. Continuing improvements in these instruments have increased their spatial resolution (now at the micrometer level), speed of analysis, and ease of operation. The performance capabilities of the instrument are very well suited to a wide variety of studies as well as to more general applications in materials sciences, chemistry, engineering, and life sciences. However, their role in modern earth science research is so important that at most universities electron microprobes are based in earth science departments.

High spatial resolution for trace elements and

isotopes has been sought primarily through the so-called ion microprobe or the secondary ionization mass spectrometer (SIMS). The super-high-resolution ion microprobe, or SHRIMP, developed at the Australian National University, is a very successful version of a SIMS instrument that emphasizes mass resolution. Successful SIMS applications include reliable radiometric determinations on a single mineral grain or even age zonations within grains, isotopic tracing in meteorites of nucleosynthesis in the solar system, and measurement of trace element distributions between minerals. High-spatial-resolution trace element analysis is useful in studies of magma formation and differentiation, in kinetic studies of metamorphic reactions, and in laboratory studies of distribution coefficients and diffusion kinetics. Another important application of the ion microprobe is determination of the abundances and isotopic compositions of trace elements in fluid inclusions of minerals, which provide information on the pressure, temperature, and composition of the fluids that form ore deposits.

High-Pressure, High-Temperature Technology

It is now possible to investigate material properties under the extreme conditions in the Earth's interior. Pressure, temperature, and chemical variables can be controlled in ways that simulate conditions even in the core. Indeed, earth scientists have pioneered techniques for achieving the highest sustained pressure and temperature conditions attainable in the laboratory. Furthermore, simultaneous developments in synchrotron radiation and geochemical analysis techniques will permit molecular structure, bulk density, elastic and viscous properties, phase transitions, and diffusion rates to be directly determined while the material is subject to the extreme conditions of the deep interior.

High-pressure, high-temperature research on samples larger than those that can be studied in diamond anvils is now carried out largely in Japan; there are a few such laboratory facilities in the United States. Large volumes (cubic millimeters or larger) are especially critical for investigating phase transitions in polyphase systems, deformation mechanisms, and interfacial or grain-boundary phenomena. Pressures up to 300 kb (30 GPa) and temperatures in excess of 2000°C can now be obtained with existing devices, making them ideal for studying the structure and processes of the mantle.

New technology is needed to reach pressures

corresponding to those of the core-mantle boundary in large samples. In situ measurements on samples of a cubic millimeter or more at temperatures of 6000°C and pressures up to 1,000 kb (100 GPa) would go a long way toward solving some of the mysteries of this fascinating region of the Earth. Development of this type of equipment will require new superhard materials in innovative configurations.

In addition to the development of super-high-pressure equipment, there is a need for a new generation of instruments capable of generating triaxial stresses for carrying out brittle and plastic deformation experiments under controlled conditions up to 50 kb (5 GPa) and 1500°C. The volume of the pressure vessel's interior should be large enough to contain pressure and internal force cells and to control the chemical environment of fluid solutions.

With regard to the study of samples at high pressures and temperatures, intense, highly collimated synchrotron x-ray sources provide earth scientists with an extraordinary new technique. X-ray diffraction, fluorescence, and absorption studies can now be carried out on microsamples. There is intense competition for beam time at existing synchrotron facilities, but opportunities do exist for earth scientists to participate in the design and operation of new beam lines at the National Synchrotron Light Source at Brookhaven, the Cornell High-Energy Synchrotron Source, and the Stanford Synchrotron Research Laboratory. It is very important that they continue to do so. In addition, a new major facility is being planned at Argonne National Laboratory—the Advanced Photon Source (APS), which will provide radiation several orders of magnitude more intense than present sources. The earth science community needs to participate in the development of the APS beam lines so that they are designed to meet the needs of the earth materials research community.

DATA GATHERING AND HANDLING

The practice of the earth sciences requires the analysis of information from many different sources. Questions about this information include:

- Is adequate information available to the earth science community to provide informed answers?
- In what forms is this information available?
- How is access to the information obtained?
- Is the information being refined and augmented?

- What resources are applied to obtaining and updating the information, both in general and for specific investigations?
- Who is responsible for obtaining, archiving, updating, and providing interpretations of this information?

Not surprisingly, there are no complete answers to these questions, because different kinds of information are handled in different ways. It is, however, quite possible to provide partial answers to the questions, to identify present successes, and to point out potential courses of action.

Enormous amounts of information are available to the earth scientist attempting to understand geology, resources, environment, and hazard potential. For example, topographic map coverage for the land surface and offshore areas of the Exclusive Economic Zone is both detailed and up to date. Maps are the traditional and useful format for presenting earth science information; geological maps and many more specialized kinds of maps are available. Basic information of this kind is handled on the national scale by federal agencies—the U.S. Geological Survey (USGS) in the case of surface topography and the National Oceanic and Atmospheric Administration (NOAA). This is also the case for much other important information.

At the state level, the geological and more specialized maps and publications of the state geological surveys represent a body of information complementary to that of the federal agencies. To compare the scale of operations, it is useful to observe that expenditures by the Geologic Division of the USGS in a recent year were about $200 million, while expenditures for all the state geological surveys together were about $133 million.

National (and international) professional societies, especially the Geological Society of America, the American Association of Petroleum Geologists, the Society of Economic Geologists, and the American Geophysical Union, as well as many local and regional societies, publish a significant proportion of the essential information available about the geology of the United States. An achievement currently nearing completion is publication by the Geological Society of America of a unique up-to-date series of assessments of the geology and regional geophysics of the North American continent and its neighboring oceans—*The Decade of North American Geology*. Field guides covering much of the country, prepared on the occasion of the 1989 International Geological Congress in Washington, D.C., represent another recent achievement. Field guides in general are a distinctive and very important part of the geoscience data base.

Fossils and rocks—including such materials as well cuttings, drill cores, deep-sea cores, meteorites, and ice cores—are vital parts of the national geological data base. All require curation and research. National and local facilities abound, but there is always a risk of deterioration and even abandonment of collections because of lack of funds and interested people.

National and local collections of aerial photography, some specialized in such features as sun-angle, wavelength, and obliquity, form an important archive. For example, aerial photographs of coasts traversed by hurricanes acquired at intervals over the past 70 years show the pace of global change to a spectacular degree.

Space-based data, which are expensive to acquire, are not in all cases as well archived as some initially cheaper data. There are cases where older data tapes can, apparently, no longer be played back on the more modern computers and the older computers cannot be maintained. A separate issue is that the cost of buying current space-acquired data (such as Landsat and SPOT imagery) is so high as to prohibit its use in routine geological research.

The petroleum industry has generated an enormous amount of information in a great variety of forms. Most significant are well logs and seismic reflection data, but other forms of geophysical data such as gravity and magnetic data also are important. Relatively little of this material is in the public domain and available to the research community. However, an open market exists within the industry, and this information is being widely used. The issue of whether and how to achieve greater public access to any of these data for social or research purposes has not been fully addressed. Drill cuttings from industry wells, on the other hand, are in many cases systematically archived by state geological surveys, and both researchers and industry make use of this material.

As major petroleum companies downsize their operations and increasingly focus their attention on foreign ventures, they have less need for domestic geological and geophysical data. Because of this, their proprietary concerns have lessened. In addition, the cost of properly maintaining the data is becoming a concern. As a result, the danger exists that much of this data, collected at enormous cost (multibillions of dollars), could be lost to both research scientists and those still actively engaged in domestic exploration and production. The industry and others would welcome innovative arrangements to preserve this data resource.

The Digital Data Revolution

The solid-earth sciences and related economic sectors are in the middle of a lasting, irreversible revolution: the digital information age is upon them. The old and new orders of analog and digital data reluctantly coexist, but the digital revolution unhaltingly progresses. In turn, the geosciences are a prime driving force, in the United States and worldwide, that compels the computer industry to develop new and powerful digital data acquisition, processing, archiving, display, and network communication technologies. U.S.-based oil companies spend $3 billion to $4 billion each year for digital data acquisition, processing, and archiving for oil exploration and related efforts. Applied geophysics, particularly, oil exploration seismology, is the single largest nonmilitary user (and buyer) of supercomputers. Thanks in part to this thrust from the geosciences and its commercial applications, the U.S. computer and information industry has attained a global leadership role that has yielded substantial economic benefits.

The transition of efforts from manual to computer-based evaluation and analysis has affected all sizes and types of organizations, from major integrated oil companies to independent consultants to universities and government agencies. With the availability of inexpensive and powerful personal computers and the emergence of sophisticated workstations and special-purpose software, the daily activities of most earth scientists now directly involve computer-processed data.

But not all subdisciplines in the solid-earth sciences have equally participated in and benefited from this unfinished digital information revolution. Many technologies still need to be developed and refined. Data format standards are badly needed. Data management and dissemination programs and centers need improvements and national policies. Economic incentives to convert to digital data usage deserve the highest priorities. National data communication networks need to be improved in capacity and geographical reach. Targeted funding is needed to bring modern digital data technologies not only to a few privileged research institutions but also to ordinary users, professional groups, public offices, and educational institutions on the broadest possible scale.

To face these challenges, there needs to be a comprehensive data-base and data analysis capability. Part of this capability is already in the making. Satellite images, digital topography, and gravity and magnetic field data are available on a global scale, but sufficient resolution is still lacking on local scales for these data to be of use to many scientists, land planners, and many other users. Geological maps are rarely available in digital form. Data formats and data bases are often geared to special purposes. Making data exchanges is difficult at best. Just to find out who has what data and how to access the data is a task for experts, not ordinary users.

These and other difficulties call for more emphasis on data management and data exchange policies. There must be the willingness and fiscal resources to create national examples and leadership roles. The earth sciences have provided excellent examples of national and global data exchange in the past; the International Geophysical Year and the Worldwide Seismic Networks are two notable examples. These kinds of cooperative data collection and dissemination efforts must now be brought into the digital age throughout the geosciences on a broad national, and often global, scale.

Improving Data Management

In the earth sciences, the computer revolution translates into solving more problems more effectively and efficiently. However, this requires ready access to an increasing quantity and variety of digital data—a situation that is only a goal, not a reality, to many earth scientists today.

One of the first problems facing the earth sciences in the management of data is capturing raw data from field observations in digital form. The magnitude of this problem is so great that any formal national effort is precluded at this time. However, mission-oriented digitizing of existing data is moving forward, and new data are increasingly being collected in digital form.

Once the data have been gathered, effective communication of data requires an environment of common public exchange among all elements of the earth science community, both nationally and internationally. The phenomenal growth of telecommunications and computer capabilities in recent years has had a major impact on the distribution of earth data. Hundreds of data bases are being continually updated with field and laboratory results generated by geoscientists from throughout the world. Various organizations have been established for the purpose of maintaining an inventory of published reports on geoscientific topics, including the "gray literature" of government papers and corporate reports.

With the explosion of data bases in academia,

government, and industry, individual research organizations have developed a large number of specific applications formats, making data exchange difficult and expensive. Limited coordination now exists within the geosciences community for the retention and distribution of data. The advent of the personal computer has brought this problem to the level of the individual, affecting researchers, data vendors, and software developers.

The present time is one of transition. In the past, data were collected, processed, and preserved in various ways but rarely digitally. In the future, digital acquisition, digital processing, and selective digital preservation will be general. The transition is accompanied by its own problems. For example, effective use of data for important purposes other than those originally intended may be inhibited or precluded. Drill-hole data obtained in mining exploration can be useful in cases of groundwater contamination. Only infrequently are a mining company's exploration records preserved and made publicly available, and the staff of a mining company might have no idea that borehole data ultimately could be useful for purposes other than mining exploration. Furthermore, decades may elapse after data are gathered before they prove useful. Companies may disappear through liquidation or merger, and records are forgotten, discarded, or lost. Even if records survive, there may be little incentive to maintain them in accessible form after exploration or mining has ceased in a region. It is only through the diligence of one individual that the valuable records and drill cores of the once mighty, and now defunct, Anaconda Corporation have been preserved.

Most government agencies and industrial organizations have specific missions, and data may be collected solely for their particular purposes. For example, petroleum companies collect data for finding and developing oil and gas fields, coal companies gather data to define minable coal seams, and metal mining companies block out ore bodies that will be profitable to mine. Often these companies pay little attention to other exploitable commodities. This is unfortunate because data useful to other companies or government agencies in exploring for other commodities could be obtained at little additional cost. For example, the "mud-logger" services widely used during the drilling of oil and gas wells during the past 30 years could also monitor drilling wells for traces of copper, zinc, lead, uranium, and other metals—information that could be useful to mining companies. The aggregate footage of oil and gas wells drilled in the United States is immense, and virtually all of the drilling has been through sedimentary rocks in which there is some potential for mineralization.

In many cases, data gathered by government agencies can be obtained by prospective users, but prospective users may not know that specific data files exist. Even if the data have been declared to be publicly accessible, and there is public knowledge of their existence and extent, the difficulties in obtaining the data are often so great that access is inhibited.

The situation in industry is somewhat different. Enormous amounts of data exist, but many data files are regarded as proprietary and are not accessible to others except through purchase or exchange. Indeed, some companies' internal policies prohibit dissemination under any circumstances. The situation is similar at the Department of Defense, where data are often classified for national security reasons.

In academia and some nonprofit research organizations, data obtained in scientific research often reside with individual researchers. Centralized archives for storing research data generally do not exist in academic institutions. Often, the data exist solely in written form, or they may be stored in machine-readable form on tapes or diskettes, but they generally remain in the possession of individuals. Exchange or dispersal commonly takes place on a person-to-person basis. When researchers leave, retire, or die, their data may be effectively lost to subsequent users.

A major concern of the research and applications community is how to improve access to the vast amount of earth science data held by government, private, and academic organizations. Currently, there is no national system charged with the management, service, retrieval, and dissemination of this exponentially growing resource. Many of these concerns are addressed in the recent *Federal Plan for Global Change Data Management*, and the committee encourages such efforts.

For some time it has been recognized that the creation of data directories would improve the practice of the earth sciences by promoting the exchange and broader use of data within the community. These directories would provide general access to, and information about, data gathered outside a scientist's own specialty or research environment and would reduce the duplicate generation of data.

Existing catalogs for data centers in federal agencies are not well coordinated, although various initiatives to increase this coordination have been undertaken. National and international data centers for the earth sciences are maintained by NOAA, USGS, and the National Aeronautics and Space Administration (NASA). All have catalogs, but they are only now beginning to be available on-line, and there is no general directory or interchange format for use in government agencies, private industry, and academia.

Questions of data management also extend internationally. A good example is global seismology. The United States has been a leader in converting data tapes into a standard format and making them available to seismologists throughout the world. An expansion of this effort could improve our understanding of the causes and frequency of major earthquakes.

One of the principal stimuli for the formation and promulgation of various international geoscience unions and congresses has been recognition of the need for systematic geological data exchange and standardization. The international dimensions of the earth sciences, including those related to data exchange, are discussed below.

GLOBAL COLLABORATION

The study of the Earth is intrinsically global. A proper understanding of paleontology, stratigraphy, mountain building, and many other solid-earth science subjects calls for basic field data from diverse regions. Geoscientists must conduct experiments on a global basis to determine the composition and dynamics of the planet.

The diverse geological provinces within North America represent field laboratories that offer a wide range of geological phenomena and processes. However, these represent only portions of a much broader spectrum, and an accurate interpretation of their significance requires examination of critical areas in other parts of the world. Study of foreign geological settings may strengthen, or may force rejection of, models that are based solely on surveys made within an individual geoscientist's regional environment. It is the recognition of this need for broader scientific backgrounds and data bases that has prompted expanded global collaboration in the earth sciences.

International Collaborative Activities

Studies in the earth sciences have been carried out on international and interdisciplinary bases for many years. By the late 1800s, international expeditions and data exchanges were increasingly common in the geosciences. Cooperative efforts eventually led to the development of more formal mechanisms for collaboration through international conferences, congresses, scientific unions, commissions, and geoscience programs.

In 1878 the first International Geological Congress (IGC) was convened in Paris. This event inaugurated a series of IGCs that have been held, with few interruptions, every 4 years up to the present time. The principal goals of the IGC are to arrange general assemblies where ideas and information can be exchanged and to provide an opportunity to examine geological features in or near the host country of the congress. Both the 28th IGC (in 1989 in Washington, D.C.) and the 29th IGC (in 1992 in Kyoto, Japan) were attended by nearly 6,000 geoscientists and representatives of related disciplines from all parts of the world.

The latter part of the nineteenth century also witnessed the development of seismological equipment capable of registering earthquake waves anywhere on Earth, and measurement stations were installed in Europe, Japan, and America. This de facto global network introduced a new activity into the solid-earth sciences by allowing essentially simultaneous observations of seismic waves produced by a single earthquake in diverse parts of the world. Similar developments were occurring in geomagnetism. As a result, agreements on geophysical measurement standards and data exchange were made on an international scale.

These activities led to formation of the organization that was the antecedent of the International Union of Geodesy and Geophysics (IUGG). The objectives of the IUGG, established in 1919, are the promotion and coordination of physical, chemical, and mathematical studies of the Earth and its environment. The IUGG at present consists of seven essentially autonomous associations, five of which are concerned with the solid earth: the International Association of Geodesy (IAG), the International Association of Seismology and Physics of the Earth's Interior (IASPEI), the International Association of Geomagnetism and Aeronomy (IAGA), the International Association of Physical Science of the Oceans (IAPSO), and the International Association of Volcanology and Chemistry of the Earth's Interior (IAVCEI).

By the early 1920s, international nongovernmental cooperation in the geosciences assumed its modern form with the creation of the International Council of Scientific Unions (ICSU), in which the IUGG and a number of other scientific unions participate. Geologists continued their international

activities through the IUGG and through the periodic IGC; in 1961 the ICSU formed the International Union of Geological Sciences (IUGS).

The objective of the IUGS is the continuing coordination of international geoscientific research activities. Particular goals of the IUGS are to encourage the study of geoscientific problems of worldwide significance, to facilitate international and interdisciplinary cooperation in geology and related sciences, and to support and provide scientific sponsorship for the IGCs. The IUGS was originally organized along disciplinary lines, but as programs developed special commissions were established in a number of fields. Thirty associations, together representing tens of thousands of geologists, are affiliated organizations of the IUGS.

International collaboration in the study of the Earth received a major boost from the International Geophysical Year (IGY) in 1957–1958. The concept of dedicating a full year to a particular scientific topic was not new, since the First Polar Year had been designated in 1882–1883 and the Second Polar Year was observed in 1932–1933. The IGY demonstrated the feasibility and effectiveness of international cooperation among essentially all countries in a scientific endeavor of common interest. Although the IGY included only limited specific activities in the solid-earth sciences, it provided the opportunity to plan new programs in the geosciences on a global scale.

In 1960 a new international project, modeled in principle on the IGY, was proposed to study "the upper mantle and its influence on development of the crust." This endeavor, referred to as the Upper Mantle Project (UMP), was carried forward by IUGG throughout the 1960s. The UMP witnessed and promoted the development of plate tectonics. This recognition of the solid earth as a dynamic system led the IUGG and IUGS to design a new cooperative earth science program for the 1970s, the International Geodynamics Project (IGP). The IGP focused on processes within the solid earth and their impact on our environment.

The International Lithosphere Program (ILP), instituted by IUGG and IUGS in 1980 as the successor to the IGPs, seeks to elucidate the origin, dynamics, and evolution of the lithosphere, with particular attention to the continents and their margins. Its full title, International Lithosphere Program: The Framework for Understanding Resources and Natural Hazards, reflects the intent to relate basic science to societal and economic issues. A specific goal of the ILP has been to strengthen interactions between basic and applied research involving geology, geophysics, geochemistry, and geodesy as these disciplines are used in the interpretation of mineral and energy resource origins, mitigation of geological hazards, and maintaining an environmental balance.

The success of the program requires active participation by many nations and their respective geoscientific organizations. Currently, 62 nations are represented by scientists serving on the working groups, related task groups, and various coordinating committees. The ILP has emphasized "key projects," including the Global Seismic Network, the Continental Scientific Drilling Program, the Global Geoscience Transects (GGT) Project (modeled on the successful North American Continent-Ocean Transects Program), and the World Stress Map.

The GGT Project consolidates geological, geophysical, and geochemical data to produce interpretative cross sections projected to the base of the crust, and deeper where data permit. Geoscientists from all countries are encouraged to compile such transects in a systematic manner and in a common format to make possible comparisons of the crust throughout the world.

An outstanding example of a recent and effective international field program is the Quebec-Maine-Gulf of Maine transect, a collaborative project including the USGS, the Geological Survey of Canada, the Maine Geological Survey, and university investigators. It involves acquisition of seismic reflection and refraction profiles, as well as gravity and magnetic data, that will be digitized and made available to the scientific community.

Since its foundation, the United Nations has been the source of various initiatives involving the earth sciences. Most of these have been implemented through the United Nations Educational, Scientific, and Cultural Organization (UNESCO) and through the United Nations Development Program (UNDP). For example, the United Nations has organized and funded initial evaluations of the potential resource bases in a number of developing nations. Such studies have involved U.S. agencies, the private sector, and in some instances members of the academic geoscientific community.

In 1987 the United Nations passed a resolution supporting the establishment of the International Decade for Natural Disaster Reduction (IDNDR). The IDNDR program, which commenced in 1990, includes five basic objectives:

■ improve all nations' capabilities to mitigate the effects of natural disasters;

■ apply existing knowledge of the causes and effects of such disasters in developing guidelines for reacting to future events;

■ encourage scientific and technical studies aimed at reducing the loss of lives and property during natural disasters;

■ disseminate existing and new data related to natural hazards in a timely fashion; and

■ develop means to assess, predict, prevent, and mitigate future disasters through appropriate technical assistance programs.

The earth science community will play a key role, both in establishing priorities for the IDNDR and implementing subsequent studies.

One of the most successful cooperative programs in the geosciences has been the International Geological Correlation Program (IGCP). This ambitious effort was conceived by IUGS in 1968 and subsequently carried forward as a joint activity of IUGS and UNESCO. The IGCP has supported over 200 cooperative projects, involving participation by thousands of scientists from over a hundred countries during its existence.

A number of U.S. geoscientists are involved in IGCP projects. Most U.S. participation has developed through initiatives taken by individual scientists who have contacted specialists in their field elsewhere in the world and organized cooperative research. Program guidelines encourage specialists from different countries to reach agreement on common standards of data acquisition and interpretation. Some IGCP projects incorporate objectives pertinent to the genesis of energy and mineral resources, and most include a component of geoscientific training involving developing nations.

The International Geosphere-Biosphere Program (A Study of Global Change) is a current major international effort modeled in part on the IGY. It involves a wide range of sciences, including components of the solid-earth sciences. The proposal for this international effort, which was put forth by the ICSU unions, focuses on a program "to describe and understand the interactive physical, chemical, and biological processes that regulate the total earth system, the unique environment that it provides for life, the changes that are occurring in this system, and the manner in which they are influenced by human actions." Over 40 nations have agreed to participate in the program, which is expected to continue for decades. U.S. scientists have joined with colleagues from other countries in planning studies of the atmosphere, biosphere, and oceans from space and from the surface.

In 1983 the secretary-general of the United Nations established an independent commission to formulate "a global agenda for change" that would propose long-term strategies for achieving sustainable environmental development by the year 2000 and beyond. To accomplish this ambitious goal, it was recognized that greater cooperation would be required among developing as well as developed countries. The resulting World Commission on Environment and Development (WCED) met from 1983 to 1987 and produced a report entitled *Our Common Future*. In the report overview, entitled "From One Earth to One World," the WCED members summarized the challenges that humankind must address to establish "a new development path . . . one that will sustain human progress not just in a few places for a few years, but for the entire planet into the distant future."

Scientific research in space, the oceans, and Antarctica extends beyond the boundaries of individual countries. As a follow-up to the IGY experience, several scientific unions called for the creation of international organizations under the ICSU to encourage continued global cooperation in these three areas. To this end, the ICSU established three scientific committees with mandates for operation of indefinite duration: Space Research (COSPAR), Oceanic Research (SCOR), and Antarctic Research (SCAR). The programs developed and monitored by these international bodies include numerous solid-earth studies and involve solid-earth scientists in interdisciplinary activities.

The Ocean Drilling Program (ODP) is a successful global geoscience venture that evolved from a project of U.S. scientists. The concept and engineering for the ODP began in the 1960s when an effort was made to drill scientific holes through thin oceanic crust and into the upper mantle (Project Mohole). Project Mohole was conceived with good intent, but the scientific enthusiasm exceeded both the financial commitments and the current technical capabilities. Nevertheless, a commitment to scientific drilling had been established, and important related technology development was begun. In 1968 the research drilling vessel *Glomar Challenger* embarked on a mission to explore the crust of the world's oceans and thus to test a number of geological concepts. This Deep Sea Drilling Project (DSDP) was succeeded by the current ODP, which is a $45 million per year venture involving the United States (through NSF) and 19 other nations. During the past 20 years, over a thousand holes have been drilled, and more than 100 km of drill core has been accumulated. The cores, correlated with geophysical survey interpretations, have been essential to scientists throughout the world in determining the ages and distribution of oceanic sediments, the structure of the crust, and the worldwide history of oceanic and climatic changes.

Although international science need not always be conducted under a mantle of formalized programs, such arrangements are useful in gaining commitments from participating countries. Productive ventures in international science are also being conducted by individual scientists—either working alone or sharing project responsibilities with colleagues from other countries.

U.S. Collaborative Activities

Besides U.S. involvement in many major international earth science organizations, events, and bilateral programs, U.S. government agencies, such as the NSF, USGS, NASA, and NOAA, have played important roles in global programs. International collaboration has also been encouraged and facilitated by various U.S.-based scientific societies, by the National Academy of Sciences, and by individual academic institutions and scientists in the United States.

Since its formation in 1950, the NSF has fostered and supported U.S. participation in international science activities that "promise significant benefit to the U.S. research and training effort." The foundation's policies encourage U.S. awareness of science and engineering developments in foreign countries, stimulate initiation of international cooperative activities, provide opportunities for scientific collaboration in developing countries, and offer support to U.S. institutions for research studies conducted abroad. Cooperative science programs currently being conducted by the NSF include studies in Australia, the People's Republic of China, the former Soviet Union, Eastern European countries, India, Japan, Korea, Argentina, Brazil, Mexico, Venezuela, New Zealand, Taiwan, and Western Europe.

The Western European program is one of the most ambitious, involving over a dozen nations. NSF support includes travel funds for U.S. scientists to visit other nations for variable periods of time, research participation grants to support individual or joint studies in host nations, and funds for international seminars addressing topics of common interest to participating countries. NSF also currently supports research by U.S. earth scientists on an exchange basis with other nations' Antarctic expeditions under the provisions of the Antarctic Treaty.

The NSF provides joint funding with other nations to form special organizations committed to scientific studies. An example is the U.S.-Israel Binational Science Foundation (BSF), an agreement signed in 1972 to establish a program of cooperative scientific research and related activities to be conducted principally in Israel. The BSF office, located in Jerusalem, coordinates projects in a variety of research areas, including the natural sciences. A similar arrangement with what was then Yugoslavia was established in 1973 to develop a program of cooperative science and technology projects.

In 1984 the NSF, in cooperation with the USGS, agreed to support the development of a global network of seismic stations that would telemeter readings to data centers around the world. Development of the full network is expected to require in excess of $100 million and involve at least 10 years of effort under currently projected funding levels. The resulting global system will provide high-quality geophysical data that can be studied by earth scientists throughout the world to further develop models of the interior, to improve our understanding of earthquake dynamics, and to continue the mapping of mantle convection patterns. This Global Seismographic Network (GSN) program is coordinated by the Incorporated Research Institutions for Seismology (IRIS), a private nonprofit corporation currently composed of 62 U.S. universities. The program's activities have been coordinated with those of the Worldwide Standardized Seismic Network (WWSSN) established under the auspices of NOAA, which is now supported and managed by NSF.

The success of the GSN program depends on the cooperation of many nations, several of which have global, regional, and/or national seismic network projects of their own. The most notable of the foreign global networks is the French GEOSCOPE project. This national program, established in 1982, now consists of 25 worldwide stations. The GEOSCOPE network will constitute the French contribution to a worldwide GSN program.

In 1985 seismologists recognized that the seismology community must coordinate its efforts if it was to develop an optimum GSN. Under the sponsorship of the Interunion Commission on the Lithosphere, representatives of 20 institutions met in Karlsruhe, Federal Republic of Germany, and founded the Federation of Digital Broad-Band Seismographic Networks. The federation, which represents 10 countries at present, has adopted standards for the system response of federation seismographic stations and for the formats to be used in the exchange of earthquake data. Future network siting plans have been made by its members, resulting in an improved collective global network. Preparations are now being made for the deployment of seismographic stations meeting federation standards. If current proposals are adequately supported, the federation's global network should comprise at least 90 stations by the early 1990s.

As stated earlier, the ODP is an example of a consortium-based international activity (operated through the Joint Oceanographic Institutions, Inc., on behalf of NSF) that continues to produce valuable earth science information. There are several other examples of consortia that have been formed (or are forming) that involve international cooperation in seismic reflection profiling and continental drilling.

Government-to-government agreements and letters of understanding have been proposed by various U.S. government agencies to foster global geoscience collaboration. The USGS uses such agreements as its principal mechanisms for carrying out overseas collaboration in both basic and applied geoscientific research. At present the USGS has 37 agreements in place. Most of the agreements provide a direct working relationship between the USGS and its counterpart bureau, agency, or department in the partner nation.

The U.S. Bureau of Mines has also negotiated a series of cooperative agreements with counterpart agencies in nations throughout the world. Each of the cooperative arrangements addresses a combination of societal, economic, and technical issues of mutual concern to the participants. Geoscientific input is a component of many of these agreements.

Cooperative agreements between the United States and other nations also can enable individual scientists to do field work in foreign countries, facilitate the transfer of funds for operating purposes between agencies, and provide for transmittal of geological data and specimens to the United States. Such provisions may be supplemented to allow future projects to be carried out under "umbrella" provisions added during the term of the agreement. In some cases, such as the Earth Sciences Protocol and the Earthquakes Studies Protocol between the United States and the People's Republic of China, several U.S. agencies may be involved.

In 1986 a memorandum of understanding for the development of scientific cooperation in the earth sciences within the framework of the topic "Evolution of Geological Processes in the History of the Earth" was executed between the United States and the former Soviet Union. This accord was developed as a component of the Agreement on Scientific Cooperation between the then Soviet Academy of Sciences and the U.S. National Academy of Sciences. The memorandum of understanding calls for identification of cooperative research projects in the earth sciences that would "contribute to our understanding of global processes and that could not be effectively pursued in the absence of such cooperation." Specific topics of study identified to date include:

- rift systems—Baikal (Russia), Rio Grande (USA);
- aleutian arc volcanism (USA)—Kurile/Kamchatka (Russia);
- collisional systems—Caucasus/Transcaucasus (Russia), Western United States;
- xenolith studies to determine chemical stratigraphy of the continental crust and mantle;
- volatile substances in igneous petrology; and
- deep drilling.

All of these projects include field and laboratory studies that will integrate geological, geophysical, and geochemical components.

In recent years the U.S. Congress has become more sensitive to environmental issues as well as political and economic factors in considering U.S. financial support of international development projects. Consequently, it has instructed the administrators of the Agency for International Development, the World Bank, and other international loan organizations to expand their evaluations of proposals soliciting U.S. support to include studies of the potential impact of new developments on the environment of the nations involved. Such evaluations are to focus on surficial geology, water resources, and potential geological hazards that may be precipitated by a proposed development. Participation of the geoscientific community in such appraisals is a requirement that will continue on a global basis.

The importance of multinational cooperation in the geosciences has also been recognized by professional societies. Among these is the Geological Society of America (GSA), in which geologists from Mexico and Canada have long played prominent roles, the GSA has allied itself with the Association of Geoscientists for International Development in seeking avenues of global collaboration. The GSA has also formed an International Geology Division to promote international cooperation, with particular emphasis on expanding collaboration between geoscientists in North America and developing countries.

The American Geophysical Union (AGU) has been closely associated with the U.S. National Committee for the IUGG since 1920. Although the AGU is a U.S. organization, it has a definite international orientation. One of its more important offices is that of the foreign secretary.

U.S. scientists and institutions play important roles in ongoing international geoscientific mapping programs. One example is the Circum-Pacific Map Project, an activity of the Circum-Pacific Council for Energy and Mineral Resources, which is supported by the USGS, the IUGS, and organizations from countries rimming the Pacific Ocean. This project, which has been active since 1974, has produced a series of maps and reports summarizing the geology and resource status of the Pacific Ocean and its margins. A similar effort, the Circum-Atlantic Project, has recently been started within the IUGS.

The American Association of Petroleum Geologists (AAPG), an organization whose members are principally involved in the applied geosciences, is becoming increasingly international in its composition and scope. Non-U.S. citizens now constitute 19 percent of its 38,000 members. A trend toward the international continues, as 32 percent of new members in the past 5 years have been non-U.S. citizens working in the United States and abroad.

The U.S. petroleum and mineral industries have recognized the benefits of global collaboration in the earth sciences for many years. This awareness and participation have been demonstrated by:

- providing the geoscientific community with timely access to geological and geophysical data produced in worldwide programs;
- participating in major data integration projects, such as the Circum-Pacific and Circum-Atlantic international mapping projects; and
- contributing to broader international understanding of new concepts in the geosciences through presentations, publications, and participation in professional societies.

Private industry has been involved in global geosciences through an assortment of individuals and organizations. These include contractors and consultants employed by industry and government agency clients to provide services outside the United States and resource companies seeking energy and mineral commodities by means of international exploration and development programs. The data acquired by these groups that involve global collaboration are of two general types:

■ Information that is classified as sensitive or proprietary and is of a nature that provides an owner with a competitive edge. This type of data is not made readily available to the geoscientific community; often it is withheld for a specific period of time. Geological, geophysical, and geochemical reports, maps, and logs in this category represent a significant potential source of international geoscientific information that eventually is released for general reference.

■ Scientific data that can be dispersed to the community as they are acquired through a number of media. They are disseminated through presentations at meetings, publication of papers in professional journals, participation in international organizations, and sharing of data with appropriate agencies of the host country, such as geological surveys and mineral development agencies.

Petroleum and mineral companies have historically contributed to the education of the geoscientists of developing nations in the course of exploring and evaluating the natural resource potential of the host countries. For many years, training was on an informal basis, but a number of companies have now established more structured procedures, including financing the college educations of students and company employees from Third World nations. As the resource-rich countries have achieved greater economic independence, exploration and development have become joint-venture operations, and the scientific and technical education of the host nation's citizens is now seen as a requirement rather than an option. This has resulted in greater independence on the part of the developing countries and in exposure of both students and instructors to other nations' geological frameworks.

The extent to which such geoscientific backgrounds are shared on a global scale depends largely on the economic situation for a private company in a foreign country. As commodity prices fluctuate, private firms can find it difficult, if not impossible, to continue to operate in certain locales and may be forced to withdraw from that particular country. In such cases, nonoperating expenses, such as funds invested in educating local residents, often are curtailed or withdrawn. Consequently, the contributions to the global geoscientific data base by the private sector have been erratic over the years.

Many U.S. multinational corporations have established overseas facilities specifically designed to keep abreast of advances in the geoscientific disciplines. Individual programs range from major investments in overseas laboratories to closer monitoring of scientific journal articles. The personnel involved in these joint efforts include both U.S. scientists and residents of the host countries. Overseas offices and facilities have allowed U.S. firms to maintain an awareness of scientific and technical advances, developments in the host country that affect a company's goals and economic success, and the availability of resources required for a firm's programs.

Over the past two decades, U.S.-based companies have been faced with a significant increase in foreign competition from Europe, the Middle East, the Far East, Latin America, and Australia. This includes a significant expansion of nationalized oil companies' efforts to compete internationally outside their home countries. Available capital no longer provides the U.S. private sector an advantage in competing for development rights. As a result, petroleum exploration and development projects have involved a growing number of joint-venture participants. Nonpetroleum-related economic interrelationships and technology transfers are often required by foreign governments as well.

The United States must remain aware of the world's energy and mineral supplies to recognize our nation's potential resource vulnerability. Resource reliance is an even more important economic issue if a nation experiences major fluctuations in its dependency on certain commodities. It is essential that geoscientists be allowed access to other geological environments if mineral and hydrocarbon resource concentration mechanisms are to be accurately interpreted.

In the course of providing the United States with its basic mineral and energy resources needs, private industry and the earth science community must be aware of the potential environmental impacts of resource exploitation. Exploration and development decisions involve a continual balancing of values if the needs of future generations are to be weighed against the needs of today. These concerns can be most effectively evaluated only within the context of the Earth as a whole.

Other Nations' Activities

Many other developed countries are making major investments in earth science programs both within and outside their respective borders. Canada, Germany, Japan, and France are among the nations that have established long-term geoscience assistance and research projects in other countries.

Support for the education of scientists from developing nations by the developed nations is a method of global collaboration that receives limited publicity. France currently subsidizes more graduate and undergraduate students from the Third World than any other nation. This includes geoscience students studying in France and other countries. This commitment to educating citizens of African and Latin American countries so that they can independently recognize and contribute to the economic, ecological, and environmental needs of their respective nations is considered to be a sound investment by the French government.

The education of scientists is an opportunity to assist developing countries that should not depend on the vagaries of individual U.S. agency budgets or on programs provided by other nations. The impact that scientists will make on their societies has been recognized, and subsidizing the education of Third World geoscientists should be considered a worthwhile investment on the part of the United States.

Proposed Programs

New solid-earth science programs are expected to be implemented in the near future. Most of these should expand U.S. participation in cooperative geoscience studies on a global scale.

A study on earth system sciences (ESS) by NASA stated as a goal "to obtain a scientific understanding of the entire earth system on a global scale by describing how its component parts and their interactions have evolved, how they function, and how they may be expected to continue to evolve on all time scales." The ESS study identified three basic approaches to accomplish this objective:

■ long-term global observations of the Earth from the surface and from space,
■ an improved data system to process existing and future information regarding the Earth, and
■ development of models on the basis of the data obtained.

Input from the solid-earth science community at an international level will be an essential requirement for the success of this program.

The NRC's Space Studies Board recommended in its report *Mission to Planet Earth* that a program be established during the period 1995–2015 with four primary goals:

■ to determine the composition, structure, and dynamics of the crust and interior;
■ to understand the dynamics and chemistry of the oceans, atmosphere, and cryosphere and their interactions with the solid earth;

■ to characterize the relationships of living organisms with their physical environments; and
■ to monitor the interaction of human activities with the natural environment.

The Mission to Planet Earth plan calls for an extensive network of land- and ocean-based observatories that will measure the physical properties of the atmosphere, lithosphere, and oceans. NASA and the Mission to Planet Earth plans have objectives that relate to other major studies, including the Global Change Program and the International Lithosphere Program.

A program to "Study the Earth's Deep Interior" (SEDI) was approved by the International Union of Geodesy and Geophysics in 1987. The principal objective of SEDI is to encourage cooperative studies of the structure, composition, and dynamics of the interior, particularly the lower mantle, and core-mantle boundary region. Specific topics included in the program are the geomagnetic dynamo and secular variation; paleomagnetism and the evolution of the deep interior; the composition, structure, and dynamics of the core; the dynamo energetics and structure of the inner core; the core-mantle boundary region; and lower-mantle structure, convection, and plumes.

A 1987 NRC report, *International Role of U.S. Geoscience*, summarized the status of U.S. participation in international geosciences and proposed changes and expansions for that role. One conclusion was that there had been a gradual decline in U.S. involvement in the geosciences on a global scale since World War II. Prior to 1945, U.S. foreign aid programs supported numerous geoscientific studies and mineral resource development projects throughout the world in cooperation with U.S. private companies. Many of these were terminated or reduced in the early 1950s. As a consequence, joint scientific research efforts between U.S. geoscientists and their foreign counterparts declined.

The committee identified three areas that could benefit from an expanded U.S. role in the global geosciences and recommended actions to address each:

■ *Basic Scientific Research*. The changing nature of geoscientific research in the world requires an expansion of U.S. involvement in international science consultation and data exchange, increased support for science and technology agreements, and additional opportunities for U.S. geoscientists to participate in field studies in foreign environments.

■ *Economic Interests*. In order to improve the competitive status of the United States, the flow and exchange of relevant geoscience information through U.S. embassies and consulates should increase, and cooperative programs involving geoscientists from the Third World and other nations should be strengthened through organizations such as the Agency for International Development.

■ *Foreign Policy*. The awareness on the part of U.S. federal government groups involved in foreign policy decisions of the importance of the geosciences in negotiating global agreements should be improved. Specific policy topics requiring scientific input include waste management, acid rain, hazard reduction, energy and mineral resource availabilities, and desertification.

The committee's report included a proposal that the United States consider the establishment of an American Office of Global Geosciences "to remedy existing deficiencies and to develop a long-term mechanism for an increased geoscience contribution to U.S. foreign policy, economic growth, and basic research." The proposed office would serve as a clearinghouse for international geoscience information and could help coordinate geoscientific projects and activities involving private industry, governments, and academia. Examples of how this could be accomplished include the assignment of additional technically qualified regional resource officers to U.S. embassies and consulates. These officers, referred to in the past as mineral attachés, could keep abreast of global mineral and energy resource availability, promote the interchange of scientific and technical data, and be aware of local developments affecting science-related activities.

RECOMMENDATIONS

The following discussion and recommendations focus on undergraduate education, instrumentation and facilities, data collection and analyses, and global collaboration. In other sections of this chapter, conclusions and recommendations are highlighted in specific sections and are not repeated here.

Education in the Solid-Earth Sciences

Significant changes are required to make the training of solid-earth scientists reflect changing societal demands on the profession. The committee believes that no single discipline's viewpoint is adequate for understanding the behavior of earth processes—even for those that are fairly well de-

fined. **The conventional disciplinary courses should be supplemented with more comprehensive courses in earth system science.** Such courses should emphasize a global perspective, interrelationships and feedback processes, and the involvement of the biosphere in geochemical cycles. This is important as application of the earth sciences is increasingly toward interdisciplinary problems.

Curricula are beginning to change in response to new and emerging fields but inevitably lag. **New courses need to be developed to prepare students for growth in both employment and research opportunities in areas such as hydrology, land use, engineering geology, environmental and urban geology, and waste disposal.** Such courses will be necessary to prepare students for changing careers in both the extractive industries and environmental areas of the earth sciences. No longer are these two areas separate, as mineral and energy resources need to be exploited in environmentally sound ways. Many of these "new" courses will cut across departmental boundaries. **Colleges and universities should explore new educational opportunities (at both the undergraduate and graduate levels) that bridge the needs of earth science and engineering departments.** This need arises from the growth of problems related to land use, urban geology, environmental geology and engineering, and waste disposal. The convergence of interests and research is striking, and the classical subject of "engineering geology" could become a significant redefined area of critical importance for society.

The need for training students to undertake basic research is fundamental; **funding for independent research at all levels of education should be increased**. The understanding of the scientific process acquired through research will serve students well whether they choose an academic, applied, or industrial career path.

As new concepts arise and fields change, there is a need, even for experienced scientists, for forms of continuing education. **The sabbatical leave concept should be enhanced to provide researchers the ability to evolve with the field**. Such a procedure would promote scientific currency and foster information exchange to the benefit of all.

Finally, **support for graduate and postdoctoral studies should be strengthened**. Supplying the basic educational training for a hundred doctoral-level and several hundred master's-level earth scientists requires between $10 million and $20 million per year in scholarships and educational research support. This is a small but extremely cost-effective investment of high importance to the nation.

Instrumentation and Facilities

The breadth of the solid-earth sciences has ensured that an extraordinary range of instruments are required for research. One way of categorizing these instruments is used in the Implementation section of Chapter 7. They have been classified according to where they are deployed. A large class is deployed in laboratories, but other instruments that are used in field studies, in the widest sense, may be deployed in space, on aircraft, at the land surface, on the surface of the sea, and on the sea bed. Still other instruments are deployed within the crust in drilled holes.

The trade-offs between costs of deployment and scientific return represent a particular challenge, because different parts of the earth science community are involved with different instruments and environments. Information summarized in Chapter 7 indicates that on balance all the priority themes outlined in this report are addressed by the funding of instrumentation and facilities under one or more federal programs. **Vigilance in pursuing priorities is needed to ensure that an appropriate balance continues to be maintained.** Many national laboratories have instrumentation and personnel that could be valuable to both university and industrial researchers. Arrangements should be made to enable qualified researchers access to this significant resource.

Data Gathering and Handling

The digital data revolution has provided exceptional opportunities for comprehensive and rapid assessment of data, but it has also created several new and complex problems that need the immediate attention of organizations and individuals involved in the solid-earth sciences. Despite the indispensability of data to the solid-earth sciences, data are not a discipline and therefore generally lack an associated constituency. Attempts to organize data bases and manage them are often construed as interfering with individual scientists or organizations or as a drain on limited scientific resources. This attitude must change if data are to be used to their greatest advantage.

Today, the retention and dissemination of data in the earth sciences are characterized by the lack of an overall national policy. Yet these data are a national resource and form the backbone of the computer revolution as applied to the solid-earth sciences. Most data have a long useful life and a potential breadth of usefulness that far transcends the purposes for which they were collected.

To maximize the effectiveness of the data that exist or are being gathered, a national earth science data policy or set of guidelines should be established. One element of this policy should be the establishment of a distributed national data management system. A data management system that will permit fast, inexpensive, and convenient retrieval of data is feasible within the foreseeable future. Modern telecommunications and computer technology allows the establishment of a national, networked, distributed data system built on existing data centers.

National policies are needed that will provide incentives for organizations and individuals to first digitize existing and future solid-earth science data and then to place those data in a national data system. To avoid misunderstanding and to maximize cooperation, all segments of the community, especially the private sector, must be strongly involved in developing these policies from the outset. Suitable incentives may be complex, but it is clear that the rights of individuals, organizations, and data brokers must be recognized and held inviolate.

A number of types of incentives can be considered. Because funds attached to a national system would probably be inadequate for the purchase of significant quantities of data, consideration should be given to providing tax incentives to encourage private industry to make data publicly available. Attention should also be given to providing incentives and guidelines for academic researchers. For example, the awarding of a grant or contract involving federal funds might be accompanied by a stipulation that data obtained in the research must be made publicly available (the NSF's Ocean Sciences Division has such a policy). Furthermore, consideration could be given to providing publication credit for placing data sets into central repositories. Special consideration should be given to data acquired by private firms or individuals when exploring public lands. For example, conditions could be established whereby the granting of exploration permits is accompanied by the requirement that data obtained in surveys be placed in a national data system with eventual public access. Already, some states have public release requirements for oil and gas well records.

Policies for firms and individuals that stake mining claims on federal lands also need to be investigated. Existing mining law, stemming from 1872, requires that claimants to federal lands do a hundred dollars worth of assessment work per year per claim (claims have prescribed maximum dimensions). The proof of assessment work might be extended to include submissions of verified data obtained in the course of that work.

Finally, a revision in policies for oil and gas exploration and exploitation on the outer continental shelf should be considered. The present policy is that most (or all) such data are filed with the U.S. Minerals Management Service, but the data do not become publicly accessible. Public access after prescribed periods might be included in requirements or incentive policies.

Ideally, a common data exchange format could be agreed to, so that any data set could be translated into the exchange format, routed across networks to any computer on the network, and finally transformed from the exchange format into the format of the receiving machine. This arrangement would permit the collectors of data to be the stewards of their data base, so that they could retain responsibility for the quality and security of their data. However, all data-generating individuals and organizations should be encouraged to make their data available to a national archive-information center as a backup to a distributed system and also to provide data to nonspecialist users.

Professional societies and government agencies have made commendable efforts toward standardized data bases and exchange formats, which are essential to a national distributed data system. Additional exchange formats are required, and data definitions and quality standards must be established and accepted for broad use in the community if data exchange is to be effective. In addition, strong liaison should be maintained with the international community in coordinating data-base and exchange formats and in standardizing data definitions.

An initial step in a distributed data system should be an on-line national data directory listing earth science data holdings of all participating organizations, with provision for continual updating of the directory. The USGS has helped to show the way by making vast sets of data available at low price on CD-ROM. Several European agencies are currently addressing the need for a single repository of information that catalogs and collates all items of potential interest to earth scientists. The United States needs to become more involved in this process. A distribution center should inventory the geoscientific data currently available from all nations.

A national archival information directory should provide an on-line catalog of data sets using standard descriptive elements, and the system should be equipped with a mechanism to encourage those generating information to document their data adequately. Arrangements to maintain the system need to be made with specialists who have the requisite education and expertise to satisfy both the subject and the computer requirements of the user. The data directory should be available to any individual. The interface between the data directory and the user should be simple enough to allow the casual user to search the system adequately.

To help fulfill the recommendation for a national data policy or set of guidelines, a national advisory committee on solid-earth science data should be set up to provide policy oversight through the entire data chain. This committee should be made up of representatives from the diverse sectors of the solid-earth community. It should concern itself with recommending national policy, identifying problem areas, suggesting mechanisms for solving the problems, and strengthening communication of solid-earth science data among government, industry, and academia. It could also review and suggest improvements to research funding policies regarding improved data management.

In the area of the educational and training needs in data handling, professional societies, appropriate federal agencies, and colleges and universities need to take a leading role in ensuring that scientific and professional personnel acquire the background needed to use the available technology. Training and experience in data management should be encouraged for students of the solid-earth sciences at an early stage in their education. Most students receive no formal training in data-base management, although most undergraduates now are required to learn one or more computer languages. With a modest adjustment of curriculum, the principles of data management could be incorporated into these programming courses. For earth science students, instruction must concentrate on the creation and manipulation of data bases typical of those they will use later. **Federal agencies, professional organizations, and private institutions should work closely with the academic community to strengthen curricula and improve facilities for educating students, educators, researchers, and practitioners in the area of data management.**

The above discussion has dealt primarily with digital data, but the discussion and recommendations apply in general to any type of earth science data. For example, a national policy or set of guidelines should be established for the acquisition, archiving, and distribution of physical samples—solid, liquid, and gaseous. A distributed national sample management system, parallel to the data management system described above, should be implemented. Where appropriate, it should build on existing sample repositories; an example of such repositories is that established for curation of materials from the Deep-Sea Drilling Program and the Ocean Drilling Program. The problems of sample capture, maintenance, and distribution are much the same as those encountered with digital data and are every bit as pressing.

Global Collaboration

Major advances in the earth sciences are most likely to originate from better understanding of how the Earth functions as a total system. Therefore, if there is to be an improved interpretation of that system, the energy fluxes that drive it, and the mechanisms that control them, **data must be collected and concepts developed at a global scale and integrated into comprehensive models**. Although we cannot expect to design an exact model of the Earth, more accurate interpretations can be expected if data sets are global in scope and concepts are tempered by discussions with geoscientists from multiple disciplines and other nations.

One of the principal stimuli for the formation and promulgation of international geoscience unions and congresses has been the recognition of a need for systematic geological data exchange and standardization. The phenomenal growth of telecommunications and computer capabilities in recent years has generated both great opportunities and great complications regarding the exchange and standardization of data. U.S. geoscientists must work closely with their foreign counterparts to make global data available on a global basis. The system of World Data Centres established through ICSU is a first step.

Cooperative earth science programs between nations must include exchanges of scientists as well as data. In recent years U.S. scientists have had many more opportunities to go to those portions of the globe that had previously been restricted. The restraints facing U.S. scientists are no longer generated principally by a foreign government's political policy or the inaccessibility of a geological terrain. They are more often due to funding limitations or delays within a bureaucratic process. For example, there have been instances of excessively delayed or limited responses by U.S. scientific groups to offers of cooperative studies by the former Soviet Union, its successors, and the People's Republic of China. After many years

of scientific isolation, these nations are currently willing to allow foreign scientists to visit and study unique geological features within their boundaries. U.S. geoscientists should be given the opportunity to take advantage of these offers to share access and scientific concepts while cooperative attitudes prevail.

U.S. scientists must also recognize that there are languages other than English. The parochial attitude that travelers from the United States have no reason to learn the language of another nation is obsolete. Geoscientists from the United States are obviously handicapped in not being able to understand scientific presentations by their non-English-speaking peers and not being able to read articles in foreign scientific journals. Both undergraduate and graduate curricula at U.S. universities should encourage that foreign-language courses be taken by students anticipating involvement in the earth sciences.

In 1982 the National Science Board stated that the United States was at a critical point in its international scientific relationships. This conclusion was based on evidence that:

■ American scientists were no longer leading in many scientific fields;

■ U.S. industry was being significantly challenged in its overall technological capabilities;

■ scientific problems were becoming more global in scope and would have greater impact on the well-being of the United States;

■ cooperation between nations in scientific endeavors would become increasingly important in the short term;

■ the expanding scales and complexities of scientific problems now required facilities and operations whose costs justified greater international cooperation in sharing the expenses, risks, and benefits;

■ foreign policy considerations were playing an increasing role in the conduct of international scientific activities; and

■ U.S. science and technology policies and programs needed to be coordinated because of their growing interdependence and impact on U.S. national security.

The board pointed out in its 1982 report that "maintaining the vigor of the U.S. research effort requires a broad worldwide program of cooperation with outstanding scientists in many nations." The report concluded that "planning for new facilities and the setting of priorities for major scientific investigations and programs should be carried out with the full recognition of the priorities of other countries, and in an environment that encourages complementary or planned supplementation, cost sharing, and coherence of the various efforts of cooperating countries."

There is an obvious need for collaboration between earth scientists on a global scale. Recommendations as to specific means by which such cooperation could be expanded will vary within disciplines and nations. Those topics that involve a broad representation of the geosciences and transcend the nebulous boundaries established between disciplines should be given particular attention.

For example, a growing concern of all nations is the impact of increasing populations on our environment. Since many of these problems are worldwide, the solutions require global collaboration. Geoscientists will play an important role in mitigating the effects of high population densities while social scientists and politicians address the causes. Earth scientists will also play increasingly important roles in assembling the crucial data required to make the correct decisions. The United States, as a member of the world community, has a responsibility to aid in resolving such problems.

BIBLIOGRAPHY

NRC Reports

NRC (1983). *Toward an International Geosphere-Biosphere Program*, National Academy Press, Washington, D.C.

NRC (1986). *Global Change in the Geosphere-Biosphere: Initial Priorities for an IGBP*, U.S. Committee for an International Geosphere-Biosphere Program, National Research Council, National Academy Press, Washington, D.C., 91 pp.

NRC (1987). *Earth Materials Research: Report of a Workshop on Physics and Chemistry of Earth Materials*, Committee on Physics and Chemistry of Earth Materials, Board on Earth Sciences, National Research Council, National Academy Press, Washington, D.C., 122 pp.

NRC (1987). *International Role of U.S. Geoscience*, Committee on Global and International Geology, Board on Earth Sciences, National Research Council, National Academy Press, Washington, D.C., 95 pp.

NRC (1988). *Space Science in the Twenty-First Century: Imperatives for the Decades 1995 to 2015: Mission to Planet Earth*, Task Group on Earth Sciences, Space Science Board, National Research Council, National Academy Press, Washington, D.C., 121 pp.

NRC (1990). *Facilities for Earth Materials Research*, U.S. Geodynamics Committee, Board on Earth Sciences and Resources, National Research Council, National Academy Press, Washington, D.C., 62 pp.

NRC (1990). *Assessing the Nation's Earthquakes: The Health and Future of Regional Seismograph Networks*, Committee on Seismology, Board on Earth Sciences and Resources, National Academy Press, Washington, D.C., 67 pp.

NRC (1990). *Spatial Data Needs: The Future of the National Mapping Program*, Mapping Science Committee, Board on Earth Sciences and Resources, National Research Council, National Academy Press, Washington, D.C., 78 pp.

NRC (1990). *Research Strategies for the U.S. Global Change Research Program*, U.S. National Committee for the IGBP, National Research Council, National Academy Press, Washington, D.C., 291 pp.

NRC (1990). *A Safer Future: Reducing the Impacts of Natural Disasters*, U.S. National Committee for the Decade for Natural Disaster Reduction, National Research Council, National Academy Press, Washington, D.C., 67 pp.

NRC (1991). *Solving the Global Change Puzzle: A U.S. Strategy for Managing Data and Information*, Committee on Geophysical Data, Board on Earth Sciences and Resources, National Research Council, National Academy Press, Washington, D.C., 52 pp.

NRC (1991). *International Global Network of Fiducial Stations: Scientific and Implementation Issues*, Committee on Geodesy, Board on Earth Sciences and Resources, National Research Council, National Academy Press, Washington, D.C., 129 pp.

NRC (1992). *A Review of the Ocean Drilling Program Long Range Plan*, Ocean Studies Board, Board on Earth Sciences and Resources, National Research Council, National Academy Press, Washington, D.C., 13 pp.

NRC (1992). *Toward a Coordinated Spatial Data Infrastructure for the Nation*, Mapping Science Committee, Board on Earth Sciences and Resources, National Research Council, National Academy Press, Washington, D.C.

Other Reports

American Geological Institute (1987, 1989). *Geoscience Employment and Hiring Survey*, AGI, Alexandria, Virginia.

World Commission on Environment and Development (1987). *Our Common Future*, Oxford University Press, 400 pp.

ESSC (1988). *Earth System Science: A Program for Global Change*, Earth Systems Sciences Committee, NASA Advisory Council, National Aeronautics and Space Administration, Washington, D.C., 208 pp.

AAAS (1989). *Science for All Americans: Project 2061*, American Association for the Advancement of Science, Washington, D.C., 217 pp.

The International Geosphere-Biosphere Programme: A Study of Global Change (IGBP) of the International Council of Scientific Unions (1990). *The Initial Core Projects*, IGBP Report No. 12, International Council of Scientific Unions, Stockholm, 232 pp. plus appendices.

Joint Oceanographic Institutions, Inc. (1990). *Ocean Drilling Program: Long Range Plan, 1989–2002*, Joint Oceanographic Institutions, Inc., Washington, D.C., 119 pp.

Inter-Union Commission on the Lithosphere (1991). *International Lithosphere Program: Annual Report*, Report No. 15, Inter-Union Commission on the Lithosphere, International Council of Scientific Unions, Stockholm, 112 pp.

NSTA (1992). *Scope, Sequence, and Coordination of Secondary School Science, Volume I: The Content Core, A Guide for Curriculum Designers*, National Science Teachers Association, Washington, D.C., 151 pp.

The International Geosphere-Biosphere Programme: A Study of Global Change (IGBP) of the International Council of Scientific Unions (1992). *The PAGES Project: Proposed Implementation Plans for Research Activities*, International Council of Scientific Unions, Stockholm, 112 pp.

Office of Science and Technology Policy (1992). *Our Changing Planet: The FY 1992 U.S. Global Change Research Program*, Committee on Earth and Environmental Sciences, Office of Science and Technology Policy, Washington, D.C., 90 pp.

Interagency Working Group on Data Management for Global Change (1992). *The U.S. Global Change and Information Management Program Plan*, Committee on Earth and Environmental Sciences, Office of Science and Technology Policy, Washington, D.C.

7

Research Priorities and Recommendations

INTRODUCTION

The solid-earth sciences offer a wealth of research opportunities. These include basic questions such as the origin of the Earth, abstract challenges such as the consequences for continental evolution of convection within the solid mantle, and provocative issues such as the disappearance of the dinosaurs or how life without sunlight is generated and sustained at submarine hydrothermal vents. Earth science research continues to solve socially relevant issues such as land-use planning and the prediction of earthquakes. And now, with the global perspective offered by earth system science and international collaboration, problems of vital concern to society's future—such as global change, contamination of water supplies, formation of mineral deposits, and prospects for future energy sources—demand contributions from earth scientists.

The survival and prosperity of humanity depend on knowledge about the earth processes that produce resources, hazards, and environments. The world's growing population needs more energy, more minerals, and more water resources and generates increasing concentrations of waste products that can pollute the air, water, and land. As more people settle in marginal regions they face increasing danger from geological hazards. Humanity will become an agent of its own destruction unless

efforts to manage all of Earth's bounty as a nonrenewable resource prevail at every level. To do so will require scientific understanding of Earth's natural processes, particularly the linkages among the geospheres, the solid earth, the hydrosphere, the atmosphere, and the biosphere. The earth sciences—spurred by a combination of innovative concepts, powerful data-handling and modeling capabilities, refined field methods, and advanced laboratory techniques—are in an era of intellectual accomplishment that will provide this understanding.

Recognition of the interconnectivity of earth processes was initiated by the plate tectonics revolution. The ocean crust is composed of materials that emerge from the interior at spreading centers, is modified as it moves along the surface, and returns to the interior in subduction zones; the continents are built and modified by processes related to the same internal processes that modify the ocean crust. The system of interconnecting influences ranges from convection in the interior and the mechanism driving plates along the surface through the interchanges with the hydrosphere and biosphere that result in long-term atmospheric, oceanic, and climatic changes, to the effects of human activity on the geological cycles. Emerging perspectives permit a synthesis of earth science data on the global scale. Supercomputers provide breathtaking opportunities to sift enormous

> *Research in the solid-earth sciences is essential for the well-being of global society and for sustaining a high quality of life in the United States.*

quantities of global data and to simulate and explain earth processes by modeling experiments. New instruments are poised for development and use in monitoring the whole Earth from space, in deducing its inner structure and workings by seismology, and in exploring the composition of its smallest particles with high-resolution analytical probes.

Distinct intellectual paths wind through the structure of the solid-earth sciences, from theoretical research to the applications that flow from it. *Boundaries between theoretical and applied earth sciences are artificial*. Although theoretical research may be defined as speculative inquiry having no practical value, all engineering programs apply pure theory as an integral foundation for design and production. Research programs designed to improve the human condition—whether they are related to resource problems with water, energy, and minerals, to hazards presented by earthquakes, volcanic eruptions, landslides, and floods, or to environmental issues of global warming, desertification, and waste contamination—are crippled without basic research aimed at understanding earth processes.

The variety of research opportunities in the earth sciences can be categorized under priority themes. Deliberate consideration can then be given to how these themes might best be supported and developed during the next decade. This brings in the difficult issue of setting priorities among first-class research opportunities and pressing societal needs, within and across scientific fields. The way in which science priorities are established will surely be influenced by the 1991 report *Federally Funded Research* by the Congressional Office of Technology Assessment.

This chapter begins with a discussion of the problems of establishing criteria for setting science priorities. Following a summary of research initiatives and recommendations made over the past decade, the *goals* and *objectives* of the solid-earth sciences, as viewed by the committee, are presented as the *Research Framework* used throughout this report. Selected groups of *research opportunities* from the wide research areas covered in Chapters 2 through 5 represent the first stage of prioritization. For each of the eight *priority themes* that arise from the Research Framework, a single *top-priority research selection* was chosen with a remarkable degree of consensus. These eight top-priority research selections are discussed along with their supporting research programs and infrastructure; in addition, two high-priority selections for each theme are presented. The last section reviews the facilities needed to implement these major programs, which leads to the *research recommendations*. Comments about present and future research funding are then followed by a set of *general recommendations*.

SETTING RESEARCH PRIORITIES

Funding scientific research and technology is an expensive enterprise. Growing numbers of individual scientists require increasing support, and the megaprojects of big scientific collaborations consume vast amounts of money. These sometimes conflicting pressures emphasize the need for development of a national science agenda. That agenda should implement a system for setting priorities within each discipline and among all the sciences.

Planning and Decision Making

In the earth sciences new research initiatives usually develop within subdisciplines and reflect the interests of individual scientists. Initiatives spawned by independent scientists or groups of scientists inevitably become involved with funding agencies at an early stage. Scientists commonly establish a consensus about research directions and priorities by active participation in national and international workshops and conferences, by communication with colleagues, and by interaction with representatives from funding agencies. Scientists with common goals form working groups that determine implementation strategies, facility requirements, and needs for technology developments. Advisory committees can provide evaluations and recommendations on the long-range objectives and priorities in their field as well as the specific needs for funding, manpower, instrumentation, and facilities.

Supporters of each new initiative make the case for their project's funding. They attempt to persuade funding agencies and government of the paramount importance of investment in what they have concluded is a key area of research. If the funding organizations are to receive the critical advice that they need to make sensible allocation decisions, it is essential that the subdisciplines remain active and responsible in developing a consensus about directions and priorities.

The selection of priority research opportunities within a subdiscipline is relatively easy compared with the next step of ranking programs, or selecting priorities, among several subdisciplines. There is an additional problem of rational evaluation when support of a particular subdiscipline is shared by more than one agency. Similar oversight evaluations and comparisons are required before the relative merits of the science initiatives can be judged against other programs often competing for the same funds.

Until recently there was no body charged with the task of establishing priorities across and among different agencies supporting the earth sciences. However, when global change was recognized as an integral part of public policy during the 1980s, the Committee on Earth Sciences (CES) was appointed to focus disparate federally funded research on the global environment and organize it into the U.S. Global Change Research Program—a focused, agency-spanning effort to coordinate scientific understanding of global change. The success of the CES provided impetus for the reorganization and revitalization of the Federal Coordinating Council for Science, Engineering, and Technology (FCCSET), an interagency group charged with orchestrating federal research and development activities that cut across the missions of more than one federal agency. One of the seven new umbrella committees established to oversee broad areas of science and technology is the Committee on Earth and Environmental Sciences (CEES), successor to CES.

The CEES is a coordinating board composed of working groups and subcommittees dedicated to relevant research topics. It has demonstrated the utility of interagency activity coordination and of planned research program development on a national scale. A partnership has evolved not only among the government agencies but also with the scientific community. This orientation is shared particularly with the U.S. National Academy of Sciences (NAS) and increasingly with international organizations. The CEES, working with NAS, has developed a framework for planning and action, founded on five basic tenets, which include guidance by a set of priorities, evaluation criteria, and agreed-upon roles for the various government agencies and the CEES.

The agencies prepare project summaries providing specific information. At a series of initial meetings, dialogue between agency and CEES representatives provides the foundation for the CEES recommendations for the annual program. This is followed by the exchange of written material, in-

> "Are the resources available for the endeavor of solid-earth science commensurate with the challenges or the available talent? Are there too many of us for the resources? Are there too few resources for the many of us?"
>
> Charles L. Drake (EOS, 1990)

terpersonal briefings and discussion, and a series of meetings between agency representatives and a working group chairman from CEES. These meetings lead to consensus. Interactive revisions of the initial proposal then end with a final agency-endorsed recommendation to the Office of Science and Technology Policy and the Office of Management and Budget (OMB).

Despite the complexity of this procedure, a similar protocol could work well to determine priorities within the whole of earth sciences, as it works for those aspects addressing questions about global change. A high-level committee, with the capability and authority to evaluate priorities within earth system science, would present its findings to the FCCSET, which could then promote earth system science to OMB as coherent national activities, rather than as a collection of agency programs.

Individual and Group Research

The committee concluded, in conformity with many other reports, that the first priority must be adequate support of the best proposals from individual investigators. The National Science Foundation's (NSF) merit review task force endorsed the principle that the individual grant for basic research is central to academic science and technological enterprise. This is also one of the three guiding principles espoused by the OMB in prioritization of agency requests. Despite these declarations of principle, individual investigators—as a group—feel threatened by inadequate support. The intellectual resources contributed by individual members of the earth science community are the most valuable asset that community can claim. Core support for individual investigators will ensure the diversity in ideas and approaches that characterizes scholarly activity in the United States.

But many problems in the earth sciences are sufficiently complex that progress can be achieved only through cooperative multidisciplinary studies. In these cases, large-scale facilities cultivate growth and success; access to expensive, innovative, and often centralized new instruments and facilities has been a key stimulus in many breakthroughs.

Clearly, not every earth science institute can have its own synchrotron, portable seismometer array, accelerator mass spectrometer, or ocean and continental drilling program. Funding bases and management practices have developed to implement these facilities and ensure their cross-disciplinary use in an efficient and effective manner. A large instrument or facility is of no use if there is no funding for the science it supports or if the technical staff and operating budget are inadequate.

Do large, cooperative, expensive research programs drain support from the small grants programs? There is no simple answer, because even meritorious large-scale projects remain unfunded or have been terminated because of insufficient funds. Either overall funding for science is inadequate or the education system has produced more research scientists than it can support. But how can there be too many scientists when some projections of current trends indicate serious shortages of Ph.D.-level scientists in the first decade of the next century?

Program science focuses on achieving specific objectives, such as resource assessment, space exploration, natural hazard reduction, or waste management. Although its emphasis is commonly on practical ends, some program science in recent years has involved the assembly of multidisciplinary research teams for projects in pure science. Examples are studies of the continental lithosphere by deep drilling, studies of structure by reflection seismology, and establishment of global seismic networks. Science of this sort presents significant challenges because current scientific knowledge is fully exploited while new fundamental science is being developed. As long as the large programs are based on scientific goals, projects by individual investigators can make valuable contributions.

Key sources of support for program science include government agencies and industry. The setting of priorities is done by these organizations or by Congress if the funding needed is very large. The importance of these large projects should be judged according to the same standards as nonprogram science efforts in order to maintain a responsibly consistent set of priorities for all scientific disciplines.

Peer Review and Evaluation

If scientists do not establish their own system of evaluation, priorities will be set for them by bureaucrats or politicians. These professionals are skilled at tailoring budgets that address conflicting needs, but they are seldom expert in science; initiation and survival of scientific projects could come to depend more on the political savvy of special-interest lobbies than on scientific merit. If political expediency were the goal, many adverse consequences could be anticipated, including a short-circuiting of the peer review process, the intellectual exchange that most scientists consider essential for maintaining the quality of research.

The committee concluded that credible evaluation should always involve some form of peer or merit review because it is effective for judging both the competence of an investigator and the merit and utility of a research project. Peer review should be the quality control point in ranking large or small proposals. Close scrutiny at this point will ensure that an excellent proposal in any area finds support and that a poor proposal—even in a very important area—is rejected. There should be no interference with or protection of programs, and reviewers should encourage innovation. At the same time, funding renewals should be reviewed as rigorously as initial proposals.

The essential criteria for peer review comprise competence of the investigators, excellence of the proposal, utility of the research, and effect on the infrastructure. The peer review system for judging merit is ideal for identification of high-quality research, but the system is overburdened: in an endless cycle of paperwork, federal funds stimulate a large academic research base, which is then required to submit proposals for review. This demands an enormous effort on the part of competent scientists who could otherwise be conducting their own research. Much time and exertion are wasted in the preparation and review of unsuccessful, unfunded proposals. In this situation, creativity and innovation are stifled because overloaded reviewers may tend to reject unfamiliar thinking.

A 1990 NSF report addressed this problem. Its recommendations suggested methods for streamlining the peer review system, but resource availability may restrict their adoption. An important factor in these reviews is recognition of the roles of both small and large projects because they complement each other. Advances are usually initiated by individuals, but fulfillment often requires large teams. Similarly, priorities are commonly realized within a particular discipline, with consequent major advances evolving from interdisciplinary activities.

Priorities change and must be updated; granting agencies must set their courses years in advance. Dramatic swings in emphasis can only lead to loss of credibility. It is essential that any evaluation process create the environment for, and be responsive to, new ideas and techniques despite the risk.

TABLE 7.1 Evaluation Criteria for Research Proposals

Scientific merit is assessed on the basis of:
- Objectives and significance
- Breadth of interest
- Conformability to specified goals
- Potential or actuality of new discoveries
- Downstream benefits
- Bottleneck breakers
- Transfer values
- Education of professionals

Societal benefits to be considered include:
- Improvement of the human condition
- Relevance for industry
- National security and advantage
- Opportunity for international cooperation
- General education

The feasibility of a proposal includes programmatic or practical concerns such as:
- Scientific logistics and infrastructure
- Community commitment
- International involvement
- Timeliness
- Probability of success
- Costs: scientific and social

The return per dollar needs to be considered.
There should be a favorable ratio of benefits (societal + scientific + security) to cost.

Evaluation Criteria and Prioritization

Priority decisions should consider the three guiding principles applied by OMB in its assessment of funding requests from federal agencies:

1. Support is required for certain programs that address national needs and national security concerns.
2. Support for basic research must be adequate: small science receives high priority in the agencies' final programs.
3. Support for the scientific infrastructure and facilities must be maintained at adequate levels.

However, criteria for evaluation are the heart of the priority-setting process. Criteria for setting priorities and evaluating proposals are similar, although implementation may differ according to the scale of the initiative and the mandate of a sponsoring agency. The criteria that should be applied to research proposals through strict peer review include scientific merit, societal benefit, feasibility, and positive cost-benefit analysis. The lists of factors to be considered under each of these major criteria can become very long; a selection is displayed in Table 7.1.

Moving from lists to a workable selection presents many problems. An example from mineral resources illustrates the problems of prioritization. Selection of the most promising research opportunities in mineral resource research could greatly accelerate scientific progress and potentially save millions of dollars in research and exploration expenditures. Presumably the most promising opportunities are those with the greatest potential for dramatic advances in scientific understanding or for providing solutions to societal problems at the lowest costs and with the greatest potential for success. Higher priority might be awarded for a variety of reasons, for example, to support a historically productive investigator or line of research, to provide seed money for a risky but promising new line of research, to test a major scientific hypothesis, or to solve a significant societal problem.

However, perceptions of research priorities in mineral resources are apt to differ at different levels within an organization and between organizations. At the national level, preference might be given to strategic minerals that are in short supply in this country. The NSF might favor research made possible through the development of a new analytical method or a recent scientific discovery. An individual state might choose projects related to its particular resources. Government departments would select projects related to their missions. At a university, research emphasis will reflect the academic interests of faculty members. Similar diversity exists within various segments of industry and between industry, government, and academic institutions. Thus, the establishment of priorities within the field of mineral resources, as in any field, is dependent on the goals of the establisher. Those goals must be clearly thought out and communicated before priorities within and between disciplines can be assessed, much less ranked.

Once the goals have been established, it is necessary to design a procedure to apply the evaluation criteria to rank research proposals. One proposal is that the criteria be formulated into a standard set of questions and that the written answers produced are compared and judged; this formalizes the common procedure of open discussion. Others argue that this approach is too qualitative; they advocate a quantitative method of evaluation, using weighted criteria, making prioritization a structured, uniform process that can be defended. The problem is that the criteria (see, e.g., Table 7.1) probably do not have equal weight, and the weighting may well vary according to the mission of a funding agency. Such methods may be no more objective than those that arrive at consensus through discussion because of the subjective nature of the selection of weighting factors.

The committee explored various methods of scoring criteria in an attempt to establish a numerical ranking of research topics. We found that a numerical system appears to offer some degree of success when similar proposals are compared but is not effective when tested for ranking priorities among disparate proposals.

PREVIOUS RECOMMENDATIONS AND INITIATIVES

The committee used a variety of materials in the preparation of this report. It established 22 panels that prepared topical working papers. In addition, there were several recent publications reviewing specific aspects of the solid-earth sciences (or issues involving the solid earth) that had been prepared mainly, but not exclusively, by advisory panels or workshops under the aegis of the National Research Council (NRC). These publications were treated from the outset as the equivalent of additional panel reports, providing recommendations reached by consensus within a particular earth science community. Similar reports have been considered as they were published. Many of the committee members had participated in the preparation of these reports and long-range plans; their experience helped to put discussions and possibilities into a realistic perspective. The selection of priorities in this volume, therefore, reflects the conclusions of many previous committees that have dealt with the earth sciences.

Perhaps the most striking aspect of research planning during the past few years has been the growing parallel perception in different research communities that their interests are part of a global system. Consequently, there has been convergence among the research plans of groups concerned nominally with solid-earth, atmospheric, space, and ocean sciences. This convergence has focused on the driving processes within the solid earth and on global change as manifested mainly in the atmosphere and oceans.

This historical development is illustrated in Table 7.2 by a sequence of selections of research topics, beginning with the 1983 report prepared by the NRC Board on Earth Sciences (now the Board on Earth Sciences and Resources) at the request of NSF. That report, *Opportunities for Research in the Geological Sciences* (ORGS), recommended the eight priorities shown in the upper left of the table.

Another 1983 report was a research briefing de-

veloped for the NRC Committee on Science, Engineering, and Public Policy (COSEPUP) for the White House Office of Science and Technology Policy and federal agencies. The five research areas listed (which were based on the ORGS recommendations) were identified as those most likely to return the highest scientific dividends as a result of incremental federal investment. Four of these areas already had operating programs or were organized promptly. The organizations were the (1) Consortium for Continental Reflection Profiling; (2) Deep Observation and Sampling of the Earth's Continental Crust; (3) Incorporated Research Institutions for Seismology (IRIS); and (4) various satellite programs, of which the Global Positioning System (GPS) in particular was relevant. The fifth area was organized later; a 1987 report on the NRC Workshop on Physics and Chemistry of Earth Materials identified three major research topics where significant advances could be expected from research on earth materials.

In 1988 the NRC Space Studies Board published *Mission to Planet Earth*, one of six volumes responding to the National Aeronautics and Space Administration's (NASA) request "to determine the principal scientific issues that the discipline and space science would face during the period from 1995–2015." The volume outlined a bold integrated program for determining the origin, evolution, and nature of our planet and its place in the solar system. The importance of combining the space-borne program with an earth-based program was emphasized. The research objectives were addressed by the four themes given in Table 7.2, which proposed to expand NASA's mission by treating the whole Earth as a solar system planet.

The long-range plan for the NSF Division of Earth Sciences, prepared by NSF's internal advisory committee in 1988, emphasized *A Unified Theory of Planet Earth*. The influence of the two previous reports, ORGS and COSEPUP, is evident from the selection of research priorities.

The 1990 *Long-Range Plan of the Ocean Drilling Program* represents a distillation of workshop and panel discussions through 4 years and the conclusions of two major international conferences. The plan is based on four high-priority research themes, with 16 objectives. These research themes extend much deeper into the Earth than was envisaged in the early phases of ocean drilling programs.

In 1989 NASA's Solid Earth Science Branch was formed by joining two previously autonomous NASA programs on geology and geodynamics. This union reflected a recognition of the need to

TABLE 7.2 Priority Development: Previous Research Recommendations and Initiatives

1983: ORGS	*1983: COSEPUP*	*1987: PACEM*
Continental lithosphere	Seismic studies, continental crust	Mantle convection
Sedimentary basin evolution	Continental scientific drilling	Material transport through fluid flow
Magmas	Physics/chemistry of geological materials	Evolution of continents
Physical and chemical properties of rocks	Global digital seismic array	
Tectonic processes	Satellite geodesy	
Convection of Earth's interior		
Evolution of life		
Surficial processes		

1988: Misson to Planet Earth

1. Composition, structure, dynamics, and evolution of the interior and crust.
2. Structure, dynamics, and chemistry of the oceans, atmosphere, and cryosphere and their interactions with the solid earth (including the global hydrological cycle, weather, and climate).
3. Characterizing the interactions of living organisms among themselves and with the physical environment (including their effects on the evolution of the environment).
4. Monitoring and understanding the interaction of human activities with the natural environment.

1988: NSF	*1990: ODP*	*1991: NASA*
Continental lithosphere	Crust and upper mantle	Global geophysical networks
Physics and chemistry of earth materials	Physical behavior of the lithosphere	Soils and surface mapping
Global change: geological reconstruction	Fluid circulation in the lithosphere	Global topographic mapping
Fluid mechanics in earth sciences	Oceanic and climatic variability	Geopotential fields
Global positioning system (active tectonics)		Volcanism and climate
Studies in the Earth's deep interior		

ORGS: *Opportunities for Research in the Geological Sciences* (NRC, 1983).
COSEPUP: *Research Briefings 1983* (NRC, 1983).
PACEM: *Earth Material Research: Report of a Workshop on Physics and Chemistry of Earth Materials* (NRC, 1987).
Mission to Planet Earth (NRC, 1988).
NSF: Long-Range Plan, NSF Division of Earth Sciences, Advisory Committee, 1988.
ODP: Long-Range Plan for the Ocean Drilling Program, NSF, 1990.
NASA: Solid Earth Sciences in the 1990s, NASA Technical Memorandum 4256 (three volumes).

understand the Earth as a whole, comprised of interacting systems. NASA sponsored an international workshop in 1989 that developed a 10-year plan of research to integrate the two programs into one. The report, published in 1991, identified the five areas shown in Table 7.2 as deserving major emphasis in the solid-earth sciences for the 1990s.

Many other reports were taken into consideration by the committee for this volume, but those mentioned above suffice to illustrate the parallelism developing in research programs specified by organizations as different as NSF, NASA, and the Ocean Drilling Program (ODP). This is due in large part to recognition of the earth system as one that is interconnected on all scales and the fact that the wide disparity of time and space scales represented by geophysics, geochemistry, geology, fluid dynamics, and biological processes can be addressed for the first time by global data sets and modeling on high-speed computers. For example, it has become clear that ODP's existence is important for other earth science initiatives that deal with global processes and interactions to achieve their goals.

ODP also will play a role in the RIDGE (Ridge InterDisciplinary Global Experiments) initiative, which developed from several workshops, initially under the guidance of the NRC Ocean Studies Board. It illustrates the trend toward multiagency support; the planning effort is now supported by NSF, the Office of Naval Research, the U.S. Geological Survey (USGS), and the National Oceanic and Atmospheric Administration (NOAA). The rift valleys that are central to the RIDGE initiative are responsible for the formation of continental margins, where 70 percent of the world's population is concentrated. MARGINS is another interdisciplinary research initiative developed from an NRC workshop jointly organized by the Ocean Studies Board and the Board on Earth Sciences and Resources. Only recently has the patchwork of diverse studies of different disciplines become interpretable in terms of comprehensive models. The workshop group concluded that a significant change in direction from current research was required, with a shift away from phenomenological descriptions to an approach focusing on process-oriented studies and modeling fundamental physical processes. This new direction requires interdisciplinary organization and funding structures.

The trend in the initiatives outlined above has been to emphasize processes, recognizing the need for attention to the properties of earth materials. Continental drilling, like ocean drilling, is a technique. The ODP is now emphasizing the determination of processes through the technique of drilling. A similar philosophy is expressed in the 1988 report *The Role of Continental Scientific Drilling in Modern Earth Sciences: Scientific Rationale and Plan for the 1990s* (Interagency Coordinating Group for Continental Scientific Drilling, 1988), based on an international conference and workshop. The report presents a comprehensive plan for a program that "should be the mechanism by which scientific drilling activities of the Department of Energy, U.S. Geological Survey, and National Science Foundation and other agencies are coordinated and focused on critical problems of national interest . . . directed at fundamental research and closely integrated with other geological and geophysical studies to address outstanding problems in the earth sciences."

The U.S. Global Change Research Program has become a central focus because it involves important and urgent political and economic issues, which require the best of scientific attention. The International Geosphere-Biosphere Program (IGBP) was addressed by a 1988 NRC report, *Toward an Understanding of Global Change*, which identified early U.S. contributions. Also in 1988, *Earth System Science: A Program for Global Change*, was prepared by the NASA Advisory Council, with the anticipation that the program recommended would become a part of the planning for the Global Change Research Program. This involves "the initiation of a new era of integrated global observations of the Earth" and "the development of new management policies and mechanisms to foster coordination among NASA, NOAA, NSF, and other federal agencies engaged in earth system science and the study of global change." The organization and operation of the Committee on Earth and Environmental Sciences, outlined earlier, illustrate the new generation of management mechanisms.

GOALS, RESEARCH AREAS, OBJECTIVES, AND RESEARCH OPPORTUNITIES

The starting point for evaluation of solid-earth science programs must be to define the goals. Recent research and discoveries in the earth sciences have brought us to the stage where we should consider the Earth as a set of interrelated systems. The theory of plate tectonics gave new emphasis to the unifying concept of planet Earth as an integrated system, with every part functioning to some degree separately but being ultimately dependent on all others. New data on the Earth's interior reinforce the notion of an internal engine driving geological processes. The dynamic Earth behaves like a thermodynamic engine that generates stresses and flows in solid and fluid materials and causes differential transfer of matter in geochemical cycles. The crustal topography is shaped by internal movement, and the near-surface chemistry involves interaction between the oceans, the atmosphere, and fluids from the crust and mantle. The detailed architecture of the surface is carved by the action of fluids driven by energy from the external heat engine—the Sun—with the aid of gravity and tidal forces. Exchange of material deep within the interior is brought about by plate subduction, slow thermal convection of the mantle, and hot-spot volcanism. The distribution of water, economically valuable minerals, and energy resources is determined by these various processes. The multidisciplinary research areas described in this report reflect this new awareness of interconnectivity.

The committee agreed that the **GOAL** of the solid-earth sciences is

to understand and to predict the behavior of the whole earth system, from interaction between the crust and its fluid envelopes of atmosphere and hydrosphere through the mantle and the outer core to the inner core. A major challenge is to understand how to maintain an environment between the solid and fluid geospheres in which the biosphere and humankind can flourish.

Reaching this goal will require an understanding of:

■ the origin and evolution of the core, mantle, and crust and
■ the interactions and linkages between the solid earth, its fluid envelopes, and the biosphere.

Such a comprehensive understanding will provide a basis for meeting the significant challenges to society and to earth scientists:

■ to provide sufficient resources—water, minerals, and fuels;
■ to cope with the hazards—earthquakes, volcanoes, landslides, and floods;
■ to avoid perturbing the geological cycles—soil erosion, water contamination, and improper mining and waste disposal; and

■ to learn how to anticipate and adjust to environmental and global change.

The committee decided to structure its priorities on the basis of four broad objectives and the major research areas that support them. This framework or matrix of objectives and research areas served as the basis for our consideration of priorities in the solid-earth sciences.

The following four **OBJECTIVES** are derived from the challenges facing society in which fundamental understanding of the solid-earth sciences plays a primary role:

A. **Understand the processes involved in the global earth system, with particular attention to the linkages and interactions between its parts (the geospheres).**
B. **Sustain a sufficient supply of natural resources.**
C. **Mitigate geological hazards.**
D. **Minimize and adjust to the effects of global and environmental change.**

The committee selected the following five **RESEARCH AREAS** that will provide the understanding needed to address the above objectives:

I. **Global paleoenvironments and biological evolution.**
II. **Global geochemical and biogeochemical cycles.**
III. **Fluids in and on the Earth.**
IV. **Dynamics of the crust (oceanic and continental).**
V. **Dynamics of the core and mantle.**

These research areas all relate to the dynamic behavior of the earth system, but they emphasize different time scales, processes, and environments, and they progress from the surface downward into the core.

These societal challenges, objectives, and research areas were selected to provide comprehensive coverage of the whole earth system. They reflect the committee's best evaluation of where the research frontiers are, and they represent a solid foundation for making predictions about areas of research that are likely to succeed (see Table 7.3). In matrix form they constitute the **RESEARCH FRAMEWORK**, which has been used to categorize the research opportunities throughout this volume.

> *Major research opportunities arise from the new, global, highly interconnected view of the whole earth system: earth system science.*

These objectives and research areas also represent a stage in prioritization based on the broad trends of community consensus, illustrated by the series of published initiatives and plans discussed above and reinforced by the committee's discussions and draft materials prepared by the committee's panels. These **PRIORITY THEMES** have the greatest promise for achieving the goals and objectives of the solid-earth sciences. They represent the first-priority scientific issues for understanding the Earth, for discovering and managing its resources, and for maintaining its habitability. (Note that Objective A—understanding the processes—is not a priority theme; it is inherent in all of the research areas and is basic to the other three objectives.)

Each priority theme embraces a very wide range of research, as outlined in the earlier chapters. In the first stage of priority selection, subsets of **RESEARCH OPPORTUNITIES** (representing significant selection and thus prioritization) from a large array of research projects were listed in research frameworks at the ends of Chapters 2 through 5. (Chapter 6 considers the requirements of education, manpower, international collaboration, and the infrastructure of facilities and equipment required to support and maintain those opportunities.) These research opportunities include frontier areas:

■ where exploration of the unknown is still under way,
■ where different processes converge or overlap,
■ where data gathering is needed,
■ where different disciplines overlap, and
■ where conditions are ripe for computer modeling.

They are compiled here into a single table, Table 7.4.

PRIORITY THEMES AND RESEARCH SELECTIONS

Selection of Top- and High-Priority Research

The question of how to prioritize the research opportunities received much attention by the committee, and it was concluded that simply producing

TABLE 7.3 Aims of Priority Themes

Research Areas

I. Global Paleoenvironments and Biological Evolution

To develop a record of how the Earth, its atmosphere, and its hydrosphere as well as life have evolved, so as to yield understanding of how its surface environment and the biosphere have changed on all time scales from the shortest to the longest. Such a record provides perspective for understanding continuing environmental change and for facilitating resource exploration.

II. Global Geochemical and Biogeochemical Cycles

To determine how and when materials have moved among the geospheres crossing the interfaces between mantle and crust, continent and ocean floor, solid earth and hydrosphere, and hydrosphere and atmosphere. Interaction between the whole solid-earth system and its fluid envelopes represents a further challenge. Cycling through the biosphere and understanding how that process has changed in time is of special interest.

III. Fluids in and on the Earth

To understand how fluids move within the Earth and its surface. The fluids include magmas rising from great depths to volcanic eruptions and solutions and gases distributed mainly through the crust but also in the mantle.

IV. Crustal Dynamics: Ocean and Continent

To understand the origin and evolution of the Earth's crust and uppermost mantle. The ocean basins, island arcs, continents, and mountain belts are built and modified by physical deformations and mass transfer processes. The tectonic products of the deformations constitute the locales for resources introduced by chemical transportation. The shapes of landform surfaces are sculpted mainly by fluids.

V. Core and Mantle Dynamics

To provide the basic geophysical, geochemical, and geological understanding as to how the internal engine of our planet operates on the grandest scale and to use such data to improve the conditions on Earth by predicting and developing theories for global earth systems.

Objectives

A. To Understand the Processes in All Research Areas

To understand the origin and evolution of the Earth's crust, mantle, and core and to comprehend the linkages between the solid earth and its fluid envelopes and the solid earth and the biosphere. We need to maintain an environment in which the biosphere and humankind can flourish without risk of mutual or shared destruction.

B. To Sustain Sufficient Supply of Natural Resources

To develop dynamic, physical, and chemical methods of determining the locations and extent of nonrenewable resources and of exploiting those resources using environmentally responsible techniques. The question of sustainability, the carrying capacity of the Earth, becomes more significant as the resource requirements grow.

C. To Mitigate Geological Hazards

To determine the nature of geological hazards, including earthquakes, volcanic eruptions, landslides, soil erosion, floods, and materials (asbestos) and to reduce, control, and mitigate the effect of these hazardous phenomena. It is important to consider risk assessment and levels of acceptable risk.

D. To Minimize and Adjust to the Effects of Global and Environmental Change

To mitigate and remediate the adverse effects produced by global changes of environment and changes resulting from modification of the environment by human beings. These latter changes may necessitate changes in human behavior. In order to predict continued environmental changes and their effects on the Earth's biosphere, we need the historical perspective given by reconstructed past changes.

a ranked list of programs or facilities would be meaningless. The needs of different sections of the community, the various federal agencies, private corporations, and state and local governments (not to mention global and international bodies) are diverse. Priorities have already been established by government agencies, industry establishes its own priorities, and strategies in petroleum and mineral resources are driven by international economic and political factors. A critical evaluation leading to prioritization is more readily accomplished given a specific list of projects and a budget. Lacking such constraints, the committee employed the matrix of priority themes as the basis for an agenda in the solid-earth sciences, an outline of how priorities

might be determined through the next decade, depending on the availability of funds.

The committee recognized that, in a field as wide as earth system science, research needs to advance on a broad front. The research opportunities summarized in Table 7.4 all merit strong support. However, their number was clearly still too large to be considered a suitable response to the charge "to establish research priorities." The problem was to generate a selected list, and the approach adopted was to identify a single top-priority item for each of the eight priority themes that have framed this report. Candidates for the top-priority selections were solicited from individual members of the committee and then debated by the committee. A high degree of consensus was attained in making the selection, which can be attributed at least in part to the earlier effort spent in defining where the main issues and outstanding problems lie at this time. (Other groups from the diverse field of earth sciences might have made a different selection, but this selection is thought to provide as firm a basis for planning the future as any other that might be proposed.) Two high-priority research subjects were also selected for each priority theme (in one priority theme there were three); in most cases they could compete strongly for the top position (see Table 7.5). There are of course supporting and supplementary research programs associated with each of the priority selections.

Because the Earth consists of numerous complex interactive systems, it is not surprising that the research programs, facilities, equipment, and data bases related to the priority themes overlap to a considerable extent. Indeed, one possible criterion for emphasis on a particular research activity is that it relates to more then one research theme or objective. For example, seismic networks are very important for understanding (Objective A) crustal dynamics (Research Area IV) and the mantle and core (Area V), as well as for hazard reduction (Objective C). Major programs such as the national, international, and state seismic networks can thus be seen as important for science and society for several reasons. They serve two objectives and two themes, although information from them alone will not provide a complete answer to any specific research priority. Similarly, the geological history of the past 2.5 million years is important for understanding (Objective A) interaction between the Earth and its fluid envelopes (Area III), environmental and biological changes (Area I), and global geochemical cycles (Area II). These understandings are critical to assessing future global change (Objective D) and contribute substantially to sustaining water and soil resources (Objective B) and somewhat to hazard mitigation (Objective C).

Major programs such as national, international,

> "Try forecasting the future of physics . . . . I looked into the previous survey to see how well it had done in my pet field of atomic physics. The performance was unimpressive. Apparently nobody noticed that the laser was about to revolutionize atomic physics. . . . [T]he lesson is that scientific discoveries invariably exceed the power of our imaginations."
>
> Daniel Kleppner (*Physics Today, December 1991*)

and state seismic networks can thus be seen as important for science and society for several reasons. This applies to such programs as ocean drilling, which contributes to the understanding of all five research areas (Objective A) as well as to global change assessment (Objective D). Similarly, *The Role of Continental Scientific Drilling in Modern Earth Sciences* (Interagency Coordinating Group for Continental Scientific Drilling, 1988) specified applications addressing problems related to many different priority themes:

- Earthquakes and crustal deformation (III, IV, V, C)
- Volcanic and magmatic processes (II, III, IV, V, B, C, D)
- Evolution of continental lithosphere (I, II, IV, B)
- Basin evolution and hydrocarbon resources (I, II, III, IV, B)
- Mineral resources (B)
- Thermal regimes and geothermal energy (II, III, IV, V, B)
- Calibration of crustal geophysics (III, IV)
- Role of fluids in crustal processes (II, III, IV, B, C)
- Lithospheric dynamics (III, IV, V, C)
- Disposal of radioactive and toxic wastes (D)
- Subterranean bacteria (B, D)

Finally, the prominence of fluids in research priorities related to the solid earth is striking. Within research areas I through V, III is concerned directly with fluid-rock interactions, I includes paleoceanography and paleoclimatology, the processes in II are accomplished dominantly through fluids, and in IV fluids influence the strength of the crust and shape its surface in landforms. Among the societal objectives, water quality is the top priority for Area B, and microbiology in a hydrous environment is the top priority for Area D. The surface and outer few kilometers are in intimate contact with the hydrosphere, and water is a most reactive phase.

TABLE 7.4 Research Opportunities

Objectives

Research Areas	A. Understand Processes	B. Sustain Sufficient Resources Water, Minerals, Fuels
I. Global Paleoenvironments and Biological Evolution	■ Soil development and contamination ■ Glacier ice and its inclusions ■ Quaternary record ■ Recent global changes ■ Paleogeography and paleoclimatology ■ Paleoceanography ■ Forcing factors in environmental change ■ History of life ■ Discovery and curation of fossils ■ Abrupt and catastrophic changes ■ Organic geochemistry	■ Mineral deposits through time
II. Global Geochemical and Biogeochemical Cycles	■ Geochemical cycles: atmospheres and oceans ■ Evolution of crust from mantle ■ Fluxes along ocean spreading centers and continental rift systems ■ Fluxes at convergent plate margins ■ Mathematical modeling in geochemistry	■ Organic geochemistry and the origin of petroleum ■ Microbiology and soils
III. Fluids in and on the Earth	■ Analysis of drainage basins ■ Mineral-water interface geochemistry ■ Pore fluids and active tectonics ■ Magma generation and migration	■ Kinetics of water-rock interaction ■ Analysis of drainage basins ■ Water quality and contamination ■ Modeling water flow ■ Source-transport-accumulation models ■ Numerical modeling of the depositional environment ■ In situ mineral resource extraction ■ Crustal fluids
IV. Crustal Dynamics: Ocean and Continent	■ Landform response to change ■ Quantification of feedback mechanisms for landforms ■ Mathematical modeling of landform changes ■ Sequence stratigraphy ■ Oceanic lithosphere generation and accretion ■ Continental rift valleys ■ Sedimentary basins and continental margins ■ Continental-scale modeling ■ Metasomatism and metamorphism of lithosphere ■ State of the crust: thermal, strain, stress ■ Convergent plate boundary lithosphere ■ History of mountain ranges: depth-temperature-time ■ Quantitative understanding of earthquake rupture ■ Rates of recent geological processes ■ Real-time plate movements and near-surface deformations ■ Geological prediction ■ Modern geological maps	■ Sedimentary basin analysis ■ Surface and soil isotopic ages ■ Prediction of mineral resource occurrences ■ Concealed ore bodies ■ Intermediate-scale search for ore bodies ■ Exploration for new petroleum reserves ■ Advanced production and recovery methods ■ Coal availability and accessibility ■ Coal petrology and quality ■ Concealed geothermal fields
V. Core and Mantle Dynamics	■ Origin of the magnetic field ■ Core-mantle boundary ■ Imaging the Earth's interior ■ Experiments at high pressures and temperatures ■ Chemical geodynamics ■ Geodynamic modeling	

TABLE 7.4 Research Opportunities (continued)

Objectives

Research Areas	C. Mitigate Geological Hazards Earthquakes, Volcanoes, Landslides	D. Minimize Global and Environmental Change Assess, Mitigate, Remediate
I. Global Paleoenvironments and Biological Evolution		■ Environmental impact of mining coal ■ Past global change ■ Catastrophic changes in the past ■ Solid-earth processes in global change ■ Global data base of present-day measurements ■ Volcanic emissions and climate modification
II. Global Geochemical and Biogeochemical Cycles	■ Seismic safety of reservoirs ■ Precursory phenomena and volcanic eruptions ■ Volume-changing soils	■ Earth-science/materials/medical research ■ Biological control of organic chemical reactions ■ Geochemistry of waste management
III. Fluids in and on the Earth		■ Isolation of radioactive waste ■ Groundwater protection ■ Waste disposal ■ In situ cleanup of hazardous waste ■ New mining technologies ■ Waste disposal from mining operations ■ Disposal of spent reactor material
IV. Crustal Dynamics: Ocean and Continent	■ Earthquake prediction ■ Paleoseismology ■ Geological mapping of volcanoes ■ Remote sensing of volcanoes ■ Quaternary tectonics ■ Densifying soil materials ■ Landslide susceptibility maps ■ Preventing landslides ■ Dating techniques ■ Real-time geology ■ Systems approach to geomorphology ■ Extreme events modifying the landscape ■ Geographic information systems ■ Land use and reuse ■ Hazard-interaction problems ■ Detection of neotectonic features ■ Bearing capacity of weathered rocks ■ Urban planning: underground space ■ Geophysical subsurface exploration ■ Detection of underground voids	
V. Core and Mantle Dynamics		

TABLE 7.5   Priority Themes and Research Selections

Top Priorities	High Priorities
I. Global Paleoenvironments and Biological Evolution **The Past 2.5 Million Years** **A coordinated thrust at understanding how the Earth's environment and biology have changed since the onset of Northern Hemisphere glaciation about 2.5 million years ago.** Numerous current federal agency research activities can be brought to bear on this issue. National and international involvement is appropriate.	■ Environmental and biological changes over the past 150 million years since the oldest preserved oceans began to evolve ■ Environmental and biological changes prior to 150 million years ago
II. Global Geochemical and Biogeochemical Cycles **Biogeochemistry and Rock Cycles Through Time** **Establishing how global geochemical cycles have operated through time is now a realistic target.** This information is an essential element in working out how the earth system operates, and the research can help coordinate a number of federal activities. National and international activities are strong.	■ Construct models of the interaction between biogeochemical cycles and the solid earth and climatic cycles ■ Establish how geochemical cycles operate in the modern world
III. Fluids in and on the Earth **Fluid Pressure and Fluid Composition in the Crust** **Understanding the three-dimensional distribution of fluid pressure and fluid composition in the crust can be appropriately taken up at this time.** Instrumental, observational, and modeling capabilities within a number of federal programs can be effectively focused on this problem. National and international involvement is important.	■ Modeling fluid flow in sedimentary basins ■ Understanding microbial influences on fluid chemistry, particularly groundwater
IV. Crustal Dynamics: Ocean and Continent **Active Crustal Deformation** **Understanding active crustal deformation is vital to solid-earth science. There is an opportunity to revolutionize current understanding by coordinated effort.** This priority theme is of importance to the missions of several federal agencies as well as state and international bodies.	■ Landform responses to climatic, tectonic, and hydrologic events ■ Understanding crustal evolution
V. Core and Mantle Dynamics **Mantle Convection** **An integrated attack aimed at understanding mantle convection is timely.** Observational, analytical, and modeling techniques are available that can be brought to bear on the issue. Several federal agencies and national and international organizations are involved.	■ Establish the origin and temporal variation of the Earth's internally generated magnetic field ■ Determine the nature of the core-mantle boundary
B. To Sustain Sufficient Natural Resources **Improve the Monitoring and Assessment of the Nation's Water Quantity and Quality** **Establishment of a dense network of water quality and quantity measurements to manage water resources and to promote scientific advances.** This task requires coordination among federal and state agencies with existing programs in the general field.	■ Sedimentary basin research, particularly for improved resource recovery ■ Improvement of thermodynamic and kinetic understanding of water-rock interaction and mineral-water interface geochemistry ■ Development of energy and mineral exploration, production, and assessment strategies
C. To Mitigate Geological Hazards **Define and Characterize Regions of Seismic Hazard** **Because many people and much property in the United States are at risk from the hazard of earthquakes, it is timely to address the problem of seismic hazard.** This issue is of recognized importance to the missions of several federal agencies as well as to state, local, and international organizations.	■ Define and characterize areas of landslide hazard ■ Define and characterize potential volcanic hazards
D. To Minimize and Adjust to the Effects of Global and Environmental Change **Develop the Ability to Remediate Polluted Groundwaters, Emphasizing Microbial Methods** **A coordinated attack on establishing the ability to remediate polluted groundwaters on both local and regional scales.** Numerous current research activities in federal, state, and local agencies and private industry can be brought to bear on this issue. National and international involvement is appropriate.	■ Secure the isolation of toxic and radioactive waste from household, industrial, nuclear plant, mining, milling, and in situ leaching sources ■ Geochemistry and human health

The selection of a single top-priority research area for each priority theme is in one sense the end of the process of prioritization, but in another sense it is the beginning of the actual implementation of research programs. Progress in each top-priority selection cannot be entirely achieved without making progress in other areas, within the same and other priority themes. The whole system is interconnected. And while the need for funding agencies to have a set of priorities to guide their programs is clear, the danger of becoming bound by such priorities is even clearer. Breakthroughs in science are not predictable, and the priority themes and priority research selections do not pretend to cover all areas where breakthroughs might occur. Many other research areas have high potential for new opportunities and novel developments; committee members know that influential discoveries could emerge from studies that include the following:

- microbiology and fossil DNA,
- bacteria on mineral surfaces and in solutions,
- quantum mechanics,
- solar physics and its variations,
- materials science,
- computer science, and
- laser technology.

Advances in instrumentation can result in dramatic progress. Earth system scientists must be aware of research in these areas as well as areas more specific to their particular subdisciplines. They will need intellectual flexibility and acute sensitivity to the frontiers of understanding to properly assess developments in rapidly evolving scientific and technological fields.

Priority Theme I: Global Paleoenvironments and Biological Evolution

The aim of this priority theme is **to develop a record of the Earth, its atmosphere and hydrosphere, and the development of life. Such a record can provide a perspective for understanding continuing environmental change and its effect on the Earth's biosphere and for facilitating resource exploration.**

This priority theme is the subject of Chapter 3. It is from the sedimentary rock record that most of our knowledge of earth history is derived. New ways of dating rocks, fossils, and surface features are opening diverse avenues of research. In addition, the deep ocean record is revealing a distinctive and hitherto poorly understood aspect of earth history. This record is a window on the evolution of the coupled ocean-atmosphere system.

On the land, new methods for studying rates at which earth-surface processes occur are expanding our ability to understand how landforms and surficial deposits, including soils, have evolved in the recent past and how they can be expected to change, with or without human influence, in the near future.

Knowledge of the relative positions of the larger parts of the present continents for the past 500 million years provides a framework for the study of paleogeography, paleobiology, paleoclimatology, and paleoceanography. In part because of the emergence of this revolutionary new framework, these fields have been rejuvenated and are essentially new disciplines.

Given the current societal concern about the greenhouse effect, the study of the geological history of atmospheric carbon dioxide and the carbon cycle is of special significance. Also to be answered is whether the concentration of oxygen in the atmosphere has fluctuated substantially during the past billion years.

In the course of earth history, changes in sea level have dramatically altered terrestrial environments, oceanographic patterns, and atmospheric chemistry. Many of the long-term changes can be attributed to plate tectonic processes. Some more rapid changes reflect fluctuations in the volume of glaciers, but others remain to be explained. It has recently become evident that Earth's orbital geometry influences climates, glaciers, sea level, and sediment accumulation, but the details remain poorly understood. Of greatest immediate concern is whether global warming will melt glacial ice and flood major cities.

Strata and the fossils that they entomb provide a unique record of life and habitats for more than 3 billion years of earth history. New methods for

> *Studies of past interactions between life on Earth and climates, oceans, and changing continental configurations will assist in understanding organic evolution, discovering resources, and predicting and dealing with future environmental changes.*

assessing rates and patterns of evolution and extinction are enlivening interpretations of this record. A new context for this work has emerged with the expansion of research on large-scale changes in ancient oceans, atmospheres, and continental configurations. Models of the coupled ocean-atmosphere system are being tested against geological data and related to the history of life. Studies of ice cores, pollen, and marine plankton are yielding detailed pictures of environmental and biotic changes during the past several thousand years—changes that can serve as models for understanding events of the future.

Global extinctions are of great current interest. A variety of evidence has convinced many geologists that the impact of one or more meteorites or comets caused the extinction of the dinosaurs, but other great extinctions have been attributed to climatic changes driven by plate tectonics. The search for evidence of large impacts continues, and the potential role of massive volcanism is also under investigation.

Top Priority, Theme I: The Past 2.5 Million Years

Environmental and biological changes since the onset of the Northern Hemisphere glaciation.

The most recent past has been chosen as a top priority for half a dozen reasons. First, the onset of Northern Hemisphere glaciation during this period represents the most radical environmental change on Earth within the past several tens of millions of years. Second, its closeness to the present means that the record is generally more complete. Methods of study not applicable to older times can be used, such as dating techniques using ^{14}C and other cosmogenic nuclides. Third, surface features such as soils and landscapes, including mountains and river systems, have largely developed or have been strongly modified within this geologically short interval. Fourth, the relatively complete understanding that is attainable makes this interval of peculiar importance as an analog of older, less readily or completely analyzable parts of the geological record. Fifth, as inhabitants of a rapidly changing environment, the human race will find it is useful to have a full appreciation of what has happened in the geologically recent past. Students of global change have emphasized the importance of the record of progressively more recent times. And, finally, the fossil record for this interval includes

many living species—for example, humans—and many extinct species with close living relatives.

Other High Priorities

For Theme I Table 7.6 lists two additional high-priority programs, effectively recommending the pursuit of similar research back through time: first, during the relatively recent interval over the past 150 million years, and then during the long interval before that back to 3.8 billion years. The background for these investigations is outlined briefly in Table 7.4 and discussed at length in Chapter 3. Important programs already support these three priorities, which aim at integrating understanding but approach the problems in diverse ways. One simple way of dividing activities is their environment: the land surface, shallow subsurface, river system, frozen ground, glacial, lacustrine, and marine environments. These are all the focus of dedicated programs in a variety of agencies, federal and local.

Programs and Infrastructure

Preeminent among single programs relevant to understanding the environment and biological change on the 2.5-million-year time scale is the Ocean Drilling Program (ODP). Although the program operates only in the two-thirds of the Earth occupied by the oceans and their margins, no other single program can rival it in comprehensive scope. Its results embody an unrivaled record of the evolution of the atmosphere-ocean system and of ocean biology and biogeochemistry, and for this reason it is accorded the highest priority. In 1992 the NRC's Board on Earth Sciences and Resources and the Ocean Studies Board reviewed the ODP program and its forward plans.

The study of how the Earth has behaved during the past 2.5 million years involves diverse agencies and numerous programs. For example, hydrology entails an understanding of rivers and landforms as well as soil development and groundwater movement on the 2.5-million-year time scale. A variety of organizations play a part, ranging from federal agencies (e.g., Department of Agriculture, USGS, Department of Energy (DOE), Environmental Protection Agency, Army Corps of Engineers, and NASA) through the state geological surveys and water authorities to individual counties and cities. Pools of relevant information have grown, such as

TABLE 7.6 Priority Theme I: Global Paleoenvironments and Biological Evolution

Aim: To develop a record of the Earth, its atmosphere and hydrosphere, and the development of life. Such a record can provide a perspective for understanding continuing environmental change and its effect on the Earth's biosphere and for facilitating resource exploration.

Top-Priority Selection	**The Past 2.5 Million Years** **Environmental and biological changes since the onset of Northern Hemisphere glaciation** Ice record, glacial record, ancient oceanography, landforms, soils, drainage basins, river systems, shore lines, sea-level change, lakes, frozen ground, deserts, climatic variation, orbital variation, faunas and floras (especially pollen), and hominid evolution. System modeling.
Other High Priorities	■ **Environmental and biological changes over the past 150 million years since the oldest preserved oceans began to evolve** Ancient geography; oceanography; climate; faunal and floral evolution, including extinctions. Sea-level changes, terrestrial and marine deposition, the change from a hothouse to an ice house Earth. Cyclical, secular, and catastrophic phenomena. ■ **Environmental and biological changes prior to 150 million years ago from the time of the earliest preserved sediments (3.8 billion years old)** Evolution of organisms from the earliest of times, ancient geography, oceanography and climate, changes in atmospheric composition. Long-term changes in the surface environment. Ancient glaciations. Cyclical, secular, and catastrophic phenomena
Requirements	Improved resolution of rock and fossil ages. Instruments: Determination of elemental and isotopic compositions. Geological and related maps. Data bases, including subsurface records and the distribution of fossil taxa in space and time. Rock, fossil, and ice core collections with adequate curation, museums, geographic information systems, and modeling capabilities.
Major Relevant Federal Programs	**NSF**: Ocean Sciences Division—marine geology and geophysics, Ocean Drilling Program; Earth Sciences Division—surficial, paleobiology; Atmospheric Sciences Division—climate dynamics. **USGS**: various programs in Geologic Division and Water Resources Division. **DOE**: Carbon dioxide program. **NOAA** and **DOD**: various marine programs. **Army Corps of Engineers**: various coastal and riverine programs. **USDA**: programs in Soil Conservation Service.
State Programs	State geological surveys: mapping and resource studies, hydrologic studies.
Industry	Hydrocarbon, mineral, and water resource activities require understanding of the rock record of environmental and biological changes. Industry scientists are active in this area of research.
International	Ocean Drilling Program, International Geosphere-Biosphere Program, PAGES: Past Global Changes, International Geological Correlation Program.
Selected Recent Reports	NRC: *Opportunities in the Hydrologic Sciences* (1991), *Research Strategies for the U.S. Global Change Research Program* (1990), *Global Surficial Geofluxes* (1993), *Sea-Level Change* (1990); NSF: *Unified Theory of Earth Sciences* (1988), *Report on Earth System History* (1991); NASA: *Solid-Earth Sciences in the 1990s* (1991); *IGBP Report No. 12* (1990).
Recommendations	**Undertake a coordinated thrust at understanding how the Earth's environment and biology have changed in the past 2.5 million years.** The current research activities of many federal agencies bear on this issue, and international involvement would be appropriate as well.

topographic and geological maps and well-log and sample stores. This diversity of involvement extends to the coasts and shallow waters of the nation. For the deep oceans, research is funded largely through the Marine Geology and Geophysics Branch of NSF's Ocean Sciences Division, although the USGS, NOAA, and the Department of Defense also are involved.

Instrumentation for both stable and radioactive isotopic analysis will be required for programs dealing with this interval, including access to tandem accelerator facilities where appropriate. On a global scale, ice core drilling and analysis of the samples represents a unique capability.

Because the kinds of data used to study the past 2.5 million years are so diverse, there is an exceptional need to build up data bases that include paleontological, depositional environment, paleogeographic, paleomagnetic, paleohydrologic, and paleotemperature data. The data can then be used to construct paleoenvironmental models, including models of oceanic and atmospheric circulation. Geographic information system approaches are beginning to prove powerful in this kind of data handling, modeling, display, and curation; the use of spatially registered data is likely to grow.

Priority Theme II: Global Geochemical and Biogeochemical Cycles

The principal aim of this priority theme is **to determine how, when, and where materials move across the interfaces between mantle and crust, continent and ocean floor, solid earth and hydrosphere, and hydrosphere and atmosphere; cycling through the biosphere is particularly important.**

The early fractionation of the Earth led to the formation of the core, mantle, oceans, and atmosphere. Geological cycles transport material across the geospheres in a variety of ways, and many geochemical and biological cycles have been identified and studied. The near-surface rapid chemical cycles involving atmosphere, oceans, soils, and biosphere (reviewed in Chapter 3) are linked with other much slower cycles through the interior (reviewed in Chapter 2), and the equations for recycled elements cannot be solved without including all transportation paths and reservoirs. There is a contrast between the information about conditions at the surface and in the ocean and atmosphere through relatively short time scales, which comes mainly from stable isotopes in sedimentary rocks, and the information about cycling within and through the mantle on long time scales, which comes mostly from long-lived radiogenic isotopes in mantle-derived rocks.

The plate tectonic cycle leads to the continuous creation of the oceanic crust at ocean ridges and its recycling at ocean trenches. Of particular interest are the fluxes of materials and volatile components

Constant recycling of the ingredients of geological materials accompanies the Earth's evolution. The dimensions and time scales of geochemical cycles within the deep interior are much greater than those of the biogeochemical cycles external to the Earth's surfaces that involve the atmosphere, hydrosphere, and biosphere.

through the solid earth in these two environments. Unanswered questions include why a planet has plate tectonics and how subduction is initiated.

The continental crust has a composition that is fundamentally different from that of the oceanic crust and, on average, that is substantially older. Questions remaining about the chemical evolution of the continental crust include how it is formed; whether its evolution depends on the hydrologic cycle, weathering, and erosion; and whether (and, if so, how) significant quantities of continental crust are recycled into the mantle. There are also large uncertainties in the estimates of how much water and carbon dioxide are cycled through the deep mantle from the surface.

Biogeochemical cycling through geological time is of special importance because life and its environment evolve together. The atmospheric oxygen that life depends on is a product of continuing biological activity (photosynthesis) that began billions of years ago. The history of life and how it has affected the chemistry of the environment can be followed using geochemical tools.

Isotopic studies have played a key role in studies of global geochemical cycles and will continue to do so. Stable isotope ratios, such as those of carbon and oxygen, provide important chemical tracers. For example, do diamonds contain organic carbon that has been recycled into the interior of the Earth? Radiogenic isotopes can quantify the size and mean age of major global reservoirs, providing information that is complementary to geophysical studies of heterogeneities in the mantle, which are being interpreted in terms of convection.

Top Priority, Theme II: Biogeochemistry and Rock Cycles Through Time

The challenge is to interface models of geochemical cycles of the surface environment with climatic models, and with models of the geochemistry of the deep-earth system, to establish how the cycles have operated. Special

TABLE 7.7 Priority Theme II: Global Geochemical and Biogeochemical Cycles

Aim: Determine how, when, and where materials are moving across the interfaces between mantle and crust, continent and ocean floor, solid earth and hydrosphere, and hydrosphere and atmosphere; cycling through the biosphere is particularly important.

Top-Priority Selection	**Biogeochemistry and Rock Cycles Through Time** **Establish how biogeochemical cycles have operated through time** Interpret elemental, mineralogical, and isotopic records of compositional variation in sedimentary rocks from a full range of times and a variety of environments, including oceanic, continental margin, and terrestrial. Interpret organic isotopic geochemistry and fluid inclusions in sediments. Infer ancient atmospheric and oceanic chemistry.
Other High Priorities	■ **Construct models of interaction between biogeochemical cycles and the solid earth and climatic cycles that generally operate on longer and shorter time scales** An integrated approach ultimately requires estimates of all fluxes across all interfaces and construction of models using such inputs as continental configurations, oceanic composition and circulation, and global volcanic fluxes. ■ **Establish how geochemical cycles operate in the modern world** Understand weathering, soil evolution, diagenesis, nutrient cycling, gas hydrate formation and destruction, volcanic and metamorphic degassing.
Requirements	High-quality material from the entire rock record, obtained partly from ocean drilling and other core collections and partly by dedicated drilling and outcrop study (e.g., drilling continental margins and basins). Instruments for elemental, mineralogical, and isotopic analyses, including fluid-inclusion measurements where needed. Continental margin carbonate records as a special need and organic geochemical data as another. Data handling systems, GIS and modeling capabilities. Maps of ancient geography, environments, and oceanography as input for modeling.
Major Relevant Federal Programs	**NSF**: Earth Sciences Division. **USGS**: Geologic Division and Water Resources Division. **DOE**: Carbon dioxide program. **IGC/CSD** (Interagency Coordinating Group/Continental Scientific Drilling program—NSF, USGS, DOE). Museums.
State Programs	State geological surveys, mapping, curation (especially cores).
Industry	Oil, gas, coal, and mineral industry interest in changes in the ancient environment relate to geochemical change.
International	Ocean Drilling Program, International Geosphere-Biosphere Program, International Geological Correlation Program.
Selected Recent Reports	NRC: *Research Strategies for the U.S. Global Change Research Program* (1990), *Global Surficial Geofluxes* (1993); NSF: *A Unified Theory of Planet Earth* (1988).
Recommendation	**Establish how global geochemical cycles have operated through time.** This information, which is essential to determining how the earth system operates, is now a realistic target that could be achieved by coordinating a number of federal programs and current national and international activities.

attention is needed to understand the fluxes of fluids (magmas and solutions) through and between various geospheres.

The sedimentary rock record contains evidence of how the surface environment has changed as the Earth has evolved. Major, trace, and isotopic variations in the compositions of sedimentary rocks help to define the properties of the ocean, atmosphere, and biosphere at the time they were deposited. Chapter 3 discusses evidence of how, for example, temperatures and atmospheric composition have changed over the past 100 million years. Instrumental capabilities now exist to ex-

tract this kind of information from sedimentary rocks of all ages. Analytical data at present relate to sporadically distributed samples and represent time intervals in disparate ways. Their quality is also very uneven. Two steps are necessary for the implementation of this priority: (1) the acquisition of samples that are representative of the geological record and (2) access to the appropriate instrumentation. Once the samples and analyses are available, the challenge will be to establish the critical state variables for times past and to work out how and why the surface environment has changed with time. Supplementary information of several kinds will be needed. For example, the extent of the flooding of the continents must be estimated from paleogeographic mapping.

The right kinds of samples are to some extent available, or at least accessible, through field work and through archives such as the Deep Sea Drilling Project and ODP core collections. Some environments, such as continental margin carbonate deposits, are underrepresented, and dedicated continental drilling will be required to complete the record for certain intervals in some kinds of sedimentary basins.

Analytical facilities need to lay emphasis on stable isotope geochemistry and the isotopic geochemistry of organic compounds, but minor element geochemistry and strontium isotopic geochemistry also are important, as is mineralogy. The ability to analyze material from special environments, such as fluid inclusions in sediments, is needed.

Other High Priorities

For Theme II Table 7.7 gives two additional high-priority items selected from the many opportunities listed in Table 7.4 and discussed in Chapters 2 and 3. The first item arises from the need to interface models of geochemical cycles of the surface environment with climatic models and models of the geochemistry of the deep-earth system. It is particularly challenging. Although there are both material and energy fluxes across the interfaces between these systems, the time scales on which the cycles operate differ by orders of magnitude. The second high-priority item involves intensive study of the key processes operating today at or near the surface, those that involve rock-fluid interactions, as a guide to and calibration for the interpretation of cycles in the past.

Programs and Infrastructure

Table 7.7 summarizes the relevant programs, along with the industry involvement and facilities required to accomplish the research. Access to high-

> *Most of the chemical exchanges within and on the Earth involve the transport of material in fluid form—in magmas, solutions, and gases. Magmas differentiate the Earth into its major components: the core, mantle, crust, and fluid envelopes. Fluids are associated with erosion and deposition, volcanic eruptions, mineral deposits, petroleum and natural gas, water resources, and waste disposal.*

speed computers and the ability to handle large data sets are needed.

Priority Theme III: Fluids in and on the Earth

The principal aim here is **to understand how fluids (magmas, lavas, solutions, gases) move both at the surface and inside the Earth.** The fluids include magmas rising from great depths to volcanic eruptions and solutions and gases distributed mainly through the crust but also in the mantle. A distinction is made between the interactions in the interior and those at the surface. Subsurface water is the dominant fluid in the shallow crust, whereas magmas become important at greater depths, although they locally reach the surface at volcanoes.

The surface processes of erosion and deposition have shaped the surface environment throughout its history. The landforms thus generated are treated in this volume as the surface of the dynamic continents; they are discussed in Chapters 3 and 5. The concentration is on fluids within the Earth.

Magmas are covered in Chapters 2 and 4, and water and other fluids receive major attention in Chapters 3 and 4. Applications of processes involving fluids are discussed in Chapters 4 and 5. The chemical differentiation of the crust, hydrosphere, and atmosphere was accomplished by partial melting at depth within the Earth and then by the transfer of magmas, followed by sedimentary processes. Magmatic processes also lead to the concentration of elements into ore deposits.

The migration of fluids through the crust plays an essential role in its chemical evolution. Fluids promote chemical reactions and transport dissolved elements. They play an important part in faulting mechanics. Many basic questions regarding fluid migration remain unanswered or only partially answered, including how deep fluids penetrate into the crust in significant quantities, whether large quantities of fluid move laterally distances of hundreds of kilometers or more, and what the driving mecha-

TABLE 7.8 Priority Theme III: Fluids in and on the Earth

Aim: To understand how fluids (magmas, lavas, solutions, gases) move both at the surface and inside the Earth.

Top-Priority Selection	**Fluid Pressure and Fluid Composition in the Crust**
	Understanding of the three-dimensional distribution of fluids in the crust, including their pressure and compositional variations
	Direct observation of fluids: groundwater, oil, gas, hydrothermal solutions. Pressure and temperature measurements. Elemental and isotopic chemistry characterizing dissolved materials as ionic or other. Indirect inferences from minerals and rocks. Remote sensing of in-ground fluids by geophysics. Fluids in the oceanic crust at spreading centers and fluids in convergent margins, especially accretionary prisms. Rock characteristics controlling permeability and dispersion. Kinetics of rock-water interaction.
Other High Priorities	■ **Modeling fluid flow in sedimentary basins**
	Groundwater, hydrothermal, oil, and gas flow.
	■ **Understanding of microbial influences on fluid chemistry, particularly groundwater**
	Observations and experiments, the effects of specialized bacteria.
Requirements	Analytical instruments, elemental and isotopic, experimental facilities, new geophysical instrumentation, drilling on continents and oceans, including hostile hot environments. Down-hole instruments for monitoring and experiment. Curation, especially of cores. Data-handling and modeling capability; high-speed computer access.
Major Relevant Federal Programs	**NSF:** Earth Sciences Division. **USGS:** Geologic Division and Water Resources Division. **DOE:** Office of Basic Energy Sciences; programs in waste isolation and cleanup. **IGC/CSD** (Interagency Coordinating Group/Continental Scientific Drilling program—NSF, USGS, DOE).
State Programs	State geological, oil, mineral, and groundwater programs. Waste isolation and cleanup programs, core storage and archival activities.
Industry	Oil, gas, groundwater, well logging, geophysical and geochemical instrumentation. Waste isolation and subsurface restoration.
International	Ocean Drilling Program, RIDGE, hydrological activities, continental drilling.
Selected Recent Reports	NRC: *Opportunities in the Hydrologic Sciences* (1991), *Rethinking Radioactive Waste Isolation* (1990), *The Role of Fluids in Crustal Processes* (1990); NSF: *A Unified Theory of Planet Earth* (1988).
Recommendation	**Take up the challenge of investigating the three-dimensional distribution of fluid pressure and fluid composition in the Earth's crust.** The instrumental, observational, and modeling capabilities that exist within various federal programs can be effectively focused on this problem. International coordination is important.

nisms are for fluid migration (possibilities include temperature, topography, and tectonic forces).

The fluids that occupy pore spaces in rocks—including hot and cold waters, steam, oil, natural gas, and partially molten rock—interact physically and chemically with their solid surroundings. Research on the behavior of these fluids represents one of the fastest-growing branches of the solid-earth sciences. Observational, experimental, theoretical, modeling, and predictive studies are all expanding. Although the different fluids represent very different environments, there are similarities in behavior in many different parts of the solid earth. The fundamental processes associated with fluid migration in the crust must be studied in situ. This requires drilling into active zones of migration.

Careful selection of drill sites, with improved downhole instrumentation, can provide the basic data set required to develop applicable theory.

Most essential nonrenewable resources involve some interaction between the solid earth and fluids. Research on materials such as water, oil, natural gas, sedimentary ore bodies, geothermal energy, hydrothermally deposited ores, coal, limestone, and gravel all find a place under this priority theme.

Transfers of energy and material from the realm dominated by internal energy to the surface where solar energy dominates, and then back into the interior, constitute a part of the biogeochemical cycles of the planet. Many of these transfers involve interactions between the solid earth and its fluids.

The interface geochemistry of mineral-fluid assemblages is of fundamental importance, both in the crust and deep within the mantle.

Top Priority, Theme III: Fluid Pressure and Composition in the Crust

Understanding of the three-dimensional distribution of fluids in the crust, including their pressure and compositional variations.

The three-dimensional distribution of fluid pressure and fluid composition in the crust can best be addressed by an integrated approach that uses information obtained in a variety of ways. Direct observation of fluids within the crust and of fluids that have been extracted from the crust is essential. This is most feasible for groundwater, oil, and natural gas, but hydrothermal fluids are now being sampled. How to sample magmas at shallow depth is a current challenge. Measurements of the chemical and isotopic composition of these fluids is required. Continental drilling in selected environments is needed; the operation of instruments in drill holes for geochemical and geophysical measurements is desirable on short- and long-term bases.

There is a long and successful tradition of inferring the properties of fluids that have flowed through rocks from the physical and chemical properties of rocks accessible at the surface or in drill holes. Inferences about the subsurface distribution of fluids have long been made from geophysical observations at the surface and in bore holes. A current challenge is to develop reliable techniques for the direct remote sensing of subsurface fluids and their permeability.

There are well-defined problems related to the distribution of fluid in the oceanic crust. At spreading centers both hydrothermal and magmatic fluids are accessible, and the RIDGE initiative includes plans for ocean drilling and submersible study. An important step has been taken in the first dedicated ODP drilling of a sedimented spreading center. Fluid fluxes through accretionary prisms at convergent plate boundaries are recognized as representing a critical element in geochemical cycling but are proving very difficult to quantify. An integrated approach, using ocean drilling with a variety of other geophysical, geological, and geochemical techniques, is likely to be needed.

With sufficient observational data, scientific understanding, and insight, the problem of the role of fluid-driven mass (and heat) transport in rocks can be defined in a manner amenable to a mathematical solution that simulates coupled flow problems and involves physical transport and chemical reactions and kinetics. Much future research will be directed toward placing confidence bands about the output of analytical models, including predictions of future system behavior.

Other High Priorities

Of the many other research opportunities cited in Chapters 3, 4, and 5, two are listed in Table 7.8 for Theme III. The first is a most important part of the top-priority selection—to model the fluid flow in sedimentary basins. The subject of sedimentary basins as an integrative theme has recurred throughout this volume. The second focuses on the details of the chemical interaction at surfaces between minerals and fluids, particularly the role of bacteria. The same topic, applied specifically to organic wastes, is the top-priority recommendation for Priority Theme D.

Programs and Infrastructure

For Theme III Table 7.8 summarizes the relevant programs along with the industry involvement and facilities required to accomplish the research. The preeminent need is for experimental facilities, for hydrothermal systems and high-temperature, large-volume experiments. Access to data-handling and high-speed computational facilities for modeling is required, as are adequate core-storage facilities.

Priority Theme IV: Crustal Dynamics—Ocean and Continent

The principal aim of this priority theme is **to understand how the Earth's crust originates and evolves, including the nature and history of the deformations and mass transfer processes responsible for building and modifying the continents, mountain belts, island arcs, and ocean basins.**

New and continuing studies regarding the origin, structure, mass transfer processes, and history of continents and continental building blocks promise excellent research returns, as discussed in Chapters 2, 3, and 4. The shaping of the land surface into landforms is treated in Chapters 3 and 5. Plate tectonics provides a basic framework for understanding how the crust is deformed. In oceanic regions the deformation is relatively simple, with creation of new crust at spreading centers, destruction or reorganization of crust at oceanic trenches, and lateral movement of crustal blocks on transform faults. Studies of these fundamental processes are

TABLE 7.9 Priority Theme IV: Crustal Dynamics—Ocean and Continent

Aim: To understand how the Earth's crust originates and evolves, including the nature and history of the deformations and mass transfer processes responsible for building and modifying the continents, mountain belts, island arcs, and ocean basins.

Top-Priority Selection	**Active Crustal Deformation** **Integrated field studies using a variety of complementary techniques for understanding the active deformation of both oceanic and continental crust** Permanent and temporary seismic networks and arrays, geodesy (especially space geodesy), paleoseismology, neotectonics, seismic profiling, heat flow, gravity, magnetic and other geophysical techniques, geological and geomorphic studies, ages of materials and surface, including cosmogenic nuclide techniques, continental and oceanic drilling. Topographic data on land and below sea level, including side-look sonar data. Laboratory deformation studies, earthquake mechanisms, thermochronology, fluid behavior, and integrated modeling of stress and temperature fields. Studies on critical areas: *United States*—Pacific northwest, San Andreas, Basin and Range, Alaska; *international*—Tibet, Himalaya, active rifts in Africa, Asia; *oceans*—spreading centers, transforms, and convergent boundaries.
Other High Priorities	■ **Landform response to climatic, tectonic, and hydrologic events** The concept of time, landform evolution, and thresholds of instability are critical for predictions of future landform responses, and the combination of space-based observations and new dating methods provide new opportunities. ■ **Understanding crustal evolution** Integrated topographic, geophysical, and geological field studies, including deep seismic reflection programs, with refraction and wide-angle reflection. Studies directed at critical examples of older oceanic, arc, and continental environments. Drilling in critical sites; laboratory studies, including elemental, mineral, and isotopic analyses. Experimental petrology, modeling.
Requirements	Seismic networks and arrays, geodetic facilities, geophysical and geological field studies, physical and chemical laboratory measurements of earth materials. Data bases, including maps. Modeling capabilities.
Major Relevant Federal Programs	**NSF**: two NSF technology centers; Earth Sciences Division—continental dynamics, geophysics. **USGS**: Geologic Division—seismology and NEHRP. IRIS (funded by **NSF, DOD**). **NOAA. Nuclear Regulatory Commission. DOD. FEMA.**
State Programs	State geological mapping and hazard programs. Archiving: cores, logs, other data.
Industry	Hazard assessments, oil, gas, coal, and mineral industries.
International	Global seismic networks and geodetic networks, International Commission on the Lithosphere, International Decade of Natural Disaster Reduction, Ocean Drilling Program.
Selected Recent Reports	NRC: *Real-Time Earthquake Monitoring* (1991), *Assessing the Nation's Earthquakes: The Health and Future of Regional Seismograph Networks* (1990), *International Global Network of Fiducial Stations* (1990), *Active Tectonics* (1986), *Geodesy in the Year 2000* (1990); NSF: *A Unified Theory of Planet Earth* (1988), *Ocean Drilling Program: Long-Range Plan* (1990); NASA: *Solid Earth Science in the 1990s* (1991).
Recommendation	**Coordinate and intensify efforts to understand active crustal deformation.** The opportunity exists to revolutionize current knowledge of this area, which is vital not only to solid-earth science but also to the missions of several federal agencies and various state and international bodies.

being refined using new high-resolution marine data and satellite altimetry.

Evolution of the lithosphere cannot be understood without better constraints on lithosphere-asthenosphere coupling and flow regimes. The geochemical evolution, tectonics, and thermal history of the crust are inextricably intertwined with the dynamics of the upper mantle. There are great opportunities for vastly improved three-dimensional mapping of the crust and upper mantle, its structure, and its properties.

Modeling of the behavior of the ocean floor was one of the earliest successes of the plate tectonics revolution; a more recent success has been modeling of the thermal and igneous behavior of active spreading centers. The complex structural patterns revealed by modern topographic studies show that there is a need for further analysis before the fault and topographic patterns at the spreading centers can be interpreted equally well.

Deformation of the continental crust is much more complex than in the oceans, and the crust is much thicker. A combination of geological mapping and seismic studies, together with data from deep drill holes, is required for determination of continental structure. Seismic networks take the pulse of the crust, providing information on where deformation is occurring and on its state of stress. Seismic reflection profiling was developed by the petroleum industry to determine the structure of

The physical response to stresses associated with mantle motions and the forceful erosion of the continents by the fluid envelopes of the atmosphere and hydrosphere shape the rocks from which humanity obtains its resources and control the stability of the surface over which the bloom of humanity spreads.

sedimentary basins. The projects of the Consortium for Continental Reflection Profiling and later programs in the United States and elsewhere have used this technique to determine the deep three-dimensional structure of active and ancient mountain belts.

Geochemical studies of minerals and rocks provide data essential for characterization of rock masses and for dating geological processes. Experimental phase equilibrium studies provide the framework for calibration of the processes, depths, and temperatures of deformation of rocks and the behavior of magmas.

Crustal dynamics may soon be interpretable in terms of a fully integrated model of the earth system. Such a system will demonstrate how the results of global topography can be related to mantle convection and how this in turn can be related to global plate tectonics and rifting of the lithosphere, as well as to the modes and rates of formation, and removal by erosion, of mountain ranges. The model will include answers to questions about how global heat flow and heat flux through ocean and continental crust have changed; how seafloor spreading rates and length of mid-ocean ridges have changed; how the rate of subduction and length of subduction zones have changed; how to reconstruct the former positions of continents, geological terranes, and passive and active margins; and how the near-surface processes are linked to energetic phenomena at the core-mantle boundary. Each of these research areas demands new theoretical models and the relating of these models to the total geological picture.

Top Priority, Theme IV: Active Crustal Deformation

Integrated field studies using a variety of complementary techniques for understanding oceanic and continental crust deformation.

A practical distinction is made between deformation of the continents and deformation of the ocean floor because different techniques are applicable in the two environments. Integrated approaches to understanding, using a variety of complementary techniques, are a strongly recommended feature of research in both environments.

Active crustal deformation in the oceans can be addressed by combination of the following techniques: (1) seismic networks to localize and characterize earthquakes (this includes networks consisting of temporary deployments of ocean-bottom seismographs); (2) other geophysical techniques, including seismic profiling and refraction, heat flow, gravity, magnetic, and electromagnetic methods of study; (3) high-resolution topographic surveys involving multibeam and near-bottom-source echo sounding and side-look sonar; (4) direct observation and sampling using submersible vessels and remotely operated vehicles; (5) ocean drilling; and (6) geodetic observations. At present these are practicable only in places like the Afar and Iceland, where active "oceanic" crustal deformation is taking place above sea level, but there exists a possibility of developing underwater geodetic techniques.

The various methods of studying active deformation of the ocean floor are of different utility for spreading centers and convergent boundaries. Active transform boundaries represent a third and slightly different environment for research. Continuing modeling needs require access to and manipulation of very large data sets.

Active crustal deformation in the continents can also be addressed by half a dozen methods: (1) seismic networks to localize and characterize earthquakes; (2) geodetic studies, including continuous monitoring to assess how strain is built up and released; (3) paleoseismology, including trenching to date how earthquakes have occurred through time on time scales of up to tens of thousands of years; (4) other geophysical techniques, including seismic profiling and refraction, heat flow, gravity, magnetic, and electromagnetic methods of study and remote sensing; (5) geological and geomorphic field studies complemented by determination of the ages of surfaces and deposits eroded from active areas (dating of minerals at outcrop and in sediments helps in determining the timing and extent of active processes); and (6) dedicated drilling. This last method can play a particularly useful role in the study of active deformation because it allows fluids in faults to be directly observed and permits direct estimates of stress distribution close to the hole. Instruments for a variety of measurements, including seismometers, can be placed in drilled holes to great effect. Integrated observations of active deformation of the kind outlined above require an ability to model deformation, stress, temperature, and compositional data on solids and fluids together. Laboratory deformation studies are needed as well.

Antarctica and Greenland present particular challenges in continental dynamics. Airborne measurements of elevation, ice thickness, gravity, and magnetics, with useful spatial resolution, are becoming feasible because of GPS navigation; these measurements could revolutionize our understanding of the dynamics of these continents.

Other High Priorities

For Theme IV Table 7.9 lists two other high-priority topics. The first is the response of landforms to climatic, tectonic, and hydrologic events—another integrated topic dealing with the surface on which humankind lives. The second is the evolution of the whole crust, which deals with the history and products of past active processes. Other priority research is described in Chapters 2, 3, and 5.

Programs and Infrastructure

Active deformation on the ocean floor is primarily a topic of research sponsored by NSF, USGS, NOAA, and the Department of Defense (DOD). Seismological programs, including the activities of IRIS and ODP, represent two of the largest activities. The RIDGE initiative embraces much research on active deformation as well as igneous, hydrothermal, and biotic studies. Convergent margins have not yet lent themselves to such a coordinated initiative.

Federal, state, and local government agencies are all involved in active deformation studies. These studies are mainly related to earthquake research and considerations of hazard. Scientific leadership has traditionally come from the USGS and NSF; the establishment of NSF science and technology centers for earthquake engineering in Buffalo and Los Angeles (the Southern California Earthquake Center, which also has substantial USGS support) shows the continuing importance to the federal government of at least the seismic aspects of the study of active deformation. The National Earthquake Hazard Reduction Program (NEHRP) is a manifestation of the same continuing interest. Other federal agencies with a special interest in active deformation that is concentrated on earthquake research include the Federal Emergency Management Agency (FEMA), DOD, the Nuclear Regulatory Commission, and DOE. The missions of these agencies are different, but they converge (or even overlap) in the area of active earthquake deformation of the continents. Special problems exist in areas such as New Madrid and Charleston, where episodes of active deformation can be separated by hundreds or even thousands of years, but substantial hazard ensues when events do occur. The NSF Division of Polar Programs is responsible for all scientific research in Antarctica, but other agencies (e.g., Naval Research Laboratory) do much in the Arctic.

Active deformation is strongly concentrated in areas close to plate boundaries and, within the continents, in broad plate boundary zones. In the United States these areas lie in the western states, Puerto Rico, the Virgin Islands, and Alaska. Current geological and geophysical studies are supported by federal agencies (principally USGS and NSF, but several others are involved, too) and state geological surveys. Outstanding advances in understanding active deformation have been made in all these areas in the past decade. The integrated approach recommended here is likely to prove critical in the future. It will, however, require an ability to model deformation, stress, temperature, and compositional data on solids and fluids together. Laboratory deformation studies also are needed.

Priority Theme V: Core and Mantle Dynamics

The principal aim of this priority theme is **to understand the internal operation of the Earth in geophysical, geochemical, and geological terms as an essential element in developing a theory of the overall earth system.**

Study of the deep interior has involved geophysical techniques such as seismic tomography as well as measurement of gravity, magnetic fields, and geodetic values. Isotopic and geochemical measurements have facilitated the evaluation of geophysical data. New ideas on the behavior of fluids in the deep interior have been tested by experimentation and computer models. Laboratory simulation using high-pressure apparatus has provided information about the properties of mantle and core materials, as well as information for the calibration of processes in terms of depth and temperature.

Plate tectonics is a kinematic description of the movement of the surface plates, and it is now generally accepted that this movement is one result of thermal convection in the mantle. Aspects of mantle convection that are poorly understood include whether the whole mantle convects or whether mantle convection is layered, whether subducted lithosphere sinks to the core-mantle boundary at a depth of 2,900 km or only to a lower thermal boundary layer at 670 km, and whether volcanism within plates (rather than at plate margins) reflects rising plumes that are generated at a 670- or 2,900-km depth.

Studies of seismic velocities give a general density structure of the mantle but are only now beginning to resolve the density differences directly associated with mantle convection. Seismic tomography infers the global density structure. The quality of the data is quite variable because of the lack of global coverage of high-resolution seismographs. In order to infer temperature distributions from density distribution, detailed knowledge of the related state relations is required. This necessitates experiments at high pressures and temperatures on the relevant minerals and rocks.

Geochemical measurements of lavas and rocks from the mantle provide data that supplement the seismic approach. Radiogenic isotopes can quantify the size and mean ages of the mantle rock reservoirs from which the lavas were melted, thus bearing on the mantle's convection history.

The gravity field, especially as determined from spacecraft in low Earth orbit, contributes to our understanding of the mass distribution, and hence the convective flow, in the mantle. Measurements of the surface heat flow show that the flux is dominated by the convective transfer of hot material to the surface at ocean spreading centers, but quantifying heat transfer from mantle plumes and discriminating between heat generated within the

> *The operation of the Earth's internal engine is the main driving force for many geological processes, some of which are simultaneously influenced by the external engine, driven by solar energy. Understanding how our planet operates on the grandest scale provides data for improving conditions on Earth by predicting and developing theories for global earth systems.*

continental crust and heat from the underlying mantle remain challenges. A further challenge is provided by electromagnetic measurements at the surface; it has not yet proved possible to infer much about mantle structure from these observations. The core exerts a major influence on the behavior of the lower mantle, and knowledge of the magnetic field and its time-dependent variation, as measured in magnetic observatories at the surface and in space, is essential to fully understand how the core works.

Changes in the structure of melts and the resultant density variations at high pressures may exert a major influence on the chemical differentiation of the Earth and other planets. Supercomputers make possible the theoretical modeling of mantle convection and of the atomic geochemistry of minerals and melts, using the first principles of quantum mechanics.

Top Priority, Theme V: Mantle Convection

Establish the variations in temperature and composition and the resulting flow structure in the mantle. This involves physical and chemical approaches, laboratory calibrations, and imaging the interior by remote sensing.

The structure of the mantle is currently being investigated by four principal techniques:

1. *Interpreting earthquake signals received at global seismic network stations.* Large numbers of these data are computed for numerous events in such techniques as "seismic tomography." This approach has allowed identification of those volumes of the mantle that have a velocity higher or lower than average. These volumes are resolved to such an extent that intriguing correlations with other phenomena are beginning to emerge. During the next decade greatly improved results can be expected from the accumulation of considerably more data, advances in the quality of instruments, deployment of instruments more widely over the surface of the Earth (in both permanent and temporary arrays), and advances in computational procedures.

2. *Interpreting the gravity field.* The gravity field at

and close to the surface depends on the local distribution of mass within the planet. At length scales of 100 km or more, this signal is dominated by mass distribution in the mantle. Resemblances between the mantle structure revealed by regional variations in the gravity field and the mantle structure determined from seismic data are beginning to emerge. Higher-resolution gravity data are needed. A satellite experiment in low Earth orbit (less than 200 km from the surface) could play a role, and airborne regional gravity is becoming a possibility with GPS positioning. Gravity data over land and ocean have been subject to military classification by many nations, but in a rapidly changing world this may not always be so.

3. *Interpreting geochemical, especially isotopic, variations in the mantle.* Interpreting these variations from rocks derived by partial melting from different parts of the mantle provides an understanding of mantle heterogeneity that is completely independent of the above two sources of information but should be compatible. For example, rock composi-

---

TABLE 7.10 Priority Theme V: Core and Mantle Dynamics

Aim: To understand the internal operation of the Earth in geophysical, geochemical, and geological terms as an essential element in developing a theory of the overall earth system.

Top-Priority Selection	**Mantle Convection** **Establish the variations in temperature and composition and the resulting flow structure in the mantle.** Heat generated inside the Earth drives plate tectonics, earthquakes, and volcanoes. Mantle convection is the deep underlying control of these and many other processes.
Other High Priorities	■ **To establish the origin and temporal variation of the earth's internally generated magnetic field** The main field is generated in the core, and its temporal variations indicate how it is generated and why it changes. ■ **To understand the nature of the core-mantle boundary** Seismic observations are beginning to reveal heterogeneities on large spatial scales at this the most prominent reactive boundary within the Earth.
Requirements	Complete a global broadband high-dynamic-range seismometer network and temporary array capability, underwater where needed, and with colocated gravity, magnetic, and space geodetic instruments, where appropriate. New instruments and adequate samples for characterizing mantle-source geochemistry in elemental and isotopic composition. High-pressure and high-temperature instruments for experimental simulation. Synchrotron facilities. Satellite and airborne gravity and magnetic measurements. Access to supercomputational facilities for simulation of mantle flow. Ancient magnetic reversal data, ancient magnetic intensity data. Drilling active plumes.
Major Relevant Federal Programs	Seismic networks (**USGS**); IRIS (funded by **NSF, DOD**). NEHRP (**NSF, USGS, FEMA**). Magnetic observatories (**USGS**); satellite magnetics, gravity, and altimetry (**NASA, DOD**); space geodesy (**NASA**); UNAVCO (**NSF** funded). Advanced geochemical instrumentation (**NSF, DOE, USGS**). High-pressure experimentation (**NSF, DOE**). Drilling: ODP (**NSF**), Interagency Coordinating Group/Continental Scientific Drilling program (**NSF, USGS, DOE**). High-speed computation (**NSF, DOE**). Various other programs in **NASA, NSF, DOD, DOE**, and **USGS**.
State Programs	Seismic networks, drilling (Hawaii).
Industry	Instrument building.
International	SEDI (Studies of Earth's Deep Interior), international seismic networks. Satellites: TOPEX/POSEIDON, ARISTOTELES, MAGNOLIA. ODP.
Selected Recent Reports	NRC: *Earth Materials Research* (1987), *Facilities for Earth Materials Research* (1990), *Assessing the Nation's Earthquakes* (1990), *International Global Network of Fiducial Stations* (1991), *Geodesy in the Year 2000* (1990), *Geomagnetic Initiative* (1993); NSF: *A Unified Theory of Planet Earth* (1988); NASA: *Solid Earth Sciences in the 1990s* (1991).
Recommendation	**Mount an integrated attack on solving the problem of understanding mantle convection.** Seismic networks, satellite data, high-pressure experiments, magnetic observatories, geochemistry, drilling, and computational modeling can all be marshaled into the fray. Again, federal, national, and international organizations will be involved.

tions indicative of a source in a region of above-average mantle temperature could be expected to correlate with a region of slow seismic velocity and mass deficiency, as indicated by the gravity field. Fuller understanding of variations in temperature and compositional variation for the mantle will necessarily involve all three sources of information. The present mantle flow structure appears to record influences of subduction, and possibly other processes such as continental rupture, over time scales of hundreds of millions of years. These changes appear to be recorded by the compositions of igneous rocks derived from the mantle.

4. *Laboratory experiments attempting to replicate the high-pressure, high-temperature conditions of the mantle.* Such experiments have greatly advanced in the past decade using diamond-anvil and split-sphere devices and dynamic shock wave apparatus. There is now a need to improve the capabilities of existing instruments and to develop their successors. Carrying out experiments at high temperature and high pressure in larger volumes than is currently possible, and measuring the physical and chemical properties of materials under the experimental conditions are needed as an adjunct to other mantle studies. For many studies in mineral physics, pure crystals larger than a few tens of micrometers are needed. The technology available for the growth of laser and electrooptic crystals can also be used to grow large crystals of synthetic minerals.

Other High Priorities

For Theme V Table 7.10 gives two high-priority topics. The first is concerned with variations in the magnetic field. The second deals with the core-mantle boundary, where all of the approaches applied to mantle convection are relevant. Other priority research opportunities in this area are discussed at the end of Chapter 2.

The magnetic field of the Earth consists of the main field generated in the core, signals from the lithosphere where magnetic minerals lie above their curie points, and the relatively rapidly varying signals of the external field. The magnetic field and its temporal variation can throw light on possible interaction at the core-mantle boundary. Continuing observations of the field at the surface in a global network of observatories and from space are needed to characterize the time-dependent behavior of the field. The lithosphere-generated part of the field is known very unevenly, and this data set needs improvement, too.

Programs and Infrastructure

Federal, state, and international programs for Theme V are listed in Table 7.10. The facilities required are numerous. A dense global digital network of modern seismographs would provide invaluable data on the Earth's structure and dynamic processes. The capability to deploy temporary seismic arrays in critical places is essential. Other important geophysical probes are needed to make gravity and magnetic measurements from the surface and satellites. Continuous monitoring of the Earth's magnetic field from orbit will be essential. Samples of ocean-floor basalt obtained by drilling will serve as a source of information on the chemical heterogeneity of the mantle. New generations of mass spectrometers and ion microprobes are needed, as well as improved experimental equipment for high-pressure, high-temperature studies in phase equilibria and the physical properties of mantle materials. Access to synchrotron radiation facilities is important. High-speed computational capabilities are critical for handling the vast amounts of data involved in mantle studies, both for processing and modeling.

Priority Theme B: To Sustain Sufficient Natural Resources

The principal aim of this priority theme is **to develop dynamic, physical, and chemical methods of determining the locations and extent of nonrenewable resources and of exploiting those resources using environmentally responsible techniques. The question of sustainability, the carrying capacity of the Earth, becomes more significant as the resource requirements grow.**

Chapter 4 covers this area, drawing on Chapters 2 and 3; the environmental aspects are considered in Chapter 5. The Research Framework for these studies involves all five priority research themes.

In many parts of the United States, water is becoming an increasingly scarce resource. Groundwater is required for both human consumption and agriculture. It is essential to develop an understanding of the rates at which groundwater is being replaced. Water-use regulations must be based on scientific understanding of the fundamental processes and basic information about water quantity and quality. Even where abundant water resources are available, the pervasive nature of pollution is only now becoming evident. A sound basis must be provided for the siting of waste repositories of all types.

As known mineral deposits are mined and consumed, the need to discover additional, often buried, deposits increases. Integrated studies involving all superposed geological processes, drawing on a wealth of new global tectonic and geochemical concepts, are generating mineral deposit models for making resource predictions. The countries with the strongest mineral resource positions a decade or two from now will be those that

> *To ensure a continuous and reasonable supply of water, energy, and mineral resources for the nation and the world, we need comprehensive understanding of the Earth's crust and its fluids.*

know in the best detail the structure of the upper 10 km of their continental crust.

It is only a matter of time until global shortages of petroleum resources develop. It is essential that the United States develop all available petroleum resources and improve the efficiency of extraction. The importance and extent of both horizontal and vertical migration must be evaluated. Methods of primary recovery extract

---

TABLE 7.11   Priority Theme B: To Sustain Sufficient Natural Resources

Aim: To develop dynamic, physical, and chemical methods of determining the locations and extent of nonrenewable resources and of exploiting those resources using environmentally responsible techniques. The question of sustainability, the carrying capacity of the Earth, becomes more significant as the resource requirements grow.

Top-Priority Selection	**Improve the Monitoring and Assessment of the Nation's Water Quantity and Quality** To manage water resources and promote scientific advances in understanding, it is essential to maintain and improve the nation's programs of monitoring and assessing the physical, chemical, and biological properties of water resources—both surface waters and groundwaters.
Other High Priorities	■ **Sedimentary basin research, particularly for improved resource recovery** Origin and evolution of sedimentary basins; how source rocks, reservoirs, and traps form and evolve; movement of fluids (oil, water, and gas) in basins; thermal and diagenetic histories; origin of petroleum. High-resolution reservoir studies. Modeling of coal, oil, gas, minerals (e.g., uranium), and water in basins embodying geological history as well as thermal, chemical, and fluid transport with time. ■ **Improvement of thermodynamic and kinetic understanding of water-rock interaction and mineral-water interface geochemistry** Application to understanding fluid circulation in the geothermal environment. ■ **Development of energy and mineral exploration, production, and assessment strategies** The strategies involve modeling, artificial intelligence, and geophysical site characterization, with particular applications to the search for totally buried ("blind") ore bodies.
Requirements	Water-sampling network and analytical facilities; field and subsurface data, including geophysical data for sedimentary basins, cores, cuttings, well logs, curation. Organic, elemental, and isotopic analytical instruments. Modeling capabilities.
Major Relevant Federal Programs	For water: **USGS**: Water Resources Division, including NAWQA (with **EPA** and **DOE**). For sedimentary basins: **USGS, NSF, DOE**. For mineral deposits: **USGS, NSF, Bureau of Mines, DOE**.
State Programs	State geological mapping and resource studies; archival logs, core cuttings, state water-resource quality programs. Local (county and city) water programs.
Industry	Oil, gas, and mineral exploration and production companies; activities of hydrogeological corporations and consultants.
International	Ocean Drilling Program.
Selected Recent Reports	NRC: *Opportunities in the Hydrologic Sciences* (1991), *The Role of Fluids in Crustal Processes* (1990).
Recommendation	**Establish a dense network of water quality and quantity measurements, including resampling at appropriate intervals.** Coordination of federal and state agencies that have programs in the field will be needed.

TABLE 7.12   Priority Theme C: To Mitigate Geological Hazards

Aim: To determine the nature of geological hazards—earthquakes, volcanoes, landslides, soil erosion, floods, hazardous materials (e.g., asbestos)—and to reduce, control, and mitigate the effect of these hazardous phenomena. It is important to consider risk assessment and levels of acceptable risk.

Top-Priority Selection	**Define and Characterize Regions of Seismic Hazard** **Hazard reduction requires better understanding of where and how earthquakes occur, and progress toward predicting when they are likely to occur can emerge from this understanding.** National and regional seismic information, real-time earthquake monitoring, seismotectonic regionalization, paleoseismology, neotectonics, active fault studies, geodesy, strong-motion seismology.
Other High Priorities	■ **Define and characterize areas of landslide hazard** Use of GIS, ancient landslides, landslide mechanisms, timing, frequency, diversity. ■ **Define and characterize potential volcanic hazards** Real-time volcano monitoring of seismicity, surface deformation, thermal and infrared measurements, gaseous emissions.
Requirements	Seismic networks; strong-motion instrumentation; fault studies, including trenching. High-resolution topographic and geological maps. Geodesy, including space geodesy; surface, airborne, and space-borne volcano monitoring. Land-based, airborne, space-borne, and marine coastline monitoring.
Major Relevant Federal Programs	**NSF:** two NSF technology centers; Earth Sciences Division—continental dynamics, geophysics. **USGS:** Geologic Division—seismology and NEHRP. IRIS (funded by **NSF, DOD**). **NOAA. Nuclear Regulatory Commission. DOD. FEMA.**
State Programs	State geological, hazard, and other programs (e.g., highways).
Industry	Engineering geology/earthquakes, landslides.
International	IDNDR, international and volcanologic and seismic programs.
Selected Recent Reports	NRC: *Real-Time Earthquake Monitoring* (1991), *Assessing the Nation's Earthquakes* (1990), *A Safer Future* (1990), *Active Tectonics* (1986).
Recommendation	**Define and characterize regions of seismic hazard.** Because many people and much property in the United States are endangered by earthquakes, improved understanding of seismic occurrences is a pressing need. This issue is important to the missions of several federal agencies and to organizations ranging from local to international.

only 25 to 30 percent of the available petroleum from a field. The development of more efficient techniques of secondary and tertiary recovery must be a high priority. This will require a fundamental understanding of transport processes in the crust, including the distribution of porosity and permeability.

Top Priority, Theme B: A National Assessment of Water Quality and Quantity

Maintain the quality and supply of water. The immediate requirement is the establishment of a comprehensive network for sampling surface waters and groundwaters, which could be used for water management and scientific understanding.

A water quality and quantity assessment pro-

gram should be developed and maintained. Surface waters and groundwaters must be sampled using a comprehensive network and must be chemically analyzed in a program that provides for regular repetition of sampling and analysis. Because we have no adequate baseline data set for the nation's water quality and quantity, it will be difficult to quantify changes that are taking place now or that may take place in the future. Acquisition of such data is an important step in the formulation of research programs dealing with the nation's water supply.

The need for this kind of network has been recognized, and the activities of the National Water Quality Assessment Program represent an important start. But establishing an adequate network has proved difficult, partly because there is no single federal agency with the lead responsibility and partly because so many state and local agencies are involved.

Other High Priorities

For Theme B Table 7.11 lists three other high-priority areas, which are discussed in Chapter 4. Related aspects relevant to the environment and waste disposal are treated in Chapter 5. The first topic, sedimentary basins, is a recurrent high-priority theme involving all of the resources considered here. An integrated multidisciplinary approach to resource assessment and exploitation is needed. A much greater emphasis must be placed on understanding and possible mitigation of the potential adverse effects of exploiting and using petroleum and mineral resources. The second topic involves the frontier of mineral-water interface geochemistry, which has relevance for all resources and many geological processes. The third topic involves development of mineral exploration and assessment strategies.

Programs and Infrastructure

Federal and state programs for Theme B are listed in Table 7.11. Industry is heavily involved in these endeavors, of course, and much of the research in energy and mineral resources is driven by economic and political factors.

Priority Theme C: To Mitigate Geological Hazards

The principal aim of this priority theme is **to determine the nature of geological hazards— earthquakes, volcanoes, landslides, soil erosion, floods, hazardous materials (asbestos)— and to reduce, control, and mitigate the effect of these hazardous phenomena. It is important to consider risk assessment and levels of acceptable risk.**

Some of the geological processes described in Chapters 2 and 3 are associated with hazards, which are reviewed in Chapter 5. New and continuing studies on land resources and geological hazards are required to develop an understanding of the global ecosystem. The relevant basic research involves the priority research themes of mantle convection, con-

> *The classic aim is to determine and control earthquake hazards, volcanic hazards, soil erosion, landslides and land subsidence, and floods, but hazards are increasingly associated with water pollution and with rocks and minerals in the human environment.*

tinental dynamics, and fluid fluxes. Floods, landslides, and hurricanes are hazards related to the interaction of the Earth's fluid envelope with its surface. Research on these topics in the solid-earth sciences is primarily directed toward understanding the environmental conditions, so that adverse consequences can be minimized in the future by an informed society.

A large fraction of the U.S. population is subject to the prospect of a destructive earthquake, the most feared natural phenomenon. The primary reason for this fear is that earthquakes occur without warning. Earthquake prediction, a major scientific goal, must be pursued both empirically by studying a variety of precursory phenomena and theoretically by developing a better understanding of the fundamental mechanics of crustal mechanics. Seismic networks provide important information on where deformation is occurring and on the state of stress in the crust. At the present time, the relative earthquake hazard can be estimated. Earthquakes will certainly occur with higher frequency in the western United States than in the eastern part. Earthquakes can be expected to occur on the San Andreas Fault, but they also occur on many other faults, mapped and unmapped. In terms of hazard assessment, important areas of research include a better understanding of earthquakes on secondary faults in an active tectonic zone and a better understanding of why earthquakes occur in plate interiors (the 1811–1812 sequence of major earthquakes near New Madrid, Missouri, is an example).

Volcanism is an essential process in the geochemical evolution of our planet. It also presents a substantial hazard to humans. Millions of people live in areas that have been devastated by recent eruptions and the associated mudflows. The 1991 eruption of Mount Pinatubo disrupted the lives of thousands of people and killed many. Improved forecasting of eruptions and improved estimates of the statistical hazard are important research challenges. At Mount Pinatubo volcanologists worked closely with civil authorities. A monitoring and alert network led to evacuations that saved lives. As demonstrated by research at Mount St. Helens, seismic and geodetic observations provide substantial warnings of impending eruptions. However, they cannot predict the severity of eruptions, so it is difficult to make evacuation plans. By quantifying the magnitude of eruptions in the geological record, it will be possible to improve estimates of the likelihood and styles of future eruptions.

Groundwater contaminated by nuclear and chemical waste represents a growing hazard. Research is growing, too, but it is not clear that the full expertise of the solid-earth science community is being involved on a large enough scale.

The potential health hazards of such materials as asbestos and radon need to be evaluated in terms of levels of acceptable risk, taking into account the expertise of mineralogists and other appropriate scientists. There is a risk that legislation based on extrapolation of unreliable data could place large (and avoidable) financial burdens on the country. This is given high priority in Priority Theme D, the environment.

There is a small but real possibility that an asteroid or comet could strike the Earth in the next century. Anything larger than the small objects that fall as meteorites could cause substantial damage, and a larger body could generate a global catastrophe.

Top-Priority, Theme C: Define and Characterize Regions of Seismic Hazard

Hazard reduction requires better understanding of where and how earthquakes occur. Progress toward predicting when they are likely to occur emerges from this understanding.

The establishment of two NSF-supported technology centers, one in earthquake research and the other in earthquake engineering research, the development of the NEHRP program, and a variety of other activities following the Loma Prieta earthquake have helped to bring seismic hazard activity to the forefront. New ideas are emerging that challenge some of the old assumptions about the very mechanisms by which earthquakes originate and propagate, particularly the role of chaotic behavior in the seismic cycle. At the same time, our experience in the probabilistic assessment of seismic risk is growing. There is a growing need for integrated approaches, including seismic networks, space geodesy, geology, neotectonics, and paleoseismology. Paleoseismology has a special need for

the excavation of trenches across fault zones to characterize past seismic activity.

Other High Priorities

For Priority Theme C Table 7.12 lists two high-priority research areas: defining and characterizing areas of landslide hazard and defining and characterizing potential volcanic hazard; both are considered in Chapter 5. Landslide hazards arise from the association, in a small area, of a variety of phenomena such as steep slopes, wet conditions, and weak surficial rocks. Understandably, the integration of disparate data using geographic information systems has been found useful. The distribution and timing of ancient landslides can be an important guide to the hazard, and this information is also important for understanding aspects of past global change.

Volcano hazard research has been stimulated by the occurrence first of the Mount St. Helens explosive events, by the substantial loss of life at Nevada de la Ruiz, and the destruction resulting from Mount Pinatubo. Volcanologists are beginning to make excellent and informed use of the opportunities provided by these and other events to focus their efforts in the directions most useful to society.

Programs and Infrastructure

Federal agencies are actively involved today in assessing these geological hazards. The USGS, the NSF (both the Division of Earth Sciences and the Directorate for Engineering), FEMA, and the National Institute of Standards and Technology have lead roles in NEHRP and represent a national commitment in this direction. Other agencies, such as DOE and the Nuclear Regulatory Commission, have mission-oriented roles in seismic hazard analyses. The existence of the two NSF centers related to earthquake studies is a good indication of the importance the federal government attaches to such hazards.

National and regional seismic networks, with real-time monitoring capabilities where appropriate, are needed. Seismotectonic regionalization is not yet adequate and should be improved using seismic, strong-motion, and paleoseismic (especially trenching) techniques.

Priority Theme D: To Minimize and Adjust to the Effects of Global and Environmental Changes

The aim of the last priority theme is **to mitigate and remediate the adverse effects produced by**

global changes and those changes resulting from modification of the environment by human beings.

The scientific background for this objective is covered mainly in Chapter 3, with some applications to global change appearing in Chapter 5. Global change is popularly associated with change in climate, but scientists recognize broader implications, extending to such phenomena as sea level, groundwater quality, and biodiversity. The record of the past is vital to an understanding of global change. We need to be able to characterize variability within the past record and to identify the causes of that variability as well as to understand system states very different from those of today. System transitions, especially abrupt ones, have much to teach us.

The past is also particularly useful for testing the credibility of new climatic models. If they cannot reproduce what has already occurred, their ability to predict conditions in the future is likely to be limited.

The influence of volcanism on climate has been documented, although its contribution to global extinctions is still debated. The implications of a massive ash eruption for climate change and agriculture may be a serious threat to global habitability, but the frequency of such eruptions is low.

The role of orbital variation in modifying temperature at the surface over the past few hundred thousand years has been demonstrated by geologists studying a deep-sea core from the Indian Ocean. Awareness is growing that extraterrestrial influence might be discerned operating in much of the earth system.

The intensity of the magnetic field has varied by more than 10 percent in the past few hundred years, and there is some evidence that much larger changes in intensity accompany episodes of magnetic reversal. Very little is known, or even guessed, about whether and how magnetic-field variation might affect the environment, especially living plants and animals, but research is warranted.

Pollution of groundwater in a variety of ways is a major and costly problem in many areas, and it is evident that humans are polluting these water resources at an ever-increasing rate. Research is growing, too, but it is not clear that the full expertise of the solid-earth science community is being involved on a large enough scale.

Top Priority, Theme D: Develop the Ability to Remediate Polluted Groundwater

The need, the opportunity, the potential for success, and the scientific challenge represented by thorough understanding of the environment and how polluted groundwater might be remediated, particularly with microbial methods.

Large and increasing quantities of organic waste material occupy unconfined areas in the United States, both at the surface and underground, including groundwater. Removal of these materials from their present sites is becoming increasingly difficult, and the need for in situ modification is clear. Research into such activities as breeding and using specialized microorganisms to modify and render harmless organic waste materials in a broad range of environments is needed and offers a unique challenge. The subsurface science programs of DOE and EPA represent prominent initiatives in this field.

Other High Priorities

Two other high priorities for Theme D are listed in Table 7.13. The first is related to the top-priority selection in the sense that secure isolation of toxic and radioactive wastes will in the future reduce the pollution of water supplies. Microbiological techniques may be used here as well, but the immediate need is to characterize geologically secure repository sites and to determine how most efficiently to convert nuclear waste into solid glass and ceramic materials. The second topic represents an opportunity for multidisciplinary research into the health risks of some minerals and elements; with proper characterization of the geological materials

> *To predict continued environmental changes and their effects on the biosphere, we need as a historical perspective the results of new and continuing studies on the reconstruction of global warming trends and other atmospheric/hydrospheric changes, as determined from geochemical cycles, paleoclimatology, evolution of life, and extinctions.*

TABLE 7.13 Priority Theme D: To Minimize and Adjust to the Effects of Global and Environmental Change

Aim: To mitigate and remediate the adverse effects produced by global changes and those changes resulting from modification of the environment by human beings.

Top-Priority Selection	**Develop the Ability to Remediate Polluted Groundwater, Emphasizing Microbial Methods** **The need, the opportunity, the potential for success, and the scientific challenge represented by thorough understanding of the environment and how polluted groundwater might be remediated.** Research into breeding and using specialized microorganisms to modify and render harmless organic waste materials in a broad range of environments is needed. The opportunity, the potential for success, and the scientific challenge represented by thorough understanding of the environment in which the microorganisms are to work.
Other High Priorities	■ **Secure the isolation of toxic and radioactive waste from household, industrial, nuclear plant, mining, milling, and in situ leaching sources** Microbiological techniques; characterization of sites especially for water distribution and flow; metal recovery and recycling from waste; conversion of nuclear waste into glass and ceramics with additional ceramic casing. ■ **Geochemistry and human health** Interdisciplinary research into geochemistry and human health involving the earth sciences, mineralogy, materials science, medical research, public education, and risk evaluation.
Requirements	A global data base of present-day surface and shallow near-surface geochemical composition and geological structure as base documentation so that future changes can be directly detected. Advanced analytical and modeling capabilities. Data bases, GIS capabilities.
Major Relevant Federal Programs	Programs in **EPA, DOE** Subsurface Science, **USGS, USDA, DOD, NSF.**
State Programs	State environmental, hydrological, and geological programs.
Industry	Much research by industry in this general area.
International	International geochemical survey.
Selected Recent Reports	NRC: *Opportunities in the Hydrologic Sciences* (1991), *Rethinking Radioactive Waste Isolation* (1990), *Policy Implications of Global Change* (1991).
Recommendation	**Establish the ability to remediate polluted groundwaters on local and regional scales, emphasizing microbial methods.** Coordination of local, industry, state, and federal activities will enhance the potential for success, and international involvement would be desirable.

and public education, the levels of risk can be weighed with the potential remediation costs. These and other research opportunities are discussed in Chapter 5.

Programs and Infrastructure

Numerous current federal, state, industry, and local agency research activities are addressing these problems, and a coordinated attack with international involvement is appropriate. Huge sums of money are spent in research and cleanup by engineering companies dealing with environmental problems. The engineering approaches would be improved in due course by stronger research programs.

RESEARCH IMPLEMENTATION: FACILITIES, EQUIPMENT, AND DATA NEEDS

For identification of the most pressing needs for instruments and the infrastructure in general, the strategy is, first, to determine the most important scientific issues. Priorities are then set by considering what resources need to be allocated to address the scientific problems. Some facilities and instruments are essential for several different scientific problems, and priority among these depends not only on how many problems can be addressed but also on which scientific issues are considered to be most important.

Significant physical facilities are needed in several

phases of research programs, to pursue exploration, to gather data, to evaluate the data, and to model the systems. Equipment ranges from large platforms (e.g., space satellites and drilling vessels) through supercomputers, large laboratory experimental equipment (e.g., large-volume, high-pressure apparatus), sensitive analytical instruments (e.g., ion microprobes), a host of smaller laboratory instruments, and field equipment (e.g., seismometers). Computational workstations for individual scientists are becoming a necessity. These facilities constitute the infrastructure of the science. They have been listed with the research opportunities at the ends of Chapters 2 through 5 and are discussed at some length in Chapter 6.

It is critical that the nation develop a scientific infrastructure that is responsive to supporting new discoveries, advanced instrumentation, and intellectual breakthroughs. Only by making relevant field observations, carrying out related laboratory studies, and integrating results in terms of comprehensive global models will we be able to understand the fundamental processes governing the behavior of our planet. The principal objective here is to ensure that solid-earth science is on the leading edge of fundamental research and that it can address public concerns, environmental crises, national requirements, and international responsibilities.

A summary of the facilities and equipment needed to advance the top-priority research projects in the solid-earth sciences appears in Table 7.14. Six categories are distinguished. Four are related to the locations in which the observing instruments are based: space, aircraft, the Earth's surface, and the sea surface. Laboratory instruments for analysis and experiment constitute the fifth category; the sixth embodies data bases in the broad sense as well as the ability to handle them using geographic information systems and to build models using computers, including the most advanced computational facilities. The research priority themes for which particular facilities and equipment are most relevant are indicated in the table.

Space-Based Instruments and Programs

Satellite-obtained electromagnetic spectral imagery has been widely used by solid-earth scientists since the first Landsat was launched in 1972. The addition of a thematic mapper to later platforms and the advent of the SPOT satellites have both broadened the way in which the imagery can

be used. Management of the Landsat program is to be under NASA and the DOD, and the data management activities will be the responsibility of the USGS EROS Data Center. Twenty continuous years of Landsat data are proving to be an important resource for land-surface and hydrologic research. The future of this essential tool for the study of the Earth appears promising, but a history of vicissitudes suggests that it is important for the user community to continue to express its interest in ongoing observation. Future developments are closely linked to the Earth Observing System (EOS). Specific instruments of EOS that are of importance to the solid-earth sciences include the Moderate-Resolution Imaging Spectrometer (MODIS), the Advanced Spaceborne Thermal Emission and Reflection Radiometer (ASTER), and the High-Resolution Imaging Spectrometer (HIRIS). The USGS's EROS Data Center is planned to be linked into the EOS Data and Information System.

Magnetic fields measured in low Earth orbit integrate the effects of the main core-generated field with temporal variations of the magnetosphere and fields related to magnetic materials in the lithosphere. In very low orbit—below 200 km, for example—fields are strongly influenced by magnetic materials in the lithosphere; measurements at this elevation are relevant to understanding the structure of both the oceanic and continental lithospheres. At higher altitudes—for example, 800 km—the field is less influenced by lithospheric variation and the main field is dominant. The European Space Agency's satellite, ARISTOTELES, which will follow such earlier satellites as NASA's MAGSAT, is planned to orbit sequentially at 200 and 800 km. Ultimately, to monitor the long-term variation of the main field, it will be necessary to maintain magnetometers continu-

TABLE 7.14 Facilities and Equipment Needed for Implementation

	Relevant Priority Theme							
	A							
Facility or Equipment and Characteristics to Be Measured	I	II	III	IV	V	B	C	D
Space-Based Instruments and Programs								
Electromagnetic spectral imagery: Landsat, SPOT, ASTER, and successors	■	■	■	■	-	■	■	■
Magnetic fields: Main and lithospheric fields: ARISTOTELES and successors	-	-	-	■	■	-	■	-
Gravity: ARISTOTELES and successors	-	-	-	■	■	-	-	-
Altimetry: GEOSAT, TOPEX successors, active space-borne radars	-	-	-	■	■	-	-	-
Geodesy: GPS system, satellite laser ranging, VLBI and successors to all, GLRS	-	-	■	■	■	-	■	-
Surface topography: in sequence—stereo-SPOT, ASTER, laser ranging, microwave interferometry	■	■	■	■	■	■	■	■
Radar mapping: SIR-A, -B, -C	■	-	-	■	-	■	■	■
Aircraft-Based Instruments and Programs								
Airborne EM spectral imagery: AVIRIS and related instruments	■	-	■	■	-	■	■	■
Airborne gravity, including gradiometers	-	-	-	■	■	■	-	-
Airborne magnetometry, regional and high-resolution	-	-	-	■	■	■	-	-
Airborne lasers, radar, other airborne geophysical instruments	■	■	-	■	■	■	■	-
Land-Surface-Based Instruments and Programs								
Seismic networks: global, national, local, temporary	-	-	■	■	■	-	■	-
Seismic reflection: shallow, conventional, and deep capabilities	■	-	■	■	■	■	■	■
Other surface geophysics: gravity, magnetics, electromagnetic, heat flow	-	-	■	■	■	■	■	■
Surface geochemistry: a data base	■	■	■	-	-	■	-	■
Field geology: specialized methods (e.g., helicopter access), GPS navigation, work in remote areas, and excavation, especially trenching	■	■	■	■	■	■	■	■
Continental drilling: shallow <1 km, deep 1 to 5 km, ultra-deep >5 km	■	■	■	■	■	■	■	-
Sea-Surface-Based Instruments and Facilities								
Research vessels: seismic, gravity, magnetic, heat flow and other instruments, side-look and multibeam sonar, dredging and other sampling	■	■	■	■	■	■	-	■
Submersibles: manned and remotely operated	■	■	■	■	-	■	-	■
Ocean drilling capability	■	■	■	■	■	■	-	■
Laboratory Instrumentation and Facilities								
Advanced instrumentation for chemical and isotopic analyses of solids, liquids, and gases, including in situ and small-particle analyses	■	■	■	■	■	■	■	■
Access to accelerator and synchroton radiation facilities	■	■	■	■	■	■	■	■
Equipment for very-high-pressure experiments	-	-	-	-	■	—	—	
Large-volume, high-pressure instrumentation	-	-	-	■	■	-	-	-
Advanced organic chemical and isotopic analyses	■	■	■	-	-	■	-	■
Data Bases, Maps, and Collections								
Global digital topographic data sets for land and sea	■	■	■	■	■	■	■	■
Subsurface data banks: seismic, potential field, bore-hole cores, cuttings, and logs	■	■	■	■	■	■	■	■
Geological and other surface maps	■	■	■	■	■	■	■	■
Physical and chemical properties of earth materials	-	■	■	■	■	■	-	-
Museum curation and storage of fossils, rocks, minerals, rock cores, ice cores, meteorites	■	■	■	■	■	■	■	■
Geographic information system capabilities	■	■	■	■	■	■	■	■
Advanced modeling capabilities and access to advanced computational capabilities	■	■	■	■	■	■	■	■

ously in orbit. Although it is possible that more magnetometers will be orbited within the decade, attainment of the ideal of continuous measurement remains a challenge.

Earth's gravity field will be able to be measured by accelerometers mounted on ARISTOTELES. Spatial resolution at an orbit of 210 km will be a few milligals over 100 km. These measurements can be expected to improve understanding of the Earth's structure. However, orbits of less than 200 km and advanced instrumentation, including gravity gradiometers, are required for the kinds of spatial resolution needed to address many problems in the solid-earth sciences.

Satellite altimetry has proved useful in improving gravity and bathymetric models of the oceans. Higher spatial resolution than is currently available would be very useful. National security considerations have restricted availability, but circumstances are changing, and there is a possibility that within the decade more information may become accessible. More advanced altimetry will become available within the decade from such systems as the joint French-NASA satellite—Ocean Topography Experiment (TOPEX/POSEIDON), which was launched in August 1992.

Space geodesy has made staggering advances over the past decade, and the coming decade promises further developments in very-long-baseline interferometry, satellite-laser ranging, and GPS utilization. Dedicated satellite systems for geodesy, which perhaps represent an ideal, are unlikely to be attained within the coming decade.

Surface topographic data of high spatial resolution both above and below sea level are essential for progress in numerous aspects of the solid-earth sciences. Their availability is very uneven, especially in digital form, and acquisition of better data is both practicable and important. Above-sea-level, space-obtained data can be used in a variety of ways. Application of different techniques, from "Stereo-SPOT" through ASTER-stereo to laser ranging and microwave interferometry, may be needed. The ultimate requirement is a digital data set for the whole Earth with a spatial resolution of a few meters and a vertical resolution of about a meter. This may prove attainable, at least for land areas, within a decade.

Aircraft-Based Instruments and Programs

The use of aerial photography in the solid-earth sciences, which dates back more than 60 years, continues to be important. Refinement is continu-

ing; for example, infrared measurements are used in mineral exploration and in the search for geothermal areas. Advanced spectroscopic systems such as NASA's Airborne Visible Infrared Imaging Spectrometer (AVIRIS) are proving powerful in research, and further developments are likely in the coming decade.

Measurement of the local gravitational field of the Earth from the air has been difficult because of the complex and rapidly varying pattern of accelerations imparted to sensors in aircraft. Systems with two or more GPS receivers mounted on an aircraft to give both location and attitude control are potentially capable of solving many problems. With this capability, gravity measurements at high spatial resolution could be obtained for much of the world within the decade. Flights to tie together existing data sets and surveys of new areas are needed. Observations over the Antarctic and Greenland ice sheets are particularly important because of the inaccessibility of subice rock. Other measurements can be made from the same aircraft at the same time.

Airborne magnetometry has been important in regional studies and in mineral exploration for most of the past 50 years. High-resolution surveys are being recognized as useful in establishing the structure of the continental basement where it is buried under gently dipping cover; this is an area where growth is likely in the coming decade. A special case is that of the ice sheets of polar regions, where magnetometry can give an unrivaled reconnaissance view of continental structure.

Other airborne geophysical instruments that are important to the solid-earth sciences include electromagnetic methods and lasers. Improved positioning and attitude control with GPS systems will help to broaden their use. Most airborne geophysical studies are flown at low elevations, and the possible role of robot aircraft in this kind of low-level flight has not been fully explored.

Land-Surface-Based Instruments and Programs

Seismometers of various kinds are important for addressing most of the priority themes recognized in this report. The deployment of global networks of very broadband seismometers is an essential step toward understanding the deep interior. The establishment of a global seismic network by the USGS and the IRIS consortium represents a major step in this direction. These groups cooperate with other nations in the Federation of Digital Seismic Networks. During the first 5 years of the current decade, deployment of over 100 stations should be attained, and comparable progress in data collection and management is expected. Real-time data collection and telemetry are envisaged for a smaller set of stations. Plans for the latter half of the decade include the placing of some seismic stations on the ocean floor; deep deployment is intended for quietness. Global network data contribute to our understanding of the crust and the active processes in the crust and are useful in hazard studies, but national, regional, and local networks of seismometers are also needed for all these purposes. A number of federal, state, and local government agencies are involved in operating networks, and the opportunities for coordinating efforts and developments in such areas as real-time seismology over the decade are great.

Seismic instruments for temporary deployment can be used in a variety of ways. The IRIS consortium is building up a set of advanced instruments for this purpose. As the decade proceeds and more instruments come into use, more and better experiments should prove feasible. Instruments of comparable sophistication will be needed worldwide.

Seismic reflection is the most widely used technique in geophysics. In the coming decade its use at shallow depths for problems in resource and environmental understanding is likely to expand; commercial and local government users are likely to be most active. At conventional depths the petroleum industry is likely to continue to lead, and developments such as three-dimensional surveys and advanced processing and data-handling techniques will extend into other communities. Deep-seismic reflection methods are being used in many parts of the world, and during the decade important developments can be expected in its use, both for extending coverage and for advanced acquisition, processing, and analysis.

Other surface techniques in geophysics are likely to profit from the increasing ability to use several methods in the same area, from improved acquisition, and from better data-handling techniques.

Surface geochemistry is a developing field. Solid-earth scientists have long known the approximate chemical composition of areas at the surface, but detailed studies have been related to mineral exploration and sometimes to environmental medicine. The new idea is that society needs baseline information on the geochemistry of the surface so that human-accelerated change can be monitored and understood and mitigative and remedial steps can be taken.

Field geology, the basic endeavor of the solid-earth scientist, is likely to continue to involve many earth scientists. The passage of the National Geological Mapping Act in 1992 (P.L.102-285) shows that the need for this activity is appreciated nationally. Roles for the USGS and state geological surveys are recognized. There are special problems in field geological work in remote areas such as much of Alaska, the Arctic, and the Antarctic as well as parts of Asia, Africa, and South America. The use of modern capabilities such as GPS navigation and the acquisition of field data in digital form should continue to develop during the coming decade. Temporary excavation, which has always played a part in field geology and is now integrated with shallow geophysics, is becoming more prominent, especially in environmental and engineering geology and paleoseismicity studies, as well as in such traditional fields as mineral exploration.

Continental scientific drilling has developed rapidly in the past decade, and its unique role in solid-earth science research is becoming well defined. Federal government efforts in the United States have responded to the Continental Scientific Drilling and Exploration Act of 1988. An interagency coordinating group (DOE, NSF, USGS) has provided an effective working mechanism in such recent successful programs as those in the Newark Basin and at Creede in Colorado. Planned activities include Katmai and on the island of Hawaii. Internationally there are active programs, especially in Europe (Germany) and the former Soviet Union, and workers from the United States are involved in experiments in some of these and other planned international drilling programs.

Sea-Surface-Based Instruments and Facilities

Solid-earth science research at sea is dominated by investigations of the sea bottom and the underlying crust. Although federal, state, local, private, and commercially owned vessels are all active in research, the strength of the federally funded research fleet and of the oceanographic institutions is critical. Specific problem-solving efforts, such as those under the RIDGE initiative, are proving a successful way of integrating research efforts; they make use of many facilities. The new initiative addressing the problems of the margins of the oceans is similar and represents comparable opportunities for research development.

The use of submersibles and especially of remotely operated vehicles of various kinds is likely to increase during the 1990s. Submarines with scientific observers have played a distinctive role in marine geology, geophysics, and geochemistry. The remotely operated vehicle represents a capability as yet largely unproven.

The Ocean Drilling Program and its predecessors have contributed as much as any facility to the rapid development of the solid-earth sciences over the past 25 years. The current program is planned through 1998. A recent NRC review of the program recognizes the scientific value of continued ocean drilling and recommends broadening participation, ensuring breadth in future activities, and focusing on the establishment of "a highly accurate geochronologic framework for studies of past ocean processes and rates as well as for an optimum approach to drilling technology."

Laboratory Instrumentation and Facilities

The development and use of advanced instruments are critical to the solid-earth sciences. The need for such instrumentation and facilities extends to every one of the priority themes developed in this report. Specialized needs have been defined throughout the report in tables and in the text. A 1990 report by the NRC Board on Earth Sciences and Resources, *Facilities for Earth Materials Research*, addressed community needs by identifying two distinct levels of implementation: one ("Schedule A") incorporates initiatives justified by present technology, manpower, and demand, and the other ("Schedule B") defines a minimum level below which research on some topics cannot be carried out at an internationally competitive level. This approach is potentially useful to those involved in implementing recommendations, and a similar approach might be extended to other areas treated in this report.

Advanced instruments for chemical and isotopic analyses of the compositions of naturally occurring solids, liquids, and gases are all needed. The earth sciences have particular requirements for determining compositions at small sites within materials (e.g., at several places in a microscopically zoned crystal), as well as compositions of tiny particles (e.g., stratospheric dust from volcanoes). Modern beam instruments are well adapted for these purposes, and future developments are likely.

Access to accelerator and synchrotron facilities will be increasingly important as their capabilities for analytical purposes become better developed and more widely appreciated in the solid-earth science community. These issues are thoroughly discussed in the *Facilities for Earth Materials Research* report,

but it is clear that implementation of the recommendations of the present report could have implications for the way in which facilities are used. For example, increased emphasis on the geology of the past 2.5 million years, which is recommended in this report, requires more dating by cosmogenic nuclide measurement. Developments of this kind could put a strain on existing facilities for accelerator mass spectroscopy.

Equipment for very-high-pressure experimentation will be extremely valuable. The ability to generate in the laboratory pressures and conditions as great as those obtained at the center of the Earth has been a significant development in the past decade. The need for the coming decade is to conduct a range of experiments that fully exploit the new capabilities represented by such instruments as diamond cells and to use those capabilities together with other forms of advanced instrumentation.

Large-volume, high-pressure instrumentation also is necessary. The ability to simulate conditions of the deepest crust and upper mantle in "large" volumes (greater than about 1 mm^3), which has been developed only in the past decade, opens up the opportunities for expanded experimental activity.

There is a growing need in the solid-earth sciences for advanced organic chemical and isotopic analyses. The study of organic geochemistry has been stimulated by such diverse interests as working out the origin and evolution of petroleum and understanding the preservation of ancient DNA. The importance of organic composition and structure in interpreting ancient environments, both at the surface and after burial, is a field that is developing fast enough for it to be singled out from among other laboratory instrumentation and facility considerations.

Data Collection and Storage

The most striking progress in the solid-earth sciences may be in the development and use of data bases. Arrays of large and diverse sets of basic research data are important for all research areas, including modern maps representing three-dimensional data bases collating geological, geophysical, geochemical, geochronological, geotechnical, and geobiological data. The data accrue too rapidly for convenient storage in hard copy, and geographic information systems must be brought into widespread use for display and analysis of these data. For much research, access to data on material and thermodynamic properties, reaction kinetics, and the nature of fluid-rock interactions (physical, chemical, and biological) is important. Hierarchical computational capabilities, with constant general use of relatively simple systems and access to progressively more complicated and advanced facilities, will be needed both for data handling and model construction and testing. Advanced instruments, themselves, require advanced computational capabilities.

In addition to collections of data, there are collections of materials that must be preserved and made available for research when needed. At a time when many geologically important localities are being overtaken by urban development and private lands or are being exhausted through mining activities, the archival curation of important collections of fossils, minerals, rocks, and ores is an increasingly important aspect of the earth sciences. There is a need to evaluate the collections of subsurface samples recovered from drill cores, for example, because these samples were collected at signifcant expense and, from an economic perspective, can be considered unique. However, as the resource industries abandon areas of active exploration, samples and records are being discarded because curation costs would be prohibitive. The need for such information is important for both scientific and applied reasons—for instance, such samples could lead to the refined understanding of reservoir heterogeneity needed for enhanced hydrocarbon recovery activities. Major national museums such as the Smithsonian Institution have become one of the important elements in the curation of a limited amount of these materials. In addition to terrestrial materials, there are unique collections of lunar rocks and meteorites that yield progressively more secrets about their inaccessible sources as time passes and instrumental techniques are improved. Other materials that need proper curating are the ice cores drilled from Antarctica and other ice sheets and the huge library of deep-sea cores.

A large proportion of the information that is important for implementation of the priority areas identified in this report is spatial. Maps present a peculiar problem, although perhaps only a temporary one. Existing geological maps, for example, embody a huge potential resource. Although optical scanners can be used to digitize information from maps, the procedure is not yet very reliable, and inordinate amounts of time-consuming verification and attribute coding are demanded. A simple distinction can therefore be made between acquisition of data in digital form and digitizing existing data sets. The coming decade will perhaps be one of

transition. That transition is likely to require substantial resource commitment, one that is concomitant with the need.

Global digital topographic data sets were discussed in the section on space-based facilities. Access to that kind of high-resolution topographic data and the ability to manipulate them are likely to prove important mainly because so much solid-earth data needs to be interpreted with topographic control.

There are large data bases of subsurface information, including seismic reflection data, well-log data, and core and well-cutting collections, as well as detailed gravity and magnetic data, all of which are relevant to the properties of the relatively shallow subsurface. In the United States the use of some of these data sets is confined to those involved in oil and gas exploration, since they are privately owned. Other sets are under the care of state and federal agencies and are in the public domain. Digitization, remote access, and the question of broadening use are important considerations, as are preservation, quality control, and other curatorial matters.

Geological maps and other maps of surface and near-surface properties are essential in the study of the solid earth. The issue of making and updating maps was considered earlier, but there is a linked question of publication, data storage, access, and availability. The USGS and various state agencies are active in addressing these issues, as identified by the National Geological Mapping Act (P.L.102-285). Compatibility and standardization are likely to remain important. Worldwide opportunities, such as access to maps of the former Soviet Union, which up to now have been secret or hard to obtain, could provide occasion to apply the lessons learned in domestic efforts.

Museum curation and storage of fossils, rocks, minerals, rock cores, rock cuttings, ice cores, and meteorites are a growing concern. Because of the value of the materials, their unique character, or the huge replacement cost, they must be stored not only where they are accessible for research but where scholars who appreciate their special value (e.g., taxonomists) can supervise preservation. Local, state, national, and even global considerations (only two nations have collected rocks on the moon, but researchers from many nations have been able to work on lunar materials) are involved. Opportunities for improving curatorial facilities are likely to provide a focus—and escalating costs to provide a challenge.

A paleontological data base is desperately needed. Hard-copy documentation of the fossil record has reached a triumphal peak in the successive editions of the *Treatise on Invertebrate Paleontology*. Researchers now need computer access to the kinds of basic information found in the treatise about duration of existence of a life form, geographical distribution, and much more. The ability to manipulate large data bases can be expected to reveal a great deal about how life has evolved; this is currently an almost indigestible mass of information. This specific data set is singled out here not only as an example of the kind of data base that can be put together from existing material but because it has unique potential for improving our understanding of the history of life.

Data bases covering the physics and chemistry of earth materials also are needed. Hard-copy editions of *The Data of Geochemistry* have been published by the USGS at intervals over the past 70 years, and the Geological Society of America has published an important handbook of physical constraints. An accessible data base that included, for example, material properties, thermodynamic data, kinetic data, and fluid-rock interaction data would significantly assist researchers.

Geographic information systems, which are widely used in land and environmental management, are relevant to all of the committee's priority themes. There will be greater use in areas where such systems are already important, as well as broad extension of their use through much of the solid-earth sciences. Advances in data-handling capability, standardization, and easy (often on-line) access to data bases are needed.

Advanced modeling capabilities and access to advanced computational facilities will complement both the data uses outlined above and the measurement needs discussed earlier. Singling out those fields in which effort will yield the most results in the coming decades is difficult. It is easier to note where activity has been great already because these are clearly promising areas for future success. Processing seismic data generated in oil and gas exploration has long been a leading computational activity, and the demands of modern three-dimensional surveys are particularly challenging. The teleseismic data of global seismic networks is now being processed to generate tomographic images of the mantle, and computational models of mantle flow can be compared with the seismic-derived images. At the other end of the spatial scale, modeling of crystal structures from ab initio calculations is a successful field likely to be more widely applied in the solid-earth sciences. Geochemical modeling, especially advances on the "box models" of early geochemical

cycle studies, is likely to prove fruitful. Paleoenvironmental models, especially where they attempt to accommodate oceanic and atmospheric circulation, are both challenging and likely to become more important. There are clearly opportunities for the greater involvement of solid-earth scientists in aspects of all four subcomponents of the federal government's initiative in high-performance computing systems.

FINANCIAL SUPPORT OF PRIORITY RESEARCH

Current Agency Expenditures

The federal funding levels for fiscal year 1990 have been categorized based on the Research Framework; detailed information on the trends is given in Appendix A. Because of the diversity of agencies and accounting methods, there is some uncertainty about what matrix box is most appropriate for some of the research funds, but the broad picture is valid. Table 7.15 summarizes research allocations among the eight priority themes and illustrates the range of support from federal agencies, state programs, and international activities. A glance at Appendix A, where details of the financial survey are given, will illustrate the difficulty of extracting this information from the different reporting formats of the various agencies. Nevertheless, this table provides a fair picture of the distribution of research support among the areas of the Research Framework and shows fields of concentration by the different agencies.

The total research expenditure for fiscal year 1990 is estimated to be $1.368 billion. This includes $153.5 million for infrastructure and education but does not include a share of the basic operating costs of DOE's national laboratories.

Within the Research Framework, the greatest percentage of support is in the sustenance of resources (Objective B; areas II, III, and IV). If some of the support in this category (e.g., soil studies, cartography, bathymetry) were to be considered peripheral to the solid-earth sciences, crustal dynamics (objectives A and B; IV) would assume the "lead" position.

Industry Support of University Research

The petroleum industry has traditionally supported hydrocarbon research that involves theoretical and, more particularly, applied geology, geo-

TABLE 7.15 Approximate Percentages of Expenditures Keyed to the Research Framework of the Federal Agencies for Fiscal Year 1990[a]

Research Areas	Objectives			
	A. Understand Processes	B. Sustain Sufficient Resources— Water, Minerals, Fuels	C. Mitigate Geological Hazards—Earthquakes, Volcanoes, Landslides	D. Minimize Global and Environmental Change—Assess, Mitigate, Remediate
I. Global Paleoenvironments and Biological Evolution	2	<1	<1	1
II. Global Geochemical and Biogeochemical Cycles	4	20	—	1
III. Fluids in and on the Earth	2	12	<1	3
IV. Crustal Dynamics: Ocean and Continent	19	22	4	6
V. Core and Mantle Dynamics	4	—	<1	—

[a]One percent of the total of $1,368 million is about $13 million (see Appendix A).

physics, and geochemistry. Likewise, so has the mining industry. It is hard to assign a meaningful dollar cost to all this research. A rough guide might be this: the seismic-exploration industry worldwide is expected to rise to about $5 billion by the mid-1990s. If about 1 percent of this sum goes to related earth science research, industry support would be about $50 million. Other estimates indicate that $100 million to $275 million is expended annually on oil and gas research in the United States in both the public and the private sectors. Although most of the research is in-house, both mining and petroleum industries historically have supported research projects conducted in university departments and have collaborated in research with federal agencies (e.g., Bureau of Mines and DOE).

Mining industry support of university research typically involves funding graduate-student field or laboratory work, summer or interim employment of graduate students, consulting arrangements with faculty, and direct grants. During the fiscal decline of the mining industry in the early and mid-1980s, this support diminished considerably as companies cut back on research and exploration activities and on geoscientific personnel. In recent years a growing proportion of the supported research has been in the area of low-temperature, heavy metal geochemistry—a reflection of concern about waste management. At the same time, support for basic research in ore-forming processes and igneous petrology has declined.

The petroleum industry currently supports uni-

versity research through granting foundations in the form of doctoral and master's fellowships, direct faculty support, and grants for equipment and laboratories. At the same time, many companies are providing support directly through their research and operating subsidiaries, either through membership in industrial consortia or direct funding of research by faculty and students. Additional research funding is handled by trade associations, such as the American Petroleum Institute and the American Gas Association. The industry-supported Petroleum Research Fund of the American Chemical Society has played an important role for decades. A wide variety of university programs have been encouraged through these means, ranging from basic research in petrology, paleontology, and sedimentology to technologies for reservoir characterization, enhanced oil recovery, and seismic signal processing. Petroleum industry support of environmental research is growing. Particular emphasis is being placed on disposal of solid and liquid wastes and groundwater management.

The main thrust of oil and gas company research is naturally toward the development of technology and science that may be directly applied to exploration for and development of oil and gas. If an application cannot be defined, support for a research project is unlikely to be granted. It should be noted, however, that a surprising number of research programs pursued by industry have led to significant bodies of fundamental knowledge that in turn have supported societal endeavors quite apart from the search for energy resources.

Suggestions for Future Funding

Although the course of events in future years can be assessed only very generally, in terms of both level of support and its fluctuation, it is clear that federal funding is substantial in relation to all of the committee's priority themes. Programs are commonly—but not always—related to a particular discipline or technique (e.g., drilling) and are often related to several priority themes. Only the activities related to the major mission of the Earth Observing System (EOS) lend themselves to crude representation of funding-time scenarios (funding wedges), and these generally show an upswing when the EOS will be getting under way.

One question to ask is whether specific recommendations—for example, the eight top-priority recommendations of this report—are addressed by the mix of programs under way and envisaged by the federal agencies. To a considerable extent, the answer is clearly yes. For example, understanding the history of the past 2.5 million years requires advanced analytical facilities, ocean drilling, Landsat and related data, geological mapping, advanced data management, and access to supercomputing. Understanding that history will increase our knowledge of the origin and nature of the surficial deposits

and of the recorded environmental and hydrological history over that interval. This is exactly the kind of understanding needed for developing sound waste isolation practices.

There is need for a more detailed assessment of the extent to which the priority themes identified in this report will be addressed in federal programs in the coming decade. Questions that could be asked include the following: Are the planned activities of the various agencies adequate? Are they complementary? Is there duplication? Are international activities integrated with those of the United States? Are there significant pieces missing? What activities are most timely? These programmatic questions are best addressed from within because the detailed information they require is usually available only for the past, not the future, and is difficult for outsiders to interpret. This report provides a background that explains why particular scientific questions have been accorded priority within the solid-earth science community. Earlier in this chapter, under the heading Planning and Decision Making, it was shown that there are broadly based mechanisms in place within the federal government for going the necessary step further: assessing at the interagency and program level such issues as whether the priority questions are being addressed, whether a better job could be done by allocating existing resources differently, or whether additional resources are needed.

RECOMMENDATIONS

Recommendations for action in areas affecting the solid-earth sciences—education, research support, and the national approach to both—are presented below. The committee's over-arching recommendation, which is basic to all its other suggestions, is that the United States make a commitment to earth system science. Knowledge of the interrelationships among the solid earth, its fluid envelopes, and the biosphere is crucial to humankind's continued well-being.

Education Recommendations

The continued vitality of the solid-earth sciences is critically dependent on a continuous supply of well-prepared geoscientists. Chapter 6 presents a number of recommendations for actions to be taken at the graduate, undergraduate, and secondary-school levels. Three recommendations for college curricula merit special attention:

EDUCATION RECOMMENDATION 1: Conventional disciplinary courses should be supplemented with more comprehensive courses in earth system science. Such courses should emphasize the whole Earth, interrelationships and feedback processes, and the involvement of the biosphere in geochemical cycles.

EDUCATION RECOMMENDATION 2: New courses need to be developed to prepare students for growth in both employment and research opportunities in areas such as hydrology, land use, engineering geology, environmental and urban geology, and waste disposal. Such courses will be necessary to prepare students for changing careers in the extractive industries and environmental areas of the earth sciences. No longer are these two areas separate, as mineral and energy resources need to be exploited in environmentally sound ways.

EDUCATION RECOMMENDATION 3: Colleges and universities should explore new educational opportunities (at both the undergraduate and graduate levels) that bridge the needs of earth science and engineering departments. This need arises from the growth of problems related to land use, urban geology, environmental geology and engineering, and waste disposal. The convergence of interests and research is striking, and the classical subject of "engineering geology" could become a significant redefined area of critical importance for society.

Research Recommendations

As mentioned earlier, the committee discovered a remarkable degree of consensus when it selected the top-priority research area for each of the priority themes. The eight top-priority research recommendations are listed below (and summarized in Table 7.5). Each has two high-priority research recommendations associated with it under the same priority theme. In many cases they were strong contenders for the top-priority position, and the choice was difficult. The high-priority selections are given below the top-priority selections.

RESEARCH RECOMMENDATION 1 (Priority Theme I): There should be a coordinated thrust at understanding how the Earth's environment and biology have changed in the past 2.5 million years. The current research activities of many federal agencies bear on this issue, and international involvement would be appropriate as well.

High-priority topics are:

■ to work out the environmental and biological changes that have taken place over the past 150 million years, since the oldest preserved oceans began to evolve and

■ to explore environmental and biological changes prior to 150 million years ago.

RESEARCH RECOMMENDATION 2 (Priority Theme II): The earth sciences need to establish how global geochemical cycles have operated through time. This information, which is essential to working out how the earth system operates, is now a realistic target that could be achieved by coordinating a number of federal programs and current national and international activities.

High-priority topics are:

■ to construct models of the interaction between biogeochemical cycles and the solid earth and climatic cycles and
■ to establish how geochemical cycles operate in the modern world.

RESEARCH RECOMMENDATION 3 (Priority Theme III): The earth sciences need to take up the challenge of investigating the three-

dimensional distribution of fluid pressure and fluid composition in the Earth's crust. The instrumental, observational, and modeling capabilities that exist within various federal programs can be effectively focused on this problem. International coordination is important.

High-priority topics are:

- to model fluid flow in sedimentary basins and
- to improve understanding of microbial influences on fluid chemistry, particularly groundwater.

RESEARCH RECOMMENDATION 4 (Priority Theme IV): There should be coordinated and intensified efforts to understand active crustal deformation. The opportunity exists to revolutionize current knowledge of this area, which is vital not only to the solid-earth sciences but also to the missions of several federal agencies and various state and international bodies.

High-priority topics are:

- to explore landform responses to climatic, tectonic, and hydrologic events and
- to increase comprehension of crustal evolution.

RESEARCH RECOMMENDATION 5 (Priority Theme V): An integrated attack on solving the problem of understanding mantle convection needs to be mounted. Seismic networks, satellite data, high-pressure experiments, magnetic observatories, geochemistry, drilling, and computational modeling can all be marshaled into the fray. Again, federal, national, and international organizations will be involved.

High-priority topics are:

- to establish the origin and temporal variation of the Earth's internally generated magnetic field and
- to determine the nature of the core-mantle boundary.

RESEARCH RECOMMENDATION 6 (Priority Theme B): A dense network of water quality and quantity measurements, including resampling at appropriate intervals, should be established as a basis for scientific advances. Coordination of federal and state agencies that have programs in the field will be needed.

High-priority topics are:

- sedimentary basin research, particularly for improved resource recovery;
- improvement of thermodynamic and kinetic understanding of water-rock interaction and mineral-water interface geochemistry; and
- development of energy and mineral exploration, production, and assessment strategies.

RESEARCH RECOMMENDATION 7 (Priority Theme C): There should be an effort to define and characterize regions of seismic hazard. Because many people and much property in the United States are endangered by earthquakes, improved understanding of seismic occurrences is a pressing need. This issue is important to the missions of several federal agencies and to organizations ranging from local to international.

High-priority topics are:

- to define and characterize areas of landslide hazard and
- to define and characterize potential volcanic hazards.

RESEARCH RECOMMENDATION 8 (Priority Theme D): The earth sciences need to develop the ability to remediate polluted groundwater on both local and regional scales, emphasizing microbial methods. Coordination of local, industry, state, and federal activities will enhance the potential for success, and international involvement would be desirable.

High-priority topics are:

■ to secure the isolation of toxic and radioactive waste from household, industrial, nuclear plant, mining, milling, and in situ leaching sources and

■ to investigate the relationship between geochemistry and human health.

General Recommendations

Recommended priorities for research will need to be developed within the existing complex structure in which federal agencies, most with highly specific missions, interact with universities, with industry, and with each other. These groups should also be interacting with professional societies, state and local agencies, other nations, and international organizations. The series of recommendations that follow is intended to provide guidance for the diverse communities involved in research and practice in the solid-earth sciences in the coming decade.

RECOMMENDATION 1. There should be a major commitment to earth system science, emphasizing interrelationships among all parts of the Earth. The recommended commitment should be akin to the space missions that

> The study of the whole earth system is essential for the solution of global problems.

have revolutionized our understanding of other planets in the past two decades. We are able for the first time to recognize the features associated with the internal evolution of our planet, the actual heterogeneities that drive the geological processes of the Earth. Thus, we are at the threshold of a new and fundamental understanding of global geological phenomena. To be effective, any "Mission to Planet Earth" must be a visionary and broad-ranging study of our entire planet, from core to crust. At least four elements are widely recognized as being crucial to this program: (1) the need for global observations, including those based on space technologies and international collaborations; (2) the development and application of novel instrumentation; (3) the utilization of new computer technologies; and (4) a commitment to support advanced training.

RECOMMENDATION 2. High priority should continue to be given to the best proposals from individual investigators. The intellectual resources represented by mem-

> Individual science is innovative science.

bers of the scientific community are our most valuable asset. The U.S. scientific and industrial population may receive less support in some areas than our international competitors, but it does not suffer from lack of imagination. Core support for individual investigators is the best way to ensure that the diversity in ideas and approaches that is at the root of American inventiveness remains a strong feature of the U.S. effort.

RECOMMENDATION 3. The newest tools for data acquisition need to be made available for use in earth science research. Advanced instrumentation is urgently needed for experiment and analysis in the laboratory and for deployment in space (on satellites), at sea (on research vessels and on the sea bottom), in aircraft, and on land (in networks and in boreholes and movable arrays).

> New instrumentation offers unparalleled opportunities for acquiring information about the Earth.

RECOMMENDATION 4. The opportunities for the integration and use of observations and measurements from advanced space-borne instruments in solid-earth geophysics and geology should continue to be made available. The opportunity for increased understanding of the continents using an integrated approach with remote sensing, field, laboratory, and other data (e.g., seismic) is extraordinary. Remote sensing data should be incorporated and used as a standard field geology tool throughout the undergraduate curriculum and especially in field geology courses. At the graduate level, research should address geological problems aided by remote sensing methods rather than consider remote sensing as a separate discipline.

> Observations and measurements made from space will inspire new concepts and Earth models.

RECOMMENDATION 5. There is an essential need for the production and availability of interactive data banks on a national level within the earth sciences. With new methods of digital acquisition, handling, and archiving, and with growth in the use of geographic information systems along with the Global Positioning System, there are major opportunities to apply the computer revolution to the solid-earth sciences. It is time to integrate the vast amounts of solid-earth science data in nondigital form, like maps, with the exponentially growing digital data sets. National coordination of data-handling services, retrieval procedures, networking, and dissemination practices is required to improve access to the wealth of data held by government, industry, and academic organizations. This will ensure the best use of data in understanding the Earth, sustaining resources, mitigating hazards, and adjusting to environmental change.

> The vast amounts of earth data on hand, together with the new data that will be acquired, must be made available to all.

RECOMMENDATION 6. Efforts need to be made to expand earth science education to all. All citizens need to understand the earth system to make responsible decisions about use of resources, avoid-

> Understanding of earth systems is essential for sustainable development of the world.

ance of natural hazards, and maintenance of the Earth as a habitat. Public school systems must respond to this need. At the university level curricula should be adjusted to meet the needs of contemporary society while maintaining excellence at the professional level.

RECOMMENDATION 7. Research partnerships involving industry-academia-government are encouraged to maximize our understanding of the Earth. Coop-erative multidisciplinary inves-

> The cooperation of industry, academia, and government in supporting research will have a synergistic effect.

tigations that pool intellectual resources residing in government, academic, and industrial sectors can produce more comprehensive research efforts. The primary objectives of the government, industry, and academic groups are diverse. The breadth of disciplines that collectively exist within groups spans our science, but each has its own primary research objectives. Each sector has much expertise to offer that would make it possible to capitalize on the complementary nature of collaboration. The solid-earth sciences stand to gain immeasurably if these three major research groups establish forward-looking cooperative programs.

RECOMMENDATION 8. Increased U.S. involvement in international cooperative projects in the solid-earth sciences and data exchange is essential. The solid-earth sciences are an intrinsically inter-

> International scientific cooperation is needed to further understanding of global earth systems.

national undertaking. Increased understanding of the Earth as a system requires that regional problems be looked at from an international perspective. Cooperative programs involving both nongovernmental international science programs and individuals should be strengthened. Groups involved in U.S. foreign policy decisions should be aware of the importance of the earth sciences in global agreements about issues such as waste management, acid rain, hazard reduction, energy and mineral resources, and desertification. Cooperation between the West, the former Soviet Union, and Eastern Europe presents a timely opportunity for U.S. scientists to join with scientists from those countries in data collection and data sharing to increase knowledge of global earth systems. Such cooperation with other countries can be an important tool in U.S. foreign policy.

BIBLIOGRAPHY

NRC Reports

NRC (1983). *Opportunities for Research in the Geological Sciences*, Committee on Opportunities for Research in the Geological Sciences, Board on Earth Sciences, National Academy Press, Washington, D.C., 95 pp.

NRC (1983). *Research Briefings 1983*, Committee on Science, Engineering, and Public Policy (COSEPUP), National Academy Press, Washington, D.C.

NRC (1986). *Studies in Geophysics—Active Tectonics*, Geophysics Study Committee, Board on Earth Sciences and Resources, National Research Council, National Academy Press, Washington, D.C., 266 pp.

NRC (1986). *Global Change in the Geosphere-Biosphere: Initial Priorities for an IGBP*, U.S. Committee for an International Geosphere-Biosphere Program, National Academy Press, Washington, D.C., 91 pp.

NRC (1987). *Earth Materials Research: Report of a Workshop on Physics and Chemistry of Earth Materials*, Committee on Physics and Chemistry of Earth Materials, Board on Earth Sciences, National Academy Press, Washington, D.C., 122 pp.

NRC (1987). *International Role of U.S. Geoscience*, Committee on Global and International Geology, Board on Earth Sciences, National Academy Press, Washington, D.C., 95 pp.

NRC (1988). *Space Science in the Twenty-First Century: Imperatives for the Decades 1995 to 2015: Mission to Planet Earth*, Task Group on Earth Sciences, Space Science Board, National Academy Press, Washington, D.C., 121 pp.

NRC (1988). *The Mid-Oceanic Ridge: A Dynamic Global System*, Ocean Studies Board, National Academy Press, Washington, D.C., 351 pp.

NRC (1989). *Margins: A Research Initiative for Interdisciplinary Studies of Processes Attending Lithospheric Extension and Convergence*, Ocean Studies Board, National Academy Press, Washington, D.C., 285 pp.

NRC (1989). *Volcanic Studies at Katmai*, U.S. Geodynamics Committee, Board on Earth Sciences and Resources, National Academy Press, Washington, D.C., 9 pp.

NRC (1990). *Facilities for Earth Materials Research*, U.S. Geodynamics Committee, Board on Earth Sciences and Resources, National Academy Press, Washington, D.C., 62 pp.

NRC (1990). *Assessing the Nation's Earthquakes: The Health and Future of Regional Seismograph Networks*, Committee on Seismology, Board on Earth Sciences and Resources, National Academy Press, Washington, D.C., 67 pp.

NRC (1990). *Spatial Data Needs: The Future of the National Mapping Program*, Mapping Science Committee, Board on Earth Sciences and Resources, National Academy Press, Washington, D.C., 78 pp.

NRC (1990). *Geodesy in the Year 2000*, Committee on Geodesy, Board on Earth Sciences and Resources, National Academy Press, Washington, D.C., 176 pp.

NRC (1990). *Rethinking High-Level Radioactive Waste Disposal: A Position Statement of the Board on Radioactive Waste Management*, Board on Radioactive Waste Management, National Academy Press, Washington, D.C., 38 pp.

NRC (1990). *A Review of the USGS National Water Quality Assessment Pilot Program*, Water Science and Technology Board, National Academy Press, Washington, D.C., 153 pp.

NRC (1990). *Studies in Geophysics—The Role of Fluids in Crustal Processes*, Geophysics Study Committee, Board on Earth Sciences and Resources, National Academy Press, Washington, D.C., 170 pp.

NRC (1990). *Studies in Geophysics—Sea-Level Change*, Geophysics Study Committee, Board on Earth Sciences and Resources, National Academy Press, Washington, D.C., 234 pp.

NRC (1990). *Research Strategies for the U.S. Global Change Research Program*, U.S. National Committee for the IGBP, National Academy Press, Washington, D.C., 291 pp.

NRC (1990). *A Safer Future: Reducing the Impacts of Natural Disasters*, U.S. National Committee for the Decade for Natural Disaster Reduction, National Academy Press, Washington, D.C., 67 pp.

NRC (1991). *Solving the Global Change Puzzle: A U.S. Strategy for Managing Data and Information*, Committee on Geophysical Data, Board on Earth Sciences and Resources, National Academy Press, Washington, D.C., 52 pp.

NRC (1991). *International Global Network of Fiducial Stations: Scientific and Implementation Issues*, Committee on Geodesy, Board on Earth Sciences and Resources, National Academy Press, Washington, D.C., 129 pp.

NRC (1991). *Policy Implications of Greenhouse Warming*, Committee on Science, Engineering, and Public Policy, National Academy Press, Washington, D.C., 127 pp.

NRC (1991). *Opportunities in Hydrology*, Committee on Opportunities in the Hydrologic Sciences, Water Science and Technology Board, National Academy Press, Washington, D.C., 348 pp.

NRC (1991). *Real-Time Earthquake Monitoring: Early Warning and Rapid Response*, Committee on Seismology, Board on Earth Sciences and Resources, National Academy Press, Washington, D.C., 52 pp.

NRC (1992). *Oceanography in the Next Decade: Building New Partnerships*, Ocean Studies Board, National Academy Press, Washington, D.C., 201 pp.

NRC (1992). *A Review of the Ocean Drilling Program: Long-Range Plan*, Ocean Studies Board, Board on Earth Sciences and Resources, National Academy Press, Washington, D.C., 13 pp.

NRC (1993). *Toward a Coordinated Spatial Data Infrastructure for the Nation*, Mapping Science Committee, Board on Earth Sciences and Resources, National Academy Press, Washington, D.C.

NRC (1993). *Geomagnetic Initiative: The Global Magnetic Environment—Challenges and Opportunities*, U.S. Geodynamics Committee, Board on Earth Sciences and Resources, National Academy Press, Washington, D.C.

NRC (1993). *Studies in Geophysics—Global Surficial Geofluxes*, Geophysics Study Committee, Board on Earth Sciences and Resources, National Academy Press, Washington, D.C.

Other Reports

Hanks, Thomas C. (1985). *The National Earthquake Hazards Reduction Program—Scientific Status*, U.S. Geological Survey Bulletin 1659, U.S. Government Printing Office, Washington, D.C., 40 pp.

NASA (1987). *From Pattern to Process: The Strategy of the Earth Observing System*, EOS Science Steering Committee, National Aeronautics and Space Administration, Washington, D.C., 140 pp.

ACES (1988). *A Unified Theory of Planet Earth: A Strategic Overview and Long Range Plan for the Division of Earth Sciences*, Advisory Committee for Earth Sciences, National Science Foundation, Washington, D.C., 48 pp.

ESSC (1988). *Earth System Science: A Program for Global Change*, Earth Systems Sciences Committee, NASA Advisory

Council, National Aeronautics and Space Administration, Washington, D.C., 208 pp.

Interagency Coordinating Group for Continental Scientific Drilling (1988). *The Role of Continental Scientific Drilling in Modern Earth Sciences: Scientific Rationale and Plan for the 1990s*, 151 pp.

Mueller, I. I., and S. Zerbini, eds. (1989). *Proceedings of the International Workshop on the Interdisciplianry Role of Space Geodesy*, Erice, Sicily, Italy, July 23–29, 1988, Springer-Verlag, Berlin.

The International Geosphere-Biosphere Programme (IGBP): A Study of Global Change (1990). *The Initial Core Projects*, IGBP Report No. 12, International Council of Scientific Unions, Stockholm, 232 pp. plus appendixes.

NSF (1990) *Report of the Merit Review Task Force*, National Science Foundation, Washington, D.C., 22 pp.

Joint Oceanographic Institutions, Inc. (1990). *Ocean Drilling Program: Long Range Plan, 1989–2002*, Joint Oceanographic Institutions, Inc., Washington, D.C., 119 pp.

Interagency Coordinating Group for Continental Scientific Drilling (1991). *The United States Continental Scientific Drilling Program, Third Annual Report to Congress*, 35 pp. + appendixes.

NSF (1991). *The Advisory Panel Report on Earth System History*, Division of Ocean Sciences, National Science Foundation, Washington, D.C., 99 pp.

NASA (1991). *Solid-Earth Science in the 1990s: Volume 1—Program Plan*, NASA Technical Memorandum 4256, National Aeronautics and Space Administration, Washington, D.C., 61 pp.

NASA (1991). *Solid-Earth Science in the 1990s: Volume 2—Panel Reports*, NASA Technical Memorandum 4256, National Aeronautics and Space Administration, Washington, D.C., 296 pp.

NASA (1991). *Solid-Earth Science in the 1990s: Volume 3—Measurement Techniques and Technology*, NASA Technical Memorandum 4256, National Aeronautics and Space Administration, Washington, D.C., 171 pp.

OTA (1991). *Federally Funded Research: Decisions for a Decade*, Office of Technology Assessment, Congress of the United States, U.S. Government Printing Office, Washington, D.C., 314 pp.

The International Geosphere-Biosphere Programme: A Study of Global Change (IGBP) of the International Council of Scientific Unions (1992). *The PAGES Project: Proposed Implementation Plans for Research Activities*, Stockholm, 112 pp.

OSTP (1992). *Our Changing Planet: The FY 1993 U.S. Global Change Research Program*, Committee on Earth and Environmental Sciences, Federal Coordinating Council for Science, Engineering, and Technology, Washington, D.C., 90 pp.

OSTP (1992). *Grand Challenges 1993: High Performance Computing and Communication*, Federal Coordinating Council for Science, Engineering, and Technology, Washington, D.C., 68 pp.

Interagency Working Group on Data Management for Global Change (1992). *The U.S. Global Change and Information Management Program Plan*, Committee on Earth and Environmental Sciences, Office of Science and Technology Policy, Washington, D.C.

A

Data Base of Federal Programs and Their Budgets for Fiscal Year 1990

IAN D. MACGREGOR
National Science Foundation

INTRODUCTION

In the management of federal science programs it is important to have a broad perspective of the range and scale of other comparable activities. This is particularly critical in the solid-earth sciences because federal agencies use a wide range of earth science subdisciplines to accomplish a variety of national goals. In order to gain a perspective and sense of the diversity I have collected a data base that provides information on the scope of the solid-earth science disciplines as they are supported and used by U.S. federal agencies in achieving their mission goals.

METHODOLOGY

A questionnaire (which appears at the end of this Appendix) was sent to all federal agencies that make use of or support the solid-earth sciences. In order to make the data as quantitative as possible, numerical data were requested for each category assigned in the questionnaire. The categories included identification of federal programs, total solid-earth science budgets, institutional distribution, subdisciplinary fields, functional goals, geographic region, and instrumentation and facilities. It is estimated that about 85 to 90 percent of the overall effort was captured.

QUALITY OF DATA

Any data base has its limitations and, correspondingly, the current effort needs to be used with caution. Problems that may be readily identified are listed below as follows.

■ *Definitions*. Interpretation of the categories listed in the data tables (see questionnaire) have been left to the perception of the program officer providing the data. Correspondingly, the data suffers from the problem that program directors who provided data may use the defined categories somewhat differently leading to unconstrained overlap in definitions. Since a uniformly unambiguous acceptance of definitions is an intractable task, it was felt, that in the compromise between rigor and the efficacy of collecting the data, that a more flexible approach was acceptable. However, it should be noted that the program staff providing data are professionals, who have a comprehensive grasp of their fields and the diversity of federal support. In addition, a fair amount of information exchange and shared programs among the different program managers assures that each manager has a reasonable grasp of the scope and style of other federal programs leading to comparable definitions.

■ *Research Category under Agency Functional Goals*. The category, "Research," listed as one of the functional goals of the agencies has not been expanded to identify the ultimate purpose of the research. Correspondingly, it is not possible to identify how the research budget relates to the agencies' missions as shown in their functional categories.

■ *Classification of "Solid-Earth Sciences."* By design the classification of "solid-earth sciences" that has been used in very broad. For example the areas of soil science, cartography, and bathymetry have been included. The data base includes all applied areas that may benefit from the application of skills derived from training in the basic disciplines of the solid-earth sciences, or use basic solid-earth science information as essential components to accomplishing mission goals. But, the data base can be used more selectively, because separate categories can be individually identified.

■ *Programs Not Included in the Data Base*. There are a few programs from which data were not collected and there is always the possibility that programs have not been identified. In terms of the total expenditures it is estimated that essentially 85 to 90 percent of the solid-earth science activities are reported. Unidentified programs are probably small because the major efforts are well known.

These points caution judicious use of the data base. One may not expect accounting accuracy and the exact figures in each pigeon hole should be assigned some error. An estimate of error is difficult but is probably reasonable to expect that values are within ten percent of the true numbers. The best use of the data is to get a qualitative to semiquantitative estimate of the scope of the federal agencies in the solid-earth sciences. In the latter sense the data are the only comprehensive accumulation of solid-earth science information that is currently available. Moreover, the information has been accumulated at the working level where there is a high degree of knowledge for the technical and scientific contributions and programmatic content of the federal solid-earth science effort.

DATA

The data are available on disk in spreadsheet format and each program or agency may be examined in terms of any of the parameters collected in the data base. As an illustration of the overall effort summary graphs of the distributions by Agency Function and Discipline are included for the total federal effort and for each agencies. The breakdown by Agency Function gives a visual description of the use of solid-earth scientists in accomplishing the missions of the agency, and the disciplinary divisions show the distribution of the types of skills that are needed. The Table gives a quantitative indication of the expenditures for each agency.

QUESTIONNAIRE

I. Program Description

1. Hierarchial position within agency (top down)
   Agency
   Organization
   Officer
   Telephone
2. Address
3. Narrative description of program (Short: 1 to 3 sentences)
4. Comments on collection/maintenance of solid-earth science data base

II. Budget Information

1. Total Budget (in thousands)
   Program level
   For FY 1989, FY 1990, and FY 1991 list

	Internal	External	Total
Element	FTE/$	$	$

2. Budget devoted to solid-earth sciences (in thousands)
   Program level
   For FY 1989, FY 1990, and FY 1991 list

	Internal	External	Total
Element	FTE/$	$	$

3. Total external funds for solid-earth sciences (from #2), distributed by institution (industry, university, federally funded research laboratory) for FY 1989, FY 1990, and FY 1991

III. Subdisciplinary Fields of the Solid-Earth Sciences (FY 1990)

Estimate percentage support of the total solid-earth science budget.

_____ Geochemistry
    _____ Analytical geochemistry
        _____ Isotopes
            _____ Stable
            _____ Radioactive
        _____ Rock/mineral/fluid major element chemistry
        _____ Rock/mineral/fluid trace element chemistry
    _____ Biogeochemistry (rock/organism interactions)
    _____ Cosmochemistry and meteoritics
    _____ Experimental geochemistry
    _____ Igneous geochemistry
    _____ Metamorphic geochemistry
        _____ High temperature (>200° C)
        _____ Low temperature (<200° C)
    _____ Organic geochemistry
    _____ Volcanology
_____ Geology
    _____ Archeology
    _____ Geomorphology
    _____ Mathematical geology and geostatistics
    _____ Quaternary geology
    _____ Sedimentology
    _____ Stratigraphy
    _____ Structural geology

_____ Surficial geology
     _____ Processes
     _____ Remote sensing
_____ Tectonics

_____ Geophysics
     _____ Geophysical modeling
     _____ Mineral physics
     _____ Physical properties of rocks
     _____ Potential field
          _____ Geodesy and gravity
          _____ Geomagnetism and paleomagnetism
          _____ Heat flow
     _____ Seismology/acoustics

_____ Glaciology

_____ Hydrology

_____ Mineralogy/crystallography

_____ Paleobiology
     _____ Paleoecology
     _____ Paleontology
          _____ Invertebrate
          _____ Paleobotany
          _____ Vertebrate

_____ Other (specify)

IV. Functional (Multidisciplinary Mission-Related) Goals of the Solid-Earth Sciences (FY 1990)

Estimate percentage support of the total solid-earth science budget.

_____ Basic Research (*Sensu strictu*)

_____ Economic geology
     _____ Mineral resources
          _____ Metals
          _____ Nonmetals
     _____ Energy resources
     _____ Hydrocarbons
          _____ Coal
          _____ Oil and gas
     _____ Geothermal
     _____ Hydrological resources
          _____ Groundwater reservoirs
          _____ Surface reservoirs

_____ Education and Human Resources

_____ Engineering Geology
     _____ Construction
     _____ Land use/urban geology
     _____ Mining engineering
     _____ Mining technology/mineral extraction
     _____ Petroleum engineering

_____ Global Change Studies (solid-earth components only)

_____ Natural Hazard Reduction

____ Earthquakes
    ____ Engineering
    ____ Geology/geophysics/geochemistry
____ Floods
____ Landslides
    ____ Engineering
    ____ Geology/geophysics/geochemistry
____ Subsidence
    ____ Engineering
    ____ Geology/geophysics/geochemistry
____ Volcanoes

____ Planetology (exclusive of Earth)
____ Planetary surfaces
____ Planetary interiors

____ Regulatory Geology
____ Environmental geology
____ Toxic wastes
    ____ Chemical
    ____ Radioactive

____ Soils
____ Resources
____ Processes (chemical, physical, biogeochemical, and mineralogical)

____ Other (specify)

V. Geographic Classification of Solid-Earth Science Support (FY 1990)
Estimate percentage support of the total solid-earth science budget.

____ Extraterrestrial
____ Meteorites
____ Planets

____ Terrestrial
____ Continental (non Polar regions)
____ Marine
    ____ Geology
    ____ Geophysics
____ Polar Regions
    ____ Arctic
    ____ Antarctic

____ Other (specify)

VI. Instrumentation and Facilities for the Solid-Earth Sciences (FY 1990)
Estimate percentage support of the total solid-earth science budget.

____ Instrumentation (Items < $500,000)
____ Geochemical
____ Geological
____ Geophysical

____ Facilities (Items > $500,000)
____ Geochemical
____ Geological
____ Geophysical

TABLE A.1 Summary of Federal Agency Expenditures in the Solid-Earth Sciences (FY 1990)

Agency Totals			Facilities and Instrumentation			Education		
	$ millions	Percentage		$ millions	Percentage		$ millions	Percentage
USDA	276.04	20.18	USDA	3.51	0.26	USDA		0
DOD	61.15	4.47	DOD	25.09	1.83	DOD		0
DOE	161.07	11.77	DOE	7.57	0.55	DOE	0.41	0.03
DOI	582.25	42.56	DOI	32.67	2.39	DOI	10.29	0.75
DOS	0.50	0.04	DOS		0	DOS	0.00	0
EPA	36.22	2.65	EPA	1.51	0.11	EPA	0.91	0.07
NASA	66.47	4.86	NASA	15.74	1.15	NASA		0
NOAA	41.85	3.06	NOAA	0.60	0.04	NOAA		0
NRC	7.77	0.57	NRC		0	NRC		0
NSF	134.70	9.85	NSF	52.46	3.83	NSF	2.75	0.20
Total	1368.01	100.00		139.16	10.17		14.37	1.05

USDA, U.S. Department of Agriculture; DOD, Department of Defense; DOE, Department of Energy; DOI, Department of the Interior; DOS, Department of State; EPA, Environmental Protection Agency; NASA, National Aeronautical and Space Administration; NOAA, National Oceanic and Atmospheric Administration; NRC, Nuclear Regulatory Agency; NSF, National Science Foundation.

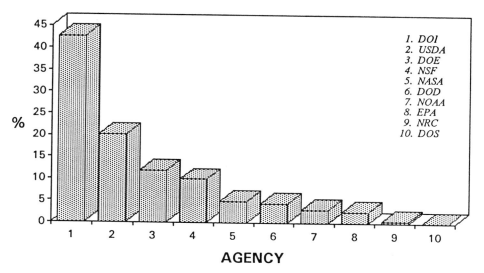

FIGURE A.1 Individual agency expenditures in the solid-earth sciences in FY 1990 expressed as a percentage of the total ($1,368 million); see Table A.1 for details.

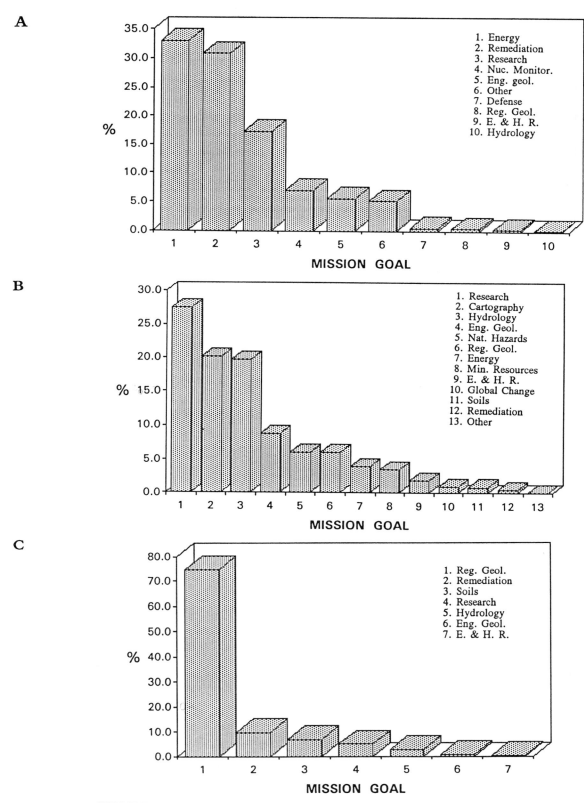

FIGURE A.2 Percentage of the solid-earth science expenditures within selected agencies as related to their mission goals. A, DOE; B, DOI; C, EPA; D, NASA; E, NSF; F, all agencies in survey. The mission goals include research, soils, cartography, energy, regulatory geology, engineering geology, remediation, defense, natural hazards, mineral resources, global change, education and human resources, nuclear monitoring, land management, planetary geology, and other.

D

E

F

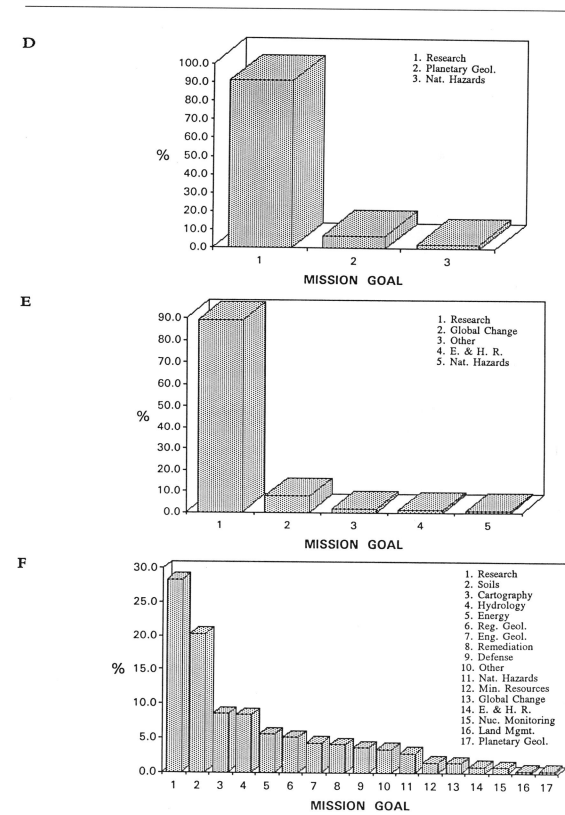

FIGURE A.2 *Continued*

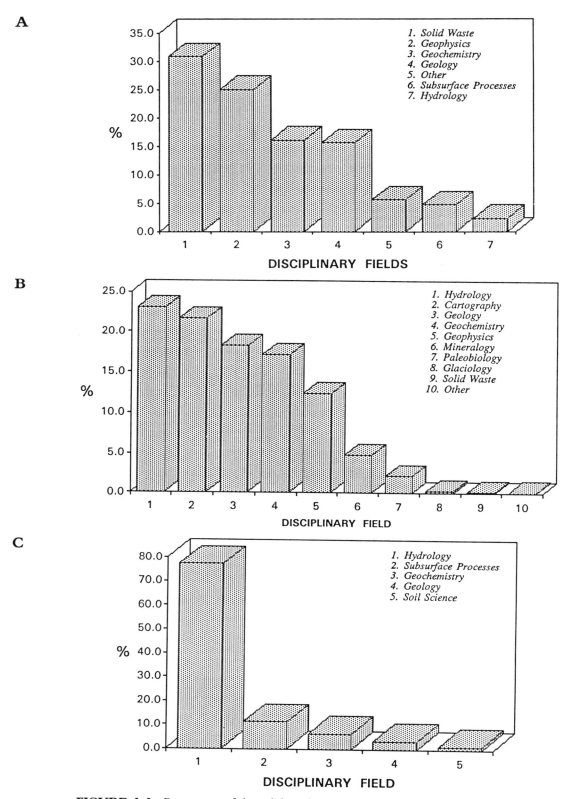

FIGURE A.3 Percentage of the solid-earth science expenditures within selected agencies subdivided according to discipline. A, DOE; B, DOI; C, EPA; D, NASA; E, NSF; F, all agencies in survey.

D

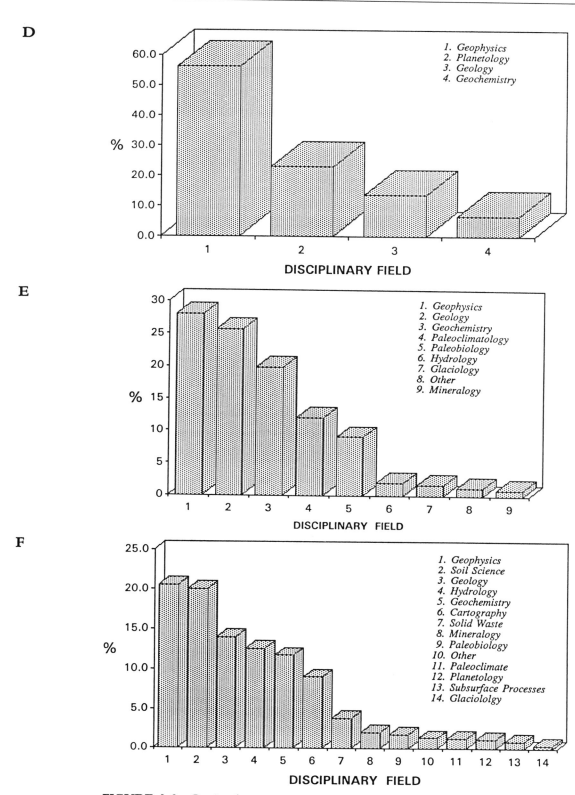

FIGURE A.3 *Continued*

APPENDIX

B

PANELS

The committee formed 21 panels to help synthesize the vast body of earth science knowledge on specific societal issues or related to a few subdisciplines. The panels and their membership are listed below; their chairmen, who also served on the overall committee, are designated by asterisks. The panels, through several individual meetings, produced draft reports that provided a major input to the report. Because of differences in approach and content among the panel reports, there are no plans to issue those draft materials.

Active Tectonics
Robert E. Wallace,* U.S. Geological Survey, Menlo Park

Data Bases and Data Management
James E. Biesecker, U.S. Geological Survey, Reston
Larry G. Carver, University of California, Santa Barbara
John C. Davis, University of Kansas
Robert M. Hamilton, U.S. Geological Survey, Reston
John W. Harbaugh, Stanford University
William J. Hinze,* Purdue University
Brian R. Shaw, BHP Petroleum (Americas'), Inc.

Dynamics and Evolution of the Core and Mantle
Don L. Anderson, California Institute of Technology
Donald J. DePaolo, University of California, Berkeley
Raymond Jeanloz,* University of California, Berkeley
Thorne Lay, University of Michigan
Ronald T. Merrill, University of Washington
Frank M. Richter, University of Chicago
John H. Woodhouse, Harvard University

Earth Surface Processes
Athol D. Abrahams, State University of New York, Buffalo
Thure E. Cerling, University of Utah
John E. Costa, U.S. Geological Survey, Vancouver
Michael D. Harvey, Water Engineering & Technology, Inc.

James C. Knox, University of Wisconsin, Madison
Mark F. Meier, University of Colorado
Stanley A. Schumm,* Colorado State University
Howard G. Wilshire, U.S. Geological Survey, Menlo Park

Energy Resources
Linda A. F. Dutcher, Consultant, Golden, Colorado
John D. Haun,* Barlow & Haun, Inc.
Edward McFarlan, Jr., Exxon (retired)
Robert J. Weimer, University of Colorado

Geochemical Cycles
Robert A. Berner,* Yale University
C. Bryan Gregor, Wright State University
John M. Hayes, Indiana University
Heinrich D. Holland, Harvard University
Antonio C. Lasaga, Yale University

Geochronology and Chronostratigraphy
James C. Brower, Syracuse University
Alexander N. Halliday, University of Michigan
T. Mark Harrison, University of California, Los Angeles
Teh-Lung Ku, University of Southern California
Robley K. Matthews, Brown University
Julie D. Morris, Carnegie Institution of Washington
George R. Tilton, University of California, Santa Barbara
Robert E. Zartman,* U.S. Geological Survey, Denver
G. Alan Zindler, Columbia University

Geologic Hazards
Clarence R. Allen,* California Institute of Technology
Kevin Coppersmith, Geomatrix Consultants
Robert W. Decker, Mariposa, California
Thomas L. Holzer, U.S. Geological Survey, Menlo Park
Robert L. Schuster, U.S. Geological Survey, Denver

Global Collaboration
G. Arthur Barber,* Consultant, Denver, Colorado
Fred Barnard, Consultant, Golden, Colorado
Charles L. Drake, Dartmouth College
Robert N. Ginsburg, University of Miami
William R. Greenwood, U.S. Geological Survey, Reston (deceased)
Priscilla C. P. Grew, Minnesota Geological Survey
Leonard E. Johnson, National Science Foundation
Peter T. Lucas, Shell Development Co.
A. Thomas Ovenshine, U.S. Geological Survey, Reston

History of Life
Andrew H. Knoll, Harvard University
Jennifer A. Kitchell, University of Michigan
Steven M. Stanley,* Johns Hopkins University
S. David Webb, University of Florida

Hydrology
Craig M. Bethke, University of Illinois, Urbana
John D. Bredehoeft,* U.S. Geological Survey, Menlo Park
Steven M. Gorelick, Stanford University

Nicholas C. Matalas, U.S. Geological Survey, Reston
M. Gordon Wolman, Johns Hopkins University

Instrumentation and Facilities
Richard W. Carlson, Carnegie Institution of Washington
Charles C. Counselman III, Massachusetts Institute of Technology
Larry W. Finger,* Carnegie Institution of Washington
Murli H. Manghnani, University of Hawaii
William H. Menke, Columbia University
Walter D. Mooney, U.S. Geological Survey, Menlo Park

Land Use and Geological Engineering
James W. Erwin, U.S. Army Corps of Engineers
Allen W. Hatheway,* University of Missouri at Rolla
George E. Heim, Leighton and Associates
Christopher C. Mathewson, Texas A&M University
Theodore R. Maynard, Department of Public Works
Susan G. Steele, Denver Water Department

Mineral Resources
Samuel S. Adams,* Minerals Consultant
Paul B. Barton, U.S. Geological Survey, Reston
Marco T. Einaudi, Stanford University
Frederick T. Graybeal, ASARCO, Inc.
Mark H. Reed, University of Oregon

Modeling
Susan W. Kieffer, Arizona State University
Peter L. Olson, Johns Hopkins University
John B. Rundle, Lawrence Livermore National Laboratory
Norman H. Sleep, Stanford University
Donald L. Turcotte,* Cornell University
David A. Yuen, University of Minnesota

Paleooceangraphy, Paleoclimatology, and Paleogeography
Michael A. Arthur, University of Rhode Island
William W. Hay, University of Colorado
James P. Kennett, University of California, Santa Barbara
John E. Kutzbach, University of Wisconsin, Madison
Peter J. McCabe, U.S. Geological Survey, Denver
Judith T. Parrish,* University of Arizona

Physics and Chemistry of Earth Materials
Thomas J. Ahrens, California Institute of Technology
R. James Kirkpatrick, University of Illinois
Bruce D. Marsh, Johns Hopkins University
William F. Murphy, Schlumberger-Doll Research
Virginia M. Oversby, Lawrence Livermore National Laboratory
Charles T. Prewitt,* Carnegie Institution of Washington

Professional Community
Gordon P. Eaton, Lamont-Doherty Geological Observatory
Marvin E. Kauffman,* National Science Foundation
Amy S. Mohler, Texas Eastern Gas Pipeline
Allison R. Palmer, Geological Society of America
David A. Stephenson, Harding Lawson Associates
Stephen H. Stow, Oak Ridge National Laboratory

Reinhard A. Wobus, Williams College
Hatten S. Yoder, Jr., Carnegie Institution of Washington

Remote Sensing
Raymond E. Arvidson, Washington University
Alan R. Gillespie, University of Washington
Alexander F. H. Goetz,* University of Colorado
G. Randy Keller, Jr., University of Texas at El Paso
Harold R. Lang, Jet Propulsion Laboratory
Jean-Bernard H. Minster, University of California, San Diego
Lawrence C. Rowan, U.S. Geological Survey, Reston
Floyd F. Sabins, Jr., Chevron Oil Field Research Company
Mark F. Settle, ARCO Oil & Gas Company
Roger J. Phillips, Southern Methodist University

Sedimentary Basins and Basin Analysis
Nicholas Christie-Blick, Columbia University
Grant Garven, Johns Hopkins University
Joseph W. Hakkinen, Marathon Oil Company
Teresa A. Jordan, Cornell University
Lisa M. Pratt, Indiana University
Lee R. Russell,* ARCO Oil & Gas Company
Leon T. Silver, California Institute of Technology
Ronald C. Surdham, University of Wyoming

Structure, Dynamics, and Evolution of the Lithosphere
Marion E. Bickford, University of Kansas
B. Clark Burchfiel, Massachusetts Institute of Technology
David S. Chapman, University of Utah
Bruce R. Doe,* U.S. Geological Survey, Reston
W. Gary Ernst, Stanford University
Bryan L. Isacks, Cornell University
David L. Jones, University of California, Berkeley

Community Input

In August 1988 and January 1989 a "Dear Colleague" letter from the Chairman of the Committee on Status and Research Objectives in the Solid-Earth Sciences: A Critical Assessment was sent to a large number of addressees. These included the earth science departments of many colleges and universities, the oil and gas industries, geological engineers and engineering geologists, some U.S. Geological Survey Branch Chiefs, State Geological Surveys, and chairmen of 21 NRC boards and committees. In March 1989 the "Dear Colleague" letter and a description of the study were distributed to approximately 120 geoscience organizations with a request that one of the items (or a combination) be printed in a forthcoming publication. The two letters and accompanying memos are reproduced on the following pages.

We would like to take this opportunity to thank the respondents to the "Dear Colleague" letter and have listed them below.

Dr. John B. Anderson, Dept of Geology & Geophysics, Rice University
Dr. John T. Andrews, Dept of Geological Sciences, University of Colorado
Dr. Subir K. Banerjee, Dept of Geology & Geophysics, University of Minnesota
Dr. Charles A. Baskerville, McLean, Virginia

Dr. Kenneth E. Bencala, U.S. Geological Survey

Prof. Charles R. Bentley, Dept of Geology & Geophysics, University of Wisconsin, Madison

Mr. Manuel G. Bonilla, Palo Alto, California

Mr. Dwain K. Butler, Waterways Experiment Station, U.S. Army Corps of Engineers

Dr. Lokesh Chaturvedi, Environmental Evaluation Group

Prof. D. L. Clark, Dept of Geology and Geophysics, University of Wisconsin

Dr. Donald M. Davidson, Jr., Dept of Geology, Northern Illinois University

Dr. Owen K. Davis, Dept of Geosciences, University of Arizona

Dr. J-Cl. De Bremaecker, Dept of Geology & Geophysics, Rice University

Mr. B. Louis Decker, Ballwin, Missouri

Prof. John R. Delaney, School of Oceanography, University of Washington

Prof. Robert S. Dietz, Department of Geology, Arizona State University

Prof. James Dorman, Center for Earthquake Research & Information, Memphis State University

Dr. A. Dreimanis, Dept of Geology, University of Western Ontario

Dr. Herbert H. Einstein, Dept of Civil Engineering, Massachusetts Institute of Technology

Prof. T. Edil, Dept of Materials Science & Engineering, University of Wisconsin, Madison

Prof. Wilfred A. Elsers, Institute of Geophysics & Planetary Physics, University of California, Riverside

Dr. Alfred G. Fischer, Dept of Geological Sciences, University of Southern California

Dr. Mel Friedman, College of Geosciences, Texas A&M University

Dr. W. S. Fyfe, Faculty of Science, University of Western Ontario

Dr. John W. Geissman, Dept of Geology, University of New Mexico

Dr. Lee C. Gerhard, Kansas Geological Survey

Dr. Charles E. Glass, College of Mines, University of Arizona

Dr. Harry W. Green II, Dept of Geology, University of California, Davis

Dr. Charles V. Guidotti, Dept of Geological Sciences, University of Maine

B. C. Haimson, Dept of Metallurgy & Mineral Engineering, University of Wisconsin, Madison

Prof. Richard L. Hervig, Center for Solid State Science, Arizona State University Dr. Mason L. Hill, Whittier, California

Dr. Lincoln Hollister, Dept of Geological & Geophysical Sciences, Princeton University

Dr. Donald M. Hoskins, Bureau of Topographic & Geologic Survey, Department of Environmental Resources, Harrisburg, Pennsylvania

Dr. Nicholas Hotton III, Department of Paleobiology, National Museum of Natural History, Smithsonian Institution

Mr. Jeffery R. Keaton, Consulting Geotechnical Engineers, Sergent, Hauskins, & Beckwith

Dr. Michael M. Kimberley, Dept of Marine, Earth, & Atmospheric Sciences, North Carolina State University

Dr. Gary L. Kinsland, Dept of Geology, University of Southwestern Louisiana

Mr. Louis Kirkaldie, Avondale, Pennsylvania

Dr. George deV. Klein, Dept of Geology, University of Illinois

Dr. Robert F. Legget, Ottawa, Ontario, Canada

Prof. Rosalie F. Maddocks, Department of Geosciences, University of Houston

Dr. James McCalpin, Department of Geology, Utah State University

Dr. Diane McKnight, U.S. Geological Survey

Dr. Robert C. Melchior, Department of Geology & Biology, Bemidji State

College

Mr. Martin O. Mifflin, Mifflin & Associates, Inc.

Dr. Brian J. Mitchell, Dept of Earth & Atmospheric Sciences, Saint Louis University

Mr. Luis Rey M. Morales, Environmental Department, H. V. Lawmaster & Company, Inc.

Dr. Paul Morgan, Department of Geology, Northern Arizona University

Dr. Ivan I. Mueller, Department of Geodetic Science, Ohio State University

Dr. Brendan Murphy, Department of Geology, St. Francis Xavier University

Mr. Norman K. Olson, South Carolina Geological Survey

Prof. M. Ostrom, Dept of Materials Science & Engineering, University of Wisconsin, Madison

Dr. Shailer S. Philbrick, Ithaca, New York

Dr. Howard J. Pincus, San Diego, California

Dr. Brian R. Pratt, Department of Geology, University of Toronto

Dr. Jonathan G. Price, Nevada Bureau of Mines & Geology

Dr. George R. Priest, Department of Geology and Mineral Industries, State of Oregon

Mr. Courtney Riordan, Office of Research and Development, Environmental Protection Agency

Dr. Eleanora I. Robbins, U. S. Geological Survey

Mr. W. H. Roberts III, Houston, Texas

Dr. W. I. Rose, Jr., Dept of Geological Engineering, Geology, & Geophysics, Michigan Technological University

Mr. Gerald P. Salisbury, San Marino, California

Mr. James L. Sampair, Geological Engineer, J. L. Sampair Associates

Dr. Roger T. Saucier, Waterways Experiment Station, U.S. Army Corps of Engineers

Dr. Steven D. Scott, Dept of Geology, University of Toronto

Mr. Jerry A. Sesco, Forest Service, Department of Agriculture

Dr. Charles W. Shabica, Earth Science Department, Northeastern Illinois University

Dr. W. Edwin Sharp, Dept of Geological Sciences, University of South Carolina

Dr. Nobu Shimizu, Woods Hole Oceanographic Institution

Dr. Rudy Slingerland, College of Earth & Mineral Sciences, Pennsylvania State University

Dr. James E. Slosson, Consulting Geologists, Slosson and Associates

Dr. Jon Spencer, Arizona Geological Survey

Mr. Donald K. Stevens, Office of Basic Energy Sciences, Department of Energy

Prof. Donald J. Stierman, Department of Geology, University of Toledo

Dr. Raphael Unrug, Dept of Geological Sciences, Wright State University

Dr. Gerald J. Wasserburg, California Institute of Technology Dr. E. G. Wermund, Bureau of Economic Geology, University of Texas, Austin

Dr. James H. Williams, Missouri Department of Natural Resources

Dr. John W. Williams, Department of Geology, San Jose State University

Dr. Clark R. Wilson, Dept of Geological Sciences, University of Texas at Austin

Dr. Robert S. Yeats, Dept of Geology, Oregon State University

Dr. Grant M. Young, Dept of Geology, University of Western Ontario

"DEAR COLLEAGUE" LETTER AND ACCOMPANYING MEMO OF AUGUST 1988

August 4, 1988

STATUS AND RESEARCH OBJECTIVES IN THE SOLID-EARTH SCIENCES: A CRITICAL ASSESSMENT

Dear Colleague:

This is a request or your assistance in a challenging opportunity for the solid-earth sciences community, which is outlined in the enclosed letter. The Committee hopes you will share its conviction that this project has substantial potential to benefit the solid-earth sciences and that you will distribute copies of the letter to all scientists in your department. Thoughtful contributions from the solid-earth sciences community will be essential to the success of the project.

Sincerely,

Peter J. Wyllie
Chairman, Committee on Solid-Earth Sciences

August 4, 1988

MEMORANDUM

To: Scientists in Departments dealing with solid earth sciences
From: Peter J. Wyllie
       Chairman of Committee on:

STATUS AND RESEARCH OBJECTIVES IN THE SOLID-EARTH SCIENCES: A CRITICAL ASSESSMENT

This is an invitation to participate in a challenging project initiated by the Board on Earth Sciences. You may have admired and perhaps envied the disciplinary surveys emphasizing opportunities and directions for future research that have been published for Chemistry, Physics, Astronomy and Astrophysics. Frank Press, President of the National Academy of Sciences, has given the solid-earth sciences an opportunity to prepare the first major assessment of its current and potential contributions to science and the nation. Generous private foundation funding has been provided for the project.

The Committee has been charged with eight tasks, including the following:

1. to identify emerging lines of research promise
2. to identify and address key scientific and societal issues
3. to identify and assess directions, changes, and contributing factors
4. to recommend long- and short-range research priorities

The Committee passes these charges to the solid earth science community. We would like input from as many earth science scientists as possible. Please send concise statements about items (1) to (4) to the address above; they will be most effective if received before the middle of September.

The Committee does not have to start from scratch; a most important basis for the assessment will be the excellent NAS/NRC reports published during recent years. Another will be provided by solicitations such as this. About 20 panels will be working through the summer and fall on topical areas in solid earth sciences. Your comments will be distributed among these panels for consideration.

I am sure you noticed #4, and that you have been reading scientific editorials. In the absence of clear advice from the Committee, priorities will be set by others who may have little interest in the field. The Committee expects to present in this report clear evidence for dazzling scientific opportunities in, and societal contributions provided by geology, geobiology, geochemistry, and geophysics. Thank you for your participation.

"DEAR COLLEAGUE" LETTER AND ACCOMPANYING MEMO OF JANUARY 1989

January 13, 1989

Dear Colleague:

This is a request for your assistance in a challenging opportunity for the solid-earth sciences community, which is outlined in the enclosed letter. The Committee hopes you will share its conviction that this project has substantial potential to benefit the solid=earth sciences and that you will distribute copies of this letter to scientists in your state survey. If you or they should wish to circulate it more widely, we would welcome that assistance. Thoughtful contributions from the solid-earth sciences community will be essential to the success of the project.

Sincerely,

Peter J. Wyllie
Chairman
Committee on the Solid-Earth Sciences

January 13, 1989

MEMORANDUM

TO: Colleagues in the Solid-Earth Sciences Community
FROM: Peter J. Wyllie, Chairman
Committee on the Solid-Earth Sciences

This is an invitation to participate in a challenging project. Frank Press, President of the National Academy of Sciences, has given the solid-earth sciences an opportunity to prepare the first major assessment of its current and potential contributions to science and the national. Generous private foundation funding has been provided for the project.

The Committee has been charged with a number of tasks, including the following:

(1) to identify emerging lines of research promise
(2) to identify and address key scientific and societal issues
(3) to identify and assess directions, change,s and contributing factors
(4) to recommend long- and short-range research priorities.

The Committee passes these charges to the solid-earth sciences community and would like input from as many individuals as possible. Please send concise statements about items (1) to (4) to the address above. Your comments can be incorporated most effectively in the Committee's deliberations if received by mid March 1989, or as soon thereafter as possible.

The Committee has not started from scratch: a most important basis for the assessment is an excellent set of reports published by the NAS/NRC and others during recent years. Another is being provided by solicitations such as this. More than 20 panels have begun working on topical areas in the solid-earth sciences and will continue into the spring. Your comments will be distributed among these panels for consideration.

I am sure you noticed item (4) and that you have been reading scientific editorials. In the absence of clear advice from this Committee, priorities will be set by others who may have little interest in the field. The committee expects to present in this report clear evidence for dazzling scientific opportunities in, and societal contributions provided by geology, geobiology, geochemistry, and geophysics.

Thank you for your participation.

Index